Principles and Practice of Astronomy

Principles and Practice of Astronomy

Edited by Audria Baldwin

SYRAWOOD
PUBLISHING HOUSE

New York

Published by Syrawood Publishing House,
750 Third Avenue, 9th Floor,
New York, NY 10017, USA
www.syrawoodpublishinghouse.com

Principles and Practice of Astronomy
Edited by Audria Baldwin

International Standard Book Number: 978-1-68286-485-2 (Hardback)

Cataloging-in-Publication Data

Principles and practice of astronomy / edited by Audria Baldwin.
    p. cm.
Includes bibliographical references and index.
ISBN 978-1-68286-485-2
1. Astronomy. 2. Space sciences. 3. Solar system. 4. Planetary systems. I. Baldwin, Audria.
QB43.3 .P75 2017
520--dc23

Printed in the United States of America.

TABLE OF CONTENTS

PREFACE

The world is advancing at a fast pace like never before. Therefore, the need is to keep up with the latest developments. This book was an idea that came to fruition when the specialists in the area realized the need to coordinate together and document essential themes in the subject. That's when I was requested to be the editor. Editing this book has been an honour as it brings together diverse authors researching on different streams of the field. The book collates essential materials contributed by veterans in the area which can be utilized by students and researchers alike.

Astronomy is a vast field of study that delves into processes and evaluation of celestial objects. It is primarily divided into two branches, namely, observational astronomy and theoretical astronomy. This book aims to shed light on some of the unexplored aspects and the recent researches in this field. It is a compilation of chapters that discuss the most vital concepts and emerging trends in the area of astronomy. Different approaches, evaluations and methodologies related to this discipline have also been included. For all those who are interested in astronomy, this book can prove to be an essential guide.

Each chapter is a sole-standing publication that reflects each author's interpretation. Thus, the book displays a multi-facetted picture of our current understanding of applications and diverse aspects of the field. I would like to thank the contributors of this book and my family for their endless support.

Editor

Graphene etching on SiC grains as a path to interstellar polycyclic aromatic hydrocarbons formation

P. Merino[1], M. Švec[2], J.I. Martinez[3], P. Jelinek[2], P. Lacovig[4], M. Dalmiglio[4], S. Lizzit[4], P. Soukiassian[5,6], J. Cernicharo[1] & J.A. Martin-Gago[1,3]

Polycyclic aromatic hydrocarbons as well as other organic molecules appear among the most abundant observed species in interstellar space and are key molecules to understanding the prebiotic roots of life. However, their existence and abundance in space remain a puzzle. Here we present a new top-down route to form polycyclic aromatic hydrocarbons in large quantities in space. We show that aromatic species can be efficiently formed on the graphitized surface of the abundant silicon carbide stardust on exposure to atomic hydrogen under pressure and temperature conditions analogous to those of the interstellar medium. To this aim, we mimic the circumstellar environment using ultra-high vacuum chambers and investigate the SiC surface by *in situ* advanced characterization techniques combined with first-principles molecular dynamics calculations. These results suggest that top-down routes are crucial to astrochemistry to explain the abundance of organic species and to uncover the origin of unidentified infrared emission features from advanced observations.

[1] Centro de Astrobiología INTA-CSIC, Carretera de Ajalvir, km.4, ES-28850 Madrid, Spain. [2] Institute of Physics, Academy of Sciences of the Czech Republic, Cukrovarnicka 10, CZ-16200 Prague, Czech Republic. [3] Instituto Ciencia de Materiales de Madrid-CSIC, c/. Sor Juana Inés de la Cruz, 3, ES-28049 Madrid, Spain. [4] Elettra-Sincrotrone Trieste S.C.p.A., Area Science Park, S.S. 14, Km 163.5, I-34149 Trieste, Italy. [5] Commissariat à l'Energie Atomique et aux Energies Alternatives, SIMA, DSM-IRAMIS-SPEC, Bât. 462, 91191 Gif sur Yvette, France. [6] Synchrotron SOLEIL, L'Orme des Merisiers, Saint-Aubin, 91192 Gif sur Yvette, France. Correspondence and requests for materials should be addressed to J.A.M.-G. (email: gago@icmm.csic.es).

The great advances experienced by radioastronomy during recent years have depicted the interstellar medium (ISM) and circumstellar envelopes (CSE) as supporting an active and rich chemistry[1]. In these cosmic regions, both simple molecules, such as H_2 or CO, and complex ones, such as C_{60} and polycyclic aromatic hydrocarbons (PAHs), are routinely detected by their characteristic infrared spectrum[2,3]. Advances in infrared spectroscopy, powered by space missions such as *ISO* and *Spitzer*, have revealed the existence of dust grains composed of silicates, SiC or oxides, among other substances, in such regions[4–6]. All this chemical complexity present in the ISM and in CSEs plays a major role in the evolution of galaxies, comets, the formation of planets and, ultimately, in the abiotic organic chemistry that preceded the origin of life on Earth[7].

PAHs contain up to 20% of the carbon in the photodissociation regions[8], and are present in any region subjected to ultraviolet radiation[9]. However, despite being so ubiquitous and playing such a crucial role in several astrophysical processes and environments, their formation in the interstellar and circumstellar media remains poorly understood. Proposed chemical routes leading to the formation of complex polyaromatic molecules, and particularly to large PAHs, in space cannot account for their abundance and thus efficient mechanisms for their formation are still an open issue. The main current theories invoke either combustion-like[10–12] or ultraviolet-driven—acetylenic—polymerization and evaporation processes in most evolved objects[13–17]. However, the efficiency of such processes is still unknown[8,9,18] and during the last 30 years an important amount of work has been performed to determine PAHs origin and physicochemical properties.

Unfortunately, a complete carbon inventory is difficult to determine, as many other important molecules—such as carbon clusters (C_2, C_3, C_n), CH_4 and C_2H_4, which could also be abundant in the inner regions of CSEs—have not yet been quantitatively observed with sufficient detail. Within this context of chemical complexity in the ISM, new bottom-up assembling mechanisms based on concepts from nanotechnology have recently come into play. The catalytic role of interstellar dust grains in certain reactions has long been proposed; in their seminal work, Watson and Salpeter[19] and later Hornekaer *et al.*[20] studied the dust grains as the preferred sites for the formation of the most abundant molecule, H_2.

In this article, we propose—and experimentally reproduce—an alternative and efficient mechanism based on a top-down approach for producing PAHs through the hydrogen processing of SiC dust grains, which is schematically shown in Fig. 1. Such SiC grains are formed in the inner regions (1–5 R^*, R^* stellar radius) of evolved stars by means of gas-phase condensation[14,18]. We show that in space, SiC crystallites possess a significant amount of segregated carbon on their surfaces, primarily organized as graphitic overlayers. These grains travel through the CSE owing to the radiation pressure induced by the infrared photons of the central star, until they reach the cold external layers of the envelope. At this point ($> 5 R^*$), they are exposed to ultraviolet radiation. PAH emission has been observed to be prominent in photodissociation regions where most molecules are photodissociated and dust grains are processed[2,8]. Here ultraviolet photons and H atoms produced by H_2 photodissociation further process the grain surfaces. We show experimentally that atomic H interaction at elevated temperature (1,000–1,300 K) on a graphene-terminated SiC surface, where graphene denotes a single layer of carbon atoms arranged in a honeycomb structure, triggers significant surface erosion even at low atomic H exposures. Thus, the graphene layer at the SiC crystallite surface is etched, leading to the formation of broken graphene flakes with sizes ranging from a few carbon rings to

large graphene areas. This provides a method to inject PAHs and other PAH-related species into the gas phase.

The potential of surface science methodology to mimic and characterize the essential characteristics of complex problems concerning molecules on surfaces has been previously demonstrated by explaining, among others, the fundamentals of catalysis[21] or the formation of H_2 on the stardust grains[20]. We simulate the conditions of interstellar space using ultra-high vacuum (UHV) chambers equipped with *in situ* atom-resolved scanning tunnelling microscopy (STM) and third-generation synchrotron radiation-based X-ray photoemission-spectroscopy (XPS), among others. The experimental results are confirmed by *ab initio* calculations. This theory–experiment combination provides a state-of-the-art morphological, structural and spectroscopic description about molecular interactions on SiC crystallites under CSE conditions. In addition, SiC has also been widely studied in condensed matter research and nowadays it is well accepted that all SiC crystalline structures are covered with graphene after annealing in vacuum[22,23].

Results

Graphitized interstellar SiC. Dust grains produced in Carbon-rich asymptotic giant branch stars are abundant in SiC crystallites[18]. Their size can range from few nm in the proximities to the star to some µm. Extra-solar particles of SiC as big as 5 µm have been found in the interior of chondritic meteorites[24,25]. The most common circumstellar SiC polytypes, as found in the Murchison meteorite are 3C-SiC (80%) and 2H-SiC (16%). The size of the SiC grains in the ISM is sufficiently large to present atomically reconstructed facets and their main chemical aspects can therefore be confidently simulated by the use of SiC single crystals. We have used as an analogue to the dust grains a 6H-SiC(0001) single crystal, and the data were taken with a base pressure in the range of 10^{-11} mbar, similar to the conditions occurring in the densest regions of the ISM[26] (see Supplementary Note 1).

Because of temperature-induced Si sublimation and subsequent C segregation, the surface of SiC crystallites is always graphitized after annealing at typical temperatures of the ISM depletion zone (Fig. 1). In the case of SiC(0001), the surface after annealing at 1,500 K consists of a combination of atomically flat terraces partially covered with single-layer graphene (Fig. 2a) and other areas with a carbonaceous termination consisting either of a mixture of Si and C atoms in sp^2/sp^3 configuration or bilayer graphene (BLG)[23,27]. The graphene termination of different SiC specimens after being submitted to high temperature has been observed in different previous studies, both for large grains or small nanoparticles[22,23]. Therefore, we conclude that the SiC grains present at CSE end with a graphene cover, as the one shown in Fig. 2a.

G/SiC exposed to atomic hydrogen. As the surface-graphitized SiC grains travel outwards into the ISM they cool (1,500–1,000 K). During this trip, they are exposed to ultraviolet-photodissociated atomic H with an approximate impinging rate of $\approx 1 \times 10^{14}$ particles cm^{-2} s. These are the conditions we use to simulate the process: a temperature of about 1,200 K while exposing the surface to an approximate dose of 4 Langmuir (L; $1 L = 10^{-6}$ mbar s $\approx 10^{15}$ particles cm^{-2} s) of atomic H. Figure 2c–e shows STM images at different length scales with atomic resolution demonstrating that the graphene-covered SiC surface has been strongly modified. The upper part of the image in Fig. 2b still contains grapheme, whereas the lower region has been completely etched and roughens. In the etched part some remaining carbonaceous pieces can be observed (brighter features

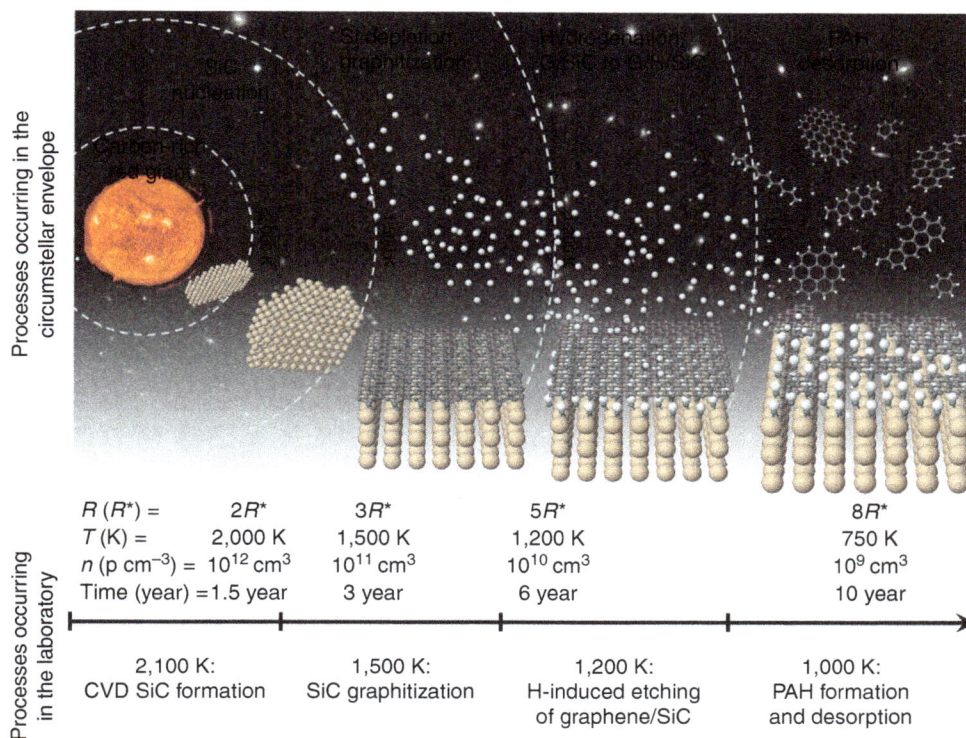

Figure 1 | Qualitative sketch of the proposed process. The steps for the formation of interstellar PAHs and aromatic-aliphatic molecules in the envelope of an evolved star can be divided into four different stages. (1) Formation of SiC in the gas phase and condensation into micrometre- and nanometre-sized grains ($T = 2,000$ K; 1-5 R^*). (2) Annealing of the SiC dust grains due to the proximity of the star and the consequent promotion of surface C-rich phases and graphene ($T = 2,000$-1,500 K, 1-5 R^*). (3) Exposure of the surface to atomic hydrogen, promoting graphitization of the C-rich surface and H passivation of the underlying buffer layer ($T = 1,500$-1,200 K; 5-20 R^*). (4) Etching of the graphene by atomic hydrogen and thermally assisted desorption of PAHs ($T = 1,200$-1,000 K, 5-20 R^*). Temperature values are taken from ref. 36. Lower row: typical temperature in laboratory experiments.

in Fig. 2d). To emphasize the erosion process, we show in Fig. 2c a height profile of an etched surface recorded on Fig. 2b after dosing the crystallite with atomic H and for comparison a height profile of an as-grown surface. The peak-to-valley difference is about 3 Å, which is the size of an atomic step in the SiC surface, suggesting that a whole layer has been disrupted. It is clear that the corrugation of the profiles strongly increases after H dosing at high temperature, indicating a strong erosion of the carbon termination.

In general, on H exposure at elevated temperature, the surface tends to reorganize from sp^3 to sp^2 configuration[27], leading to a cracked and decoupled graphene layer from the SiC surface. Figure 2d shows some formed carbonaceous structures of different sizes remaining weakly bound to the surface. High-resolution images reveal some features that can be assigned to large PAHs (see Fig. 2e,f). These features show a molecular orbital-like electronic density and a periodicity of about 2.4 Å between bright lobes[28]. This distance corresponds to the lattice constant of graphene, indicating that these structures consist of a concatenation of benzene rings (see Supplementary Note 2 and Supplementary Figs 1 and 2). This situation leaves small patches of graphene weakly bound to a H-saturated substrate, ready to be ejected by thermal or photoexcitation processes. These molecules could be the origin of the recently proposed top-down interstellar carbon chemistry[29] and the reason why SiC is abundant in evolved stars but normally gone in the ISM[30].

Figure 3a,b shows some regions of the surface that have been heavily etched. Figure 3a displays an area where the graphene cover has been strongly shrunk to small patches. Only graphene islands remain in the centre of the image. The most probable

scenario is that the original flake covered the whole upper terrace (brighter area at the left side of Fig. 3a) and continued over another graphene layer into the lower terrace (darker area at the right side of Fig. 3a). The etching process has attacked the graphene in contact with the H-passivated substrate, and the process has stopped at the step edge where the overlayer becomes BLG. Thus, the presence of BLG can be considered as a protective film with respect to H attack[31,32]. This could be the reason why this process does not disrupt the whole SiC grain. Figure 3b shows a graphene area where a small flake has been blown out from the surface and the contour of the etched region is visible. The flakes exhibit rounded smooth contours, and high-resolution images show heavily distorted electronic states at the edges, which could originate from passivation through hydrogenation of the edge atoms[33]. In both images of Fig. 3 can be seen that the edge between graphene and eroded parts is strongly modified, indicating that H erosion takes place through edges and defects of the graphene layer. These structural defects are spots to intercalate H atoms into the graphene structure. In some parts of the surface, we have found evidences of H inserted between the SiC interface and the graphene layer. The Fig. 3c shows three bumps identified by their apparent height as H atoms adsorbed on the lower basal plane of graphene.

DFT and MD theoretical analysis. To shed some light on the etching of the graphene by atomic hydrogen, and the subsequent thermally assisted PAH formation and desorption at $T = 1,000$ K and beyond 5 R^* (see Fig. 1), we have carried out density functional theory (DFT)-based atomistic molecular dynamics (MD)

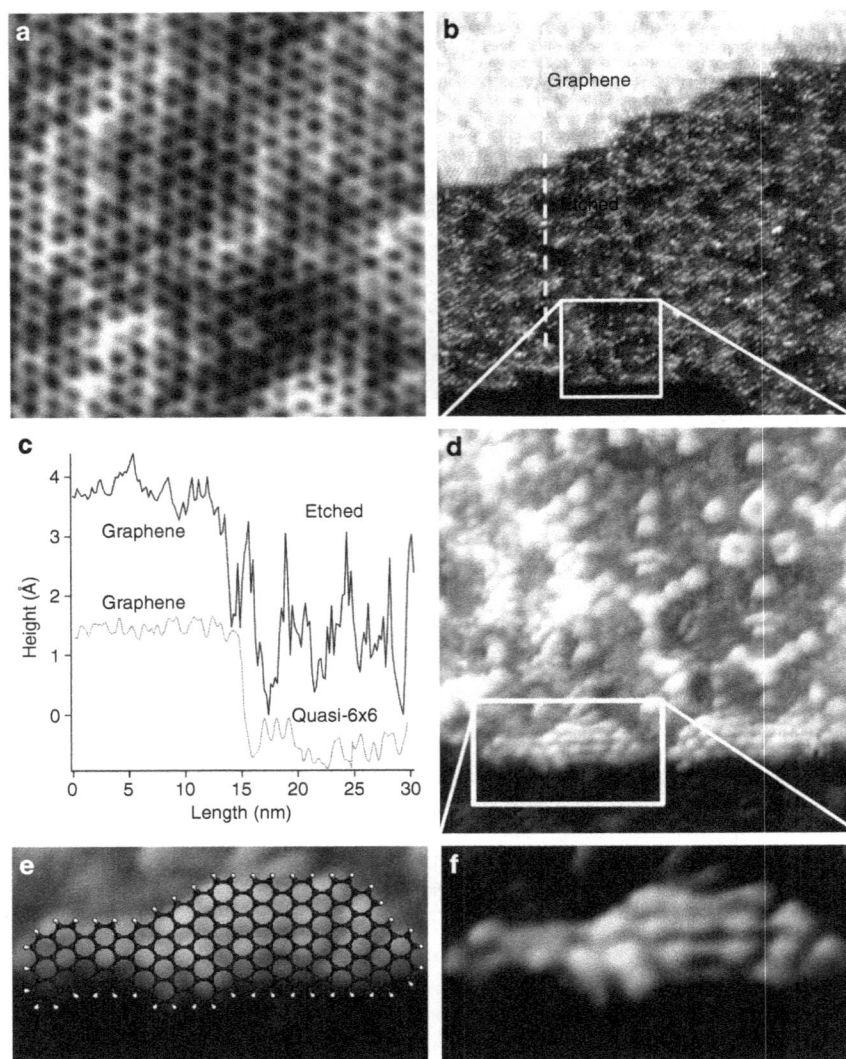

Figure 2 | Graphene etching on high-temperature and atomic H dose. (**a**) STM image (5×5 nm^2) showing a graphene region $V = 100$ mV; the small mesh corresponds to the atomic honeycomb atomic lattice of graphene. (**b**) The image shows at the upper side a graphene plane, whereas at lower part the surface has strongly roughened. 40×40 nm^2 $V = -100$ mV. (**c**) Profile on STM images recorded before (blue) and after (black) being exposed to atomic H with the surface kept at 1,200 K, taken along the dotted line indicated in. The total roughness (r.m.s.) of the surface increases from 0.22 to 0.69 Å. (**b,d-f**) Series of STM images showing the graphene surface eroded after atomic H exposure at high temperature. (**d**) STM image (10×10 nm^2) recorded on the most eroded part that shows the formation on the surface of small molecules and nanostructures. $V = -200$ mV (**f**) STM image (2.2×4 nm^2) zooming in on a detail of the surface showing protrusions corresponding to localized molecular orbitals of a PAH-like molecule. (**e**) Overlaid optimized DFT model of a proposed PAH formed.

simulations. To fully understand the mechanism underlying the PAH formation and desorption, we have designed three heuristic theoretical procedures: first, to mimic the behaviour of BLG, we have performed MD simulations ($T = 1,000$ K) of two parallel 6×6 H-saturated graphene nanoribbons (with 132 atoms each) distanced by 3.35 Å, see Supplementary Movies 1–3; second, to quantify the influence of interlayered atomic hydrogen in the PAH formation and desorption process, we have repeated the MD calculations of the nanoribbon by including around 0.5 ML atomic hydrogen content in between the interlayer region, see Supplementary Movies 4–6; and finally, to simulate the thermally assisted behaviour of a graphene single layer, we have performed MD calculations ($T = 1,000$ K) of a 6×6 H-saturated graphene nanoribbon on a (fix) fully pre-optimized fragment of on-SiC graphene buffer layer (at a distance of 3.35 Å), with the same interlayer atomic H content than in previous case (around 0.5 ML), see Supplementary Movies 7–9. All these molecular

dynamics calculations have been carried out up to a total time of 250 fs (in steps of 1 fs) at 1,000 K. The fourth set of movies (Supplementary Movies 10–12) is calculated with the same conditions than the last set; this is with a pre-optimized fragment of fixed buffer layer but the timescale is larger, up to 1,000 fs.

From this exhaustive theoretical analysis, we can conclude that the highest PAH formation and desorption efficiency is observed for the graphene single layer on the fix on-SiC graphene buffer layer with a high atomic hydrogen content in between the interlayer region, as evidenced in the STM experiments. For this case, after 250 fs, MD process reveals a C–C broken bond ratio of around 60%, to be compared with the 30 and 5% for the hydrogenated and clean graphene bilayer, respectively.

The lower PAH formation and desorption efficiency for the hydrogenated BLG may be explained in terms of that the two free-to-move graphene layers forming the bilayered system can structurally follow, adapt and accommodate the vibrations

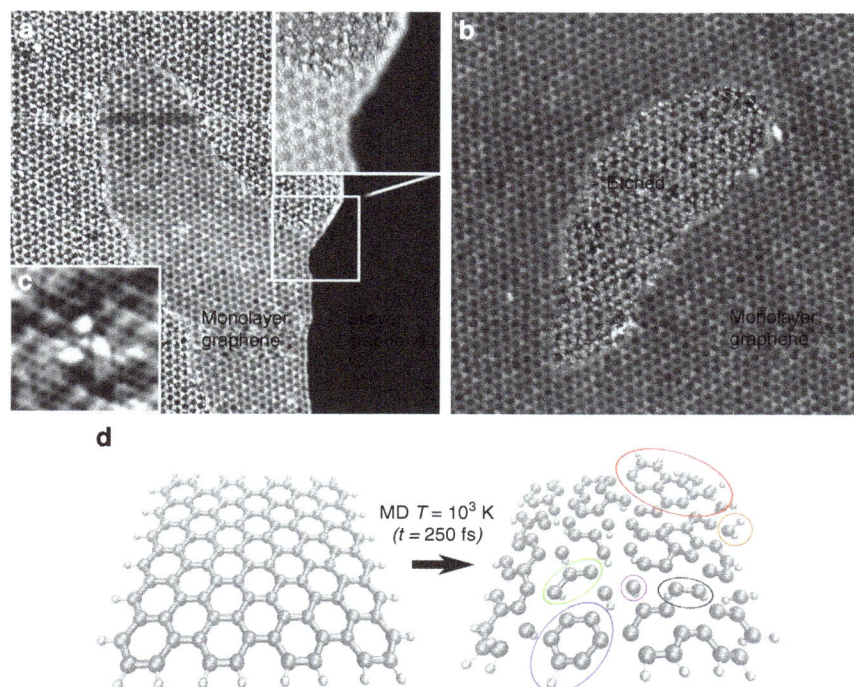

Figure 3 | STM images after H dosing at 1,100 K of G/SiC and computer simulation. (**a**) 100×100 nm^2, $-1,100$ mV. Wide view of a region where the graphene has been reduced to a stripe with rounded edges on the centre of the image. On the right part of the image, a region of BLG seems unaltered. The upper inset shows a high-resolution image of a kink where single-layer graphene, BLG and the modified etched substrate coexist. (**b**) STM image of a hole in the middle of a graphene terrace. These holes are observed after H treatment at high temperatures and they evidence etching through edges and defects. 70×70 nm^2, -1.1 V. (**c**) Atomically resolved STM image of a subsurface trimmer 2.5×2.5 nm^2, -400 mV. (**d**) Initial and final frames of the graphene surface extracted from a movie of DFT molecular dynamics calculations of the G/H/SiC system. The distance between graphene layers expand from 7% at room temperature to about 62% at 1,000 K, and finally, after 250 fs (MD time), the topmost layer disrupt in the form of small carbonaceous species. In this case, benzene rings (blue oval), acetylene molecule (black oval), methylene radical (orange oval) or PAHs (red oval) can be identified.

thermally induced in each other, reducing in this way the efficiency (with respect to the single-layer graphene). Graphene on SiC crystallites is supported on a strongly bonded (almost unaltered) buffer layer, which is unable to accommodate the thermally induced vibrations in the supported graphene single layer. Lower etching efficiency in BLG has been experimentally confirmed in Fig. 3.

Figure 3d shows the structure of the starting graphene flake on the on-SiC graphene buffer layer before (left panel, $t=0$), and after (right panel, $t=250$ fs) the MD calculation, showing the initial stages of PAH formation and desorption. The MD simulations reveal that, at $T=1,000$ K, the atomic hydrogen located in between the interlayer region starts to attack the already H-saturated edges of the nanoribbon, favouring the rupture of the external C–C bonds, and the creation of some CH$_2$ groups. As the time increases, the atomic hydrogen located in the interlayer region starts to attach to more internal C–C bonds in the ribbon, inducing their rupture, as well as the formation of CH and C$_2$H$_2$ groups, and small aliphatic C$_3$H$_n$ chains. From around 200 fs, all the functional groups tend to separate from each other and desorbs by expansion, and some larger structures can be observed, such as some aromatic rings and larger PAHs.

In addition, to theoretically quantify the graphene-decoupling pattern by the intercalated hydrogen with temperature, we have performed an additional optimization calculation (at $T=0$ K), by including the van der Waals interaction in a perturbative way (see ref. 34 and references therein), of an infinite graphene bilayer with a 0.5-ML of intercalated atomic hydrogen content. The result of this optimization process shows the intercalated hydrogen atoms attaching to internal (basal) on-top carbon

positions leading to a net interlayer distance dilatation from 3.35 Å—for the pristine bilayered graphene—to an interlayer distance of 3.60 Å—for the bilayered graphene—which can be quantified in a net graphene decoupling of 7% with respect to the clean system. This value can be easily compared with the decoupling rates extracted from the molecular dynamics simulations (at $T=1,000$ K) of the previously mentioned calculations (the pristine bilayered graphene and its H-intercalated form) after a molecular dynamics time of 250 fs. For these systems, the decoupling rates can be estimated to be around 19% for the pristine graphene bilayer at $T=1,000$ K (an average interlayer distance dilatation from 3.35 to 3.98 Å) and in around 62% for the graphene bilayer with intercalated atomic hydrogen at $T=1,000$ K (an average interlayer distance dilatation from 3.35 to 5.42 Å). This behaviour clearly manifests that both the temperature and the intercalated atomic hydrogen content improve the graphene-decoupling efficiency. Nevertheless, according to the previous numbers, the influence of the intercalated atomic hydrogen is substantially higher (around three times higher) than the temperature, showing its important catalytic role in this graphene-decoupling reaction.

XPS C1s intensity on H exposition at high temperature. The above-proposed mechanism is based on topographical information obtained from high-resolution STM images and DFT calculations. To confirm them, and to acquire quantitative information, we explore the relevant chemistry by analysing high-resolution synchrotron radiation XPS spectra before and after H exposures. The C1s core level peak gives an indication of the atomic density of C at the surface region of a sample and the particular electronic environment (that is, bonding configuration)[35]. Figure 4 shows the *C1s* spectrum of the SiC before and after exposure to 550 l of atomic H at 1,250 K. This peak, as well as the Si2*p* peak (see Supplementary Fig. 3), can be decomposed into different components in good agreement with the scientific literature[35,36] (see Supplementary Note 3 for Si2*p* curve components decomposition). The component appearing at 284.8 eV can be assigned to the graphene *sp*[2] configuration and is the biggest contribution to the C spectra (58.2% of the total), consistent with the area estimation from the STM topography. The components located at lower binding energies (about 284.0 eV) are attributed to bulk carbides and the components at the left-hand side of the spectra to the buffer layer[35]. Figure 4 shows that the whole intensity of the *C1s* peak decreases on H exposure treatment, indicating that there is about a 30% of carbon desorption, which left the surface with lower density than the original one. Although this number cannot be directly extrapolated to the CSE, the high efficiency of the presented erosion process makes these events highly probable to occur at the CSE regions.

If we perform the intensity analysis of the components of the *C1s* peak before and after H exposition, while keeping the sample at high temperatures, we can study the modification degree of every component. The *C1s* core level peak and its components indicate the quantity and density of carbon present on the sample. By tracking the changes in the intensity of every component, we can follow the modification of the carbon species upon H exposition. Therefore, in case the intensity of some particular component decreases, we can assume the number of carbon atoms in such bonding configuration is diminishing. The components of the *C1s* peak are divided in two ranges of modification after hydrogen treatment at high temperatures. The components at energies 284.8, 286.4 and 284.0 eV decrease,

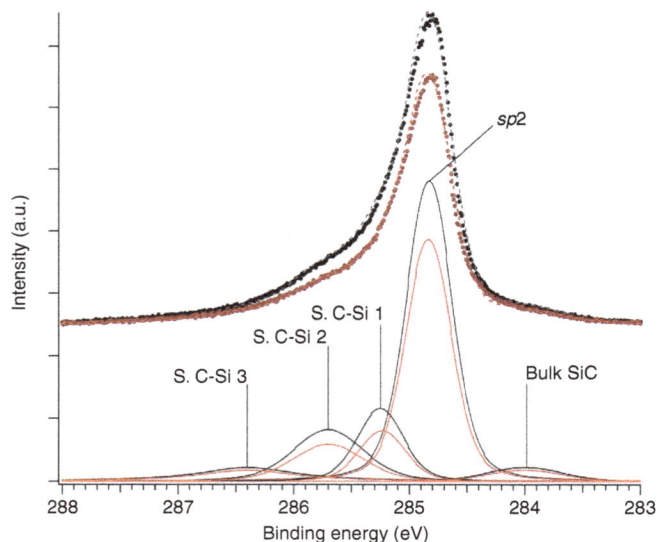

respectively, by −19.4, −19.4 and −20.7% while the components at energies 285.25 and 285.7 eV change by −30.2 and −29.8%. This evidences a loss of carbon atoms on the surface in a disordered manner that leaves the surface with a smaller C density than before the hydrogen treatment. If the carbon atoms would leave the surface in a layer-by-layer mechanism, we would find an XPS spectra modification that eventually would return the intensity and shape to its original spectra, indicating that a full layer has been eroded and a new one is exposed to the X-rays.

As the *C1s* peak intensity decreases only in two different values: approximately −20% for the graphene, bulk and S. C-Si 3 components, and approximately −30% for the surface S. C-Si 1 and S. C-Si 2 components, three values in the Si2*p* (∼13.5, ∼+4 and ∼20%) indicates certain relation between the components sharing the same degradation. The less altered components pertain to the most buried electronic configurations as the hydrogen will mostly attack the atoms located at the surface–vacuum interface. Therefore, the C bulk component is less altered than the surface S. C-Si 1 and S. C-Si 2 components. The exceptions are the graphene and the S. C-Si 3 components that are less altered than the other two surface components. This might arise because of a lower efficiency of the etching process at the carbon atoms in such configurations.

Figure 5a shows the evolution of the lineshape of the *C1s* peak after subsequent different treatments involving exposition to H atmospheres at different temperatures. The original graphitization degree of the sample is around one monolayer. After a first cycle, 50 L of atomic hydrogen while keeping the sample at 1,000 K, the *C1s* intensity decreased 5%. Subsequently, we exposed it to a second cycle, 90 L with a sample temperature of 1,200 K, that resulted in an extra 7% intensity loss. The third subsequent cycle was performed keeping the sample at room temperature (300 K) and submitting it to a hydrogen dose of

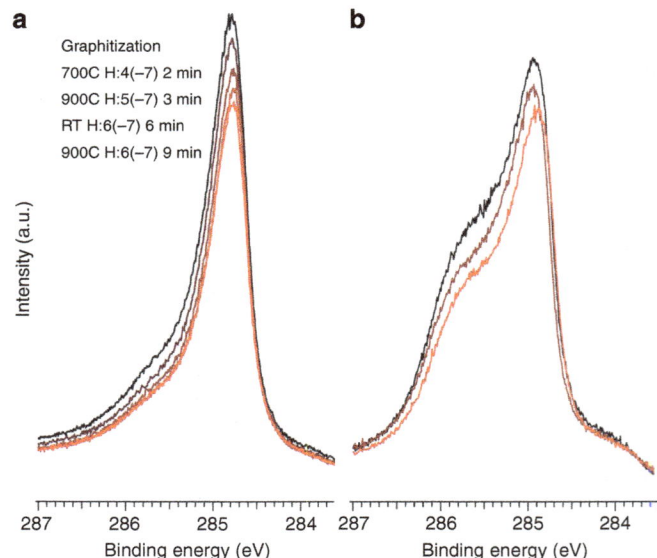

Figure 4 | Synchrotron-based XPS spectra of C1s. The spectra are shown before (black curve) and after (red curve) H treatment (550 L) at 1,200 K. Beneath the experimental data, we show their decomposition into separate curve components. A decrease in the C peak of about 28% in the surface-related components indicates that those C species have been etched away through H processing. The component appearing at 284.8 eV can be assigned to the graphene *sp*[2] configuration. The components located at lower binding energies (about 284.0 eV) are attributed to bulk carbides and the components at the left-hand side of the spectra to the buffer layer (between graphene and SiC). Photon energy: 400 eV.

Figure 5 | High-resolution evolution of the XPS core level peak of C1s. (**a**) Spectra corresponding to the exposure of a graphene covered SiC (1 ML) surface to atomic H at different temperatures during different times. (**b**) Spectra corresponding to the exposure of a partially graphene-covered SiC surface, surface to atomic H at different temperatures and doses: black curve (as graphitized), brown curve (first two steps of a figure, that is, 2 min at 700 C and 4 × 10[−7] mbar and later 3 min at 900 C and 5 × 10[−7] mbar) and red curve (second two steps of **a**, that is, 6 min at 30 C and 6 × 10[−7] mbar followed by 9 min at 900 C and 6 × 10[−7] mbar. The photon energy for both spectra is 400 eV.

220 L, and it resulted in an extra intensity loss of 3%. The last cycle performed on this sample consisted of 325 L of atomic hydrogen while the sample was kept at 1,200 K, and it resulted in an extra 3% decrease of the photoelectron intensity. In total, we dosed around 685 L of hydrogen on the sample, which caused a total intensity decrease of 19%. Importantly, in all the steps of the experiment using low hydrogen pressure, the XPS intensity was reduced after high-temperature dose. This experiment is an indication of the low-pressure etching efficiency on SiC samples with graphene coverage around one monolayer.

The etching process also takes place on samples with a graphene layer covering less than one monolayer. In Fig. 5b, we see the C1s XPS core level spectra of a SiC sample whose degree of graphitization is lower than one monolayer. The intensity after submitting the sample to cycles of hydrogen treatment at high sample temperatures (1,000–1,200 K) diminishes around 14%. The degree of modification of the hydrogen etching is dependent, in a non-trivial manner, on the hydrogen dose, the temperature of the sample during the exposition and the original degree of graphitization of the sample.

Discussion

Surface science aims at studying highly controlled systems and tracing the influence of particular parameters in the final product of a chemical transformation. This approach has produced outstanding results in different fields, such as solid-state chemistry or catalysis[21]. The environmental parameters in our simulation chambers, such as temperature, vacuum and gas composition, were as close as possible to the conditions occurring in the CSE. Of course, real CSEs have a much complex chemistry than the one gathered in our experimental ideal environmental simulation chambers. Molecules such as H_2, CO or C_2H_2 are present in the proximities of evolved carbon-rich stars[37], and electromagnetic radiations, covering large regions of the spectrum, have non-negligible intensities in the vicinities of the stars. These parameters might play a role in the mechanism proposed here for the formation of PAHs. However, in the case of the mechanism that we are proposing, the role of radiation and small molecules is not so important. First, because graphene absorbs electromagnetic radiation through excitation of vibrational phonon modes, which later decay through radiation emission, this process does not alter the intrinsic properties of graphene[38]. Second, because graphene is inert to most of the molecules present in the CSE, such as H_2, CO, C_2H_2. We have performed exactly the same experiment using H_2 instead of atomic H and no etching (not even molecular hydrogen adsorption) takes place. We have obtained exactly the same STM images as for the clean SiC surface (Fig. 2a).

This accurate methodology allows us to test the influence of many parameters in the process. Our simulation experiments and calculations indicate that the size of the resulting PAHs strongly depends on the hydrogen concentration and the formation temperature. If atomic hydrogen is included in the system and temperature is much lower than 900 K, we just have adsorption of H on the surface. If the temperature exceeds 1,500 K, we have new graphene formation at the surface as it is etched away (the surface regenerates). To have efficient graphene erosion, temperature in the 900–1,200 K range and high doses of atomic hydrogen are required. Thus, in our experiment performed in UHV chambers, we have been able to mimic the most important conditions for SiC dust grains in the CSE, and this kind of research opens the door to use UHV-model systems to understand the precise role of co-absorption of molecules, other than hydrogen, and radiation in dust-processing mechanisms (for further understanding of the UHV equivalence of the CSE, see Supplementary Note 1).

SiC is produced close to the evolved star as small-sized nanoparticles (typically few nm) and it has been proven that the crystallization rates of SiC grains can be very fast ($\approx \mu m\,h^{-1}$) at high temperatures[18]. In our experiment, we have used single-crystal surfaces to simulate the process, and one might think that the surface chemical properties of large extended crystals and nanoparticles could be completely different. However, this is not the case. Just a few crystallographic unit cells are required for recovering the bulk electronic structure of SiC. There are several high-resolution transmission electron microscopic studies using SiC nanoparticles of diameter about 5 nm revealing that the graphitization process takes place efficiently on high-temperature annealing[39,40] all around the SiC nanoparticle[22]. Therefore, our single-crystal approach can be seen as a suitable atomic scale description of the astrochemical processes taking place on the surface of both, micrometre- and nanometre-sized, SiC grains populating the CSE.

The usual size of PAHs in the ISM is estimated to be in the range of hundreds of C atoms, in good agreement with the one that we have indentified in Fig. 2d, which has around 180 C atoms. Although Fig. 3d might give the erroneous impression that only small PAHs can be produced, we would like to remark that the PAH formed by our proposed mechanism can be quite large. This misleading idea is due to the reduced number of atoms that can be handled into the complex ab initio MD calculations we have performed.

Moreover, one might think that in our model, the etching process will last forever cracking the formed PAHs into smaller and smaller pieces as time goes on. There are several factors promoting a self-stopping of the etching of graphene into PAHs. The efficiency of the process is strongly dependent on the underlying interface, as it is shown in the DFT calculated movies (see Supplementary Movies 3, 6 and 9). In other words, the graphene etching takes place efficiently because it is supported on a SiC grain. Once the PAH is decoupled from the SiC, etching is not an efficient mechanism. The hydrogen concentration and the temperature decrease as the PAH travel outwards in the CSE. PAHs arrive to regions of ISM where the H concentration can be considered negligible. The same occurs with the temperature, which rapidly decrease below 900 K. We have recorded STM images of H deposited on graphene at 300 K where H clusters are spotted on top of C atoms. This H adsorption process is also known to occur on PAHs and it is called hyperhydrogenation. We have not seen any adsorption or erosion when H_2 is used. Also, the smaller PAHs have less probability of interacting with the atomic hydrogen and a higher hydrogen adsorption energy. DFT calculations show that as you decrease the size of the PAH, the atomic hydrogen capture probability decreases. In this sense, there are two factors influencing the self-stopping hydrogenation of small PAHs. First, the size of the absorption area; the smaller the PAH, the lower the probability of scattering with an impinging adsorbate. Second, energetic considerations; the smaller a PAH, the higher the adsorption energy for an extra hydrogen (see Supplementary Note 4 and Supplementary Fig. 4). Therefore, graphene etching in the CSE is a self-limiting process yielding to significantly large PAHs.

Finally, based on these experiments, we should not expect the formed ISM PAHs to have the identifiable structures. Mixtures of PAHs, aromatic-aliphatic molecules and hyperhydrogenated PAHs molecules have been proposed to account for the astronomical infrared observations[41]. All these aromatic chemical species can be produced out of the above-presented processing mechanism of the SiC dust. We are aware that in our experiments we observe the surface of the SiC grains instead of the produced PAHs. However, the latter is difficult as a priori our mechanism does not favour a particular size of the produced species, and a global increase in mass detection experiments can just be expected.

In conclusion, we propose an alternative PAH top-down formation mechanism through graphene etching on the surface of SiC dust grains at ISM. This model is supported by a combination of different state-of-the-art experimental techniques with advanced computational methods. The good match between the available astrochemical information and the advanced surface science experimental methodology suggests that the problems of the former can be successfully understood through the techniques of the latter and opens the door to the investigation and modelling of other processes related to dust particles in ISM.

Methods

Scanning tunnelling microscope. STM experiments were performed in a UHV system with a base pressure 1×10^{-10} mbar using an Omicron variable temperature scanning tunnelling microscope, which enables us to perform measurements while keeping the sample in the 40–300 K range. Tungsten tips were used to acquire topography images in constant current mode. The chamber is equipped with an atomic H flux source that produces hydrogen by H_2 thermal cracking (99% efficiency, no protons). The gas passes through a tungsten capillary kept at 2,500 K by electron bombardment. The sample was placed at 150 mm from the capillary. The chamber contains an annealing stage that heats the sample through direct heating. After hydrogen deposition with the SiC sample at high temperature, it is transferred to the STM for local measurements.

Photoemission experiments. Synchrotron radiation XPS measurements were performed at the SuperESCA beamline of Elettra, the third-generation synchrotron light source in Italy. Here the top-up mode of operation of the storage ring generating synchrotron radiation provides a stable photon flux in a wide range of photon energies. The end station of SuperESCA consists of an UHV chamber with base pressure 1×10^{-10} mbar. It is specifically designed for high-flux and high-resolution core level spectroscopy, particularly of adsorbates on single-crystal surfaces using soft X-ray synchrotron radiation. The atomic H doser of the STM apparatus was mounted in the chamber. The main chamber is equipped with a hemispherical electron analyser using an home-made delay line electron detection system. The overall energy resolution, comprising the analyser and the photon beam resolution, was better than 50 meV.

C1s core level peak decomposition into curve components. The XPS spectra analysis of the results obtained at the SuperESCA beamline was performed using the FITT program after a Shirley background subtraction. The high resolution of the experimental system permits us to decompose the peaks into its basic curve components; every component reflects a particular electronic environment (that is, bonding configuration) and results from the convolution of a purely quantum Lorentzian energy distribution (full width at half maximum, 0.12–0.2 eV) with a Gaussian distribution (full width at half maximum, 0.4–0.7 eV)[42]. However, the carbon chemistry is very rich and the interpretation of the C1s spectra is not straightforward. The components corresponding to sp^2- and sp^3-bonding configurations measured for pure graphite and pure diamond, respectively, appear at very similar binding energies (284.68 and 285.0 eV, respectively)[43,44].

We have decomposed the C1s XPS peak of the as-grown G/SiC sample (Fig. 4) in five different basic components without using any asymmetry parameter. SiC is a wide-gap semiconductor and therefore we use symmetrical peaks for the fitting. In Supplementary Fig. 1, we show the C1s and Si2p photoelectron peaks decomposed in its basic components before and after H treatment. The component appearing at 284.83 eV can be directly assigned to the graphene contribution (or C atoms in sp^2 configuration), and is the biggest contribution to the C1s core level peak, with its 58.2% of the total area. This component is also important for calibration of the energy width, since it is narrow enough to test the Lorentzian and Gaussian widths with a high precision. The components located at lower binding energies than the main peak (in our case 283.98 eV) are attributed normally to carbides, and we assign them to the C in the SiC bulk crystal[35,36]. There are three other components at higher binding energies, which can in principle be assigned to different configurations of C atoms in the superficial rearrangement that takes place on surface reconstruction[35]. We assign these XPS components to surface-related peaks and name them with an S. in the way that follows: S. C-Si 1 at 285.26 eV, S. C-Si 2 at 285.73 eV and S. C-Si 3 at 286.4 eV, respectively. We will later discuss the validity of this assumption.

Theoretical spatially distributed density of states. The molecular orbitals of the isolated PAHs have been calculated using the DFT calculation package FIREBALL. To resemble the experimental STM images of free PAHs, they have been taken as spatially distributed orbital densities, density of states, with suitable values of the orbital density three-dimensional isosurfaces to mimic the experimental morphology.

Molecular dynamics. The atomistic MD simulations have been performed by using the fast and efficient local-orbital DFT code FIREBALL. The temperature in these calculations is simulated by giving to each non-fixed atom a random initial velocity. In particular, to mimic the physical conditions for the etching of the graphene by atomic hydrogen, the dynamical processes were simulated giving to all the systems an initial temperature $T = 1,000$ K. Then, the atoms were free to move using free MD up to a total time of 250 fs, corresponding to 250 MD steps. Three different graphenic configurations have been simulated by molecular dynamics: (a) two parallel 6×6 H-saturated flat graphene nanoribbons (with 132 atoms each) distanced by 3.35 Å; (b) the same two parallel 6×6 H-saturated flat graphene nanoribbons (with 132 atoms each) distanced by 3.35 Å as in previous case (a), but this time by including around 0.5 ML atomic hydrogen content in between the interlayer region, which initially forms an homogenous atomic H-intercalated layer; and, finally, (c) a 6×6 H-saturated flat graphene nanoribbon on a (fix) fully pre-optimized 6×6 fragment of a H-saturated on-SiC graphene buffer layer (at a distance of 3.35 Å), with the same interlayer atomic H content than in previous case (b).

Atomic hydrogen adsorption potential energy curves. The adsorption energy curves of atomic hydrogen in interaction with PAHs of different sizes have been obtained by using the accurate planewave DFT parametrization included in the QUANTUM ESPRESSO simulation package. In addition, to properly account the non-negligible dispersion forces appearing within this kind of interactions, we have used the DFT + D approach also implemented in the mentioned planewave simulation code. For this purpose, we have used a generalized gradient-corrected approximation for the electronic exchange-correlation description, and an empirical efficient van der Waals R^{-6} correction to add dispersive forces to conventional density functionals (DFT-D). Two different adsorption configurations have been analysed in the calculations: chemisorption and physisorption paths of atomic hydrogen approaching towards an 'on-top' PAH carbon atom adsorption site.

References

1. Decin, L. et al. Warm water vapour in the sooty outflow from a luminous carbon star. Nature **467**, 64–67 (2010).
2. Tielens, A. Interstellar polycyclic aromatic hydrocarbon molecules*. Annu. Rev. Astron. Astrophys. **46**, 289–337 (2008).
3. Cami, J., Bernard-Salas, J., Peeters, E. & Malek, S. E. Detection of C60 and C70 in a young planetary nebula. Science **329**, 1180–1182 (2010).
4. Peeters, E., Allamandola, L., Hudgins, D., Hony, S. & Tielens, A. in Astrophysics of Dust, Vol. 309 (eds A. N. Witt, G. C. Clayton & B. T. Draine), 141–162 (Astronomical Society of the Pacific, 2004).
5. Knacke, R. Carbonaceous compounds in interstellar dust. Nature **269**, 132–134 (1977).
6. Gail, H.-P., Zhukovska, S., Hoppe, P. & Trieloff, M. Stardust from asymptotic giant branch stars. Astrophys. J. **698**, 1136–1154 (2009).
7. Caro, G. M. et al. Amino acids from ultraviolet irradiation of interstellar ice analogues. Nature **416**, 403–406 (2002).
8. Joblin, C. & Tielens, A. G. G. M. PAHs and the Universe: A Symposium to Celebrate the 25th Anniversary of the PAH Hypothesis in EAS Publications Series 46 (Cambridge University Press, 2011).
9. Tielens, A. The molecular universe. Rev. Mod. Phys. **85**, 1021 (2013).
10. Kwok, S., Volk, K. & Bernath, P. On the origin of infrared plateau features in proto-planetary nebulae. Astrophys. J. Lett. **554**, L87–L90 (2001).
11. Frenklach, M. & Feigelson, E. D. Formation of polycyclic aromatic hydrocarbons in circumstellar envelopes. Astrophys. J. **341**, 372–384 (1989).
12. Cherchneff, I., Barker, J. R. & Tielens, A. G. Polycyclic aromatic hydrocarbon formation in carbon-rich stellar envelopes. Astrophys. J. **401**, 269–287 (1992).
13. José, C. et al. Infrared space observatory's discovery of C4H2, C6H2, and benzene in CRL 618. Astrophys. J. Lett. **546**, L123–L126 (2001).
14. Fonfría, J. P., Cernicharo, J., Richter, M. J. & Lacy, J. H. A detailed analysis of the dust formation zone of IRC + 10216 derived from mid-infrared bands of C2H2 and HCN. Astrophys. J. **673**, 445–469 (2008).
15. Pilleri, P., Montillaud, J., Berné, O. & Joblin, C. Evaporating very small grains as tracers of the UV radiation field in photo-dissociation regions. Astron. Astrophys. **542**, A69 (2012).
16. Contreras, C. S. & Salama, F. Laboratory investigations of polycyclic aromatic hydrocarbon formation and destruction in the circumstellar outflows of carbon stars. Astrophys. J. Suppl. Ser. **208**, 6 (2013).
17. Cernicharo, J. The polymerization of acetylene, hydrogen cyanide, and carbon chains in the neutral layers of carbon-rich proto-planetary nebulae. Astrophys. J. Lett. **608**, L41–L44 (2004).
18. Frenklach, M., Carmer, C. & Feigelson, E. Silicon carbide and the origin of interstellar carbon grains. Nature **339**, 196–198 (1989).
19. Watson, W. & Salpeter, E. Molecule formation on interstellar grains. Astrophys. J. **174**, 321–340 (1972).

20. Hornekær, L., Baurichter, A., Petrunin, V., Field, D. & Luntz, A. Importance of surface morphology in interstellar H2 formation. *Science* **302**, 1943–1946 (2003).

21. Sachs, C., Hildebrand, M., Völkening, S., Wintterlin, J. & Ertl, G. Spatiotemporal self-organization in a surface reaction: from the atomic to the mesoscopic scale. *Science* **293**, 1635–1638 (2001).

22. Ostler, M. *et al.* Direct growth of quasi-free-standing epitaxial graphene on nonpolar SiC surfaces. *Phys. Rev. B* **88**, 085408 (2013).

23. Soukiassian, P. G. & Enriquez, H. B. Atomic scale control and understanding of cubic silicon carbide surface reconstructions, nanostructures and nanochemistry. *J. Phys. Condens. Matter* **16**, S1611–S1658 (2004).

24. Daulton, T. *et al.* Polytype distribution in circumstellar silicon carbide. *Science* **296**, 1852–1855 (2002).

25. Daulton, T. *et al.* Polytype distribution of circumstellar silicon carbide: Microstructural characterization by transmission electron microscopy. *Geochim. Cosmochim. Acta* **67**, 4743–4767 (2003).

26. Caro, G. M. *et al.* New results on thermal and photodesorption of CO ice using the novel InterStellar Astrochemistry Chamber (ISAC). *A&A* **522**, A108 (2010).

27. Riedl, C., Coletti, C., Iwasaki, T., Zakharov, A. A. & Starke, U. Quasi-free-standing epitaxial graphene on SiC obtained by hydrogen intercalation. *Phys. Rev. Lett.* **103**, 246804 (2009).

28. Repp, J., Meyer, G., Stojković, S. M., Gourdon, A. & Joachim, C. Molecules on insulating films: scanning-tunneling microscopy imaging of individual molecular orbitals. *Phys. Rev. Lett.* **94**, 026803 (2005).

29. Berné, O. & Tielens, A. G. Formation of buckminsterfullerene (C60) in interstellar space. *Proc. Natl Acad. Sci.* **109**, 401–406 (2012).

30. Whittet, D., Duley, W. & Martin, P. On the abundance of silicon carbide dust in the interstellar medium. *Mon. Not. R. Astron. Soc.* **244**, 427–431 (1990).

31. Zhang, Y., Li, Z., Kim, P., Zhang, L. & Zhou, C. Anisotropic hydrogen etching of chemical vapor deposited graphene. *ACS Nano* **6**, 126–132 (2011).

32. Diankov, G., Neumann, M. & Goldhaber-Gordon, D. Extreme monolayer-selectivity of hydrogen-plasma reactions with graphene. *ACS Nano* **7**, 1324–1332 (2013).

33. Park, C. *et al.* Formation of unconventional standing waves at graphene edges by valley mixing and pseudospin rotation. *Proc. Natl Acad. Sci.* **108**, 18622–18625 (2011).

34. Dappe, Y. J. & Martínez, J. I. Effect of van der Waals forces on the stacking of coronenes encapsulated in a single-wall carbon nanotube and many-body excitation spectrum. *Carbon N.Y.* **54**, 113–123 (2013).

35. Emtsev, K. V. *et al.* Towards wafer-size graphene layers by atmospheric pressure graphitization of silicon carbide. *Nat. Mater.* **8**, 203–207 (2009).

36. Virojanadara, C., Yakimova, R., Zakharov, A. A. & Johansson, L. I. Large homogeneous mono-/bi-layer graphene on 6H–SiC(0 0 0 1) and buffer layer elimination. *J. Phys. D. Appl. Phys.* **43**, 374010 (2010).

37. Agúndez, M. & Cernicharo, J. Oxygen chemistry in the circumstellar envelope of the carbon-rich star IRC+ 10216. *Astrophys. J.* **650**, 374–393 (2006).

38. Bonaccorso, F., Sun, Z., Hasan, T. & Ferrari, A. Graphene photonics and optoelectronics. *Nat. Photon.* **4**, 611–622 (2010).

39. Peng, T., Lv, H., He, D., Pan, M. & Mu, S. Direct transformation of amorphous silicon carbide into graphene under low temperature and ambient pressure. *Sci. Rep.* **3**, 01148 (2013).

40. Oku, T., Niihara, K. & Suganuma, K. Formation of carbon nanocapsules with SiC nanoparticles prepared by polymer pyrolysis. *J. Mater. Chem.* **8**, 1323–1325 (1998).

41. Bauschlicher, Jr C. *et al.* The NASA Ames polycyclic aromatic hydrocarbon infrared spectroscopic database: the computed spectra. *Astrophys. J. Suppl. Ser.* **189**, 341–351 (2010).

42. Zampieri, G. *et al.* Photoelectron diffraction study of the 6H-SiC(0001) sqrt[3] × sqrt[3] R30° reconstruction. *Phys. Rev. B* **72**, 165327 (2005).

43. Diaz, J., Paolicelli, G., Ferrer, S. & Comin, F. Separation of the sp^{3} and sp^{2} components in the C1s photoemission spectra of amorphous carbon films. *Phys. Rev. B* **54**, 8064–8069 (1996).

44. Wilson, J., Walton, J. & Beamson, G. Analysis of chemical vapour deposited diamond films by X-ray photoelectron spectroscopy. *J. Electron. Spectrosc. Relat. Phenom.* **121**, 183–201 (2001).

Acknowledgements

We acknowledge funding from the following Spanish Grants: CSD2007-41 (NANOSELECT), CSD2009-38 (ASTROMOL), MAT2011-26534. M.S. and P.J. acknowledge GAAWM100101207. Access to the Elettra facility has been granted by the European Community's Seventh Framework Programme (FP7/2007-2013) under Grant agreement number 226716. The research leading to these results has received funding from the European Union Seventh Framework Programme under Grant agreement number 604391 Graphene Flagship. P.M. acknowledges financial support from the Rafael Calvo Rodés programme. J.I.M. acknowledges CSIC-JAE fellowship programme, co-founded by the European Social Funding. M.S. and P.J. acknowledge GAAV project n. M100101207. We thank O. Berné for fruitful discussions. Background images for artistic assembly of Fig. 1 are from NASA.

Author contributions

All experiments were performed by P.M. who also co-wrote the manuscript, made the data analysis and discussed the main results. M.S. collaborated with STM measurements; DFT and MD calculations were performed by P.J. and J.I.M.; P.L., M.D. and S.L. contributed in the synchrotron radiation XPS set-up and measurements. P.S. participated in the project conception using SiC and helped with the XPS measurements. J.C. designed the main astrochemical idea, co-wrote the manuscript and discussed the results. J.A.M.-G. conducted the research, discussed all data and results and co-wrote the manuscript. All authors commented on the final version of the manuscript.

Additional information

2

Extreme ultraviolet imaging of three-dimensional magnetic reconnection in a solar eruption

J.Q. Sun[1], X. Cheng[1], M.D. Ding[1], Y. Guo[1], E.R. Priest[2], C.E. Parnell[2], S.J. Edwards[3], J. Zhang[4], P.F. Chen[1] & C. Fang[1]

Magnetic reconnection, a change of magnetic field connectivity, is a fundamental physical process in which magnetic energy is released explosively, and it is responsible for various eruptive phenomena in the universe. However, this process is difficult to observe directly. Here, the magnetic topology associated with a solar reconnection event is studied in three dimensions using the combined perspectives of two spacecraft. The sequence of extreme ultraviolet images clearly shows that two groups of oppositely directed and non-coplanar magnetic loops gradually approach each other, forming a separator or quasi-separator and then reconnecting. The plasma near the reconnection site is subsequently heated from ~ 1 to ≥ 5 MK. Shortly afterwards, warm flare loops (~ 3 MK) appear underneath the hot plasma. Other observational signatures of reconnection, including plasma inflows and downflows, are unambiguously revealed and quantitatively measured. These observations provide direct evidence of magnetic reconnection in a three-dimensional configuration and reveal its origin.

[1] School of Astronomy and Space Science, Nanjing University, Nanjing 210093, China. [2] School of Mathematics and Statistics, University of St Andrews, Fife, KY16 9SS Scotland, UK. [3] Department of Mathematical Sciences, Durham University, Durham DH1 3LE, UK. [4] School of Physics, Astronomy and Computational Sciences, George Mason University, Fairfax, Virginia 22030, USA. Correspondence and requests for materials should be addressed to X.C. (email: xincheng@nju.edu.cn) or to M.D.D. (email: dmd@nju.edu.cn).

Magnetic reconnection plays an important role in various astrophysical, space and laboratory environments[1] such as γ-ray bursts[2], accretion disks[3,4], solar and stellar coronae[5,6], planetary magnetospheres[7,8] and plasma fusion[9,10]. In the classic two-dimensional (2D) model, reconnection occurs at an X-point where anti-parallel magnetic field lines converge and interact. As a consequence, free energy stored in the magnetic field is rapidly released and converted into other forms of energy, resulting in heating and bulk motions of plasma and acceleration of non-thermal particles[11]. In the past decades, much attention has been paid to validate this picture. One piece of direct evidence is from *in situ* solar wind measurements at the magnetosheath and magnetotail of the Earth[12]. Most observational evidence is from remote sensing observations of solar flares, including cusp-shaped flare loops[13], plasma inflows/outflows[14,15], downflows above flare arcades[16], double hard X-ray coronal sources[17], current sheets[18] and changes in connectivity of two sets of EUV loops during a compact flare[19,20]. With many of these observations, researchers were trying to reveal the 2D aspects of reconnection. However, reconnection is in reality a process in 3D that occurs in places where magnetic connectivity changes significantly, namely, at null points[21], separators[22,23] or quasi-separators[24].

Recently launched spacecraft Solar Terrestrial Relations Observatory (STEREO) and Solar Dynamics Observatory (SDO) provide us an unprecedented opportunity to observe reconnection in a 3D setting. Utilizing stereoscopic observations from these two spacecraft, 3D configurations of various solar phenomena have been reconstructed[25,26]. Here we study

reconnection through its reconstructed 3D magnetic topology as well as many other signatures. The Extreme Ultraviolet Imager (EUVI)[27] on board STEREO and the Atmospheric Imaging Assembly (AIA)[28] on board SDO provide the necessary observational data; in particular, the AIA has an unprecedented high spatial resolution (0.6 arcsec per pixel), high cadence (12 s) and multi-temperature imaging ability (10 passbands).

Results

Overview of the reconnection event. The event of interest occurred on 27 January 2012, when STEREO-A and SDO were separated in space by 108 degrees along their ecliptic orbits (Fig. 1a). From ~00:00 to 03:00 UT (universal time), a pre-existing large-scale cavity, which refers to the dark region in the EUV or soft X-ray passbands and is usually interpreted to be the cross-section of a helical magnetic flux rope[29–31], appears above the western solar limb as seen from the Earth. The reason why the cavity is dark may be that the density has decreased or that the plasma temperature has increased to a value outside the effective response of the lower temperature passbands. The cavity, mostly visible in the AIA 171 Å passband (sensitive to a plasma temperature of ~0.6 MK), starts to expand and rise from ~01:40 UT, and finally results in a coronal mass ejection (CME) that is well observed by the AIA and the Large Angle and Spectrometric Coronagraph (LASCO)[32] on board Solar and Heliospheric Observatory (SOHO; Fig. 1b). The slow rise of the cavity causes its two legs, which are made of cool loops, to approach each other and form an X-shaped structure near 03:00 UT (Fig. 1c,d and

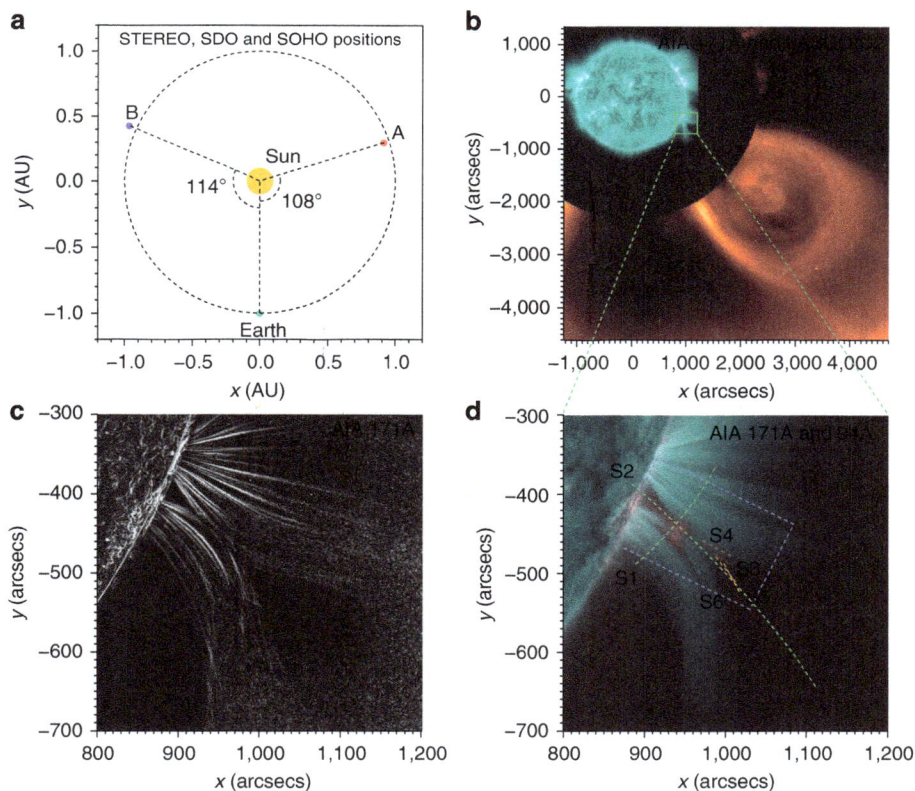

Figure 1 | Overview of the 17 January 2012 solar flare and CME reconnection event. (a) The positions of the Sun, Earth and STEREO-A/B satellites (SOHO is at L1 point and SDO is in the Earth orbit). **(b)** A composition of the AIA 171 Å passband image (cyan) and the LASCO C2 white-light image (red). The green box indicates the main flare region. **(c)** The enhanced AIA 171 Å image showing a clear X-shaped structure. **(d)** A composite image of the AIA 171 Å (cyan) and 94 Å (red) passbands. Cyan (red) indicates coronal loops with a temperature of ~0.6 MK (~7.0 MK). Six dashed lines denote six slices (S1–S6) that are used to trace the evolution of various reconnection features with time.

Supplementary Movies 1 and 2). Following the disappearance of the cool loops (cyan in Fig. 1d), a hot region (~ 7 MK; visible in the AIA 94 Å passband) immediately appears near the X-shaped structure, indicating the initial heating of a solar flare (red in Fig. 1d). Unfortunately, as the flare soft X-ray emission is very weak and submerged in the emission from the decay phase of a previous flare, the accurate magnitude of the flare is not recorded by Geostationary Operational Environmental Satellite. We also note that there are no X-ray observations from the Reuven Ramaty High Energy Solar Spectroscopic Imager because of annealing.

3D topology and origin of magnetic reconnection. Observations from SDO (the AIA 171 Å passband; Fig. 2a), in combination with STEREO-A observations (the EUVI 171 Å passband; Fig. 2b), enable us to reconstruct the 3D topology of the reconnection and its evolution. Owing to the high magnetic Reynolds number of the ionized corona, the plasma is frozen to the magnetic field; and so

the loop-like plasma emission is reasonably assumed to outline the geometry of the magnetic field[33]. We select two magnetic loops (cyan and green dashed lines in Fig. 2a,b) that can most clearly exhibit the reconnection process. With images from two perspective angles, the 3D structure of the loops is reconstructed (Fig. 2c and Supplementary Movie 3). The results display a clear picture of how the connectivity of the loops changes as the reconnection proceeds. Before reconnection, two nearly oppositely directed loops are anchored, respectively, at each side of the filament in the active region (left panel of Fig. 2b). The plasma between their legs has been heated to a moderate temperature (left panel of Fig. 2c).

With the rise of the cavity, the underlying loops of opposite polarities gradually approach each other. As the inward movements of the loops are not coplanar, an apparent separator or quasi-separator appears at ~ 04:14 UT (middle panel of Fig. 2b). We calculate the 3D global magnetic field on January 26 using the potential field assumption[34] and find an absence of pre-existing null points and separators in the reconnection region. However,

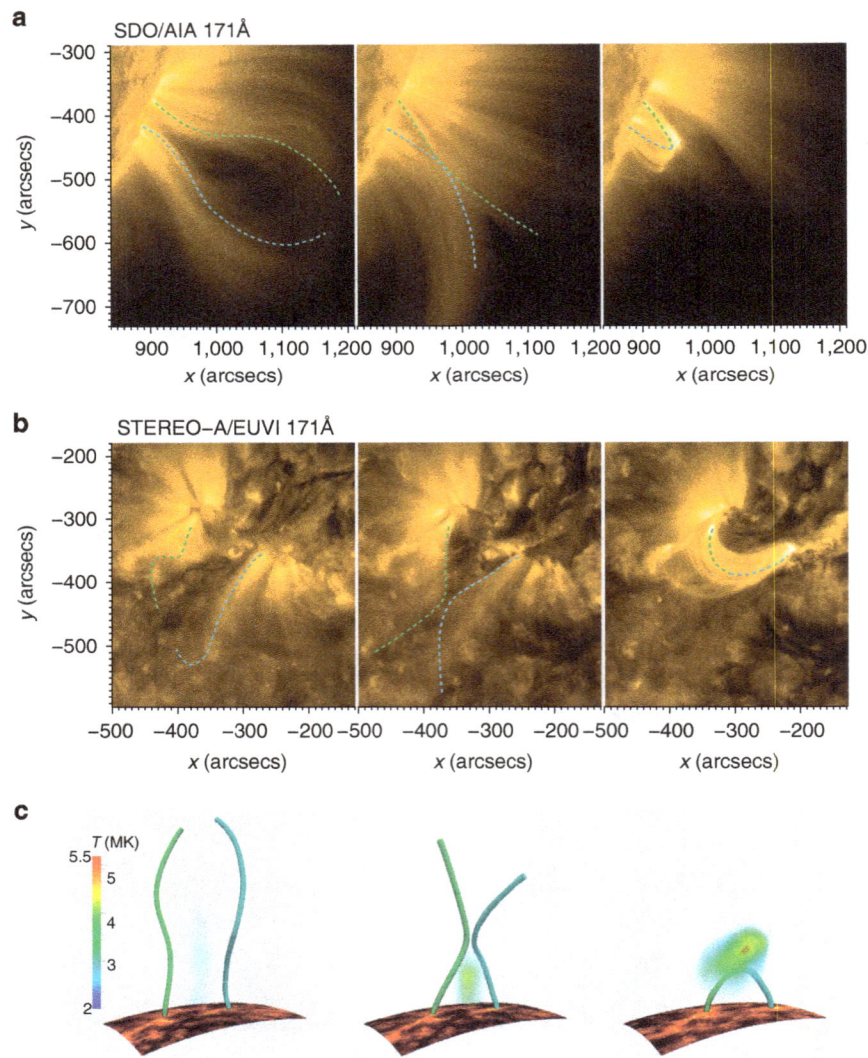

Figure 2 | Plasma and magnetic configurations during the reconnection process. (**a**) The AIA 171 Å images at 02:14 UT (left), 04:14 UT (middle) and 08:14 UT (right) displaying the side view of the evolution of two sets of coronal loops. The cyan and green dashed curves show selected coronal loops representing two magnetic field lines involved in the process. (**b**) The EUVI 171 Å images showing the top view of the reconnection. The cyan and green dashed curves give another view of the same loops as in panel **a**. (**c**) The reconstructed 3D magnetic topology (cyan and green curves) and heated regions (cloud-like structures) before, during and after the reconnection. The bottom boundaries are the projected EUVI 304 Å images showing the footpoints of the flare and the separation of two flare ribbons.

the simple magnetic field in the original bipolar source region is strongly sheared from January 21 as shown by the long-existing filament/prominence at the bottom of the cavity (Fig. 2b). It suggests that a new separator or quasi-separator is formed with the prominence taking off (middle panel of Fig. 2c). As the reconnection initiates, free magnetic energy starts to be released, the most obvious consequence of which is to form a hotter region underneath the reconnection site.

Topologically, the reconnection between the two groups of loops forms poloidal field lines above the reconnection site, increasing the twist of the erupted flux rope. At the same time, a cusp-shaped field below the reconnection site quickly shrinks into a semicircular shape to form flare loops[35] (right panel of Fig. 2c). With the acceleration of the CME, more plasma is heated to temperatures up to ∼5 MK, suggesting an enhanced reconnection. However, the heated region is still confined between the reconnection site and the flare loop top but with a spatial extension.

Quantitative properties of magnetic reconnection. AIA observations with high spatial and temporal resolution successfully capture evidence for reconnection including bilateral inflows, instantaneous heating of plasma near the reconnection site, and downflows that are related to the reconnection outflows[16]. To quantitatively investigate the inflows, we select an oblique slice in the 171 and 94 Å composite images (S1 in Fig. 1d). The time-distance plot (Fig. 3a) clearly shows that the

bilateral cool loops (cyan) keep converging to the middle (reconnection) region from ∼00:00 UT. Once the visible innermost loops come into contact at ∼03:10 UT, they immediately disappear; meanwhile, hot plasma (red) appears at the reconnection site. Several trajectories for inflows are tracked. The velocities of the inflows vary from 0.1 to 3.7 km s^{-1} (Fig. 3b). Moreover, for each trajectory, the velocity tends to increase towards the reconnection site, indicating that an inward force exists on both sides of the current sheet to accelerate the inflows.

The time-distance plot for the vertical slice (S2) shows the eruption of the CME and the downflows above the flare loops (Fig. 3c). In the early phase, only the slow rise of the CME cavity is detectable. However, along with the fast eruption of the CME, magnetic reconnection is initiated, which causes the plasma at the bottom to be rapidly heated. During this process, many dark voids intermittently appear above the heated region, rapidly falling, propagating a distance of ∼20–30 Mm, and finally disappearing. The time-distance plots (Fig. 3d) for four selected slices (S3–S6) show that the different downflows have almost similar trajectories. Their velocities range from ∼100 to 200 km s^{-1} initially, but quickly decrease to tens of km s^{-1} (Fig. 3e).

Role of magnetic reconnection in the flare and CME. The 2D temperature maps (Fig. 4a) and the time-distance plot of the temperature along the vertical slice (Fig. 4b) reveal the detailed temperature evolution of the heated region and flare loops. Before

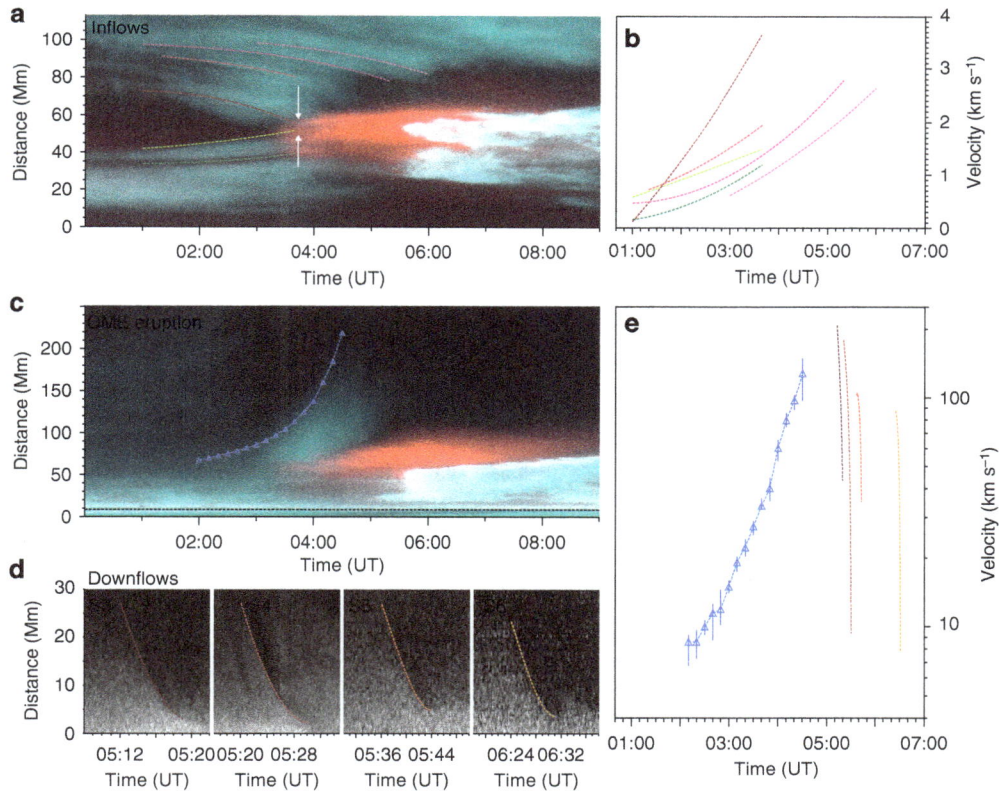

Figure 3 | Temporal evolution of plasma inflows and downflows during the reconnection. (a) A time-distance plot of the composite AIA 171 Å (cyan) and 94 Å (red) images along the direction of the inflows (denoted by S1 in Fig. 1d) showing the approach of oppositely directed loops (two white arrows). The dashed lines with different colours (green to pink) denote the height-time measurements of the inflow at different locations. **(b)** The velocities of inflows, which are derived by cubic-fitting to the height-time data. **(c)** A time-distance plot of the composite AIA 171 Å (cyan) and 94 Å (red) images along the rising direction of the CME (denoted by S2 in Fig. 1d). The blue dashed line denotes the height-time measurement of the CME bubble. **(d)** The time-distance plots of the AIA 94 Å images along the direction of four selected downflows (S3–S6 in Fig. 1d). The dashed lines are the height-time measurements of the downflows. **(e)** The velocities of the CME (blue) and four downflows (brown to yellow). The error in the velocity of the CME (marked by the vertical symbol size) mainly comes from the uncertainty of the height, which is taken as the s.d. of 10 measurements.

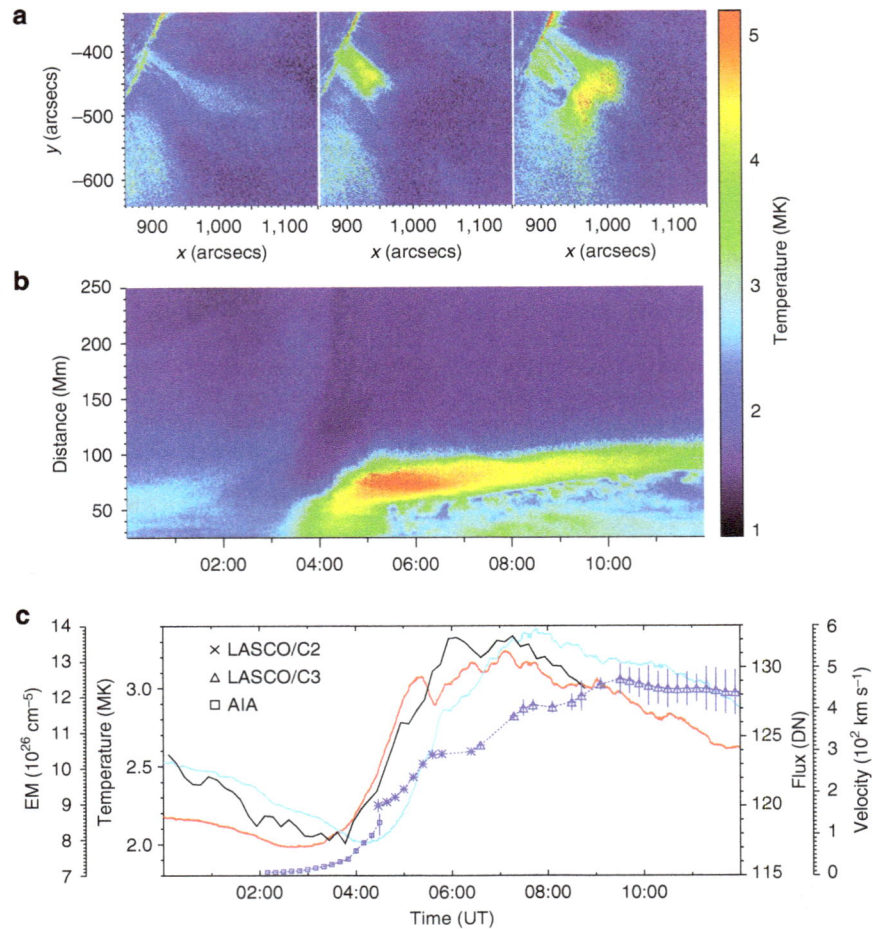

Figure 4 | Temporal evolution of the flare heating and the CME acceleration. (a) The DEM-weighted temperature map at three instants (02:14, 04:14 and 08:14 UT) showing the location of the region heated by the reconnection. **(b)** A time-distance plot of the temperature map along the rising direction of the CME (denoted by S2 in Fig. 1d) illustrating the temperature evolution of the CME bubble and the flare region. **(c)** The temporal evolution of the CME velocity (blue), the flare emission intensity in the EUVI 304 Å passband (black, a proxy of the flare soft X-ray flux), mean temperature (red) and total EM (cyan). The error in the velocity (marked by the vertical symbol size) mainly comes from the uncertainty of the height, which is taken as the s.d. of 10 measurements.

the onset of the flare, the CME cavity is actually hotter than the surrounding coronal plasma, supporting the recent argument that it is most likely a pre-existing hot magnetic flux rope[30,31]. Owing to the rise motion of the hot cavity, the plasma underneath the reconnection region is quickly heated, forming a hot region with an average temperature of $\sim 4\,MK$. In the hot region, some flare loops are discernible. As the reconnection continues, the hot region further ascends and extends, and its temperature rises to $>5\,MK$. In contrast, the originally formed hot flare loops gradually cool down to $\sim 2\,MK$ and become more distinctive in the temperature maps. Meanwhile, newer hot flare loops are formed, stacked over the cool ones. The above process continues until the reconnection stops.

We deduce in detail the evolution of the emission intensity, DEM-weighted temperature, total emission measure (EM) of the flare region and the velocity of the CME (Fig. 4c). The CME velocity is mostly synchronous with the intensity increase and plasma heating of the flare but precedes the total EM by tens of minutes. From $\sim 04{:}00$ to 05:30 UT, the CME velocity, the flare intensity and the temperature rapidly increase. From $\sim 05{:}30$ to 08:00 UT, the CME is still being accelerated, but the intensity and temperature of the flare reach a maximum with only some minor fluctuations. Afterwards, the intensity and temperature start to

decrease; the CME velocity also decreases to a nearly constant value, most likely as a result of the interaction with the background solar wind[36]. On the basis of the velocity of the inflows and downflows, a lower limit on the reconnection rate (the inflow Alfvén Mach number) is estimated to be 0.001–0.03, which is able to produce the observed weak solar flare and the long-accelerating slow CME.

Discussion

We have reconstructed the 3D magnetic topology of the fast reconnection in a solar eruption and quantify the properties of the reconnection and its role in the flare and CME. The method of direct imaging overcomes the disadvantage of magnetic field extrapolations based on non-linear force-free field modelling when studying a dynamic process[37]. The excellent observations provide much needed elucidation of the physical processes involved in a flare/CME in a 3D configuration. In a traditional 2D flare model[6], the initial magnetic configuration consists of two sets of oppositely directed field lines. A current sheet formed between them is an essential ingredient for flare occurrence. Once the reconnection begins, magnetic energy is released to produce an enhanced flare emission, and post-flare loops are formed, mapping to two flare ribbons on the chromosphere. The discovery of a pre-existing flux rope makes a crucial addition to the standard paradigm[31]. It suggests that the reconnection is associated with the eruption of the flux rope. Although the flux rope is a 3D structure, observations made thus far of the reconnection are mostly restricted to 2D, in which the flux rope often appears as a hot plasma blob when viewed along the axis, and the reconnection site underneath the blob is apparently manifested as a thin and long sheet[30]. In a 3D case, the magnetic topology becomes much more complex and there are different regimes of magnetic reconnection[38]. The event studied here does reveal some new features. First, the reconnection site is more likely a separator or quasi-separator. The fact that the two sets of loops that are obviously non-coplanar are approaching each other does imply the presence of a separator or quasi-separator between them. Second, in the 2D case, reconnection forms an isolated closed field (a section of a flux rope) above, in addition to a flare loop below; while in the 3D case, reconnection supplies poloidal flux to the flux rope whose two ends are still anchored on the solar photosphere. As the reconnection proceeds, more and more poloidal flux is added to the flux rope, further accelerating the CME and in turn strengthening the flare emission. Our results are consistent with and lend observational support to models of flux rope-induced solar eruptions[39]. The primary trigger of the eruption may be torus or kink instability of a pre-existing flux rope[40], while magnetic reconnection, which occurs at a newly formed separator[22,23] or quasi-separator[24], releases free magnetic energy and helps accelerate the eruption.

Methods

3D reconstruction and visualization. Using the Interactive Data Language (IDL) program 'scc_measure.pro' in the Solar SoftWare (SSW) package, we reconstruct the 3D coordinates of the magnetic loops. This routine allows us to select a point in, for example, the AIA image. A line representing the line-of-sight from the AIA perspective is then displayed in the image from other perspectives, such as EUVI. According to the emission characteristics, we identify the same point at this line. 3D coordinates of the selected point (heliographic longitude, latitude and radial distance in solar radii) are then determined. With the same manipulation, the 3D coordinates of all the points along the magnetic loops are derived. For each loop, the reconstruction is repeated 10 times, the most optimal one of which is chosen as the result, thus ensuring the accuracy of the reconstruction. To trace the evolution of the magnetic loops, we keep their footpoints fixed. The 3D visualization is realized by the software Paraview.

Differential emission measure reconstruction. The differential emission measure (DEM) is recovered from six AIA passbands including 94 Å (Fe X, $\sim$1.1 MK; Fe XVIII, $\sim$7.1 MK), 131 Å (Fe VIII, $\sim$0.4 MK; Fe XXI, $\sim$11 MK), 171 Å (Fe IX, $\sim$0.6 MK), 193 Å (Fe XII, $\sim$1.6 MK), 211 Å (Fe XIV, $\sim$2.0 MK) and 335 Å (Fe XVI, $\sim$2.5 MK) through the regularized inversion method[41]. The observed flux F_i for each passband can be written as:

$$F_i = \int DEM(T)R_i(T)\mathrm{d}T + \delta F_i, \qquad (1)$$

where $R_i(T)$ is the temperature response function of passband i, $DEM(T)$ indicates the plasma DEM in the corona, and δF_i is the error of the observational intensity for passband i. The temperature range in the inversion is chosen as $5.5 \leq \log T \leq 7.5$. With the derived DEM, the DEM-weighted (mean) temperature and the total EM are calculated as:

$$T_{\mathrm{mean}} = \frac{\int_{T_{\min}}^{T_{\max}} DEM(T)T\mathrm{d}T}{\int_{T_{\min}}^{T_{\max}} DEM(T)\mathrm{d}T} \qquad (2)$$

and

$$EM = \int_{T_{\min}}^{T_{\max}} DEM(T)\mathrm{d}T. \qquad (3)$$

The temperature range of integration is set to be $5.7 \leq \log T \leq 7.1$, within which the EM solutions are well constrained as shown in Supplementary Fig. 1. Finally, with the mean temperature at each pixel, the 2D temperature maps are constructed.

Enhancement of EUV images. To display the fine structures of the EUV images, we enhance the contrast by the routine 'aia_rfilter.pro' in SSW. This program first sums five images and divides the summed image into a number of rings. Each ring is then scaled to the difference of the maximum brightness and the minimum one. The final images are obtained by performing the Sobel edge enhancement taking advantage of the IDL program 'sobel.pro'.

3D magnetic field extrapolation and singularity calculation. With a potential field model, we extrapolate the 3D global magnetic field structure using the Helioseismic and Magnetic Imager[42] daily updated synoptic maps of the radial magnetic field component on 26 January 2012 as the lower boundary. We further calculate the locations of all the null points and separators in the hemisphere containing the reconnection region. However, we cannot find any null points or separators in the source region of the reconnection event. It implies that the null point or separator responsible for the reconnection event is probably formed during the initial stages of the eruption. Note that, as the magnetic data were measured 1 day before the event, a possible evolution of the photospheric magnetic field could change this conclusion.

Uncertainty analysis. The errors of the DEM-weighted temperature and total EM depend on the errors of the DEM results, which come mainly from uncertainties in the temperature response functions of AIA including non-ionization equilibrium effects, non-thermal populations of electrons, modifications of dielectronic recombination rates, radiative transfer effects and even the unknown filling factor of the plasma[43]. Three representative DEM curves are shown in Supplementary Fig. 1b, from which one can find that the DEM solutions are well constrained in the temperature range $5.7 \leq \log T \leq 7.1$. To ensure the accuracy of the regularized inversion method, we also calculate the DEM with the forward fitting method[44] and find that the two results are very similar.

A possible deviation in the 3D reconstruction of magnetic topology mainly comes from the uncertainty in identifying the same feature from two different perspectives. However, this does not affect qualitatively the global 3D topology. The uncertainty in displaying the heated region is mostly from the assumption that the filling factor is 1 and the 3D temperature distribution in the hot region is of cylindrical symmetry with the cross-section of the cylinder corresponding to the DEM-weighted 2D temperature map.

Code availability. The codes 'scc_measure.pro', 'aia_rfilter.pro', and 'sobel.pro' used in the above analysis are available at the website http://www.lmsal.com/solarsoft/.

References

1. Priest, E. & Forbes, T. *Magnetic reconnection: MHD theory and applications* (Cambridge Univ. Press, 2000).
2. Dai, Z. G., Wang, X. Y., Wu, X. F. & Zhang, B. X-ray flares from postmerger millisecond pulsars. *Science* **311**, 1127–1129 (2006).
3. Balbus, S. A. & Hawley, J. F. Instability, turbulence, and enhanced transport in accretion disks. *Rev. Mod. Phys.* **70**, 1–53 (1998).
4. Yuan, F., Lin, J., Wu, K. & Ho, L. C. A magnetohydrodynamical model for the formation of episodic jets. *Mon. Not. R. Astron. Soc* **395**, 2183–2188 (2009).
5. Shibata, K. *et al.* Chromospheric anemone jets as evidence of ubiquitous reconnection. *Science* **318**, 1591–1594 (2007).
6. Sturrock, P. A. Model of the high-energy phase of solar flares. *Nature* **211**, 695–697 (1966).
7. Phan, T. D. *et al.* A magnetic reconnection X-line extending more than 390 Earth radii in the solar wind. *Nature* **439**, 175–178 (2006).
8. Xiao, C. J. *et al.* Satellite observations of separator-line geometry of three-dimensional magnetic reconnection. *Nat. Phys.* **3**, 609–613 (2007).
9. Yamada, M., Kulsrud, R. & Ji, H. Magnetic reconnection. *Rev. Mod. Phys.* **82**, 603–664 (2010).
10. Zhong, J. *et al.* Modelling loop-top X-ray source and reconnection outflows in solar flares with intense lasers. *Nat. Phys.* **6**, 984–987 (2010).
11. Priest, E. *Magnetohydrodynamics of the Sun* (Cambridge Univ. Press, 2014).
12. Paschmann, G. *et al.* Plasma acceleration at the earth's magnetopause-Evidence for reconnection. *Nature* **282**, 243–246 (1979).
13. Masuda, S., Kosugi, T., Hara, H., Tsuneta, S. & Ogawara, Y. A loop-top hard X-ray source in a compact solar flare as evidence for magnetic reconnection. *Nature* **371**, 495–497 (1994).
14. Li, L. & Zhang, J. Observations of the magnetic reconnection signature of an M2 flare on 2000 March 23. *Astrophys. J.* **703**, 877–882 (2009).
15. Yokoyama, T., Akita, K., Morimoto, T., Inoue, K. & Newmark, J. Clear evidence of reconnection inflow of a solar flare. *Astrophys. J. Lett.* **546**, L69–L72 (2001).
16. McKenzie, D. E. & Savage, S. L. Quantitative examination of supra-arcade downflows in eruptive solar flares. *Astrophys. J.* **697**, 1569–1577 (2009).
17. Sui, L. & Holman, G. D. Evidence for the formation of a large-scale current sheet in a solar flare. *Astrophys. J. Lett.* **596**, L251–L254 (2003).
18. Lin, J. *et al.* Direct observations of the magnetic reconnection site of an eruption on 2003 November 18. *Astrophys. J.* **622**, 1251–1264 (2005).

19. Su, Y. *et al.* Imaging coronal magnetic-field reconnection in a solar flare. *Nat. Phys.* **9**, 489–493 (2013).
20. Yang, S., Zhang, J. & Xiang, Y. Magnetic reconnection between small-scale loops observed with the new vacuum solar telescope. *Astrophys. J. Lett.* **798**, L11 (2015).
21. Priest, E. R. & Pontin, D. I. Three-dimensional null point reconnection regimes. *Phys. Plasma* **16**, 122101 (2009).
22. Longcope, D. W., McKenzie, D. E., Cirtain, J. & Scott, J. Observations of separator reconnection to an emerging active region. *Astrophys. J.* **630**, 596–614 (2005).
23. Parnell, C. E., Haynes, A. L. & Galsgaard, K. Structure of magnetic separators and separator reconnection. *J. Geophys. Res.* **115**, A02102 (2010).
24. Aulanier, G., Pariat, E. & Démoulin, P. Current sheet formation in quasi-separatrix layers and hyperbolic flux tubes. *Astron. Astrophys.* **444**, 961–976 (2005).
25. Byrne, J. P., Maloney, S. A., McAteer, R. T. J., Refojo, J. M. & Gallagher, P. T. Propagation of an Earth-directed coronal mass ejection in three dimensions. *Nat. Commun.* **1**, 74 (2010).
26. Feng, L. *et al.* Morphological evolution of a three-dimensional coronal mass ejection cloud reconstructed from three viewpoints. *Astrophys. J.* **751**, 18 (2012).
27. Howard, R. A. *et al.* Sun earth connection coronal and heliospheric investigation (SECCHI). *Space Sci. Rev.* **136**, 67–115 (2008).
28. Lemen, J. R. *et al.* The atmospheric imaging assembly (AIA) on the solar dynamics observatory (SDO). *Solar Phys.* **275**, 17–40 (2012).
29. Gibson, S. E. *et al.* Three-dimensional morphology of a coronal prominence cavity. *Astrophys. J.* **724**, 1133–1146 (2010).
30. Cheng, X., Zhang, J., Liu, Y. & Ding, M. D. Observing flux rope formation during the impulsive phase of a solar eruption. *Astrophys. J. Lett.* **732**, L25 (2011).
31. Zhang, J., Cheng, X. & Ding, M. D. Observation of an evolving magnetic flux rope before and during a solar eruption. *Nat. Commun.* **3**, 747 (2012).
32. Brueckner, G. E. *et al.* The large angle spectroscopic coronagraph (LASCO). *Solar Phys.* **162**, 357–402 (1995).
33. Priest, E. R. & Forbes, T. G. The magnetic nature of solar flares. *Astron. Astrophys. Rev.* **10**, 313–377 (2002).
34. Sakurai, T. Computational modeling of magnetic fields in solar active regions. *Space Sci. Rev.* **51**, 11–48 (1989).
35. Forbes, T. G. & Acton, L. W. Reconnection and field line shrinkage in solar flares. *Astrophys. J.* **459**, 330–341 (1996).
36. Gopalswamy, N. Properties of interplanetary coronal mass ejections. *Space Sci. Rev.* **124**, 145–168 (2006).
37. Cheng, X. *et al.* Formation of a double-decker magnetic flux rope in the sigmoidal solar active region 11520. *Astrophys. J.* **789**, 93 (2014).
38. Pontin, D. I. Theory of magnetic reconnection in solar and astrophysical plasmas. *Phil. Trans. R. Soc. A* **370**, 3169–3192 (2012).
39. Shibata, K. *et al.* Hot-plasma ejections associated with compact-loop solar flares. *Astrophys. J. Lett.* **451**, L83–L85 (1995).
40. Kliem, B. & Török, T. Torus instability. *Phys. Rev. Lett.* **96**, 255002 (2006).
41. Hannah, I. G. & Kontar, E. P. Differential emission measures from the regularized inversion of Hinode and SDO data. *Astron. Astrophys.* **539**, A146 (2012).
42. Schou, J. *et al.* Design and ground calibration of the helioseismic and magnetic imager (HMI) instrument on the Solar Dynamics Observatory (SDO). *Solar Phys.* **275**, 229–259 (2012).
43. Judge, P. G. Coronal emission lines as thermometers. *Astrophys. J.* **708**, 1238–1240 (2010).
44. Cheng, X., Zhang, J., Saar, S. H. & Ding, M. D. Differential emission measure analysis of multiple structural components of coronal mass ejections in the inner corona. *Astrophys. J.* **761**, 62 (2012).

Acknowledgements

SDO is a mission of NASA's Living With a Star Program, STEREO is the third mission in NASA's Solar Terrestrial Probes Program, and SOHO is a mission of international cooperation between ESA and NASA. X.C., J.Q.S., M.D.D., Y.G., P.F.C. and C.F. are supported by NSFC through grants 11303016, 11373023, 11203014 and 11025314, and by NKBRSF through grants 2011CB811402 and 2014CB744203. C.E.P. and S.J.E. are supported by the UK STFC. J.Z. is supported by US NSF AGS-1249270 and AGS-1156120.

Author contributions

X.C. and J.Q.S. analysed the observational data and contributed equally to the work. X.C. initiated the study. M.D.D. supervised the project and led the discussions. Y.G. joined part of the data analysis. C.E.P. and S.J.E. performed the magnetic field extrapolation. X.C. wrote the first draft. M.D.D. and E.R.P. made major revisions of the manuscript. The remaining authors discussed the results and commented on the manuscript.

Additional information

3

Quantum dynamics of CO–H$_2$ in full dimensionality

Benhui Yang[1], P. Zhang[2], X. Wang[3], P.C. Stancil[1], J.M. Bowman[3], N. Balakrishnan[4] & R.C. Forrey[5]

Accurate rate coefficients for molecular vibrational transitions due to collisions with H$_2$, critical for interpreting infrared astronomical observations, are lacking for most molecules. Quantum calculations are the primary source of such data, but reliable values that consider all internal degrees of freedom of the collision complex have only been reported for H$_2$-H$_2$ due to the difficulty of the computations. Here we present essentially exact, full-dimensional dynamics computations for rovibrational quenching of CO due to H$_2$ impact. Using a high-level six-dimensional potential surface, time-independent scattering calculations, within a full angular momentum coupling formulation, were performed for the de-excitation of vibrationally excited CO. Agreement with experimentally determined results confirms the accuracy of the potential and scattering computations, representing the largest of such calculations performed to date. This investigation advances computational quantum dynamical studies representing initial steps towards obtaining CO–H$_2$ rovibrational quenching data needed for astrophysical modelling.

[1]Department of Physics and Astronomy and the Center for Simulational Physics, The University of Georgia, Athens, Georgia 30602, USA. [2]Department of Chemistry, Duke University, Durham, North Carolina 27708, USA. [3]Department of Chemistry, Emory University, Atlanta, Georgia 30322, USA. [4]Department of Chemistry, University of Nevada Las Vegas, Las Vegas, Nevada 89154, USA. [5]Department of Physics, Penn State University, Berks Campus, Reading, Pennsylvania 19610, USA. Correspondence and requests for materials should be addressed to P.C.S. (email: stancil@physast.uga.edu).

Quantum mechanical studies of inelastic processes in molecular collisions began with the development of the nearly exact, close-coupling (CC) method for rotational transitions in atom–diatom collisions by Arthurs and Dalgarno[1] and Takayanagi[2] in the 1960s. Over the past five decades tremendous advances in computational processing power and numerical algorithms have allowed high-level computations of inelastic processes[3,4], as well as reactive collisional dynamics of large molecular systems, the latter using primarily time-dependent approaches[5–7]. Until recently, however, the largest full-dimensional inelastic studies for a system of two colliding molecules have been limited to H_2-H_2 collisions in six-dimensions (6D), which were performed with both time-independent[8] and time-dependent[9,10] approaches for solving the Schrödinger equation. However, to make these computations possible, the authors resorted to various angular-momentum decoupling approximations with uncertain reliability. It was only recently[11–13] that these decoupling approximations were relaxed and essentially exact full-dimensional CC computations for the H_2-H_2 system became feasible. Despite the large internal energy spacing of H_2, which allows the basis sets to be relatively compact, these calculations are computationally demanding. Replacing one H_2 molecule with a molecule that has smaller internal energy spacing, such as carbon monoxide, would further increase the computational demands. Whether the CC method could in practice be used to describe such a diatom–diatom system in full dimensionality remains an open question.

First detected in the interstellar medium in 1970 (ref. 14), carbon monoxide is the second most abundant molecule, after H_2, in most astrophysical environments. CO has been the focus of countless theoretical astrophysical studies and observations, being detected in objects as distant as high redshift quasars[15] to cometary comae in our solar system[16] to the atmospheres of extrasolar giant planets[17]. Most studies have focused on pure rotational transitions observed in the far infrared to the radio or electronic absorption in the near ultraviolet. Over the past decade, however, near infrared emission of CO due to the fundamental vibrational band near 4.7 μm has been detected in a variety of sources including star-forming regions in Orion with the Infrared Space Observatory[18] and protoplanetary disks (PPDs) of young stellar objects[19–21] with the Gemini Observatory and the Very Large Telescope. In addition, pure rotational transitions, but in the first two vibrationally excited states ($v_1 = 1$ and 2), were detected by the Submillimeter Array in the circumstellar shell of the much-studied evolved star IRC + 10216 (ref. 22). In particular, high-resolution observations of the CO fundamental band probe the physical conditions in the inner disk region, ∼ 10–20 astronomical units (AU), the site of planet formation. Detailed modelling of such environments requires state-to-state inelastic rovibrational excitation rate coefficients for CO due to H_2 collisions, but current simulations are limited to approximate scaling methods due to the dearth of explicit data[18,23].

In this Article, we address the two issues outlined above by advancing the state-of-the-art for inelastic quantum dynamics with a full-dimensional investigation using an accurate potential energy surface (PES) relevant to this scattering process, with a particular emphasis on the important region of the van der Waals complex. This was made possible through the accurate computation and precise fitting of a 6D CO–H_2 PES in the relevant region of the formaldehyde tetra-atomic system and the further development of the inelastic diatom–diatom time-independent scattering code, TwoBC[24], which performs full angular momentum coupling, the CC formalism[1], including vibrational degrees of freedom. We first briefly describe the new CO–H_2 PES and its testing through comparison of rotational excitation

calculations using the 6D PES and 4D PESs (neglecting vibrational motion) and available experimental results. The full-dimensional (6D), essentially exact, computations for rovibrational quenching of CO($v_1 = 1$) due to H_2 collisions are presented resolving a two-orders-magnitude discrepancy between earlier 4D calculations, which adopted various approximations[25–28]. Finally, the current results are shown to be consistent with the rovibrational quenching measurements for the CO–H_2 system, performed at the Oxford Physical Chemistry Laboratory from 1976–1993, which no prior calculation has yet been able to adequately explain[29–31].

Results

The CO–H_2 PES. The CO–H_2 interaction has been of considerable interest to the chemical physics community for many decades with one of the first 4D surfaces for the electronic ground state constructed by Schinke et al.[32], which was later extended by Bačić et al.[25] The group of Jankowski and Szalewicz[33] performed accurate 5D and 6D electronic energy calculations, but averaged over monomer vibrational modes, and performed several fits to obtain a series of 4D rigid-rotor surfaces, referred to as the V98 (ref. 33), V04 (ref. 34) and V12 (ref. 35) PESs. A 6D PES for formaldehyde was constructed earlier by Zhang et al.[36], but it was developed for reactive scattering applications and, consequently, limited attention was given to the long-range CO–H_2 van der Waals configuration. Therefore, as a prerequisite to 6D inelastic dynamics studies, we carried out an unprecedented potential energy calculation including over 459 756 energy points (see Methods for details). The potential energy data were then fit using the invariant polynomial method with Morse-type variables in terms of bond-distances[37,38]. The resulting 6D PES, referred to as V6D, is shown in Fig. 1 for one sample configuration. Some features of V6D are also illustrated in Supplementary Figs 2–4.

Cross-sections and rate coefficients. Time-independent quantum scattering calculations were performed using the CC formulation of Arthurs and Dalgarno[1] as implemented for diatom–diatom collisions in the 4D rigid-rotor approximation in MOLSCAT[39] and extended to full-dimensional dynamics as described by Quéméner, Balakrishnan, and coworkers[12,13] in TwoBC. In the first set of scattering calculations, the new 6D PES was tested for pure rotational excitation from CO($v_1 = 0, j_1 = 0, 1$), where v_1 and j_1 are the vibrational and rotational quantum numbers, respectively. The crossed molecular beam experiment of Antonova et al.[40], who obtained relative state-to-state rotational inelastic cross-sections, is used as a benchmark. The experimental cross-sections were determined at three centre of mass kinetic energies (795, 860 and 991 cm^{-1}), but with an initial state distribution of CO estimated to be 75 ± 5% for $j_1 = 0$ and 25 ± 5% for $j_1 = 1$. Antonova et al.[40] normalized the relative cross-sections to rigid-rotor calculations done with MOLSCAT using the V04 PES of Jankowski and Szalewicz[33]. Comparison of the experiment with new 4D rigid-rotor calculations on V12 and full-dimensional calculations on V6D are shown in Fig. 2. We find no difference in the excitation cross-sections when using V04 or V12, while the root mean square (RMS) cross-section errors between the normalized experiment and the V12 and V6D calculations range from 0.56–0.89 × 10^{-16} cm^2 to 0.55–0.95 × 10^{-16} cm^2, respectively (See Supplementary Table 1 and Supplementary Note 1). Clearly, a 4D rigid-rotor treatment of the dynamics is sufficient for describing rotational excitation at these relatively high energies.

The importance of full dimensionality for rotational excitation becomes more evident as the collision energy is reduced (see

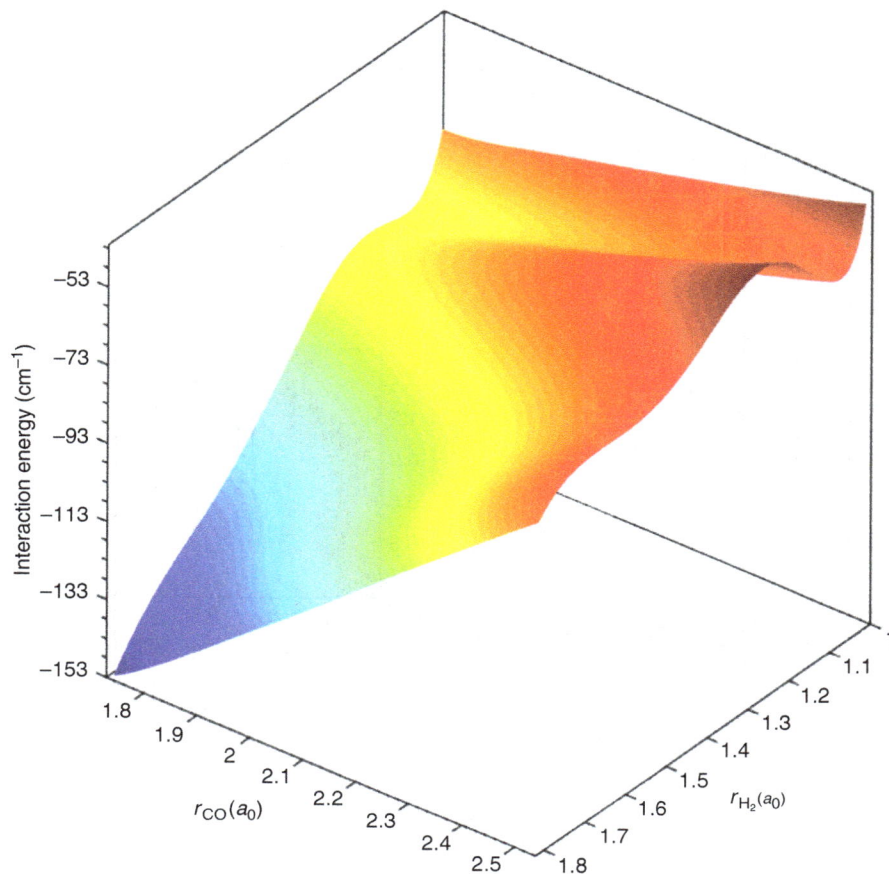

Figure 1 | The CO–H$_2$ interaction potential energy surface V6D. The potential surface is constructed in the 6D diatom–diatom Jacobi coordinates $(R, r_1, r_2, \theta_1, \theta_2, \phi)$, r_1 and r_2 are bond lengths, R the internuclear distance between CO and H$_2$ centre of masses, θ_1 and θ_2 the angles between $\mathbf{R}$ and $\mathbf{r}_1$ and $\mathbf{r}_2$, and ϕ the dihedral or twist angle. See Supplementary Figure 1. Here the dependence of the potential surface on bond lengths $r_{CO} = r_1$ and $r_{H2} = r_2$ is shown with $R = 8$ a$_0$, $\theta_1 = 180°$, $\theta_2 = 0$ and $\phi = 0$. Note that the CO(r_1) and H$_2$(r_2) diatom potentials have been subtracted.

also Supplementary Figs 5 and 6). Low-energy excitation cross-sections for the process

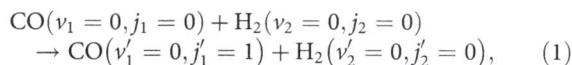

$$CO(v_1 = 0, j_1 = 0) + H_2(v_2 = 0, j_2 = 0)$$
$$\rightarrow CO(v_1' = 0, j_1' = 1) + H_2(v_2' = 0, j_2' = 0), \quad (1)$$

or $(0000) \rightarrow (0100)$, using the notation defined in Methods, were measured by Chefdeville et al.[41] in a crossed-beam experiment. They obtained the excitation cross-section for centre of mass kinetic energies from 3.3 to 22.5 cm^{-1}. Although their energy resolution was limited, three broad features were detected near 6, 13 and 16 cm^{-1} attributed to orbiting resonances. Computed cross-sections for process (1) using the 4D V12 PES and the full-dimensional V6D PES are presented in Fig. 3a. While both computations reveal numerous resonances, the resonances are generally shifted by 2–3 cm^{-1} between calculations. The energy and magnitude of the resonances are very sensitive to the details of the PESs, but differences may also be due to the relaxation of the rigid-rotor approximation with the use of V6D in TwoBC. In Fig. 3b, the two calculations are convolved over the experimental energy resolution and compared with the measured relative cross-sections. Except for the peak near 8 cm^{-1}, the current 6D calculation appears to reproduce the main features of the experiment. RMS errors are found to be 0.355 and 0.228 × 10^{-16} cm^2 for the V12 and V6D PESs, respectively. Further details on the rotational excitation calculations can be found in Supplementary Note 1.

Now that the improved performance of the V6D potential for pure rotational excitation is apparent, we turn to rovibrational

transitions. As far as we are aware, previous experimental[29–31,42,43] and theoretical[25–28] studies are limited to the total quenching of CO($v_1 = 1$). In the scattering calculations of both Bačić et al.[25,26] and Reid et al.[27], the 4D potential of Bačić et al.[25] was adopted, in which two coordinates were fixed ($r_2 = 1.4$ a$_0$, $\phi = 0$), and various combinations of angular-momentum decoupling approximations for the dynamics were utilized (for example, the infinite order sudden, IOS, and coupled-states, CS, approximations; see Supplementary Note 2). More recently, Flower[28] performed CC calculations on a parameterization of the 4D PES of Kobaysashi et al.[44] These four sets of 4D calculations for quenching due to para-H$_2$ ($j_2 = 0$) are compared in Fig. 4 with the current 6D/CC calculations on the V6D surface for the case of $j_2 = j_2' = 0$. Corresponding state-to-state and total cross-sections for collisions with ortho-H$_2$ are given in Supplementary Figs 7–9 with further details in Supplementary Note 2.

A cursory glance at Fig. 4 reveals a more than two-orders-of-magnitude discrepancy among the various calculations. The large dispersion for the previous calculations is due to a combination of reduced dimensionality and decoupled angular momentum, which makes it difficult to assess the reliability of each approximation. The current results, however, remove these uncertainties by utilizing (i) a full-dimensional (6D) PES, (ii) full-dimensional (6D) dynamics and (iii) full angular momentum coupling. The sharp peaks in the cross-sections over the 1–10 cm^{-1} range in the 6D/CC results are due to resonances[45,46] supported by the CO–H$_2$ van der Waals potential well. These resonances, in the rovibrational quenching

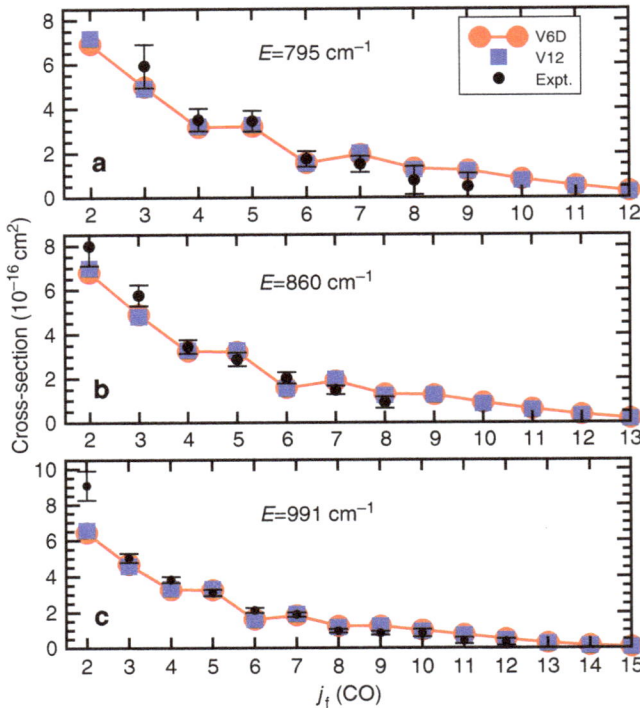

Figure 2 | State-to-state cross-sections for rotational excitation of CO($v_1 = 0, j_1 = 0, 1$) by collisions with H$_2$. Theoretical results of full-dimensional calculations on V6D and 4D rigid-rotor calculations on V12 are compared with normalized experimental results[40] for collision energies of (**a**) 795 cm^{-1}, (**b**) 860 cm^{-1} and (**c**) 991 cm^{-1}. The calculations were performed for H$_2$($v_2 = 0, j_2 = 0$), but the experimental H$_2$ rotational distribution was undetermined. The error bars correspond to twice the estimated s.d. in the weighted means of the measurements[40].

Figure 3 | Low-energy excitation cross-sections. $j_1 = 0 \to 1$ cross-sections for CO($v_1 = 0$) due to collisions with H$_2$($v_2 = 0, j_2 = 0$) are shown as a function of collision energy. (**a**) Computed cross-sections using the 4D V12 and 6D V6D PESs. (**b**) Computed cross-sections convolved over the experimental beam energy spread (lines) compared with the relative experiment of Chefdeville et al.[41] (circles with error bars). The error bars on the experimental cross-sections of Chefdeville et al. represent the statistical uncertainty at a 95% confidence interval.

of CO by H$_2$, are reported here for the first time and their prediction is made possible through our high-level treatment of the dynamics. Computations for transitions involving other initially excited j_2 states and inelastic H$_2$ ($j_2 \neq j_2'$) transitions are presented and compared in Supplementary Figs 10 and 11 and Supplementary Note 2.

Using the current 6D/CC cross-sections and the 4D/IOS-CS results of Reid et al., rate coefficients for a Maxwellian velocity distribution are computed and compared in Fig. 5 with total de-excitation measurements[29–31] reported by Reid et al.[27] The comparison is not straight forward because (i) the measurements correspond to an initial thermal population of H$_2$ rotational states, (ii) the initial rotational population of CO($v_1 = 1, j_1$) was unknown, (iii) the experimental rate coefficients for ortho-H$_2$ are estimated from para- and normal-H$_2$ measurements and (iv) the contribution from a quasi-resonant channel,

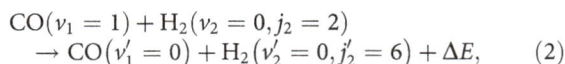

$$\text{CO}(v_1 = 1) + \text{H}_2(v_2 = 0, j_2 = 2)$$
$$\to \text{CO}(v_1' = 0) + \text{H}_2(v_2' = 0, j_2' = 6) + \Delta E, \quad (2)$$

dominates the para-H$_2$ case for $T \gtrsim 50$ K. Figure 5a displays the CO($v_1 = 1$) rovibrational de-excitation rate coefficients from the current 6D/CC calculations for collisions of ortho-H$_2$ for $j_2 = 1$ and 3, separately. These rate coefficients are summed over all j_1' and $j_2' = 1$, 3 and 5. Assuming a Boltzmann average at the kinetic temperature T of the H$_2$ rotational levels $j_2 = 1$ and 3, presumed to be representative of the experimental conditions, rate coefficients are computed and found to be in good agreement with the measurements above 200 K. From Fig. 5a, the contribution from $j_2 = 3$ is seen to be only important above 150 K. The remaining

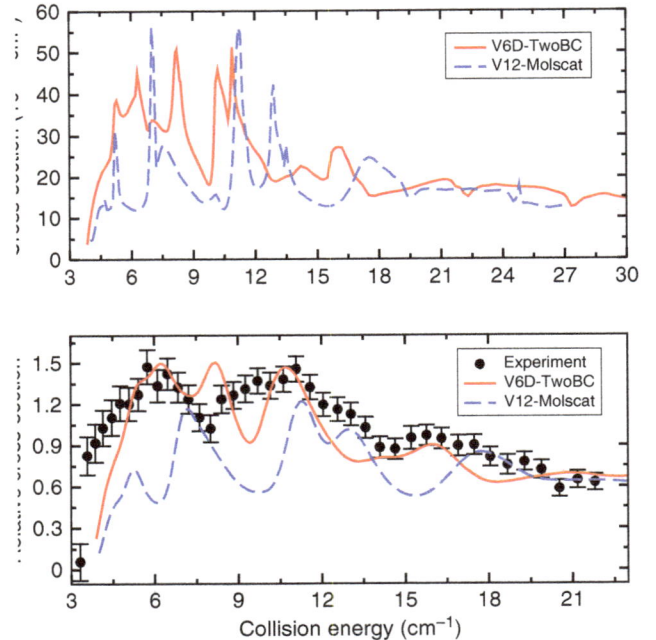

difference with experiment at low temperatures may be due to the fact mentioned above that the ortho-H$_2$ rate coefficients are not directly measured, but deduced from normal-H$_2$ and para-H$_2$ experiments. In particular, Reid et al.[27] assume that the ortho/para ratio in the normal-H$_2$ measurements is 3:1, that is, statistical, and independent of temperature. Further, as stated above, the experimental CO rotational population distribution in $v_1 = 1$ is unknown. Nevertheless, the current 6D/CC computations are a significant advance over the 4D results of Reid et al.[27], which also correspond to a Boltzmann average of rate coefficients for $j_2 = 1$ and 3.

As indicated above, the situation for para-H$_2$ collisions is more complicated due to the quasi-resonant contribution (2), a mechanism not important for ortho-H$_2$. Boltzmann-averaged rate coefficients are presented in Fig. 5b including $j_2 = 0$ and 2 summed over $j_2' = 0$, 2 and 4, with and without the quasi-resonant contribution, $j_2 = 2 \to j_2' = 6$. While the current 6D/CC results and the 4D calculations of Reid et al.[27] are in agreement that the quasi-resonant contribution becomes important for $T \gtrsim 50$ K, the relative magnitude compared with the non-resonant transitions from the 6D/CC calculation is somewhat less than obtained previously with the 4D potential. This is partly related to the fact that the 6D/CC rate coefficients for $j_2 = 0$ are significantly larger than those of Reid et al.[27] (see also the corresponding cross-sections in Fig. 4 and in the Supplementary Fig. 9). Compared with the experiment, we obtain excellent agreement for $T \lesssim 150$ K, but are somewhat smaller at higher temperatures. This small discrepancy may be related to the unknown CO($v_1 = 1, j_1$) rotational population in the measurement. Nevertheless, it is only the 6D/CC computations, that is, dynamics in full dimensionality with full angular

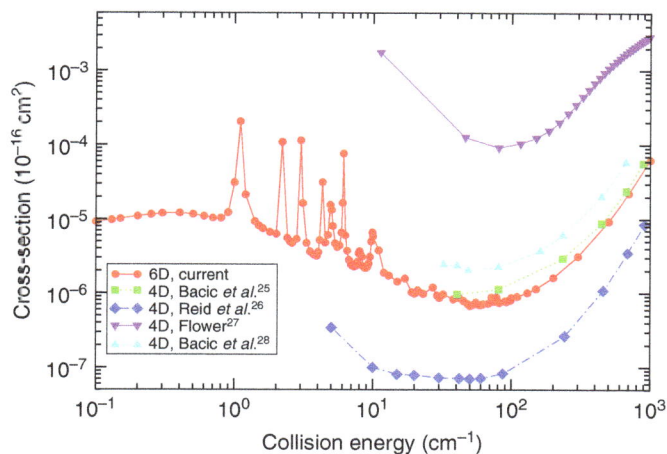

Figure 4 | Total theoretical cross-sections for the vibrational de-excitation of CO($v_1 = 1$) by para-H$_2$. Current 6D/CC results are compared with previous 4D calculations. The 4D results of Bačić et al.[26] and Reid et al.[27] do not distinguish CO rotational states, while the 4D results of Bačić et al.[25], 4D results of Flower[28], and the current 6D/CC results are for initial $j_1 = 0$ summed over all final CO($v_1' = 0, j_1'$). In every case, the H$_2$ rovibrational state remains unchanged, $v_2, j_2 = v_2', j_2' = 0, 0$.

momentum coupling, that can reproduce the measurements for both ortho- and para-H$_2$. Further, the computations for the quasi-resonant process (2) were the most challenging reported here due to the requirement of a very large basis set (see Supplementary Table 2), which resulted in long computation times, a large number of channels and usage of significant disk space (~ 0.5 TB per partial wave). In total, the cross-sections given here consumed $> 40,000$ CPU hours.

Discussion

The current investigation of the CO–H$_2$ inelastic collision system has been performed with the intent of minimal approximation through the computation of a high-level PES, robust surface fitting and full-dimensional inelastic dynamics with full angular momentum coupling. That is, within this paradigm for studying inelastic dynamics, we have advanced the state-of-the-art for diatom–diatom collisions through this unprecedented series of computations. The approach has been benchmarked against experiment for pure rotational and rovibrational transitions giving the most accurate results to date within the experimental uncertainties and unknowns. The accuracy and long-range behaviour of the 6D PES is found to be comparable to previous, lower-dimensional surfaces. The agreement of the current computation for the CO($v_1 = 1$) rovibrational quenching with

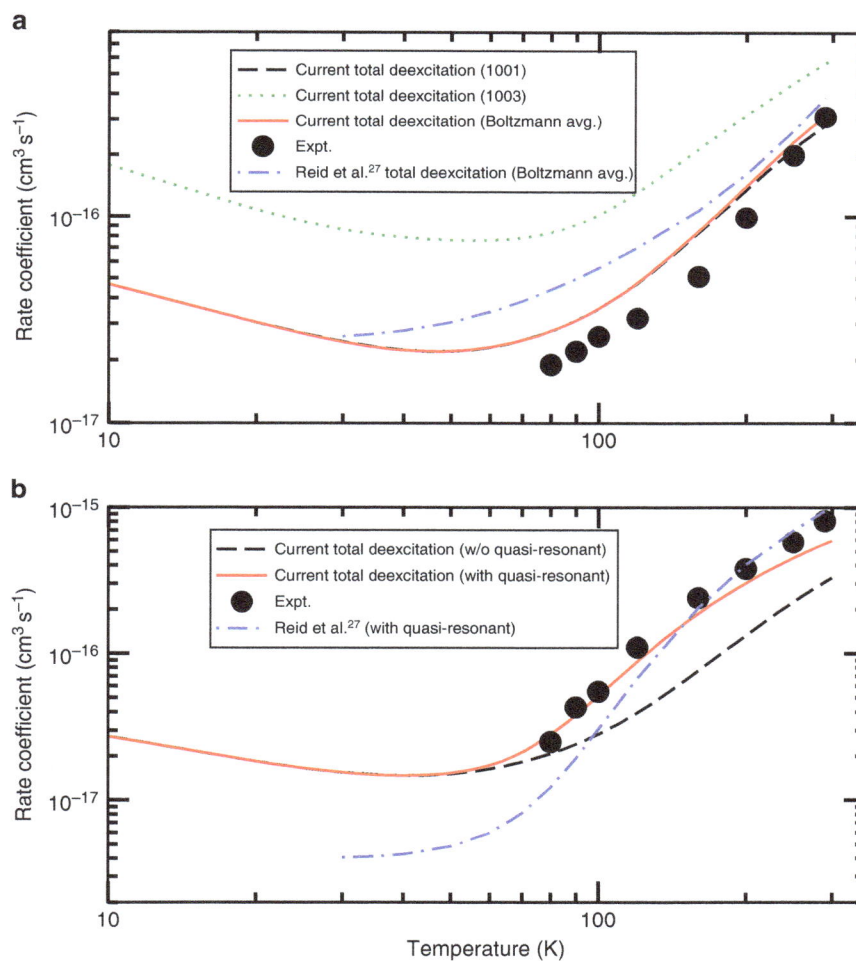

Figure 5 | Rate coefficients for the vibrational de-excitation of CO($v_1 = 1$) due to H$_2$. Current 6D/CC (solid, dashed and dotted lines) and Reid et al.[27] 4D (dot-dashed line) calculations are compared with the total CO($v_1 = 1$) rovibrational quenching experiment (symbols)[29-31]. (**a**) Ortho-H$_2$ rate coefficients for initial state-resolved, $(100 j_2) \rightarrow (0 j_1' 0 j_2')$, summed over j_1' and j_2' (dashed and dotted lines) and for a Boltzmann average of initial H$_2$($j_2 = 1,3$) (solid and dot-dashed lines). (**b**) Para-H$_2$ rate coefficients for a Boltzmann average of initial H$_2$($j_2 = 0,2$) with and without the quasi-resonant process (2). Note that the experimental uncertainties are smaller than the symbol sizes.

measurement resolves a long-standing (more than two-decades) discrepancy and justifies the requirement of a full-dimensional approach. This methodology can now, though with significant computational cost, be applied to a large range of initial rotational levels for $v_1 = 1$ and for higher vibrational excitation to compute detailed state-to-state cross-sections unobtainable via experiment.

This advance in computational inelastic scattering is particularly timely as ground-based (for example, the Very Large Telescope (VLT)) observations have focused on CO rovibrational emission/absorption in a variety of astrophysical objects, while related observations are in the planning stages for future space-based telescopes (for example, NASA's James Webb Space Telescope). In particular, we are in an exciting era of investigation into the properties of PPDs around young stellar objects[47]. PPDs provide the material for newly forming stars and fledgling planets. The CO fundamental band ($|\Delta v_1| = 1$) is a tracer of warm gas in the inner regions of PPDs and, with appropriate modelling, gives insight into disk-gas kinematics and disk evolution in that zone where planets are expected to be forming. A recent survey of 69 PPDs with the VLT[19] detected CO vibrational bands in 77% of the sources including $v_1 = 1 \rightarrow 0$, $v_1 = 2 \rightarrow 1$, $v_1 = 3 \rightarrow 2$, and even $v_1 = 4 \rightarrow 3$ in a few cases. Remarkably, rotational excitation as high as $j_1 = 32$ was observed. However, the modelling of PPDs, and other astrophysical sources with CO vibrational excitation, is hindered by the lack of rate coefficients due to H_2 collisions. We are now in an excellent position to provide full-dimensional, state-to-state CO–H_2 collisional data, which will not only have a profound impact on models characterizing these intriguing environments that give birth to planets, but also aid in critiquing current theories used to describe their evolution.

Methods

Potential energy computations.
The potential energy computations were performed using the explicitly correlated coupled cluster (CCSD(T)-F12b) method[48,49], as implemented in MOLPRO2010.1 (ref. 50). The cc-pcvqz-f12 orbital basis sets[51] that have been specifically optimized for use with explicitly correlated F12 methods and for core-valence correlation effects have been adopted. Density fitting approximations[49] were used in all explicitly correlated calculations with the AUG-CC-PVQZ/JKFIT and AUG-CC-PWCVQZ-MP auxiliary basis sets[52,53]. The diagonal, fixed amplitude 3C(FIX) ansatz was used, which is orbital invariant, size consistent, and free of geminal basis set superposition error (BSSE)[54,55]. The default CCSD-F12 correlation factor was employed in all calculations, and all coupled cluster calculations assume a frozen core (C:1 s and O:1 s). The counterpoise (CP) correction[56] was employed to reduce BSSE. Although the explicitly correlated calculations recover a large fraction of the correlation energy, the CP correction is still necessary, mainly to reduce the BSSE of the Hartree–Fock contribution. Benchmark calculations at the CCSD(T)-F12b/cc-pcvqz-f12 level were carried out on selected molecular configurations and results were compared with those from the conventional CCSD(T) method using aug-cc-pV5Z and aug-cc-pV6Z basis sets. Results showed that the CP-corrected interaction energy agrees closely with those derived from CCSD(T)/aug-cc-pV6Z.

To construct the PES, the computations were performed on a six-dimensional (6D) grid using Jacobi coordinates as shown in Supplementary Fig. 1. R is the distance between the centre of mass of CO and H_2. r_1 and r_2 are the bond lengths of CO and H_2, respectively. θ_1 is the angle between $\mathbf{r}_1$ and $\mathbf{R}$, θ_2 the angle between $\mathbf{r}_2$ and $\mathbf{R}$, and ϕ the out-of-plane dihedral or twist angle. In the potential energy computations, the bond lengths are taken over the ranges $1.7359 \leq r_1 \leq 2.5359$ a_0 and $1.01 \leq r_2 \leq 1.81$ a_0, both with a step-size of 0.1 a_0. For R, the grid extends from 4.0 to 18.0 a_0 with step-size of 0.5 a_0 for $R < 11.0$ a_0 and 1.0 a_0 for $R > 11.0$ a_0. All angular coordinates were computed with a step-size of 15° with $0 \leq \theta_1 \leq 360°$ and $0 \leq \theta_2, \phi \leq 180°$. Additional points were added in the region of the van der Waals minimum.

The PES fit.
The CO–H_2 interaction PES has been fitted in 6D using an invariant polynomial method[37,38]. The PES was expanded in the form,

$$V(y_1 \cdots y_6) = \sum_{n_1 \cdots n_6}^{N} c_{n_1 \cdots n_6} y_1^{n_1} y_6^{n_6} [y_2^{n_2} y_3^{n_3} y_4^{n_4} y_5^{n_5} + y_2^{n_5} y_3^{n_4} y_4^{n_3} y_5^{n_2}], \quad (3)$$

where $y_i = e^{-0.5d_i}$ is a Morse-type variable. The internuclear distances d_i between two atoms are defined as $d_1 = d_{HH'}$, $d_2 = d_{OH'}$, $d_3 = d_{CH'}$, $d_4 = d_{CH}$, $d_5 = d_{OH}$ and $d_6 = d_{CO}$. The total power of the polynomial, $N = n_1 + n_2 + n_3 + n_4 + n_5 + n_6$, was restricted to a maximum of 6. Expansion coefficients $c_{n_1 \cdots n_6}$ were obtained using

weighted least-squares fitting for potential energies up to 10,000 cm^{-1}. The RMS error in the fit of the PES is 14.22 cm^{-1}, which included 398,218 different geometries. This RMS error can be compared with that of 277 cm^{-1} for the 6D reactive surface of Zhang et al.[36] which, however, extended to more than 30,000 cm^{-1}. From the computed energy points, the global minimum of the total potential corresponds to the collinear arrangement H-H-C-O ($\theta_1 = 0, \theta_2 = 0, \phi = 0$) with a depth of -85.937 cm^{-1} at $R = 8.0$ a_0 with r_1 and r_2 at their, respectively, equilibrium positions. This compares with the global minimum of the interaction potential obtained by Jankowski et al.[35] from their fitted V12 PES averaged over r_1 and r_2: $R = 7.911$ a_0 and -93.651 cm^{-1}. Note that this comparison is only suggestive, as the global minimum in the total and interaction potentials coincide only when bond lengths (r_1 and r_2) are fixed at their equilibrium values as illustrated in Fig. 1 and that is not the case for V12.

Scattering theory and computational details.
The quantum scattering theory for a collision of an S-state atom with a rigid-rotor was developed[2,57–59], based on the close-coupling (CC) formulation of Arthurs and Dalgarno[1]. Details about its extension to diatom–diatom collisions with full-vibrational motion can be found in refs 12,13. In this approach, the interaction potential $V(R, r_1, r_2, \theta_1, \theta_2, \phi)$ is expanded as,

$$V(R, r_1, r_2, \theta_1, \theta_2, \phi) = \sum_{\lambda_1 \lambda_2 \lambda_{12}} A_{\lambda_1, \lambda_2, \lambda_{12}}(r_1, r_2, R) Y_{\lambda_1, \lambda_2, \lambda_{12}}(\hat{r}_1, \hat{r}_2, \hat{R}), \quad (4)$$

with the bi-spherical harmonic function expressed as,

$$Y_{\lambda_1, \lambda_2, \lambda_{12}}(\hat{r}_1, \hat{r}_2, \hat{R}) = \sum_{m_{\lambda_1} m_{\lambda_2} m_{\lambda_{12}}} \langle \lambda_1 m_{\lambda_1} \lambda_2 m_{\lambda_2} | \lambda_{12} m_{\lambda_{12}} \rangle \\ \times Y_{\lambda_1 m_{\lambda_1}}(\hat{r}_1) Y_{\lambda_2 m_{\lambda_2}}(\hat{r}_2) Y_{\lambda_{12} m_{\lambda_{12}}}^*(\hat{R}), \quad (5)$$

where $0 \leq \lambda_1 \leq 10, 0 \leq \lambda_2 \leq 6$ was used in the scattering calculations. Owing to the symmetry of H_2, only even values of λ_2 contribute.

For convenience, the combined molecular state (CMS) notation is applied to describe a combination of rovibrational states for the two diatoms. A CMS represents a unique quantum state of the diatom–diatom system before or after a collision. The CMS will be denoted as ($v_1 j_1 v_2 j_2$). v and j are the vibrational and rotational quantum numbers.

The rovibrational state-to-state cross-section as a function of collision energy E is given by,

$$\sigma_{v_1 j_1 v_2 j_2, v_1' j_1' v_2' j_2'}(E) = \frac{\pi}{(2j_1 + 1)(2j_2 + 1)k^2} \\ \times \sum_{j_{12} j_{12}' ll' J \varepsilon_l} (2J + 1) | \delta_{v_1 j_1 v_2 j_2 l, v_1' j_1' v_2' j_2' l'} - S_{v_1 j_1 v_2 j_2 l, v_1' j_1' v_2' j_2' l'}^{J \varepsilon_l}(E) |^2, \quad (6)$$

where ($v_1 j_1 v_2 j_2$) and ($v_1' j_1' v_2' j_2'$) are, respectively, the initial and final CMSs of CO–H_2, the wave vector $k^2 = 2\mu E/\hbar^2$ and S is the scattering matrix. l is the orbital angular momentum and J the total collision system angular momentum, where $\mathbf{J} = \mathbf{l} + \mathbf{j}_{12}$ and $\mathbf{j}_{12} = \mathbf{j}_1 + \mathbf{j}_2$.

Thorough convergence testing was performed in the scattering calculations by varying all relevant parameters. The CC equations were propagated for each value of R from 4 to 18.0 a_0 using the log-derivative matrix propagation method of Johnson[60] and Manolopoulos[61], which was found to converge for a radial step-size of $\Delta R = 0.05$ a_0. The convergence tests of the $v_1 = 1 \rightarrow 0$ vibrational quenching cross-section of CO with respect to the number of $v_1 = 1$ rotational channels found that at least 13–15 channels have to be included in the $v_1 = 1$ basis set, especially for low-energy scattering. On the basis of convergence tests with respect to the adopted maximum R for the long-range part of the PES, we found that the cross-sections are converged down to the lowest collision energy of 0.1 cm^{-1}. This value also guarantees that the rate coefficients are converged for temperatures > 1 K. The number of discrete variable representation points N_{r1} and N_{r2}; the number of points in θ_1 and θ_2 for Gauss-Legendre quadrature, $N_{\theta1}$ and $N_{\theta2}$; and the number of points in ϕ for Gauss-Hermite quadrature, N_ϕ, which were applied to project out the potential expansion coefficients, were all tested for convergence with the final adopted values given in Supplementary Table 2. The basis sets and the maximum number of coupled channels are also presented in Supplementary Table 2.

The resulting integral cross-sections were thermally averaged over a Maxwellian kinetic energy distribution to yield state-to-state rate coefficients as function of temperature T,

$$k_{v_1 j_1 v_2 j_2, v_1' j_1' v_2' j_2'}(T) = \left(\frac{8}{\pi m \beta}\right)^{1/2} \beta^2 \int_0^\infty E \sigma_{v_1 j_1 v_2 j_2, v_1' j_1' v_2' j_2'}(E) \exp(-\beta E) dE, \quad (7)$$

where m is the reduced mass of the CO–H_2 complex, $\beta = (k_B T)^{-1}$ and k_B is Boltzmann's constant.

References
1. Arthurs, A. M. & Dalgarno, A. The theory of scattering by a rigid rotor. *Proc. R. Soc.* A **256**, 540–551 (1960).

2. Takayanagi, K. The production of rotational and vibrational transitions in encounters between molecules. *Adv. Atom. Mol. Phys.* **1**, 149–194 (1965).

3. Dubernet, M.-L. *et al.* BASECOL2012: A collisional database repository and web service within the Virtual Atomic and Molecular Data Centre (VAMDC). *Astron. Astrophys.* **553**, A50 (2013).

4. Roueff, E. & Lique, F. Molecular excitation in the interstellar medium: recent advances in collisional, radiative, and chemical processes. *Chem. Rev.* **113**, 8906–8938 (2013).

5. Althorpe, S. C. & Clary, D. C. Quantum scattering calculations on chemical reactions. *Annu. Rev. Phys. Chem.* **54**, 493–529 (2003).

6. Bhattacharya, S., Panda, A. N. & Meyer, H.-D. Multiconfiguration time-dependent Hartree approach to study the OH + H_2 reaction. *J. Chem. Phys.* **132**, 214304 (2010).

7. Bhattacharya, S., Kirwai, A., Panda, A. N. & Meyer, H.-D. Full dimensional quantum scattering study of the CN + H_2 reaction. *J. Chem. Sci.* **124**, 65–73 (2012).

8. Pogrebnya, S. K. & Clary, D. C. A full-dimensional quantum dynamical study of vibrational relaxation in $H_2 + H_2$. *Chem. Phys. Lett.* **363**, 523–528 (2002).

9. Panda, A. N., Otto, F., Gatti, F. & Meyer, H.-D. Rovibrational energy transfer in ortho-H_2 + para-H_2 collisions. *J. Chem. Phys.* **127**, 114310 (2007).

10. Lin, S. Y. & Guo, H. Full-dimensional quantum wave packet study of collision-induced vibrational relaxation between para-H_2. *Chem. Phys.* **289**, 191–199 (2003).

11. Quéméner, G., Balakrishnan, N. & Krems, R. V. Vibrational energy transfer in ultracold molecule-molecule collisions. *Phys. Rev. A* **77**, 030704(R) (2008).

12. Quéméner, G. & Balakrishnan, N. Quantum calculations of H_2-H_2 collisions: from ultracold to thermal energies. *J. Chem. Phys.* **130**, 114303 (2009).

13. dos Santos, S. F. *et al.* Quantum dynamics of rovibrational transitions in H_2-H_2 collisions: Internal energy and rotational angular momentum conservation effects. *J. Chem. Phys.* **134**, 214303 (2011).

14. Wilson, R. W., Jefferts, K. B. & Penzias, A. A. Carbon monoxide in Orion nebula. *Astrophys. J.* **161**, L43–L44 (1970).

15. Narayanan, D. *et al.* The nature of CO emission from $z \sim 6$ quasars. *Astrophys. J. Suppl. Ser.* **174**, 13–30 (2008).

16. Lupu, R. E., Feldman, P. D., Weaver, H. A. & Tozzi, G.-P. The fourth positive system of carbon monoxide in the Hubble Space Telescope spectra of comets. *Astrophys. J.* **670**, 1473–1484 (2007).

17. Swain, M. R. *et al.* Molecular signatures in the near-infrared dayside spectrum of HD 189733b. *Astrophys. J. Lett.* **690**, L114–L117 (2009).

18. González-Alfonso, E. *et al.* CO and H_2O vibrational emission toward Orion Peak 1 and Peak 2. *Astron. Astrophys.* **386**, 1074–1102 (2002).

19. Brown, J. M. *et al.* VLT-CRIRES survey of rovibrational CO emission from protoplanetary disks. *Astrophys. J.* **770**, 94 (2013).

20. Brittain, S. D., Najita, J. R. & Carr, J. S. Tracing the inner edge of the disk around HD 100546 with rovibrational CO emission lines. *Astrophys. J.* **702**, 85–99 (2009).

21. Bertelsen, R. P. H. *et al.* CO ro-vibrational lines in HD 100546: A search for disc asymmetries and the role of fluorescence. *Astron. Astrophys.* **561**, A102 (2014).

22. Patel, N. A. *et al.* Detection of vibrationally excited CO in IRC + 10216. *Astrophys. J.* **691**, L55–L58 (2009).

23. Thi, W. F. *et al.* Radiation thermo-chemical models of protoplanetary discs IV. Modelling CO ro-vibrational emission from Herbig Ae discs. *Astron. Astrophys.* **551**, A49 (2013).

24. Krems, R. V. *TwoBC - quantum scattering program* (University of British Columbia, 2006).

25. Bačić, Z., Schinke, R. & Diercksen, G. H. F. Vibrational relaxation of CO ($n=1$) in collisions with H_2. I. Potential energy surface and test of dynamical approximations. *J. Chem. Phys.* **82**, 236–244 (1985).

26. Bačić, Z., Schinke, R. & Diercksen, G. H. F. Vibrational relaxation of CO ($n=1$) in collisions with H_2. II. Influence of H_2 rotation. *J. Chem. Phys.* **82**, 245–253 (1985).

27. Reid, J. P., Simpson, C. J. S. M. & Quiney, H. M. The vibrational deactivation of CO($v=1$) by inelastic collisions with H_2 and D_2. *J. Chem. Phys.* **106**, 4931–4944 (1997).

28. Flower, D. R. Rate coefficients for the rovibrational excitation of CO by H_2 and He. *Mon. Not. R. Astron. Soc.* **425**, 1350–1356 (2012).

29. Andrews, A. J. & Simpson, C. J. S. M. Vibrational relaxation of CO by H_2 down to 73 K using a chemical CO laser. *Chem. Phys. Lett.* **36**, 271–274 (1975).

30. Andrews, A. J. & Simpson, C. J. S. M. Vibrational deactivation of CO by n-H_2, by p-H_2 and by HD measured down to 77 K using laser fluorescence. *Chem. Phys. Lett.* **41**, 565–569 (1976).

31. Wilson, G. J., Turnidge, M. L., Solodukhin, A. S. & Simpson, C. J. S. M. The measurement of rate constants for the vibrational deactivation of $^{12}C^{16}O$ by H_2, D_2 and ^{4}He in the gas phase down to 35 K. *Chem. Phys. Lett.* **207**, 521–525 (1993).

32. Schinke, R., Meyer, H., Buck, U. & Diercksen, G. H. F. A new rigid-rotor H_2-CO potential-energy surface from accurate ab initio calculations and rotationally inelastic-scattering data. *J. Chem. Phys.* **80**, 5518–5530 (1984).

33. Jankowski, P. & Szalewicz, K. Ab initio potential energy surface and infrared spectra of H_2-CO and D_2-CO van der Waals complexes. *J. Chem. Phys.* **108**, 3554–3565 (1998).

34. Jankowski, P. & Szalewicz, K. A new ab initio interaction energy surface and high-resolution spectra of the H_2-CO van der Waals complex. *J. Chem. Phys.* **123**, 104301 (2005).

35. Jankowski, P. *et al.* A comprehensive experimental and theoretical study of H_2-CO spectra. *J. Chem. Phys.* **138**, 084307 (2013).

36. Zhang, X., Zou, S., Harding, L. B. & Bowman, J. M. A global ab initio potential energy surface for formaldehyde. *J. Phys. Chem. A* **108**, 8980–8986 (2004).

37. Braams, B. J. & Bowman, J. M. Permutationally invariant potential energy surfaces in high dimensionality. *Int. Rev. Phys. Chem.* **28**, 577–606 (2009).

38. Bowman, J. M., Czakó, G. & Fu, B. N. High-dimensional ab initio potential energy surfaces for reaction dynamics calculations. *Phys. Chem. Chem. Phys.* **13**, 8094–8111 (2011).

39. Hutson, J. M. & Green, S. MOLSCAT computer code, Version 14 http://www.giss.nasa.gov/tools/molscat/ (1994).

40. Antonova, S., Tsakotellis, A. P., Lin, A. & McBane, G. C. State-to-state rotational excitation of CO by H_2 near 1000 cm^{-1} collision energy. *J. Chem. Phys.* **112**, 554–559 (2000).

41. Chefdeville, S. *et al.* Appearance of low energy resonances in CO-para-H_2 inelastic collisions. *Phys. Rev. Lett.* **109**, 023201 (2012).

42. Hooker, W. J. & Millikan, R. C. Shocktube study of vibrational relaxation in carbon monoxide for the fundamental and first overtone. *J. Chem. Phys.* **38**, 214–220 (1963).

43. Millikan, R. C. & Osburg, L. A. Vibrational relaxation of carbon monoxide by ortho- and para-hydrogen. *J. Chem. Phys.* **41**, 2196–2197 (1964).

44. Kobayashi, R., Amos, R. D., Reid, J. P., Quiney, H. M. & Simpson, C. J. S. M. Coupled cluster ab initio potential energy surfaces for CO…He and CO…H_2. *Mol. Phys.* **98**, 1995–2005 (2000).

45. Chandler, D. W. Cold and ultracold molecules: Spotlight on orbiting resonances. *J. Chem. Phys.* **132**, 110901 (2010).

46. Casavecchia, P. & Alexander, M. H. Uncloaking the quantum nature of inelastic molecular collisions. *Science* **341**, 1076–1077 (2013).

47. Henning, T. & Semenov, D. Chemistry in protoplanetary disks. *Chem. Rev.* **113**, 9016–9042 (2013).

48. Adler, T. B., Knizia, G. & Werner, H.-J. A simple and efficient CCSD(T)-F12 approximation. *J. Chem. Phys.* **127**, 221106 (2007).

49. Werner, H.-J., Adler, T. B. & Manby, F. R. General orbital invariant MP2-F12 theory. *J. Chem. Phys.* **126**, 164102 (2007).

50. Werner, H.-J. *et al.* MOLPRO, version 20101. A package of ab initio programs http://www.molpro.net.

51. Hill, J. G., Mazumder, S. & Peterson, K. A. Correlation consistent basis sets for molecular core-valence effects with explicitly correlated wave functions: The atoms B-Ne and Al-Ar. *J. Chem. Phys.* **132**, 054108 (2010).

52. Hättig, C. Optimization of auxiliary basis sets for RI-MP2 and RI-CC2 calculations: Core-valence and quintuple-zeta basis sets for H to Ar and QZVPP basis sets for Li to Kr. *Phys. Chem. Chem. Phys.* **7**, 59–66 (2005).

53. Weigend, F. A fully direct RI-HF algorithm: Implementation, optimised auxiliary basis sets, demonstration of accuracy and efficiency. *Phys. Chem. Chem. Phys.* **4**, 4285–4291 (2002).

54. Tew, D. P. & Klopper, W. A comparison of linear and nonlinear correlation factors for basis set limit Moller-Plesset second order binding energies and structures of He₂, Be₂, and Ne₂. *J. Chem. Phys.* **125**, 094302 (2006).

55. Feller, D., Peterson, K. A. & Hill, J. G. Calibration study of the CCSD(T)-F12a/b methods for C₂ and small hydrocarbons. *J. Chem. Phys.* **133**, 184102 (2010).

56. Bernardi, F. & Boys, S. F. Calculation of small molecular interactions by differences of separate total energies - some procedures with reduced errors. *Mol. Phys.* **19**, 553 (1970).

57. Green, S. Rotational excitation in H_2-H_2 collisions - close-coupling calculations. *J. Chem. Phys.* **62**, 2271–2277 (1975).

58. Alexander, M. H. & DePristo, A. E. Symmetry considerations in quantum treatment of collisions between two diatomic-molecules. *J. Chem. Phys.* **66**, 2166–2172 (1977).

59. Zarur, G. & Rabitz, H. Effective potential formulation of molecule-molecule collisions with application to H_2-H_2. *J. Chem. Phys.* **60**, 2057–2078 (1974).

60. Johnson, B. R. Multichannel log-derivative method for scattering calculations. *J. Comp. Phys.* **13**, 445–449 (1973).

61. Manolopoulos, D. E. An improved log derivative method for inelastic-scattering. *J. Chem. Phys.* **85**, 6425–6429 (1986).

Acknowledgements

Work at UGA and Emory was supported by NASA grant NNX12AF42G from the Astronomy and Physics Research and Analysis Program, at UNLV by NSF Grant No. PHY-1205838, and at Penn State by NSF Grant No. PHY-1203228. This study was supported in part by resources and technical expertise from the UGA Georgia Advanced Computing Resource Center (GACRC), a partnership between the UGA Office of the Vice President for Research and Office of the Vice President for Information Technology.

We thank Shan-Ho Tsai (GACRC), Jeff Deroshia (UGA Department of Physics and Astronomy) and Gregg Derda (GACRC) for computational assistance.

Author contributions

B.H.Y. performed the potential energy surface (PES) fitting and scattering calculations. P.Z. calculated the PES. X.W. and J.M.B. developed the PES fitting code. N.B., R.C.F., and P.C.S. extended and modified the TwoBC code for $CO-H_2$ rovibrational scattering calculations, while N.B. assisted with the scattering calculations. B.H.Y., P.C.S. and P.Z. wrote the article with contributions from all other authors.

Additional information

Competing financial interests: The authors declare no competing financial interests.

4

Witnessing magnetic twist with high-resolution observation from the 1.6-m New Solar Telescope

Haimin Wang[1,2], Wenda Cao[1,2], Chang Liu[1,2], Yan Xu[1,2], Rui Liu[3,4], Zhicheng Zeng[2], Jongchul Chae[5] & Haisheng Ji[6]

Magnetic flux ropes are highly twisted, current-carrying magnetic fields. They are crucial for the instability of plasma involved in solar eruptions, which may lead to adverse space weather effects. Here we present observations of a flaring using the highest resolution chromospheric images from the 1.6-m New Solar Telescope at Big Bear Solar Observatory, supplemented by a magnetic field extrapolation model. A set of loops initially appear to peel off from an overall inverse S-shaped flux bundle, and then develop into a multi-stranded twisted flux rope, producing a two-ribbon flare. We show evidence that the flux rope is embedded in sheared arcades and becomes unstable following the enhancement of its twists. The subsequent motion of the flux rope is confined due to the strong strapping effect of the overlying field. These results provide a first opportunity to witness the detailed structure and evolution of flux ropes in the low solar atmosphere.

[1] Space Weather Research Laboratory, New Jersey Institute of Technology, University Heights, Newark, New Jersey 07102-1982, USA. [2] Big Bear Solar Observatory, New Jersey Institute of Technology, 40386 North Shore Lane, Big Bear City, California 92314-9672, USA. [3] Department of Geophysics and Planetary Sciences, CAS Key Laboratory of Geospace Environment, University of Science and Technology of China, Hefei 230026, China. [4] Collaborative Innovation Center of Astronautical Science and Technology, China. [5] Department of Physics and Astronomy, Astronomy Program, Seoul National University, Seoul 151-747, Korea. [6] Purple Mountain Observatory, Chinese Academy of Sciences, Nanjing 210008, China. Correspondence and requests for materials should be addressed to H.W. (email: haimin.wang@njit.edu).

Magnetic flux ropes are a group of twisted magnetic fields writhing about each other and rotating around a common axis. The importance of these helical, current-carrying magnetic flux systems was demonstrated long ago in plasma physics laboratories, such as in the experimental and modelling research with the tokamak[1,2]. As the toroidal plasma current in a tokamak produces a hoop force that expands the twisted fields along the major radius, the tokamak and similar experiments were designed to constrain this expansion by adding a strapping field, reducing the effect of the torus instability. Twisted flux ropes are also subject to kink instability[3], where the rope axis itself can evolve into a helical structure when the flux rope twist reaches a certain threshold[4]. The structure of magnetic flux ropes associated with solar eruptions and interplanetary activities have received significant attention with the application of basic plasma physics. For example, the linkage between the solar plasma and tokamaks was discussed recently[5]. In the large scale, magnetic flux ropes are found in the interplanetary magnetic clouds[6], which may interact with Earth's magnetic field to generate geomagnetic storms. Near the solar surface, there are evidences that different branches of magnetic loops could reconnect and form a flux rope during the eruption process[7–9]. On the other hand, some studies also suggest that flux ropes are built gradually and may already exist before eruptions[10,11].

Solar magnetic configurations prone to eruptions have been of great importance for space weather, which increasingly affects the modernization of human activities. There are a number of models that describe the initiation mechanism of solar eruptions, such as catastrophe or instability[12] and magnetic breakout[13]. One of the eruptive magnetic field configuration related to flux ropes is the so-called sigmoid[14]: an S- or inverse S-shaped magnetic field structure as seen in soft X-rays[15] and extreme ultraviolet[16]. In a qualitative cartoon picture[17], the onset of a sigmoid eruption begins when the highly sheared inner legs of the two branches of the sigmoidal fields are joined via magnetic reconnection[18]. This initial stage of eruption produces a low-lying shorter loop across the magnetic polarity inversion line (PIL) and a longer twisted flux rope linking the two far ends of the sigmoid. The next stage proceeds when the formed flux rope becomes unstable (due to, for example, the torus instability) and erupts outward to become a coronal mass ejection. However, the insufficient spatial resolution of coronal images predominantly used by previous observational studies of flux ropes seriously hampers an unambiguous and detailed identification of flux ropes and their relation to eruptions. Moreover, the evolution of flux ropes has not been observed in the low solar atmosphere such as the chromosphere.

Here we report on clear observational evidence of a flaring twisted flux rope, taking advantage of H-alpha and He I 10,830 Å images at high resolution (60 and 100 km, respectively) obtained by the recently built 1.6 m New Solar Telescope[19,20] (NST) at Big Bear Solar Observatory (BBSO). NST achieves the high-spatial resolution observation in the chromospheric wavelengths at a

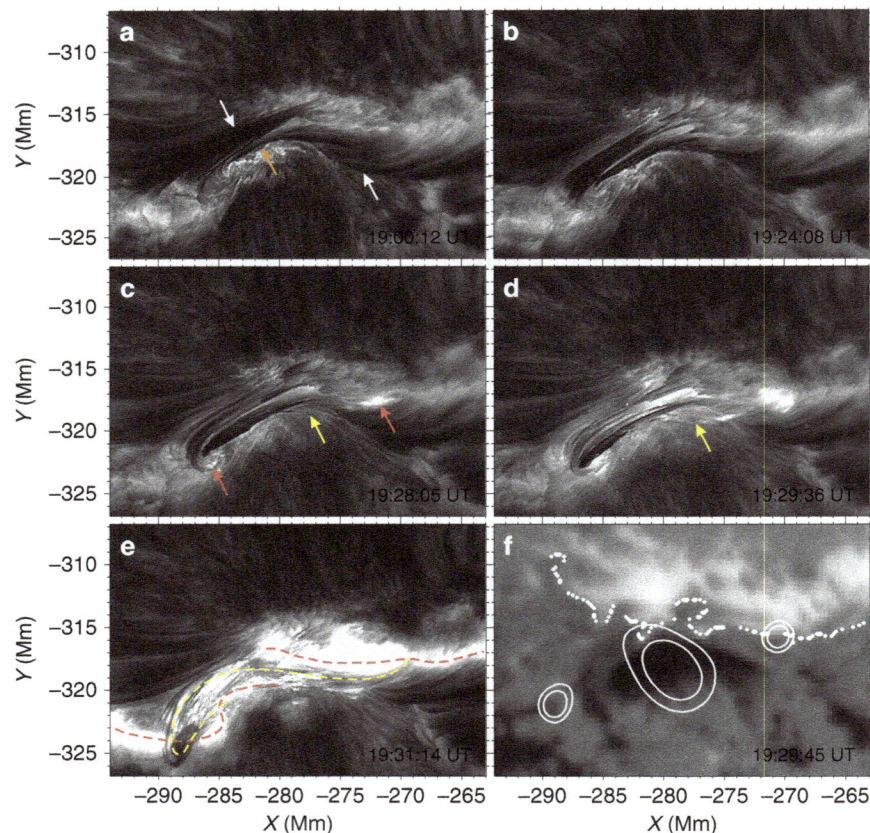

Figure 1 | Magnetic flux rope evolution in H-alpha line centre and surface magnetic field structure. The observation is for NOAA Active Region 11817 on 11 August 2013, with image time shown in the bottom right of each panel. (a–e) BBSO/NST VIS images at the H-alpha line centre. The white (orange) arrows in a indicate the weakened (enhanced) arcade loops before the flux rope activity. The red arrows in c point to footpoint brightenings. The yellow arrows in c and d and the yellow dashed line in e trace the active flux rope. The red dashed line in e delineate the two flare ribbons. (f) A corresponding line-of-sight magnetic field observed with Solar Dynamic Observatory/HMI, with the white (black) colour representing positive (negative) field (scaled from −2,000 to 2,000 G) and the dotted line indicating the PIL. The 25–50 keV HXR map is integrated from 19:29:45 to 19:30:45 UT, showing a coronal (the central contours) and two chromospheric footpoint-like (the left and right smaller contours) HXR-emitting sources of the associated flare. The full sequence of BBSO/NST H-alpha centre-line images is provided in Supplementary Movie 1.

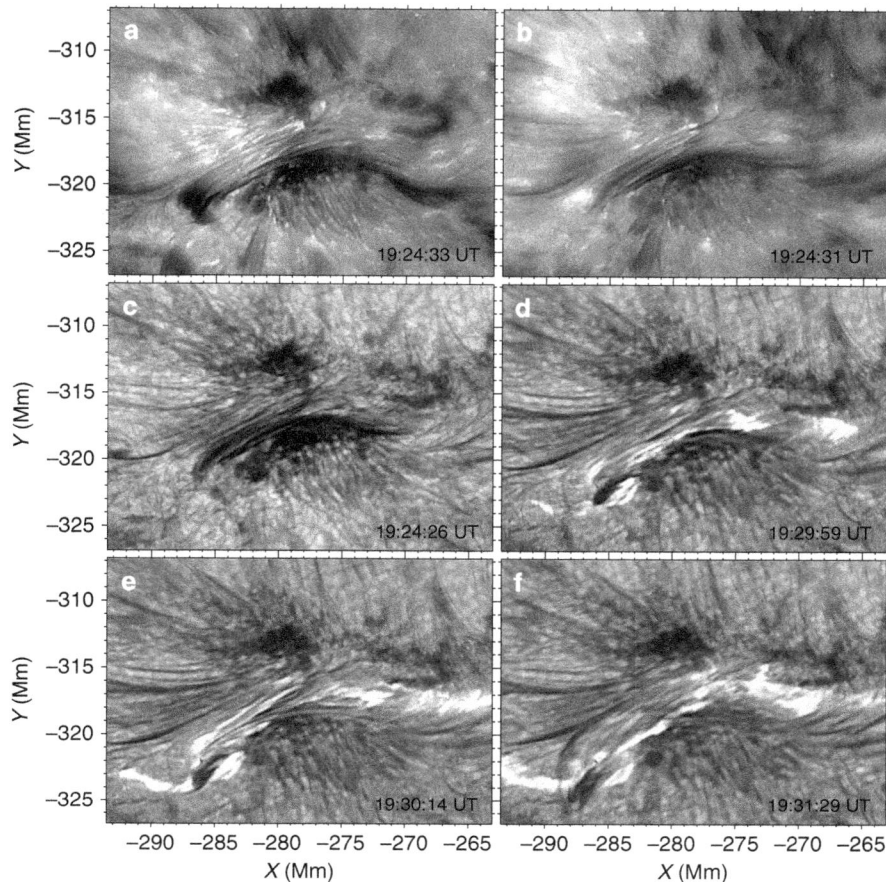

Figure 2 | Magnetic flux rope evolution in H-alpha off-bands and He I 10830 Å. BBSO/NST observation on 11 August 2013, including VIS images at the H-alpha red-wing (+0.6 Å) (**a**) and blue-wing (−0.6 Å) (**b**) together with infrared imaging magnetograph (IRIM) images at the He I 10830 Å line (**c-f**). These are observed at a lower height than the H-alpha line centre images shown in Fig. 1. It is notable that both H-alpha off-bands (**a,b**) and He I 10830 Å (**c**) images right before the event onset exhibit an inverse S-shaped flux rope structure at the centre, bearing a possible similarity to the modelled low-lying red field lines in Fig. 3. The active flux rope on top of the extending flare ribbons are clearly seen in (**d-f**). Note that some linear features running at an angle across the centre of the flux rope in **c-f** are artifacts produced by the infrared camera. The full sequences of H-alpha off-band images are provided in Supplementary Movies 2 and 3.

high time cadence (∼15 s), thus providing a unique opportunity for a fine-scale assessment of evolving structures in the low atmosphere. An overall inverse S-shaped flux bundle (SFB) is initially observed to lie along the magnetic PIL, from which a set of loops appear to peel off and grow upward into a multi-stranded twisted flux rope within about 2 min. In the meantime, compact brightenings at the footpoints of the loop threads extend to form two elongated flare ribbons. These observations are supplemented by a nonlinear force-free field (NLFFF) extrapolation model[21] based on the surface vector magnetogram taken by the Helioseismic and Magnetic Imager (HMI) telescope[22] on board the Solar Dynamic Observatory. Hard X-ray (HXR) observation of the flare recorded by the Reuven Ramaty High Energy Solar Spectroscopic Imager (RHESSI) telescope[23] is also used to reveal the likely energy release sites. We suggest that in this event, a low-lying single flux rope is embedded in sheared arcades. Following the enhancement of its twists, possibly driven by sub-surface motion, the flux rope becomes unstable. It is also possible that the two groups of the closely lying sheared arcades could reconnect to contribute to the active flux rope. The subsequent motion of the flux rope is, however, confined because of the strong strapping effect of the overlying field. These significant results provide a first opportunity to witness the detailed structure and evolution of flux ropes in the low solar atmosphere.

Results

BBSO/NST observation. On 11 August 2013, we observed an active magnetic flux rope and its associated GOES-class C2.1 flare around 19:30 UT in NOAA Active Region 11817. We present in Fig. 1 the H-alpha line-centre images at selected instances and a corresponding magnetic field on the surface. Owing to the unprecedented high resolution of NST, great details can be visualized. The full dynamics of evolution can be better seen in Supplementary Movie 1. It can be easily recognized that initially there is an overall inverse SFB lying along the PIL (the white dotted line in Fig. 1f). This SFB mainly consists of two branches of sheared arcade loops that seemingly cross and join near the centre (Fig. 1a). From about 19:15 UT (12 min before the event), the outside of these two branches of arcades (especially the eastern one; pointed to by the white arrows) appear to become weakened, while the inner loops (pointed to by the orange arrow) are enhanced (cf. Fig. 1a,b). At the beginning of the event, a bunch of long loop threads start to peel off from the SFB (pointed to by the yellow arrows in Fig. 1c,d). These threads seem to wrap around the entire SFB, with obvious brightenings seen at their two footpoints (pointed to by the red arrows) near the two ends of the SFB. In about 2 min, the threads rapidly expand and rise upward, and grow into a multi-stranded field structure with apparent twist and writhe clearly representing a flux rope (traced by the yellow dashed line in Fig. 1e). As clearly witnessed by

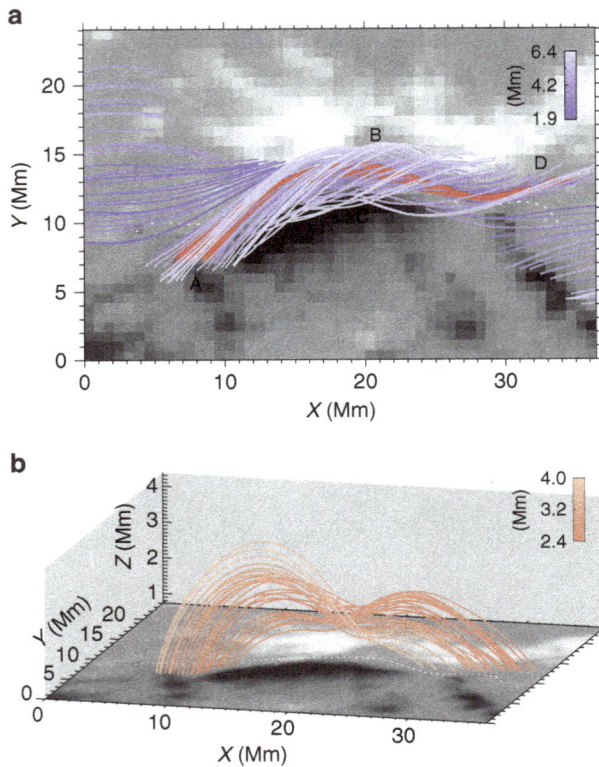

Figure 3 | Modelled of magnetic field structure. (**a**) Selected field lines from a NLFFF extrapolation model at 11 August 2013 19:22:24 UT, before the flux rope activity and the associated C2.1 flare. The model is based on the HMI vector magnetic field data (2 × 2 rebinned) in cylindrical equal area coordinate. The blue field lines have a maximum height ≲ 6.4 Mm and portray the overall inverse SFB. The long red field lines have a maximum height ≲ 4.0 Mm and run through the SFB. (**b**) A side view of the red field lines. The white dotted line in both the panels is the portion of the PIL above which we compute the average profile of decay index (see Fig. 5).

BBSO/NST, we can estimate that this flux rope possesses a twist of at least one turn from end to end. In the meantime, the initial compact brightenings in H-alpha have extended in the horizontal direction to form two bright ribbon-like patches (delineated by the red dashed line in Fig. 1e) near the two ends of the rope. A similar evolution of the SFB and the flux rope can be observed with the He I 10,830 Å images (see Fig. 2c–f). Very interestingly, compared with the H-alpha line-centre wavelength, the SFB in He I 10,830 Å and also the H-alpha offbands (± 0.6 Å; see Fig. 2a,b and Supplementary Movies 2 and 3) appears more like an integrated single structure right before the event onset.

Magnetic field structure. Disentangling the topology of the SFB as observed with BBSO/NST in projection is crucial for understanding the nature and properties of the flux rope. However, vector magnetic fields of higher atmosphere (for example, chromosphere and corona) are not routinely measured with the present instrument. To shed light on the three-dimensional magnetic field structure, we thus construct a NLFFF model at the pre-event time and plot some selected representative field lines in Fig. 3. Obviously, the extrapolated field lines coloured blue bear a reasonable resemblance to the morphology of the SFB seen in H-alpha (see Fig. 1b), showing in general two branches of arcade loops (AB and CD) that are highly sheared when passing each other along the PIL (we note that there may not exist a clear separation between the AB and CD domains). Importantly, we also find that below the blue field lines, there lies a group of

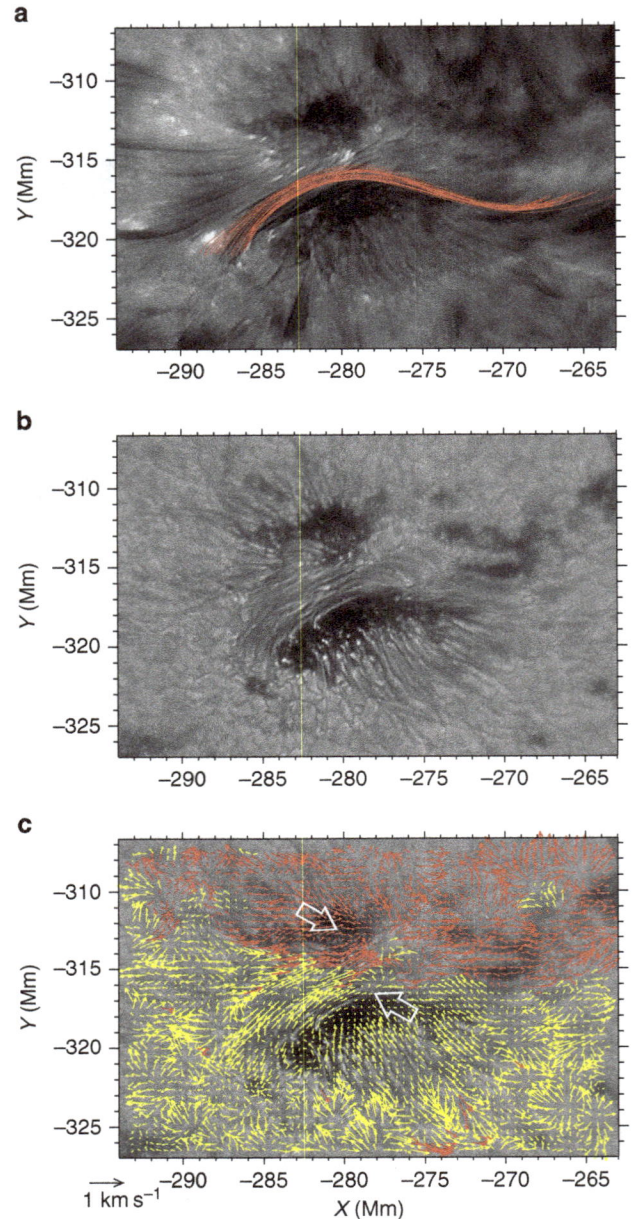

Figure 4 | Magnetic flux rope and photospheric flow field. (**a**) The same red field lines as in Fig. 3a but transformed to the image coordinate and overplotted onto a cotemporal pre-flare BBSO/NST H-alpha blue-wing image at 19:22:30 UT, showing a reasonable agreement. (**b**) A BBSO/NST TiO image of the active region at 19:12:04 UT. (**c**) The same TiO image as (**b**) but superimposed with arrows illustrating horizontal plasma flows tracked with DAVE (differential affine velocity estimator) and averaged for the time period 19:00:04–19:23:49 UT. The length of arrows denotes the flow magnitude, and for clarity, arrows in positive and negative magnetic fields are coded in red and yellow, respectively. The two white hollow arrows indicate the overall converging and shearing motions of the opposite polarity sunspot regions. The full TiO image sequence is provided in Supplementary Movie 4.

longer field lines (coloured red) that directly connect the two ends of the SFB from A to D. These field lines AD might correspond to the loops observed in He I 10,830 Å as well as H-alpha offbands at a lower height (see Fig. 2a–c), as also clearly demonstrated in Fig. 4a. A side view of the red field lines AD is presented in Fig. 3b. On the basis of the force-free field extrapolations, we can

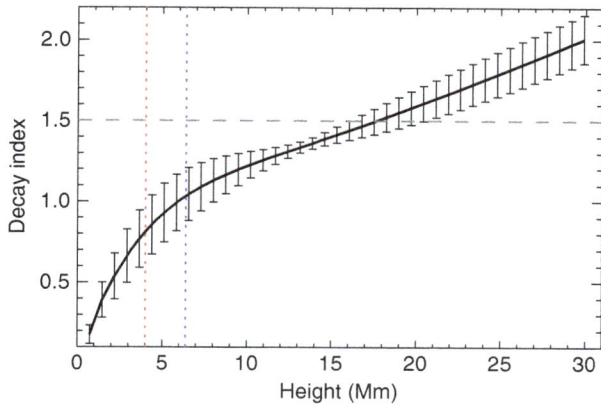

Figure 5 | Decay profile of overlying field. We average the decay index at a certain height above the portion of the PIL beneath the flux rope (plotted as the white dotted line in Fig. 3), and plot the mean value versus height as the black line with the error bars representing 1 s.d. The red and blue dotted lines correspond to the maximum height of the red (flux rope) and blue (sheared arcades) field lines plotted in Fig. 3, respectively. The grey dashed line denotes the critical decay index value of 1.5 related to the torus instability condition.

further derive the magnetic twist number of a field line as $\frac{1}{4\pi}\alpha L$, where α is the force-free parameter and L is the field line length. It is obtained that the mean twist number of loops AB and CD is about 0.8 turns and that of AD is about 1.1 turns. Therefore, we speculate that the configuration of the SFB may represent a transition from twisted loops or flux ropes (for example, AD) to sheared arcades (for example, AB and CD). This may also be implied by the high-resolution BBSO/NST H-alpha images. We caution, however, that the NLFFF approach has limitations. These include that it does not take account of the plasma, and that it cannot be used to model the process of magnetic reconnection that involves non-force-free field.

Confined motion. It is worth mentioning that the aforementioned flux rope activity was not observed to develop into a coronal mass ejection. This suggests that the flux rope motion may be confined or failed. To shed light on this phenomenon, we calculate the decay index, defined by $n = -\text{dlog}(B)/\text{dlog}(h)$, which can gauge the likelihood of eruption in the torus instability model. Here B is the strength of the horizontal component of the overlying potential field, h is the height above the surface and $n > 1.5$–2.0 is the critical instability condition[24]. The calculation is performed above locations of the central portion of the PIL (the white dotted line in Fig. 3) underlying the flux rope. From the result plotted in Fig. 5, we can see that a flux rope would experience a decay index of the critical value at about 18 Mm, which is much higher than the maximum height (~ 4 Mm) of the present flux rope of interest. This means that the magnetic field overlying the flux rope may have a strong strapping effect that prevents a full flux rope eruption. Nevertheless, the twisted flux rope may reconnect with the ambient field and reform as a less twisted field[25,26].

Discussion

On the basis of the BBSO/NST images and the NLFFF model result, we infer possible scenarios for this flux rope activity. The most likely explanation is that the twisted flux rope (modelled as loops AD underlying the SFB structure) may become unstable due to kink instability. Justifications for this view are discussed as follows. First, we show in Fig. 6 the maps with a larger field of view of twist number obtained by integrating the force-free

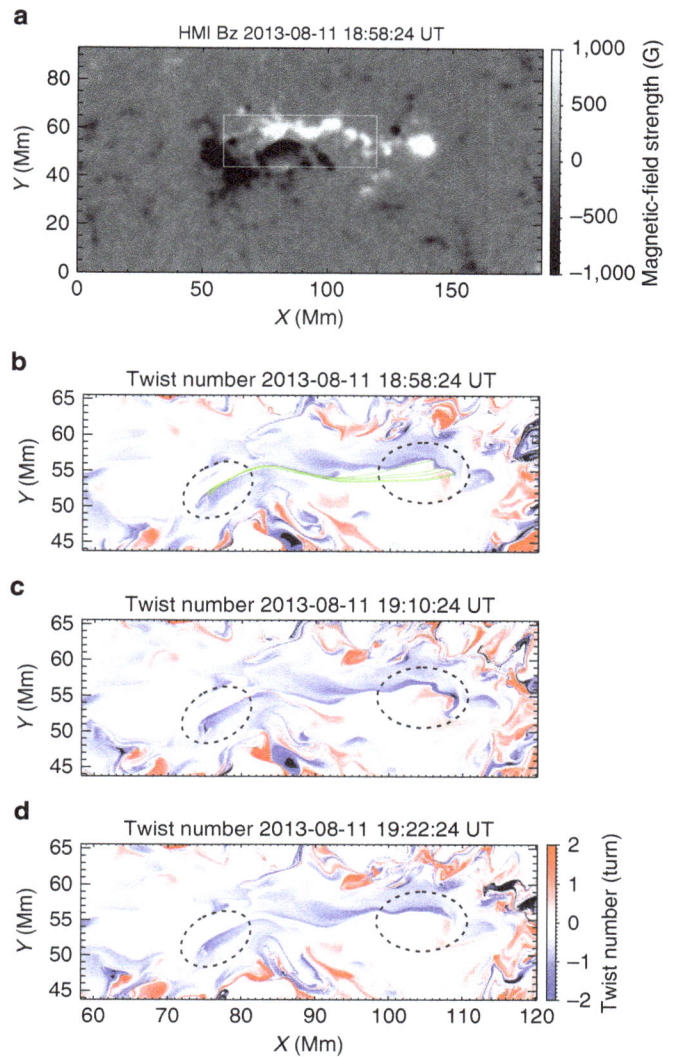

Figure 6 | Evolution of magnetic twist. Maps of twist number are obtained by integrating the force-free parameter α along field lines divided by 4π. Only considered are field lines with both footpoints located within the field of view shown in (**a**). The white rectangle in **a** indicates the region in which the maps of twist number (**b**–**d**) are calculated. Green curves show exemplary twisted field lines in the map at 18:58:24 UT. Note a transient enhancement in the twist number at 19:10:24 UT at the footpoint regions of these field lines within the dotted ellipses.

parameter α along field lines. Note that a transient enhancement in the twist number at the footpoint regions (within the dotted ellipses) of the field lines similar to AD occurred at 19:10:24 UT, for a period before the onset of the event. Therefore, the sub-photospheric motion may play a significant role in enhancing the twist and destabilizing the flux rope. Second, the NLFFF model indicates the existence of such a flux rope, with a twist number comparable to that visualized in the high-resolution BBSO/NST observation. Third, it is found that on the surface, the opposite magnetic polarities of this active region exhibit an overall converging and shearing motions, which may be favourable for the gradual build-up of flux ropes[27]. These motions are obvious when observing the TiO time sequence images (see Fig. 4b and Supplementary Movie 4) and are clearly shown in the derived flow field (see Fig. 4c). Fourth, the interaction between this rising flux rope and the SFB can explain the brightenings of the flare ribbons with the standard two-ribbon flare model.

However, it is also possible that the magnetic reconnection between the sheared arcades contributes to the active flux rope. Specifically, a magnetic reconnection may occur between sheared arcades[17,27] loosely denoted here as AB and CD. The main evidence is that the site of coronal magnetic reconnection, although not directly visible, may be manifested by a strong HXR-emitting source (the central contours in Fig. 1f) imaged by RHESSI around the flare peak. This source is located apparently between AB and CD, consistent with the speculated occurrence of reconnection. In the meantime, compact HXR emissions (left and right contours in Fig. 1f) are also produced near the two footpoints A and D, indicating enhanced precipitating electrons around the flare maximum. The weakening of the arcade structure before the flare, as shown in Fig. 1a,b, is also consistent with this view. Nevertheless, the sheared arcades still have a maximum height ($\sim$6 Mm) well below the torus instability regime (see Fig. 5).

In conclusion, high-resolution images obtained by BBSO/NST make it possible to witness, for the first time, the detailed structure and evolution of a twisted flux rope in the low solar atmosphere. Observations with such a high resolution are invaluable in the studies of flux ropes, a magnetic structure important for energetic solar and interplanetary phenomena.

Methods

Observations and data processing. The essential observational evidence of the magnetic flux rope is provided by BBSO/NST. The telescope has gone through a few upgrades for its focal plane instrument and adaptive optics system. The latest upgrade of the high-order adaptive optics system uses 308 sub-apertures, which means that the telescope aperture is divided into 308 small telescopes so as to compensate and correct the atmospheric disturbances. Together with the speckle image reconstructions[28], diffraction-limited imaging can thus be achieved. Here we use data (as shown in Fig. 1a–e, Fig. 2a,b, Fig. 4a and Supplementary Movies 1–3) from the Visible Imaging Spectrometer (VIS) at BBSO, which saw its first light with NST in the summer of 2013. This Fabry–Pérot filter-based system scans cross spectral lines in the range of 5,500–7,000 Å with millisecond steps. The H-alpha images have a spatial resolution of $\sim$0.068″ (50 km), and the cadence is 15 s for VIS to complete a scan at a 0.2 Å step from -1.0 to $+1.0$ Å around the H-alpha line centre. In addition, we also use He I 10830 Å images (as shown in Fig. 2c–f) taken by the InfraRed Imaging vector Magnetograph[29,30] (IRIM) at BBSO and the broadband TiO (a proxy for the photosphere at 7,057 Å) images (as shown in Fig. 4b and Supplementary Movie 4). The cadence is 15 s and the spatial resolution is about 120 km for 10,830 Å and 65 km for the TiO data. While H-alpha off-bands and line-centre images cover a wide height range ($\sim$1,500–3,000 km), the He I 10,830 Å line is formed around 2,000 km above the photosphere. The time sequence of TiO images can help to examine the photospheric flows (as shown in Fig. 4c), which are traced with the differential affine velocity estimator[31], with the window size set to 19 pixels for feature tracking. Moreover, context information on the magnetic field structure of the solar photosphere (as shown in Fig. 1f) is provided by the vector magnetograms taken by the HMI instrument on the Solar Dynamic Observatory.

Signatures of magnetic reconnection can be revealed using the HXR emission due to the accelerated electrons, which was recorded for this flare event by RHESSI. The image of HXR sources, presented as contours in Fig. 1f, is reconstructed with the PIXON algorithm[32] using detectors 1 and 3–8, and is integrated from 19:29:45 to 19:30:45 UT in the 25–50 keV energy band.

Magnetic field modelling. The magnetic field model, as presented in Figs 3 and 4a, is constructed using the NLFFF extrapolation technique[21]. To obtain a chromospheric-like boundary data satisfying the force-free condition, a pre-processing procedure[33] is first applied to the 2 × 2 rebinned Space-Weather HMI Active Region Patches[34] data. Then the three-dimensional magnetic field is extrapolated using a 'weighted optimization' method[35,36]. The extrapolation calculation is performed within a box of 584 × 248 × 256 uniform grid points, which corresponds to about 426 × 181 × 187 Mm^3.

References

1. Shafranov, V. D. Plasma equilibrium in a magnetic field. *Rev. Plasma Phys.* **2**, 103 (1966).
2. Bateman, G. *MHD Instabilities* (MIT Press, 1978).
3. Kruskal, M. D., Johnson, J. L., Gottlieb, M. B. & Goldman, L. M. Hydromagnetic Instability in a Stellarator. *Phys. Fluids* **1**, 421–429 (1958).
4. Hood, A. W. & Priest, E. R. Critical conditions for magnetic instabilities in force-free coronal loops. *Geophys. Astrophys. Fluid Dynamics* **17**, 297–318 (1981).
5. Browning, P. K., Stanier, A., Ashworth, G., McClements, K. G. & Lukin, V. S. Self-organization during spherical torus formation by flux rope merging in the mega ampere spherical tokamak. *Plasma Phys. Controlled Fusion* **56**, 064009 (2014).
6. Burlaga, L. F. *et al.* A magnetic cloud and a coronal mass ejection. *Geophys. Res. Lett.* **9**, 1317–1320 (1982).
7. Liu, R., Liu, C., Wang, S., Deng, N. & Wang, H. Sigmoid-to-flux-rope transition leading to a loop-like coronal mass ejection. *Astrophys. J. Lett.* **725**, L84–L90 (2010).
8. Green, L. M., Kliem, B. & Wallace, A. J. Photospheric flux cancellation and associated flux rope formation and eruption. *Astron. Astrophys.* **526**, A2 (2011).
9. Cheng, X., Zhang, J., Liu, Y. & Ding, M. D. Observing flux rope formation during the impulsive phase of a solar eruption. *Astrophys. J. Lett.* **732**, L25 (2011).
10. Zhang, J., Cheng, X. & Ding, M.-D. Observation of an Evolving Magnetic Flux Rope Before and During a Solar Eruption. *Nat. Commun.* **3 (2012)**.
11. Cheng, X. *et al.* Tracking the evolution of a coherent magnetic flux rope continuously from the inner to the outer corona. *Astrophys. J.* **780**, 28 (2014).
12. Kliem, B., Lin, J., Forbes, T. G., Priest, E. R. & Török, T. Catastrophe versus Instability for the eruption of a toroidal solar magnetic flux rope. *Astrophys. J.* **789**, 46 (2014).
13. Antiochos, S. K., DeVore, C. R. & Klimchuk, J. A. A model for solar coronal mass ejections. *Astrophys. J.* **510**, 485–493 (1999).
14. Canfield, R. C., Hudson, H. S. & Pevtsov, A. A. Sigmoids as precursors of solar eruptions. *IEEE Trans. Plasma Sci.* **28**, 1786–1794 (2000).
15. Rust, D. M. & Kumar, A. Evidence for helically kinked magnetic flux ropes in solar eruptions. *Astrophys. J. Lett.* **464**, L199–L202 (1996).
16. Liu, C. *et al.* The Eruption from a Sigmoidal Solar Active Region on 2005 May 13. *Astrophys. J.* **669**, 1372–1381 (2007).
17. Moore, R. L., Sterling, A. C., Hudson, H. S. & Lemen, J. R. Onset of the magnetic explosion in solar flares and coronal mass ejections. *Astrophys. J.* **552**, 833–848 (2001).
18. Parker, E. N. The solar-flare phenomenon and the theory of reconnection and annihiliation of magnetic fields. *Astrophys. J. Suppl. Ser.* **8**, 177–211 (1963).
19. Goode, P. R. & Cao, W. in *Society of Photo-Optical Instrumentation Engineers (SPIE) Conference Series*, Vol. 8444 (2012).
20. Cao, W. *et al.* Scientific instrumentation for the 1.6m New Solar Telescope in Big Bear. *Astron. Nachr.* **331**, 636 (2010).
21. Wiegelmann, T. *et al.* How should one optimize nonlinear force-free coronal magnetic field extrapolations from SDO/HMI vector magnetograms? *Sol. Phys.* **281**, 37–51 (2012).
22. Schou, J. *et al.* Design and ground calibration of the Helioseismic and Magnetic Imager (HMI) instrument on the Solar Dynamics Observatory (SDO). *Sol. Phys.* **275**, 229–259 (2012).
23. Lin, R. P. *et al.* The Reuven Ramaty High-Energy Solar Spectroscopic Imager (RHESSI). *Sol. Phys.* **210**, 3–32 (2002).
24. Kliem, B. & Török, T. Torus instability. *Phys. Rev. Lett.* **96**, 255002 (2006).
25. Gibson, S. E. & Fan, Y. The partial expulsion of a magnetic flux rope. *Astrophys. J. Lett.* **637**, L65–L68 (2006).
26. Kliem, B., Linton, M. G., Török, T. & Karlický, M. Reconnection of a kinking flux rope triggering the ejection of a microwave and hard X-Ray source II. Numerical modeling. *Sol. Phys.* **266**, 91–107 (2010).
27. van Ballegooijen, A. A. & Martens, P. C. H. Formation and eruption of solar prominences. *Astrophys. J.* **343**, 971–984 (1989).
28. Wöger, F. & von der Lühe, O. Field dependent amplitude calibration of adaptive optics supported solar speckle imaging. *Appl. Opt.* **46**, 8015–8026 (2007).
29. Denker, C. *et al.* Imaging magnetographs for high-resolution solar observations in the visible and near-infrared wavelength region. *Astron. Nachr.* **324**, 332–333 (2003).
30. Cao, W. *et al.* Diffraction-limited polarimetry from the infrared imaging magnetograph at Big Bear Solar Observatory. *Publ. Astron. Soc. Pac.* **118**, 838–844 (2006).
31. Schuck, P. W. Tracking Magnetic footpoints with the magnetic induction equation. *Astrophys. J.* **646**, 1358–1391 (2006).
32. Hurford, G. J. *et al.* The RHESSI imaging concept. *Sol. Phys.* **210**, 61–86 (2002).
33. Wiegelmann, T., Inhester, B. & Sakurai, T. Preprocessing of vector magnetogram data for a nonlinear force-free magnetic field reconstruction. *Sol. Phys.* **233**, 215–232 (2006).
34. Bobra, M. G. *et al.* The Helioseismic and Magnetic Imager (HMI) vector magnetic field pipeline: SHARPs--Space-weather HMI Active Region Patches. *Sol. Phys.* **289**, 3549–3578 (2014).
35. Wheatland, M. S., Sturrock, P. A. & Roumeliotis, G. An optimization approach to reconstructing force-free fields. *Astrophys. J.* **540**, 1150–1155 (2000).

36. Wiegelmann, T. Optimization code with weighting function for the reconstruction of coronal magnetic fields. *Sol. Phys.* **219,** 87–108 (2004).

Acknowledgements

We thank the BBSO staff for the NST observations, the SDO/HMI team for the vector magnetic field data, the RHESSI team for the X-ray data, T. Wiegelmann for the NLFFF extrapolation code and P. W. Schuck for the flow tracking code. This work was supported by NSF under grants AGS 0847126, 1153226, 1153424, 1250374, 1250818, 1348513 and 1408703, and by NASA under grants NNX13AG13G and NNX13AF76G. H.J. was supported by CAS Xiandao-B with grant No. XDB09000000, NSFC grants 11333009, 11173062, 11221063 and the 973 program with grant No. 2011CB811402.

Author contributions

H.W. developed the ideas for this study, coordinated the efforts of data analysis, wrote the first draft and led the discussion. W.C. developed instruments and led the observation at BBSO. C.L. contributed to the ideas, carried out the main data analysis and made major revisions of the manuscript. Y.X. contributed to the RHESSI data analysis. R.L. studied the magnetic twist and contributed to the scientific ideas. Z.Z. participated in the observations. J.C. contributed to the physical interpretation of the results. H.J. contributed the instrumentation of the 10830 observation. All authors participated in discussions, read and commented on the manuscript.

Additional information

5

Wave energy budget analysis in the Earth's radiation belts uncovers a missing energy

A.V. Artemyev[1,†], O.V. Agapitov[2,†], D. Mourenas[3], V.V. Krasnoselskikh[1] & F.S. Mozer[2]

Whistler-mode emissions are important electromagnetic waves pervasive in the Earth's magnetosphere, where they continuously remove or energize electrons trapped by the geomagnetic field, controlling radiation hazards to satellites and astronauts and the upper-atmosphere ionization or chemical composition. Here, we report an analysis of 10-year Cluster data, statistically evaluating the full wave energy budget in the Earth's magnetosphere, revealing that a significant fraction of the energy corresponds to hitherto generally neglected very oblique waves. Such waves, with 10 times smaller magnetic power than parallel waves, typically have similar total energy. Moreover, they carry up to 80% of the wave energy involved in wave–particle resonant interactions. It implies that electron heating and precipitation into the atmosphere may have been significantly under/over-valued in past studies considering only conventional quasi-parallel waves. Very oblique waves may turn out to be a crucial agent of energy redistribution in the Earth's radiation belts, controlled by solar activity.

[1] LPC2E/CNRS, 3A, Avenue de la Recherche Scientifique, 45071 Orleans Cedex 2, France. [2] Space Sciences Laboratory, University of California, 7 Gauss Way, Berkeley, California 94720, USA. [3] CEA, DAM, DIF, F-91297 Arpajon Cedex, France. † Present addresses: Space Research Institute (IKI) 117997, 84/32 Profsoyuznaya Street, Moscow, Russia. (A.V.A.); Astronomy and Space Physics Department, National Taras Shevchenko University of Kiev, 2 Glushkova Street, 03222 Kiev, Ukraine (O.V.A.). Correspondence and requests for materials should be addressed to A.V.A. (email: ante0226@gmail.com).

Since whistler-mode waves regulate fluxes of trapped electrons[1,2] and their precipitation rate[3-5] in the upper atmosphere[6-8], accurately determining the wave energy budget in the outer radiation belt has lately become an outstanding challenge for the scientific community[9]. Owing to the sparse wave data obtained by early satellites, their poor coverage of high latitudes and mainly one-component field measurements, and as linear theory was showing much higher parallel wave growth, scientists have commonly relied on the assumption that chorus waves were mainly field-aligned, that is, their propagation was weakly oblique with respect to the geomagnetic field[10,11]. Moreover, crucial theoretical works[12,13] in this area have demonstrated early on that the most important wave field component determining the wave–particle coupling efficiency was generally the magnetic field one, at least over a reasonably large range of wave obliquity. As a result, previous wave statistics focused on the sole magnetic field component of the full wave energy—showing indeed a clear prevalence of parallel waves in the equatorial region sampled by most satellites[14-17].

No global study on the basis of satellite measurements since then has led to a real revision of this conventional picture. Although some studies[16-19] and ray-tracing simulations[20] have recently hinted at both the possible presence and potential importance of very oblique whistler-mode waves, they failed to grasp the full extent of the implications, owing either to their continuing focus on statistics of the sole magnetic field component or to their use of statistical averages over such wide ranges of geomagnetic conditions that the effects of oblique waves have become blurred. Here, we study the full wave energy distribution of whistler waves, including both magnetic and electric field components. Our work suggests that the unexpected presence of a very large electrostatic energy, hitherto missing in past statistics of wave intensity and stored in very oblique waves, may profoundly change the current understanding of both the actual wave generation mechanisms and the processes of wave-induced electron scattering, acceleration and loss in the magnetosphere.

Results

Statistics of wave energy. To compare the impact of oblique and parallel waves in the formation and evolution of keV to MeV

electron fluxes in the inner magnetosphere, a reasonable approach consists in first estimating the energy density of both wave populations. Such a global survey is presented in Fig. 1. Here, we make use of 10 years of wave measurements performed by Cluster satellites[16] to evaluate the wave energy distribution throughout much of the Earth's inner magnetosphere as a function of wave obliquity and L-shell (the equatorial distance to the centre of the Earth normalized to Earth's radius). The energy density W of whistler-mode waves is determined by wave electric $\mathbf{E}$ and magnetic $\mathbf{B}$ field vectors through a complex relationship involving the tensor of absolute permittivity (see equation (1) in Methods section). Using the cold plasma dispersion relation for whistlers, W depends only on wave characteristics such as magnetic amplitude B, frequency ω, wave-normal angle θ (which defines the wave obliquity with respect to the geomagnetic field) and refractive index $N = kc/\omega$ (with k is the wave vector and c is the velocity of light).

Figure 1 with two-dimensional maps of wave energy W clearly shows that the proportion of very oblique waves, propagating near the resonance cone angle (that is, near 90°), is generally similar to or even larger than the proportion of quasi-parallel waves for L = 3 to 6 during moderate geomagnetic activity (defined by index $K_p < 3$) on the dayside. On the nightside or during more disturbed periods such that $K_p > 3$ (that is, geomagnetic storms or substorms), the amount of very oblique waves is sensibly reduced. The latter reduction stems probably from the presence of higher fluxes of hot ($\sim 100\,\text{eV}$ to $1\,\text{keV}$) plasmasheet electrons injected in the midnight region during disturbed periods[21]. Numerical ray-tracing simulations have shown that such hot electrons can damp very oblique waves propagating near their resonance cone[19,20,22].

The present results therefore challenge the conventional assumption of predominantly quasi-parallel whistler-mode waves in the outer radiation belt. A big, missing slice of the wave energy appears to be stored in very oblique waves—which are mainly made up of electrostatic energy[23]. Although most oblique waves are observed away from the equator, significant amounts moreover exist close to it. It strongly suggests that the widely accepted theory of parallel wave generation near the equator by

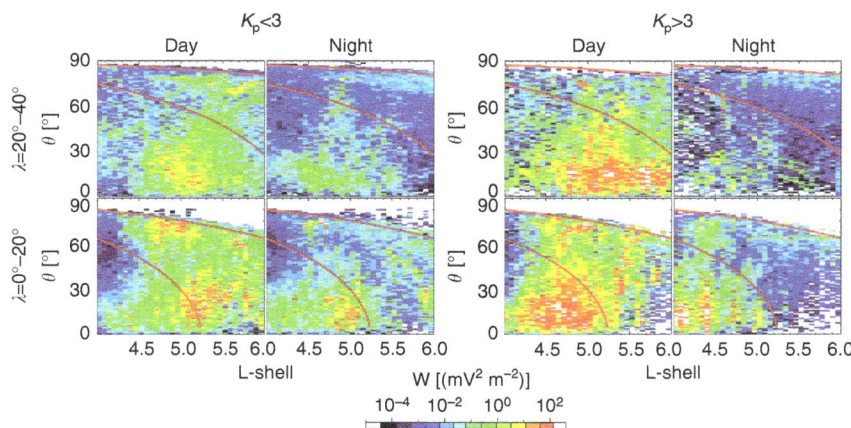

Figure 1 | Distribution of the energy of whistler waves in the Earth radiation belts. The distribution of the density of whistler wave energy W (in $\text{mV}^2\,\text{m}^{-2}$) is displayed in the (L,θ) space. Data are shown for two ranges of magnetic latitude (the near-equator region with $|\lambda| \in [0°,20°]$ and the high-latitude region with $|\lambda| \in [20°,40°]$), for day and night sectors, and for low ($K_p < 3$) and high ($K_p > 3$) geomagnetic activity. Red curves show the position of Gendrin θ_g and resonance cone θ_r angles (where $\cos\theta_g \approx 2\omega/\Omega_c$, $\cos\theta_r \approx \omega/\Omega_c$ and Ω_c is the local electron gyrofrequency). Both angles are calculated with the mean frequency of spacecraft observations, making use of precise plasma density measurements from IMAGE[39]. In the present figure, the wave refractive index has been limited to <100 in agreement with rough but conservative upper bounds due to Landau damping by average levels of hot electrons[19]. Three frequency channels have been taken into account: 2,244.9, 2,828.4 and 3,563.6 Hz, covering almost the full range from 2 to 4 kHz. Each channel is used in the corresponding L-shell range to measure only waves in whistler-mode frequency range.

an unstable electron population exhibiting a temperature anisotropy[11,24,25] might need to be supplemented with some new mechanism allowing the direct generation of very oblique waves there. This could require the presence of additional energetic electron populations differing subtly from the commonly assumed ones.

What are the consequences of the large energy of oblique waves on the dynamics of energetic particles? As their name suggests, wave–particle resonant interactions are controlled not only by wave intensity, but also by the actual efficiency of the resonant interactions. Waves must be in resonance with particles, implying that a certain relationship must be fulfilled between particle energy, pitch angle, wave frequency and obliquity (see equation (2) in Methods section). As a result, only a small portion of the total wave energy density actually corresponds to resonant waves[12]. The wave–particle coupling efficiency Φ, which depends also on cyclotron or Landau resonance harmonic number and on wave field components, provides the exact portion of wave energy interacting resonantly with particles[13,26], finally yielding the resonant wave energy $\Theta^2 = B^2\Phi^2/N^2$. Figure 2 shows the wave energy density Θ^2 of resonant waves plotted in the same manner as the wave energy density previously, using measured wave field components from Cluster. Figure 2 demonstrates that the resonant wave energy density at high θ-values (between the Gendrin and resonance cone angles) is 5–10 times larger than for parallel waves throughout the region L = 3–6. Hence, very oblique waves are expected to play a crucial role in the scattering of electrons in this region of space.

Electron lifetimes during geomagnetic storms. The remarkable effectiveness of the resonant interaction of very oblique waves with keV to MeV electrons can modify particle scattering and energization processes substantially in the radiation belts as compared with conventional theoretical estimates obtained for quasi-parallel waves alone. This effect should be most pronounced during moderately disturbed periods where oblique waves are more ubiquitous. To estimate the effects of oblique waves on resonant electron scattering during the course of a geomagnetic storm, we use here parameterizations of lower-band chorus wave magnetic intensity and θ distributions as functions of D_{st} devised on the basis of the same wave data set[18,27]. The

disturbance storm time D_{st} index is widely used to study the magnitude and internal variability of geomagnetic storms[28,29]. Two typical profiles $D_{st}(t)$ are considered (see top panel in Fig. 3), corresponding to storm types #1 and #2 (refs 28,30). The storm type #1 has a relatively long ($\sim$1.5 day) early recovery phase between $D_{st} = -100$ and -75 nT followed by a rapid increase of D_{st} back to -10 nT, while the second type has a much shorter early recovery phase followed by a much more prolonged stay ($\sim$3 days) around -50 nT.

The evolution of the lifetime τ_L of energetic electrons during the course of these two storms has been calculated numerically for various energies ranging from 100 eV to 1 MeV. Figure 3 first demonstrates the important variations of τ_L with D_{st}. Such strong variations can be explained by the combination of two main effects: (1) lifetimes increase when wave intensity decreases (both with and without oblique waves)[10,31] and (2) the wave–particle coupling Φ is significantly stronger for very oblique waves than for quasi-parallel waves over a very wide energy range (see Supplementary Fig. 2), leading to a reduction of lifetimes as the amount of very oblique waves increases during not-too-disturbed geomagnetic conditions[18,19,27]. The number of contributing resonances can moreover increase up to 10-fold for very oblique waves (see discussion of Supplementary Fig. 2 in Methods section).

When considering a realistic wave-normal angle distribution, the first clear consequence of the additional presence of very oblique waves is a general reduction of lifetimes during the storm. Most remarkably, however, such a reduction is much less significant during the early recovery period corresponding to $D_{st} < -75$ nT. The latter range actually corresponds to high parallel wave amplitudes. Very oblique waves are then almost absent, probably due to their quick damping by intense injections of hot electrons during the main phase of strong storms. Thus, an extended storm phase such that $D_{st} < -75$ nT, with intense waves and small losses, is particularly propitious for the strong energization of electrons. Later on, the competition between the opposite effects of a rapidly decreasing wave intensity and an increasing amount of oblique waves as D_{st} increases, results in a local minimum of τ_L near $D_{st} \sim -60$ nT during the early recovery phase. Finally, during nearly quiet periods with $D_{st} \sim -10$ nT at the end of storms, electron losses to the atmosphere are significantly increased by oblique waves, especially at very low energy. The remarkable difference between τ_L calculated for

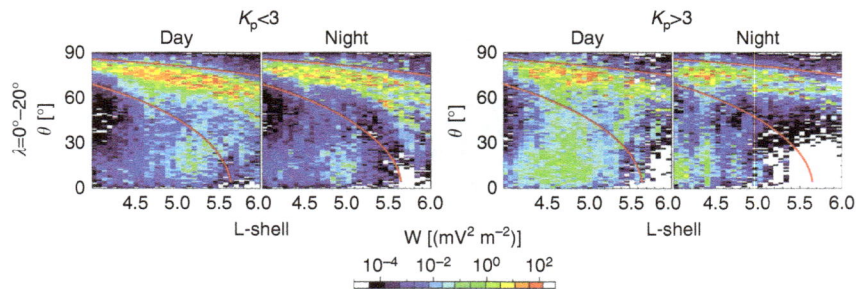

Figure 2 | Distribution of the energy of resonant whistler waves in the radiation belts. The distribution of the wave energy density Θ^2 of resonant waves (in mV^2 m^{-2}) is displayed in the (L,θ) space. The most effective resonant wave–particle interaction corresponds to a condition tan α tan $\theta \approx 1$ (where α is the particle pitch angle) for electron energy <2 MeV (ref. 31). This condition has been used to plot Θ^2 in this figure. Data are shown for one range of magnetic latitude $|\lambda| \in [0°,20°]$, for day and night sectors, and for low ($K_p < 3$) and high ($K_p > 3$) geomagnetic activity. Red curves show the position of Gendrin θ_g and resonance cone θ_r angles. Both angles are calculated with the mean frequency of spacecraft observations, making use of precise plasma density measurements from IMAGE[39]. In the present figure, the wave refractive index has been limited to <100 in agreement with rough but conservative upper bounds due to Landau damping by average levels of hot electrons[19]. Three frequency channels have been taken into account: 2,244.9, 2,828.4 and 3,563.6 Hz, covering almost the full range from 2 to 4 kHz. Each channel is used in the corresponding L-shell range to measure only waves in whistler-mode frequency range.

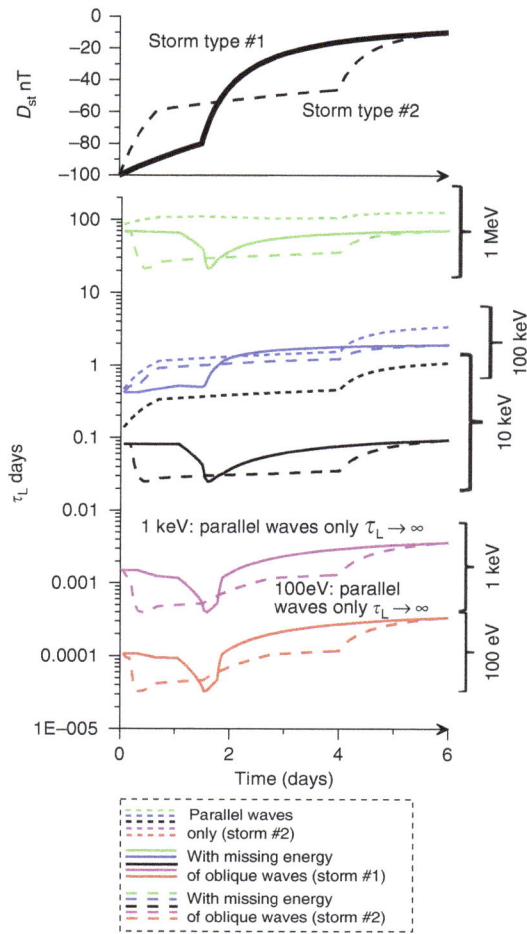

Figure 3 | Variation of electron lifetimes in the course of geomagnetic storms. Two different temporal profiles of the D_{st} index are shown in the top panel, corresponding to two different types of storms. Bottom panels show the corresponding variation of electron lifetimes τ_L calculated for $L \sim 5$ at different energies (each colour corresponding to one energy). Dotted lines correspond to τ_L for parallel chorus whistler-mode waves (τ_L is infinite for 100 eV and 1 keV in this case) in the case of storm type #2. Solid (storm type #1) and dashed (storm type #2) lines show τ_L when using a realistic distribution of wave-normal angle. Five energies are shown: 100 eV, 1 keV, 10 keV, 100 keV and 1 MeV (for $E \le (>)$ 1 keV, lifetimes are numerically calculated for electrons of equatorial pitch angle $<60°(85°)$).

parallel waves alone and with a realistic θ-distribution reaches indeed one order of magnitude for 10 keV electrons, while at lower energies (≤ 1 keV), only very oblique waves are still able to resonantly scatter electrons towards the loss-cone.

Discussion

Such results definitely show that a precise knowledge of the actual distribution of wave energy as a function of propagation angle θ is a key factor for accurately modelling the evolution of relativistic as well as low-energy electron fluxes under the influence of resonant wave–particle interactions. As noted above, this θ distribution is tightly controlled by the density and temperature of hot electrons. The large energy stored in very oblique waves can be readily tapped by sufficiently hot electrons newly injected from the outer magnetosphere and lead to their further heating via Landau damping. It, therefore, represents an accelerating factor of change for this important population[25] of particles.

More generally, the intrinsic variability of hot electron injections with geomagnetic activity[21] probably explains the observed variation of wave obliquity[18,27]. As the latter is able to fine-tune the precipitation of very low energy (especially 0.1 to 10 keV) electrons, the presence of a large amount of very oblique waves could have unexpected and major consequences on the ionospheric conductivity and on the nightside upper atmosphere ionization level at various altitudes, potentially affecting the whole magnetosphere–ionosphere coupling[7,8,32].

Beside determining electron precipitations into the atmosphere, whistler-mode waves are also responsible for the rapid energization of ~ 10 keV to 1 MeV electrons to multi-MeVs in the radiation belts during geomagnetic storms[3,5,33]. To first order, the effective energization depends mainly on the dimensionless product $D_{EE}\tau_L$ of the energy diffusion rate D_{EE} and lifetime, because a longer τ_L leaves more time for electron acceleration to proceed[18,27,34]. Moreover, the important dependence of the energy diffusion rate on the wave magnetic intensity (strongly increasing with $-D_{st}$) is almost fully compensated in this factor $D_{EE}\tau_L$ by the inverse dependence of the lifetime on the wave intensity. Since D_{EE} varies also much more weakly with wave obliquity than τ_L[35], it is the important variation of the lifetime with wave obliquity that should mainly determine the variation of the effective energization level of electrons. Thus, the comparison of lifetimes calculated with and without very oblique waves in Fig. 3 directly demonstrates the often dramatic change in energization level between these two cases.

Furthermore, the results displayed in Fig. 3 suggest that two storms with the same maximal strength but with different temporal profiles may lead to different effects on energetic electron fluxes, because of the different lifetime reductions dictated by the varying amount of very oblique waves. A storm (close to type #1) with a prolonged early recovery phase at $D_{st} < -75$ nT followed by a quick return to $D_{st} > -20$ nT should take advantage of high parallel wave intensity and weak overall losses to strongly energize electrons. Conversely, another storm (close to type #2) with a shorter initial period at $D_{st} < -75$ nT followed by a much slower recovery back to $D_{st} > -20$ nT should generally involve much stronger electron losses induced by larger amounts of very oblique waves during the early recovery phase (up to 1.5 day in Fig. 3), associated with a smaller magnetic wave intensity—efficiently reducing electron energization during that period. Later, significant losses to the atmosphere combined with modest wave intensity should nearly prevent any substantial acceleration. This could help to answer one outstanding question in radiation belt physics—why some geomagnetic storms correspond to global electron energization, while other storms with the same magnitude of D_{st} variation do not[29].

Excluding oblique waves from consideration would actually make the dimensionless energization factor $D_{EE}\tau_L$ almost constant and independent of the D_{st} time profile, as it does not depend on the bounce-averaged wave intensity. Only the consideration of an additional dimension of the system, corresponding to wave obliquity, gives a chance to obtain a significant variation of particle acceleration efficiency with D_{st} and, as a result, immediately produces a difference in particle acceleration for different $D_{st}(t)$ profiles. This effect allows to separate precipitation-dominated storms with a fast early recovery further slowing down, from acceleration-dominated storms with a slow initial recovery later on speeding up.

The surprisingly high level of very oblique wave energy discovered in Figs 1 and 2 and the strong concomitant increase of the wave–particle coupling strength have revealed that the wave obliquity, regulated by low-energy electrons injected from the plasmasheet, represents a new and important lever governing the variations of energetic electron fluxes. It indicates one possible answer to the problem of often-noted discrepancies between modern radiation belt models and observations[9,36]. The consideration of only parallel waves mostly restricts the space of wave model parameters to a single parameter—the wave amplitude. However, the distributions of wave amplitudes with geomagnetic activity and spatial location are well documented[10,16] and included in modern codes. In this study, we have clearly shown that there exists at least one additional model parameter—wave obliquity, which can control both the energization of electrons, their precipitation into the atmosphere and even the energy range of precipitated particles. The revelation of this hidden parameter and of the corresponding missing energy of very oblique waves should provide new opportunities to better understand and forecast the observed variations of energetic electron fluxes in the radiation belts as well as the global dynamics of the magnetosphere–ionosphere coupling.

Methods

Evaluation of the wave energy density. The energy density W of whistler-mode waves in the Earth's magnetosphere is determined by wave electric field $\mathbf{E}$ and magnetic field $\mathbf{B}$ vectors through the relation:

$$W \approx \frac{1}{8\pi} \left(\mathbf{E}^* \frac{d(\omega \hat{\varepsilon})}{d\omega} \mathbf{E} + \mathbf{B}^2 \right) \qquad (1)$$

where $\mathbf{E}^*$ is the conjugate vector to $\mathbf{E}$, ω is the wave frequency and $\hat{\varepsilon}(\omega)$ the tensor of absolute permittivity. Using the dispersion relation for electromagnetic whistler-mode waves in a cold magnetized plasma, electric field components can be further expressed as a function of $\mathbf{B}$. One gets $W = B^2(1 + W_E)/8\pi$ where W_E depends only on the wave characteristics: its frequency ω, wave-normal angle θ (which defines the wave obliquity with respect to the geomagnetic field), and refractive index $N = kc/\omega$ (where k is the wave vector). W steeply increases with N, which is itself a rapidly growing function of θ. The refractive index N (as well as θ) can be determined either solely from full three-component wave magnetic field measurements on a given spacecraft, or else by complete wave magnetic and electric field measurements. The dominant contributions to the wave energy distribution can be further assessed on the basis of either method. However, wave electric field measurements on Cluster satellites are often noisy, at least much more than magnetic field measurements, limiting their use in practice to some case studies. Therefore, we have chosen to resort to the just-discussed method of determination of the full wave energy density on the basis of measurements of the wave magnetic components alone. Nevertheless, the accuracy and reliability of this method must first be demonstrated.

To this aim, we have compared the crucial N values obtained by the two methods in a series of Cluster observations of chorus waves displayed in Supplementary Fig. 1. The comparison of panels (a) and (b) shows clearly that wave activity can be identified not only in magnetic field fluctuations, but also in the concomitant variations of the electric field. Most of the wave-power is concentrated around ~3 kHz—the ratio of wave frequency to electron equatorial gyrofrequency is $\omega/\Omega_{c0} \sim 0.35$. Waves can be considered as very oblique when θ is comprised between the Gendrin angle $\theta_g \approx \arccos(2\omega/\Omega_{c0})$ (which corresponds to wave propagation at a group velocity independent of frequency[37]) and the so-called resonance cone angle $\theta_r \sim \arccos(2\omega/\Omega_{c0})$ (the upper bound on θ where the cold plasma refractive index N of whistler waves goes to infinity[38]). For events in Supplementary Fig. 1, we have $\theta_g \sim 55$–$65°$ and $\theta_r \sim 75$–$85°$. Most observed whistler-mode waves are such that $\theta \in [60°, 85°]$ and can, therefore, be classified as very oblique chorus waves. A substantial part of the wave energy density (see panel (d)) consists of such oblique waves. The large ratio $W/W_B \gg 1$ shows that most of the energy density then comes from the wave electric field. More importantly, evaluations of the wave refractive index N from three-component measurements of the sole wave magnetic field yield values very similar to calculations making use of both magnetic and electric field components, attesting the reliability of the former method (compare panels (e) and (f)). The discrepancy does not exceed 25% on average, showing that this method can be safely used for evaluating the wave energy density.

However, only some part of the total wave energy density can actually interact resonantly with trapped electrons[12]. This resonant part is determined by the wave–particle coupling efficiency Φ (ref. 2) which depends on resonance harmonic number n, electron energy and pitch angle, as well as on the wave field components[13,26]. The resonance condition

$$\omega \gamma - ck\sqrt{\gamma^2 - 1} \cos\theta \cos\alpha = -n\Omega_c \qquad (2)$$

provides the necessary relation between particle energy (Lorentz factor γ), pitch angle α and wave obliquity θ. As a result, one gets a normalized estimate $\Theta^2 = \Phi^2 B^2/N^2$ of the resonant wave energy[13].

Evaluation of wave-particle coupling and diffusion rates. To demonstrate the peculiarities of electron resonant interaction with very oblique waves, additional numerical calculations of the wave–particle coupling efficiency Φ (averaged over latitude) have been performed as a function of wave propagation angle θ and geomagnetic activity index D_{st}, for various electron energies ranging from 100 eV to 1 MeV. Here, as well as for Fig. 3, we use usual values of the mean frequency $\omega_m/\Omega_{c0} \sim 0.35$ and frequency width $\Delta\omega/\Omega_{c0} \sim 0.2$ of lower-band chorus waves[19] and a ratio $\Omega_{pe}/\Omega_{ce} = 5$ corresponding to $L \sim 5$. Supplementary Fig. 1 shows that during not-too-disturbed geomagnetic conditions ($D_{st} > -60$ nT), wave–particle coupling Φ is clearly stronger for very oblique waves than for quasi-parallel waves over a wide energy range. For a given level of wave intensity, the available range of variation of the wave–particle coupling efficiency Φ at small equatorial pitch angles (near the loss-cone where particles are precipitated in the atmosphere) is so large that it could presumably explain any observed fluctuations of electron flux by fluctuations of the wave obliquity only and associated variations of electron scattering. In addition to the increase of Φ for a given resonance, the number of such contributing resonances can moreover increase 10-fold for oblique waves (see Supplementary Fig. 2).

The efficiency of charged particles resonant interaction with waves is determined by diffusion rates proportional to the weighting factor $\Phi_n^2 = \Theta^2 g_\theta(\theta) g_\omega(\omega)$ where g_θ and g_ω are normalized distributions of θ and wave frequency. To calculate the g_θ normalization, one should determine resonant k and ω for given particle pitch angle and energy. Then, an integration over θ must be performed. The upper limit of this integration is determined by the maximum value of the refractive index N_{Max}. The latter is imposed by the presence of both thermal effects in the dispersion relation and Landau damping by 100–500 eV suprathermal electrons of oblique waves near the resonance cone angle[19,22]. Using typically observed parameters for the thermal and suprathermal electron population at $L \sim 5$, it has been shown that one could take $N_{Max} \sim 120$ to 300 for lower-band chorus waves from low to high latitudes during periods of quiet to moderately disturbed geomagnetic activity, with N_{Max} varying as the inverse of the frequency ω and increasing with latitude[19]. It led us to use here (in Supplementary Fig. 2, and Fig. 3 in main text) a rough but realistic limit $N_{Max} \sim \min(300, 36\Omega_{ce}/\omega)$ corresponding to a predominant effect of Landau damping.

References

1. van Allen, J. A. & Frank, L. A. Radiation around the Earth to a radial distance of 107,400 km. *Nature* **183**, 430–434 (1959).
2. Schulz, M. & Lanzerotti., L. J. *Particle Diffusion in the Radiation Belts* (Springer, 1974).
3. Horne, R. B. Plasma astrophysics: acceleration of killer electrons. *Nat. Phys.* **3**, 590–591 (2007).
4. Horne, R. B., Lam, M. M. & Green, J. C. Energetic electron precipitation from the outer radiation belt during geomagnetic storms. *Geophys. Res. Lett.* **36**, L19104 (2009).
5. Thorne, R. M. et al. Rapid local acceleration of relativistic radiation-belt electrons by magnetospheric chorus. *Nature* **504**, 411–414 (2013).
6. Thorne, R. M. Energetic radiation belt electron precipitation—a natural depletion mechanism for stratospheric ozone. *Science* **195**, 287–289 (1977).
7. Brasseur, G. P. & Solomon., S. *Aeronomy of the Middle Atmosphere: Chemistry and Physics of the Stratosphere and Mesosphere,* (Springer, 2005).
8. Seppälä, A., Lu, H., Clilverd, M. A. & Rodger, C. J. Geomagnetic activity signatures in wintertime stratosphere wind, temperature, and wave response. *J. Geophys. Res.* **118**, 2169–2183 (2013).
9. Horne, R. B. et al. Space weather impacts on satellites and forecasting the Earth's electron radiation belts with SPACECAST. *Space Weather* **11**, 169–186 (2013).
10. Shprits, Y. Y., Meredith, N. P. & Thorne, R. M. Parameterization of radiation belt electron loss timescales due to interactions with chorus waves. *Geophys. Res. Lett.* **34**, L11110 (2007).
11. Omura, Y., Katoh, Y. & Summers, D. Theory and simulation of the generation of whistler-mode chorus. *J. Geophys. Res.* **113**, A04223 (2008).
12. Kennel, C. F. & Wong, H. V. Resonant particle instabilities in a uniform magnetic field. *J. Plasma Phys.* **1**, 75 (1967).
13. Lyons, L. R., Thorne, R. M. & Kennel, C. F. Electron pitch-angle diffusion driven by oblique whistler-mode turbulence. *J. Plasma Phys.* **6**, 589–606 (1971).
14. Hayakawa, M., Yamanaka, Y., Parrot, M. & Lefeuvre., F. The wave normals of magnetospheric chorus emissions observed on board GEOS 2. *J. Geophys. Res.* **89**, 2811–2821 (1984).
15. Meredith, N. P. et al. Global model of lower band and upper band chorus from multiple satellite observations. *J. Geophys. Res.* **117**, A10225 (2012).
16. Agapitov, O. et al. Statistics of whistler mode waves in the outer radiation belt: Cluster STAFF-SA measurements. *J. Geophys. Res.* **118**, 3407–3420 (2013).
17. Li, W. et al. Characteristics of the Poynting flux and wave normal vectors of whistler-mode waves observed on THEMIS. *J. Geophys. Res.* **118**, 1461–1471 (2013).
18. Artemyev, A. V., Agapitov, O. V., Mourenas, D., Krasnoselskikh, V. & Zelenyi, L. M. Storm-induced energization of radiation belt electrons: effect of wave obliquity. *Geophys. Res. Lett.* **40**, 4138–4143 (2013).

19. Li, W. *et al.* Evidence of stronger pitch angle scattering loss caused by oblique whistler-mode waves as compared with quasi-parallel waves. *Geophys. Res. Lett.* **41**, 6063–6070 (2014).

20. Chen, L., Thorne, R. M., Li, W. & Bortnik, J. Modeling the wave normal distribution of chorus waves. *J. Geophys. Res.* **118**, 1074–1088 (2013).

21. Li, W. *et al.* Global distributions of suprathermal electrons observed on THEMIS and potential mechanisms for access into the plasmasphere. *J. Geophys. Res.* **115**, A00J10 (2010).

22. Horne, R. B. & Sazhin., S. S. Quasielectrostatic and electrostatic approximations for whistler mode waves in the magnetospheric plasma. *Planet. Space Sci.* **38**, 311–318 (1990).

23. Ginzburg, V. L. & Rukhadze, A. A. *Waves in Magnetoactive Plasma* 2nd revised edition (Nauka, 1975).

24. Trakhtengerts., V. Y. A generation mechanism for chorus emission. *Ann. Geophys.* **17**, 95–100 (1999).

25. Fu, X. *et al.* Whistler anisotropy instabilities as the source of banded chorus: Van Allen probes observations and particle-in-cell simulations. *J. Geophys. Res.* **119**, 8288–8298 (2014).

26. Glauert, S. A. & Horne, R. B. Calculation of pitch angle and energy diffusion coefficients with the PADIE code. *J. Geophys. Res.* **110**, A04206 (2005).

27. Mourenas, D., Artemyev, A. V., Agapitov, O. V. & Krasnoselskikh, V. Consequences of geomagnetic activity on energization and loss of radiation belt electrons by oblique chorus waves. *J. Geophys. Res.* **119**, 2775–2796 (2014).

28. McPherron, R. L. in *Magnetic Storms* 98 (eds Tsurutani, B. T., Gonzalez, W. D., Kamide, Y. & Arballo, J. K.) 131–147 (American Geophysical Union Geophysical Monograph Series, 1997).

29. Reeves, G. D., McAdams, K. L., Friedel, R. H. W. & O'Brien, T. P. Acceleration and loss of relativistic electrons during geomagnetic storms. *Geophys. Res. Lett.* **30**, 1529 (2003).

30. Denton, M. H. *et al.* Geomagnetic storms driven by ICME- and CIR-dominated solar wind. *J. Geophys. Res.* **111**, A07S07 (2006).

31. Mourenas, D., Artemyev, A. V., Ripoll, J.-F., Agapitov, O. V. & Krasnoselskikh, V. V. Timescales for electron quasi-linear diffusion by parallel and oblique lower-band chorus waves. *J. Geophys. Res.* **117**, A06234 (2012).

32. Gkioulidou, M. *et al.* Effect of an MLT dependent electron loss rate on the magnetosphere-ionosphere coupling. *J. Geophys. Res.* **117**, A11218 (2012).

33. Mozer, F. S. *et al.* Direct observation of radiation-belt electron acceleration from electron-volt energies to megavolts by nonlinear whistlers. *Phys. Rev. Lett.* **1130**, 035001 (2014).

34. Horne, R. B. *et al.* Wave acceleration of electrons in the Van Allen radiation belts. *Nature* **437**, 227–230 (2005).

35. Mourenas, D., Artemyev, A., Agapitov, O. & Krasnoselskikh, V. Acceleration of radiation belts electrons by oblique chorus waves. *J. Geophys. Res.* **117**, A10212 (2012b).

36. Kim, K.-C., Shprits, Y., Subbotin, D. & Ni, B. Relativistic radiation belt electron responses to GEM magnetic storms: Comparison of CRRES observations with 3-D VERB simulations. *J. Geophys. Res.* **117**, A08221 (2012).

37. Gendrin, R. Le guidage des whistlers par le champ magnetique. *Planet. Space Sci.* **5**, 274 (1961).

38. Helliwell., R. A. *Whistlers and Related Ionospheric Phenomena* (Stanford Univ. Press, 1965).

39. Denton, R. E. *et al.* Distribution of density along magnetospheric field lines. *J. Geophys. Res.* **111**, A04213 (2006).

Acknowledgements

The work of F.S.M. and O.V.A. was supported by NASA Grant NNX09AE41G. The work of V.V.K. was supported by CNES grant Cluster Co-I DW.

Author contributions

All the authors contributed to all the aspects of this work.

Additional information

6

Abundance of live ^{244}Pu in deep-sea reservoirs on Earth points to rarity of actinide nucleosynthesis

A. Wallner[1,2], T. Faestermann[3], J. Feige[2], C. Feldstein[4], K. Knie[3,5], G. Korschinek[3], W. Kutschera[2], A. Ofan[4], M. Paul[4], F. Quinto[2,†], G. Rugel[3,†] & P. Steier[2]

Half of the heavy elements including all actinides are produced in r-process nucleosynthesis, whose sites and history remain a mystery. If continuously produced, the Interstellar Medium is expected to build-up a quasi-steady state of abundances of short-lived nuclides (with half-lives ≤100 My), including actinides produced in r-process nucleosynthesis. Their existence in today's interstellar medium would serve as a radioactive clock and would establish that their production was recent. In particular ^{244}Pu, a radioactive actinide nuclide (half-life = 81 My), can place strong constraints on recent r-process frequency and production yield. Here we report the detection of live interstellar ^{244}Pu, archived in Earth's deep-sea floor during the last 25 My, at abundances lower than expected from continuous production in the Galaxy by about 2 orders of magnitude. This large discrepancy may signal a rarity of actinide r-process nucleosynthesis sites, compatible with neutron-star mergers or with a small subset of actinide-producing supernovae.

[1] Department of Nuclear Physics, Australian National University, Canberra, Australian Capital Territory 0200, Australia. [2] VERA Laboratory, Faculty of Physics, University of Vienna, Währinger Strasse 17, A-1090 Vienna, Austria. [3] Physik Department, Technische Universität München, D-85747 Garching, Germany. [4] Racah Institute of Physics, Hebrew University, Jerusalem 91904, Israel. [5] GSI Helmholtz-Zentrum für Schwerionenforschung GmbH, Planckstrasse 1, 64291 Darmstadt, Germany. † Present addresses: Institute for Nuclear Waste Disposal (INE), Hermann-von-Helmholtz-Platz 1, D-76344 Eggenstein-Leopoldshafen, Germany (F.Q.); Helmholtz-Zentrum Dresden-Rossendorf, Helmholtz Institute Freiberg for Resource Technology, Halsbruecker Strasse 34, 09599 Freiberg, Germany (G.R.). Correspondence and requests for materials should be addressed to A.W. (email: anton.wallner@anu.edu.au).

About half of all nuclides existing in nature and heavier than iron are generated in stellar explosive environments. Their production requires a very short and intense burst of neutrons (rapid neutron capture or r-process)[1–3]. The nuclides are formed via successive neutron captures on seed elements, following a path in the very neutron-rich region of nuclei. However, the relevant astrophysical sites, with supernovae (SN)[1,2] and neutron-star (NS-NS) mergers[3,4] as prime candidates, and the history of the r-process during the Galactic chemical evolution are largely unknown. The interstellar medium (ISM) is expected to become steadily enriched with fresh nucleosynthetic products and may also contain continuously produced short-lived nuclides (with half-lives ≤ 100 My (ref. 5)), including actinides produced in r-process nucleosynthesis.

Recent r-process models within SNe-II explosions, based on neutrino wind scenarios[6,7], suffer difficulties on whether heavy elements can really be produced in these explosions. An alternative site is NS ejecta, for example, NS-NS or NS black-hole (NS-BH) mergers. Candidates of such neutron-star binary systems have been detected[8,9]. Estimations of an NS-NS merger event rate of about $(2–3) \times 10^{-5}$ per year in our galaxy would allow for such mergers to account for all heavy r-process matter in our Galaxy[3,10,11].

It was pointed out by Thielemann et al.[4] that observations of old stars indicate a probable splitting of the r-process into (i) a rare event that reproduces the heavy r-process abundances including actinides always in solar proportions, and (ii) a more frequent event responsible for the lighter r-process abundances. Galactic chemical evolution models[10,12,13] show that NS mergers, occurring at late time in the life of a galaxy, cannot account for all the r-process nuclei found in very old stars[12]. Thus, recent models suggest different r-process scenarios (similar to s process), which might occur at different nucleosynthesis sites[3,13].

To summarize, very few hints on astrophysical sites and galactic chemical evolution exist. First, the relative abundance distribution observed spectroscopically in a few old stars for r-process elements between barium and hafnium is very similar to that of the Solar System (SS)[1,14], pointing to an apparently robust phenomenon; a large scatter for the r-process elements beyond Hf and also below barium is, however, observed[3,11,15]. Second, the early SS (ESS) is known to have hosted a set of short-lived radioactive nuclides ($t_{1/2} < \sim 100$ My)[5,16–18], among them pure r-process nuclei such as ^{244}Pu (half-life $= 81$ My) and ^{247}Cm (15.6 My) clearly produced no more than a few half-lives before the gravitational collapse of the protosolar nebula[19–23].

We report here on a search for live ^{244}Pu (whose abundance in the ESS relative to ^{238}U was $\sim 0.8\%$ (see refs 5,19–21,23–25)) in deep-sea reservoirs, which are expected to accumulate ISM dust particles over long time periods. Our findings indicate that SNe, at their standard rate of ~ 1-2/100 years in the Galaxy, did not contribute significantly to actinide nucleosynthesis for the past few hundred million years. A similar conclusion is drawn, when related to the recent SNe history in the local interstellar environment: we do not find evidence for live ^{244}Pu that may be locked in the ISM in accumulated swept-up material and that was transported to Earth by means of recent SNe activity. Our results suggest that actinide nucleosynthesis, as mapped through live ^{244}Pu, seems to be very rare.

Results

Experimental concept. ISM dust particles[26,27], assumed to be representative of the ISM, are known to enter the SS and are expected to reach and accumulate on Earth in long-term natural depositories such as deep-sea hydrogenous iron-manganese (FeMn) encrustations and sediments. Such a process is confirmed by inclusion in these archives of meteoritic ^{10}Be, cosmogenic ^{53}Mn and live ^{60}Fe, the latter attributed to the direct ejecta of a close-by SN (refs 28–30). ^{244}Pu-detection would be the equivalent for r-process nuclides of the γ-ray astronomy observations of live radioactivities[17] produced by explosive nucleosynthesis in single SN events (for example, ^{56}Ni (6.1 d), ^{56}Co (77.3 d), ^{44}Ti (60.0 y) or diffuse in the Galactic plane such as ^{26}Al (0.72 My) and ^{60}Fe (2.62 My), owing to their longer half-life).

Several models, based on the frequency of SN events, the nucleosynthesis yield and the radioactive half-life, were developed to calculate the abundance of ^{244}Pu in quasi-secular equilibrium between production and radioactive decay rates. These models together with the flux and average mass of ISM dust particles into the inner SS measured by space missions in the last decade (Galileo, Ulysses, Cassini)[31] are used here to estimate the corresponding influx of ^{244}Pu nuclei onto Earth.

We compare our results also with a possible imprint of recent actinide nucleosynthesis (<15 My) from the SNe history of the Local Bubble (LB, a cavity of low density and hot temperature of ~ 200 pc diameter). Recent ISM simulations suggest about 14–20 SN explosions within the last 14 My[32–34] that were responsible for forming the local ISM structure and the LB. ^{244}Pu decay can be considered negligible during this period. The SN ejecta shaped the ISM and also accumulates swept-up material including pre-existing ^{244}Pu from nucleosynthesis events prior to the formation of the LB[35,36].

With a growth rate of a few millimetres per million years[37], hydrogenous crusts will strongly concentrate elements and particles present in the water column above. The higher accumulation rate of deep-sea sediments (millimetre per thousand years) results in a better time resolution but requires much larger sample volumes. With regard to other potential ^{244}Pu sources, we note that natural ^{244}Pu production on Earth is negligible and the ESS abundance has decayed to 10^{-17} of its pre-solar value[22,23]. Anthropogenic production from atmospheric nuclear bomb tests and from high-power reactors is restricted to the last few decades, localized in upper layers and can easily be monitored through the characteristic isotopic fingerprint of the other co-produced $^{239–242}$Pu isotopes. In fact the detection of anthropogenic ^{239,240}Pu in deep-sea sediments[38–40] and crusts[41] provides an excellent proxy for the ingestion efficiency of dust from the high atmosphere into these reservoirs, together with their chemical processing towards the final analyzed samples (Methods).

Selected terrestrial archives for extraterrestrial ^{244}Pu. Terrestrial archives like deep-sea FeMn crust and sediment archives extend over the past tens of million years. Large dust grains entering Earth's atmosphere have also been observed by radar detections[42]. Extraterrestrial dust particles, cosmogenic nuclides and terrestrial input sink to the ocean floor and are eventually incorporated into the FeMn crust or sediment. For actinide transport through the latter stages, the observed deposition of global fallout[41] from atmospheric nuclear bomb testing[38,39] in deep marine reservoirs after injection to the stratosphere serves as a proxy to extraterrestrial particles.

We chose two independent archives: a large piece (1.9 and 0.4 kg samples) from a deep-sea manganese crust (237 KD from cruise VA13/2, collected in 1976) with a growth rate between 2.5 mm per My (refs 29,37) and 3.57 mm per My (ref. 43). It originates from the equatorial Pacific (location 9°180′N, 146°030′W) at a depth of 4,830 m and covers the last ~ 25 My (refs 30,43–45). In the very same crust, the live ^{60}Fe signal mentioned above was found at about 2.2 My before present (BP)[28,29]. Our second sample, also from the Pacific Ocean, is a piston-core deep-sea sediment (7P), extracted during the TRIPOD expedition as part of the Deep-Sea

Drilling Project (DSDP) at location 17°30′ N, 113°00′ W at 3,763 m water depth and covers a time period of ~0.5–2.1 My BP (W. Smith, Scripps Geological Collections, USA, personal communication). The crust sample, covering a total area of 227.5 cm² and a time range of 25 My, was split into four layers (1–4) representing different time periods in the past (see Table 1). Each layer was subdivided into three vertical sections (B, C and D) with areas between 70 and 85 cm², totalling 12 individually processed samples. The surface layer (layer 1, with a time range from present to 500,000 years BP) contains also the anthropogenic Pu signal originating from global fallout of atmospheric weapons testing[38,39]. Next, layer 2 spans a time period from 0.5–5 My BP, layer 3 5–12 My and layer 4 12–25 My (ref. 30). We note, the age for samples older than 14 My, where no ^{10}Be dating is possible[29,37], is more difficult to establish; different age models suggest a time period of 12 to ~18–20 My (ref. 44), another model up to ~30 My (ref. 45) for layer 4. Finally, sample X, the bottom layer of hydrothermal origin (Fig. 1) served as background sample.

For archives accumulating millions of years, the expected ^{244}Pu abundance range (see discussion) is well within reach of accelerator mass spectrometry (AMS), an ultra-sensitive method[46–48] of ion identification and detection. Based on the ingestion efficiency of Pu into deep-sea manganese crusts (21%) and on the AMS ^{244}Pu-detection efficiency (1×10^{-4}, see Methods), we calculate a measurement sensitivity expressed as a ^{244}Pu flux onto Earth of the order of 0.1 to 1 atom per cm² per My ^{244}Pu from ISM deposition. Thus, for the crust with a 25 My accumulation period and with 200 cm² surface area ~500–4,000 ^{244}Pu-detection events are expected, and about a factor 100 less for the sediment sample (1.64 My time period and 4.9 cm² surface area).

AMS experimental data of ^{244}Pu abundances in Earth archives. We have developed the capability to detect trace amounts of ^{244}Pu in terrestrial archives by AMS[46] and our technique provides background-free ^{244}Pu detection with an overall efficiency (atoms detected/atoms in the sample) of ~1×10^{-4} (see Methods and Supplementary Tables 1–4). The AMS measurements determine the atom ratio ^{244}Pu/^{A}Pu where ^{A}Pu (A = 236 or 242) is a spike of known amount (added during the chemical processing of the

sample) from which the number of ^{244}Pu nuclei in the sample is obtained (see Methods). In addition to ^{244}Pu counting, we also measured the shorter lived ^{239}Pu ($t_{1/2} = 24.1$ ky) content as an indicator of anthropogenic Pu signature.

The results for the four crust layers and the blank sample, obtained from the AMS measurements on 11 individual crust samples, are listed in Tables 1 and 3 (see also Supplementary Tables 1 and 2; identification spectra are plotted in Fig. 2). We observed one single event in each of the two crust subsamples, namely layer 3, section B (B3), and layer 4, section D (D4). No ^{244}Pu was registered in the other seven crust subsamples or the blank sample (X). A clear anthropogenic ^{239}Pu and ^{244}Pu signal, originating from atmospheric atomic-bomb tests from ~1950 to 1963, was observed in the top layer (16 events of ^{244}Pu detected).

Figure 1 | Crust sample 237 KD. This FeMn crust (with a total thickness of 25 cm) was sampled in 1976 from the Pacific Ocean at 4,830 m water depth: large samples used in this work were taken from one part of the crust (hydrogenous crust, layers 1–4, left in the figure)) and from the bottom (hydrothermal origin, layer X, crust started to grow ~65 My ago[43], see also refs 44,45).

Table 1 | ^{244}Pu detector events and corresponding ISM flux compared with galactic chemical models assuming steady state.

Deep-sea archive	Time period (My)	Sample area (cm²)	Sample mass (g)	Integral sensitivity (eff. × area × time period) (cm² My)	^{244}Pu detector events (2σ limit)*	^{244}Pu flux into terrestrial archive (atoms per cm² per My)	^{244}Pu flux ISM at Earth orbit (atoms per cm² per My)†
Crust_modern	0–0.5	227.2	80	0.006	16	—	—
Layer X	Blank	~100	364	—	0	—	—
Layer 2	0.5–5	227.2	473	0.016	0 (<3)	<188	<3,500
Layer 3	5–12	227.2	822	0.075	1 (<5)	13^{+53}_{-12} (<66)	$247^{+1,000}_{-235}$
Layer 4	12–25	142.2	614	0.060	1 (<5)	17^{+66}_{-16} (<83)	$320^{+1,250}_{-300}$
Crust	**0.5–25**	**182**	**1,909**	**0.151**	**2 (<6.7)**	13^{+31}_{-11} (<44)	250^{+590}_{-205}
Sediment	0.53–2.17	4.9	101	0.0013	1 (<5)	$750^{+3,000}_{-710}$	$3,000^{+12,000}_{-2,850}$
Model and satellite data‡	Steady-state model and ISM flux data at 1 AU from satellite Cassini						20,000–160,000

eff., efficiency; ISM, interstellar medium.
The FeMn crust sample was split into four layers 1–4 (see Methods). The top layer (1 mm, 'crust modern') was removed for measuring the anthropogenic Pu content. In total two ^{244}Pu detector events were registered using AMS in all older crust samples over a 72 h counting time (column 6). We calculate from our data an extraterrestrial ^{244}Pu flux and a 2σ limit from <6.7 extraterrestrial ^{244}Pu events[49]. The sediment sample also gave one ^{244}Pu detector event and none were registered in any of the blank samples. The term 'integral sensitivity' represents a quantity that combines the overall measurement eff., the flux integration area and the time period covered by the individual samples.
*Because of the low ^{244}Pu event rate, we also display 2σ upper levels (95% confidence levels) applying low-level statistics[49].
†Using an incorporation efficiency ε = (21 ± 5)% for the crust and 100% for the sediment sample (Methods). The mean area for the crust sample is 182 cm² (accounting for the different time periods) and 4.9 cm² for the sediment sample. For calculating the ISM flux at Earth orbit, the measured ^{244}Pu flux into the terrestrial archives was corrected for the incorporation efficiency and was multiplied by a factor of 4 to account for the ratio of Earth's surface to its cross-section (that is, assuming a unidirectional and homogeneous ISM flux relative to the Solar System).
‡the steady-state ^{244}Pu flux is based on the actinide (U and Th) abundances measured in meteorites, and on present-day Pu/U and Pu/Th ISM concentrations deduced from galactic chemical evolution models. The Pu flux at 1 AU (Earth orbit) is corrected for the filtration of interstellar dust particles when entering the heliosphere of our Solar System (3–9%, see Methods).

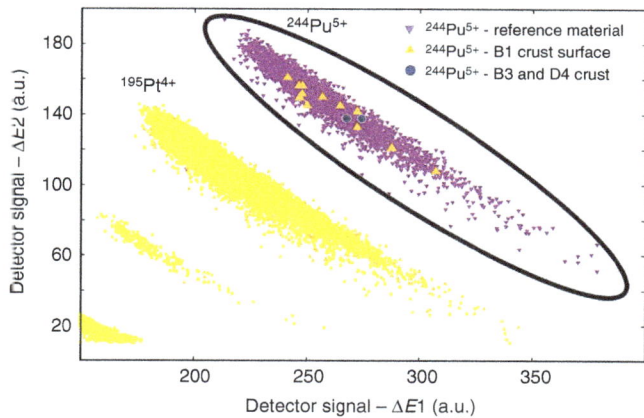

Figure 2 | ^{244}Pu detection with AMS. Identification spectra obtained in the AMS measurements with a particle detector (two independent differential energy-loss signals ($\Delta E1$ and $\Delta E2$) are plotted in x- and y-axis). Parasitic (or background) particles of different energy (for example, ^{195}Pt^{4+}) and different mass were clearly separated and do not interfere. Displayed is an overlay of ^{244}Pu^{5+} events obtained in a series of measurements for a ^{244}Pu reference material (purple triangle), the 12 events registered for one of the surface layer samples, B1 (yellow triangle) and the 2 events measured for the deeper layers B3 and D4, respectively (blue circle).

Measurements of samples from deep layers (>0.5 My) show also some events during the ^{239}Pu measurement (compared with the top layer, the ^{239}Pu count rate in the deep crust layers were a factor of ~ 100 lower, and the one in the sediment and blank sample were a factor of $\sim 1,000$ lower). Since naturally produced ^{239}Pu in these older layers is considered negligible, its presence is attributed to ^{238}U still present in the final AMS sample at about 8 to 9 orders of magnitude higher than ^{239}Pu and mimicking ^{239}Pu detector events (see Methods); we also note that the ^{236}Pu spike added for tracing the measurement efficiency was found to contain some ^{239}Pu, which we corrected for, see Supplementary Table 4). We conclude from the ^{239}Pu data that anthropogenic contamination did not add any significant ^{244}Pu detector events for all older layers (using the anthropogenic ^{244}Pu/^{239}Pu ratio obtained from the top layer, 1×10^{-4}, see Methods, Table 3). In the following, we calculate for all cases 2σ upper limits, that is, 95% confidence levels for which 0 (1 or 2) ^{244}Pu events corresponds to an upper limit of <3 (5 or 6.7, respectively) ^{244}Pu events (applying statistics for small signals[49]).

^{244}Pu flux deduced from measured terrestrial concentrations. The crust data for all sections and for all three deeper layers are compatible (for details see also Supplementary Table 1). Owing to a higher chemical yield (integral sensitivity, column 5, Table 1) layers 3 and 4 provide lower limits. The measured ^{244}Pu concentration in these layers can be converted into a ^{244}Pu particle flux using chemical yield, detection efficiency, the incorporation efficiency of Pu into the crust (21 ± 5 %, see Methods and Table 3), and the area and time period covered. We also assume that the extraterrestrial ^{244}Pu flux through Earth's cross-section is homogeneously distributed over the Earth's surface. Hence, the interstellar flux is calculated by multiplying the measured flux into the crust by a factor of $4/0.21 = 19$. We thus derive a 2σ limit[49] for the ^{244}Pu-ISM-flux at Earth orbit from data of the three layers $<3,500$, $<1,300$ and $<1,560$ ^{244}Pu atoms per cm^2 per My, and the one ^{244}Pu event in layers 3 and 4 corresponds to a flux of $247^{+1,000}_{-235}$ and $320^{+1,250}_{-300}$ atoms cm^{-2} per My, respectively. Combining all samples (2 ^{244}Pu events) a flux of

250^{+590}_{-205} and a 2σ-limit on the ^{244}Pu flux of <840 ^{244}Pu atoms per cm^2 per My is obtained. The data are plotted in Fig. 2. The single ^{244}Pu event measured for the deep-sea sediment converts to a flux of 3,000 ($<15,000$) atoms per cm^2 per My. Both archives give consistent ^{244}Pu flux limits with a higher sensitivity for the crust samples.

Discussion

First, we estimate the expected ^{244}Pu flux from ISM dust particles penetrating the SS, and their incorporation into terrestrial archives. Our experimental results are then compared with these estimations. Based on a uniform production model[18] or an open-box model[5,16,50] (see also ref. 24), taking into account Galactic-disk enrichment in low-metallicity gas, the present-day ISM atom ratio ^{244}Pu/^{238}U from SN events is calculated to be between 0.017 and 0.044 (Table 2). We further assume that the abundance of ^{238}U in ISM dust is the same as that of chondrite meteorites[51] (corrected for the SS age), 1.7×10^{-8} g per g meteorite, and derive a steady-state ^{244}Pu abundance of $(2.8–7.5) \times 10^{-10}$ g Pu per g ISM (that is, $(0.8–2.2) \times 10^{-14}$ atoms per cm^3); similar values are obtained if normalized to ^{232}Th.

For interstellar dust particles (ISDs) entering the SS[27], we have to take into account filtering when penetrating the heliosphere. ISDs were observed by the Ulysses, Galileo and Cassini[31,52] space missions over more than 5 years, for distances from the Sun between 0.4 and >5 AU. Measurements of the Cassini space mission[27,31] determine a mean flux of ISM dust of $(3–4) \times 10^{-5}$ particles per m^2 per s at a distance of 1 AU, that is, at Earth's position, with a mean particle mass of $(3–7) \times 10^{-13}$ g (0.5–0.6 µm average particle size). These particles show a speed distribution corresponding to the flow velocity of the ISM (26 km s^{-1}) and constitute 3–9% of the dust component of the ISM intercepted by the SS (see Table 2). The direct collection of a few particles identified as ISD, very recently reported[53], although of low statistical significance, supports the scenario of penetration of large ISD particles into the inner SS and may be consistent with the satellite data. It should be noted that Galactic cosmic-rays penetrate the SS and recent observations clearly demonstrate therein the presence of Th and U, and tentatively of ^{244}Pu (ref. 54).

Within the assumptions described above, the expected flux of ^{244}Pu atoms from the ISM reaching the inner SS (at Earth orbit) is $(2.5–21) \times 10^{-31}$ g Pu per cm^2 per My or 20,000–160,000 ^{244}Pu atoms per cm^2 per My. If evenly distributed over the Earth's surface (that is, assuming a unidirectional ISM flux) the ^{244}Pu flux into terrestrial archives becomes 5,000–40,000 ^{244}Pu atoms per cm^2 per My.

Our experimental results (Table 1) provide for the first time a sensitive limit of interstellar ^{244}Pu concentrations reaching Earth, integrated over a period of 24.5 My. Our data are a factor of 80–640 lower than the values expected under our constraints on ISM grain composition from a SN derived steady-state actinide production (the 2σ upper limit of ~ 840 atoms per cm^2 per My is still a factor of 25–200 lower). The lifetime of ^{244}Pu is comparable to the complete mixing time scales of the ISM[32–34]. The deep-sea crust sample integrates a ^{244}Pu flux over a time period of 24.5 My ($\sim 1/10$ of the SS rotation period in the Galaxy) corresponding to a relative travel distance of the SS of 650 pc (taking the mean speed of the measured ISM dust particles of 26 km s^{-1} (ref. 52), $\sim 1/10$ of the galactic orbital speed, as a proxy for ISM reshaping and for motion differences relative to the co-rotating local neighbourhood). These results, consistent with previous studies on extraterrestrial ^{244}Pu in crust[41,55] and sediment samples[40,56], are more sensitive by a factor of >100 and provide for the first time stringent experimental constraints on actinide nucleosynthesis in the last few hundred million years (see Fig. 3).

Table 2 | Expected ^{244}Pu fluxes at 1 AU from models and satellite data.

Meteoritic U and Th abundance data and models for the present Pu/U and Pu/Th ISM ratio

Dust mass density in the local galactic environment (g cm^{-3})[27]	1.2×10^{-26}	
U abundance in dust (from meteorite data) (g g^{-1})[51]	17×10^{-9} g U per g meteorite*	
Pu abundance in Early SS (measured from fissiogenic Xe)	^{244}Pu/^{238}U = 0.008 atom/atom[19,20,24]	
Galactic chemical evolution model	UP Wasserburg et al.[18]	Open-box model (Clayton[50], Meyer and Clayton[5], Dauphas[24,70])
Model prediction for present ISM steady-state ratio ^{244}Pu/^{238}U (atom/atom)	0.0165	0.044
Present expected ^{244}Pu concentration in ISM dust (using meteoritic data as proxy)	2.8×10^{-10} g per g	7.5×10^{-10} g per g
^{244}Pu ISM concentration	0.8×10^{-14} atom per cm^3	2.2×10^{-14} atom per cm^3
Range of ^{244}Pu ISM concentration	**$(0.8-2.2) \times 10^{-14}$ ^{244}Pu atoms per cm^3**	
Cassini: flux data at Earth orbit[31]	$(3-4) \times 10^{-5}$ particles per m^2 per s	
Cassini data: mean particle mass at Earth orbit[31,52]	$(3-7) \times 10^{-13}$ g (0.5-0.6 µm radius, $\rho = 2.5$ g cm^{-3})	
Cassini: mean mass flux	$(9-28) \times 10^{-22}$ ISM g cm^{-2} s^{-1}	
Cassini mass flux × ^{244}Pu conc. in ISM dust (^{244}Pu concentration in ISM dust × mean mass flux at 1 AU)	$(2.5-20) \times 10^{-31}$ ^{244}Pu g cm^{-2} s^{-1}	
^{244}Pu flux at Earth orbit	**20,000-160,000 ^{244}Pu atoms per cm^2 per My**	
Fraction of ISM dust found at 1 AU (mean mass flux at 1 AU versus dust mass density; using a peak velocity of 26 km s^{-1})	3-9%	

^{244}Pu fluence through a surface of 75 pc radius in swept-up material of the Local Bubble

Pre-LB intermediate dust mass density (required to form the Local Bubble, seven particles per cm^3 (ref. 34) with 1% of mass in dust)	7.6×10^{-26} g cm^{-3}
ISM dust column density over 75 pc (radius of LB)	2×10^{-5} g cm^{-2}
Total ^{244}Pu fluence from swept-up ISM dust into SS at Earth orbit (ISM dust column density × ^{244}Pu concentration in ISM dust × fraction of ISM dust found at 1 AU)	**$(0.4-3) \times 10^6$ ^{244}Pu atoms per cm^2**

ISM, interstellar medium; LB, Local Bubble; UP, uniform production.
1 solar mass = 1.99×10^{30} kg, 1 pc = 30.857×10^{15} m (1 pc^3 = 2.94×10^{55} cm^3). The ^{244}Pu fluxes at 1 AU were calculated from steady-state concentrations of ^{244}Pu from galactic chemical evolution models and Cassini satellite measurements[28,52] of the interstellar particle flux at 1 AU (Earth orbit). The U abundance in dust, the present ISM steady-state ^{244}Pu/^{238}U ratio, the Cassini data as well as the pre-LB intermediate dust mass density are literature values used for the comparison with our measured ^{244}Pu flux.
*The measured U concentration in present meteorites of (8.4 ± 0.8) p.p.b. (Lodders[51]) was adjusted for decay (~ one half-life of ^{238}U).

Table 3 | Anthropogenic Pu at the surface and the incorporation efficiency into the manganese crust.

	Surface layer 1				Blank
Time period	0-0.5 My Contains anthropogenic Pu (top 1 mm)				—
					Hydrothermal ~100 cm^2
Subsample	B1	C1	D1	Total	X
Mass (g)	32	20	28	80	364
Time period (My)	0-0.5	0-0.5	0-0.5	0-0.5	—
Total meas. eff. (10^{-4})	0.82	0.45	0.18	<0.51>	0.93
Measuring time	3.8 h	3.8 h	2.6 h	10.2 h	3.4 h
^{236}Pu atoms spike	3×10^8	3×10^8	3×10^8	9×10^8	3×10^8
^{244}Pu atoms (10^4)	18.5	5.3	13.3	37.1	**<1.7**
^{239}Pu atoms (10^8)	13.9	15.6	9.9	39.4	**0.3**
^{244}Pu/^{239}Pu (10^{-4})	1.3	0.3	1.3	1.0 ± 0.3	—
^{239}Pu atoms per cm^2 measured	1.6×10^7	2.2×10^7	1.4×10^7	1.76×10^7	—
^{239}Pu atoms per cm^2 reaching deep-sea floor* in 1976 (refs 38,39)	—	—	—	8.2×10^7	—
^{239}Pu incorporation eff. (crust)	—	—	—	**(21 ± 5)%**	

eff., efficiency; meas., measurement.
237 KD (VA13/2) deep-sea crust measurement: detailed data for the surface layer 1 (anthropogenic Pu) and the hydrothermal blank sample and determination of the Pu incorporation eff. into the deep-sea manganese crust by comparison of the known amount of atomic bomb-produced Pu at the crust's location with the measured Pu in the top layer 1.
*The amount of ^{239}Pu atoms per cm^2 reaching the deep-sea floor at the time of crust sampling (1976) is derived from the ratio (2.1 %) of the ^{239,240}Pu fluence measured in deep-sea sediments[39] (assumed to incorporate 100% of precipitated material) and the overall ^{239,240}Pu fallout fluence measured for the location of the crust[41].

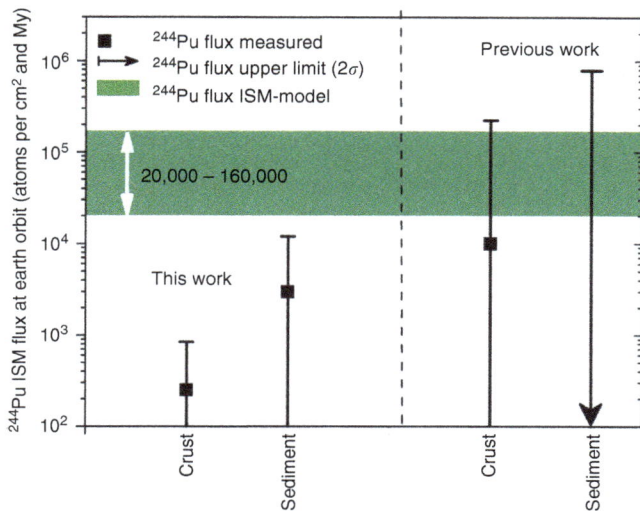

Figure 3 | Comparison of the measured ^{244}Pu flux at Earth orbit with models. The ISM ^{244}Pu flux at Earth orbit was determined from the concentrations measured in a deep-sea crust and a deep-sea sediment sample (note the logarithmic scale). Our results are compared with previous measurements (deep-sea crust[41] and sediment[40]) and to models of galactic chemical evolution[18,50] assuming steady-state conditions and taking into account filtration of dust particles when entering the heliosphere[31]. The arrows and error bars represent upper levels (2σ, 95% confidence levels) from the measurements. The green area indicates the data range deduced from the steady-state models. The crust data suggest a ^{244}Pu flux, which is a factor between 80 and 640 lower than inferred from the models.

A simple steady-state scenario might represent a simplified assumption within our local ISM environment. Compared with the typical size of ISM substructures of $\sim$50–100 pc (for example, LB)[27,32–34,57] and life-times of some 10 My, the crust sample probes, however, the equivalent of $\sim$10 such cavities (the ^{244}Pu life-time coupled with the spatial movement of the SS during the 24.5 My accumulation). Thus, we expect existing ISM inhomogeneities largely smeared out in our space- and time-integrated samples, confirming the significance of a ratio $<1/100$ between measured and expected ^{244}Pu abundance.

Further, we can relate our result to actinide nucleosynthesis during the recent SN history of the LB[32–34,58–59] in which the SS is embedded now. ISM simulations suggest the LB was formed by $\sim$14–20 SN explosions within the last 14 My (refs 32–34) with the last one $\sim$0.5 My BP (refs 32,58). To reproduce size and age of the LB, an intermediate density of seven particles per cm^3 ($\sim$10 times the mean density of the local environment now) before the first SN explosion took place, is required[34]. The mean mass density of the LB has since transformed to 0.005 particles per cm^3. The series of SNe explosions has generated the void inside the LB and has continuously pushed material into space forming an ISM shell. The SS is now placed inside the LB and thus has passed or passes the front of accumulated swept-up material including possible pre-existing ^{244}Pu from nucleosynthesis events prior to the formation of the LB[35,36].

We can distinguish three different scenarios for the recent LB history: (i) the SN activity transformed the local ISM from a dense to a low-density medium (LB), and pre-existing ISM material containing (steady state) old ^{244}Pu was swept-up and passed the SS[35,60]; (ii) direct production of ^{244}Pu in the 14–20 SNe and their expected traces left on Earth[35]; and (iii) independently, we can compare our data for ^{244}Pu with recent AMS data of ^{60}Fe influx[28,29].

In a simple first order estimate for scenario (i), we assume that the swept-up material is distributed over a surface with a radius of 75 pc. Using the pre-LB density of seven particles per cm^3 (ref. 34) with 1% of this ISM mass locked into dust, we calculate with our assumptions of Pu concentration in dust (Table 2) and a dust penetration efficiency of $\sim$6 $\pm$ 3% into the SS to Earth orbit, a ^{244}Pu fluence from swept-up material of $(0.4$–$3) \times 10^6$ ^{244}Pu atoms per cm^2 (see Table 2).

Our experimental data give a flux of 200^{+800}_{-200} ^{244}Pu atoms per cm^2 per My for the last 12 My (layers 2 and 3) at Earth orbit corresponding to a fluence of 2,300 ($<$12,000) ^{244}Pu atoms per cm^2 during this period. This experimental value for the fluence is a factor of $\sim$170–1,300 lower than the value calculated above (see Table 2) assuming swept-up material of about half the diameter of the LB is moved across the SS. We deduce approximately the same discrepancy as found for a simple steady-state actinide production scenario.

For LB scenario (ii), in a first order estimation, we take the SN-rate of 1.1–1.7 SNe/My within the LB[34,58] and a mean distance to the SS for these SN events of 100 pc. From our measured value of 200^{+800}_{-200} ^{244}Pu atoms per cm^2 per My at Earth orbit (with 6% penetration efficiency into 1 AU), this corresponds to $\sim$3,000 ($<$17,000) ^{244}Pu atoms per cm^2 per My unfiltered ISM flux, spread over a surface area with a radius of 100 pc. We deduce an average ^{244}Pu yield per SN of $(0.6^{+2.4}_{-0.6}) \times 10^{-9}$ M$_{solar}$ for the last 12 My.

Finally for LB scenario (iii), Knie et al.[28] and Fitoussi et al.[29] measured a clear ^{60}Fe signal of possible SN origin $\sim$2.2 My in the past in exactly the same crust material (237 KD) as we have used in this work for the search of ^{244}Pu (using a sample $\sim$50 cm distant; a SN origin for ^{60}Fe is being questioned by some authors[61,62], while several recent studies on ^{60}Fe in deep ocean sediments[63,64] and in lunar samples[65] confirm the results of Knie et al.[28]). Thus we can directly compare the measured fluences of ^{60}Fe and ^{244}Pu for the same event (using layer 2, 0.5–5 My). These fluence values can be converted into an atom ratio that is independent of the SS penetration efficiency and we assume the same incorporation efficiency for Fe and Pu (refs 63–66). We deduce a ^{244}Pu/^{60}Fe isotope ratio for this event of $<6 \times 10^{-5}$ (similarly, we obtain an upper limit from the sediment of $<10^{-4}$). Clearly, this ratio depends strongly on the type of explosive scenario. Literature values for this ratio are highly varying also due to large uncertainties in the r-process yields.

Our experimental results indicate that SNe, at their standard rate of $\sim$1 to 2 per 100 years in the Galaxy, did not contribute significantly to actinide nucleosynthesis for the past few hundred million years and actinide nucleosynthesis, as mapped through live ^{244}Pu, seems to be very rare. Our data may be consistent with a predominant contribution of compact-object mergers, which are 10^2 to 10^3 less frequent than core-collapse SNe[1]. A recent observation indicates indeed that such mergers may be sites of significant production of heavy r-process elements[10,11]. Our experimental work is also in line with observations of low-metallicity stars[12,14], indicating splitting into a rare and a more frequent r-process scenario allowing an independent evolution of the r-process elements Eu/Th over time[3,4]. In addition, we must conclude from our findings that, given the presence of short-lived actinide ^{244}Pu (and ^{247}Cm) in the ESS, it must have been subject to a rare heavy r-process nucleosynthesis event shortly before formation.

Methods
Details on the chemistry of the crust samples. Quantitative extraction of Pu was required from the 2.3 kg crust sample. The sample potentially contained some 10^6–10^7 atoms of ^{244}Pu, which correspond to an atom-concentration of $(0.5$–$5) \times 10^{-19}$ relative to the bulk material. No stable isobar to ^{244}Pu exists in nature and molecular interference in the measurements is excluded. The FeMn

crust sample was split into four layers 1–4 and three sections B, C and D. The top layer (1 mm, 'crust modern') was removed for measuring the anthropogenic Pu content originating from atmospheric atomic-bomb tests from ~ 1950 to 1963. Four different vertical layers represent different time periods in the past, while three different horizontal sections were chosen to identify possible lateral variations (B, C and D). The 12 individual pieces had masses between 30 and 360 g. Ten samples were measured by AMS.

The individual parts of the crust were dissolved in aqua regia and H_2O_2, and a spike of $\sim 3 \times 10^8$ atoms of a ^{236}Pu reference material was added to the leached solutions. After removal of the undissolved SiO_2 fractions, the solutions were brought to dryness and successively redissolved in concentrated HNO_3 and H_2O_2. At this stage, the sample solutions contain the actinides, but also the dominant fraction of the matrix elements of the crust, in particular Mn (~ 14 to 28%) and Fe (~ 16 to 28%). To separate the actinides from the Mn and Fe fraction, a pre-concentration step involving the selective co-precipitation of the actinides with CaC_2O_4 at pH ~ 1.7 was performed. After centrifugation, the precipitated CaC_2O_4 was converted to $CaCO_3$ in a muffle furnace at a temperature of 450 °C for several days. The $CaCO_3$ was dissolved in 7.2 M HNO_3 and the oxidation state of Pu was adjusted quantitatively to (IV)Pu by the addition of $NaNO_2$. These solutions were then loaded onto pre-conditioned anion-exchange columns containing Dowex 1×8, from which after the separation of the Ca, Am, Cm and the Th fractions Pu was eluted by reduction to (III)Pu with a solution of HI. To purify the obtained Pu fraction, two additional successive anion-exchange separations similar to the one described above were performed on the eluted Pu solutions. The Pu fractions were further purified from the organic residues by fuming with HNO_3 and H_2O_2. Successively an $Fe(OH)_3$ co-precipitation of Pu was performed in 1 M HCl by adding 2 mg of Fe powder. After centrifugation and drying of the precipitate, the $Fe(OH)_3$ was converted to iron oxide by combustion in a muffle furnace at 800 °C for 4 h. The plutonium oxide embedded in a matrix of iron oxide was then mixed with 2 mg of high-purity Ag powder and pressed in the sample holders suitable for the subsequent AMS measurement.

Owing to the massive matrix component of the crust for some samples a low chemical yield was observed (in the AMS measurements via the ^{236}Pu spike). In these cases the procedure for Pu extraction was repeated, that is, the solutions containing the Mn and Fe fraction underwent again a CaC_2O_4 co-precipitation procedure and the resulting actinide fractions were mixed with the rest of the remaining fractions originating from the first threefold anion-exchange separation. These solutions underwent a chromatographic column separation employing Tru-resin in 5 M HCl, from which after the elution of the Ca, Am and Cm and the Th fractions Pu was stripped out with a solution of 0.03 M $H_2C_2O_4$ in 0.5 M HCl.

Chemical processing of the TRIP deep-sea sediment. Similarly, Pu was extracted from two deep-sea sediment samples of 43 and 58 g mass provided by the Scripps Oceanographic Institute, the University of California at San Diego. It was a piston core (7P), extracted during the TRIPOD expedition (1966) as part of the DSDP at location 17°30′ N 113°00′ W (Pacific Ocean) at 3,763 m water depth. Two main sections were sampled with sediment depths 0–80 cm and 80–230 cm, from which the top 3 cm (containing the anthropogenic Pu) were removed. One quarter of the total cross-section throughout the ~ 230 cm length of the core was used in this study (4.9 cm^2). The samples, shipped in sealed polyethylene, were chemically processed at the Hebrew University, Jerusalem.

The physical and chemical processing of the sediments is as follows: the processing involved brief milling and calcination of the sample, alkali fusion of the sediment using NaOH at 750 °C, liquid-phase extraction of Fe and other main elements and liquid ion-chromatography to extract the Pu fraction. Prior to the alkali fusion steps, an isotope ^{242}Pu marker and a chemical ^{230}Th marker were added to the sediment. ^{242}Pu was used to monitor the efficiency of Pu detection by measuring the ^{242}Pu content of the final AMS sample (analogous to the ^{236}Pu spike in the crust samples), while ^{230}Th served as additional indicator of the chemical efficiency of actinide extraction by measuring the alpha activity of an electroplated deposition prepared from a separate fraction. The final AMS samples were obtained by co-precipitation of the Pu fraction with Fe in an ammonia solution, centrifugation and ignition to obtain a Fe_2O_3 matrix containing the Pu marker and traces. Finally, 2 mg of high-purity Ag powder was added and the powder pressed in the sample holders suitable for the subsequent AMS measurement.

AMS-measurement procedure. We have applied the most sensitive technique—AMS[46–48] to detect minute amounts of ^{244}Pu. This technique provides the complete suppression of any interfering background, for example, molecules of the same mass, by the stripping process in the terminal of a tandem accelerator, which is crucial for such experiments where only a few counts are expected. The ^{244}Pu measurements were performed at the Vienna Environmental Research Accelerator (VERA) facility in the University of Vienna[46,67–69]. This set up has been optimized for high-measurement efficiency and offered an exceptional selectivity.

The individual crust samples were spiked with a well-known amount of ^{236}Pu atoms (^{242}Pu for the sediment). ^{244}Pu and ^{239}Pu measurements were performed relative to the ^{236}Pu (^{242}Pu) spike, that is, the total efficiency and the chemical yield of Pu (when compared with the theoretical measurement efficiency) were monitored with the ^{236}Pu (^{242}Pu) spike, that was counted in short time intervals

before and after the ^{244}Pu runs in the AMS measurements. The chemical yield varied between 5 and 70% largely depending on the sample matrix.

In a sputter source, the Fe/Ag matrix, containing the Pu atoms, was bombarded with Cs ions and negative PuO ions were extracted, and energy and mass were analyzed before injection into the tandem accelerator. The negative ions were accelerated to the 3 MV tandem terminal and stripped there to positive ions in the gas stripper (O_2). ^{244}Pu^{5+} ions were accelerated to a final energy of 18 MeV and selected with a second analyzing magnet. They then had to pass an additional energy and another mass filter (electrostatic and magnetic dipoles, respectively) and were finally counted in an energy-sensitive particle detector. The system was optimized with a ^{238}U pilot beam and monitored during a measurement with reference samples containing ^{242}Pu. The measurement set up for ^{244}Pu and ^{236}Pu counting was scaled from the tuning set up. At the end of a measurement series, reference samples containing a well-known isotope ratio of ^{244}Pu/^{242}Pu were measured. The measured ratios reproduced the nominal values within 4% and confirmed the validity of scaling between the different masses in the measurement.

This set up suppresses adjacent masses (for example, ^{238}U from ^{239}Pu) by 8–9 orders of magnitude. This suppression factor was sufficient for ^{244}Pu counting as no interference from neighbouring masses is expected for ^{244}Pu (and additional isotopic suppression, for example, via time-of-flight identification, would have been at the cost of lower particle transmission). However, ^{238}U was abundant at levels of ~ 10 p.p.m. in the crust, and U separation from Pu in the chemical preparation of these samples was not 100%. Thus, the detector events registered for ^{239}Pu counting, by 2–3 orders of magnitude lower compared with modern samples, are attributed to leaking ^{238}U atoms injected as ^{238}UOH$^-$ ions together with ^{239}PuO$^-$ mimicking ^{239}Pu. To summarize, our detector event rate for ^{239}Pu suggests no significant anthropogenic ^{244}Pu contamination.

The measurement procedure was a sequence of alternating counting periods of ^{236}Pu, ^{239}Pu, ^{244}Pu and again ^{236}Pu. All samples were repeatedly measured until they were completely exhausted. The sputtering time per sample was between 5 and 20 h. Three measurement series were required to fully consume all the samples. The overall yields for the 12 crust samples were between 0.06×10^{-4} and 1.54×10^{-4}, that is, one ^{244}Pu detector event would represent correspondingly between 6,500 and 1.7×10^5 ^{244}Pu atoms in the analyzed sample. A similar procedure was followed for the sediment samples where ^{242}Pu was used as a spike.

Due to the low number of expected detector events, the machine and measurement backgrounds were carefully monitored with samples of pure Fe and Ag powders. They were sputtered identical to the samples containing the crust fractions. In addition, one crust sample (X), the lowest layer of the crust material, was of hydrothermal origin where no extraterrestrial ^{244}Pu could accumulate. This sample was chemically prepared and measured in the same way as the other crust samples and served as a process blank for potential chemistry and machine background.

Incorporation efficiency of Pu into the deep-sea crust. The incorporation efficiency of bomb-produced Pu into the crust was determined from the anthro-pogenic ^{239}Pu content in the top layer of the crust and deep-sea sediment data (details are given in Table 3). The total measurement efficiency is calculated from the total number of ^{236}Pu registered and normalized by the time fraction of ^{236}Pu AMS counting and divided by the number of ^{236}Pu atoms added as spike to the sample (3×10^8 atoms each). The number of ^{244}Pu atoms per sample is calculated from the number of ^{244}Pu events registered with the particle detector, scaled by the time fraction of AMS ^{244}Pu counting time and normalized with the measurement efficiency; the same procedure was used for ^{239}Pu atoms per sample. The ^{239}Pu detector events were corrected for a well-known contribution when adding the ^{236}Pu spike, which also contains ^{239}Pu (see Supplementary Table 1 for more details).

The average over 18 sediment cores from the Pacific measured between 1974 and 1979 gave a ^{239,240}Pu sediment inventory of 2.15 Bq m^{-2} (ref. 38). When compared with the well-known surface activity of 2.8 mCi km^{-2} (104 Bq m^{-2})[39], at the time of sampling the crust in 1976, 2.1% of anthropogenic Pu from the bomb tests was incorporated into deep-sea sediments with sediments having an incorporation efficiency of 100% (this compares well with the ratio of the time that had passed since the peak in atmospheric bomb-testing (15 years) and the mean residence time of Pu in the ocean of ~ 440 years; this ratio is, 3.4%). Taking the total anthropogenic Pu-inventory at the location of the crust (between 60 and 78 Bq m^{-2} and a ^{240}Pu/^{239}Pu atom ratio of 0.20; refs 40,41); and the fraction of 2.1% measured in sediments, we found that 8.2×10^7 ^{239}Pu atoms per cm^2 have reached the crust in the year 1976. We measured from the three crust subsamples from the top layer a ^{239}Pu surface density of $(1.76 \pm 0.44) \times 10^7$ cm^{-2}, and thus deduce an incorporation efficiency into the crust of (21 ± 5)%. We assume that the ISM-Pu is incorporated like the bomb-produced Pu.

References

1. Qian, Y.-Z. The origin of the heavy elements: recent progress in the understanding of the r-process. *Prog. Part. Nucl. Phys.* **50**, 153–199 (2003).
2. Woosley, S. E. & Weaver, T. A. The evolution and explosion of massive stars. II. Explosive hydrodynamics and nucleosynthesis. *Astrophys. J. Suppl. Ser.* **101**, 181–235 (1995).

3. Arnould, M., Goriely, S. & Takahashi, K. The *r*-process of stellar nucleosynthesis: astrophysics and nuclear physics achievements and mysteries. *Phys. Rep.* **450,** 97–213 (2007).
4. Thielemann, F. K. *et al.* What are the astrophysical sites for the *r*-process and the production of heavy elements? *Prog. Part. Nucl. Phys.* **66,** 346–353 (2011).
5. Meyer, B. S. & Clayton, D. D. Short-lived radioactivities and the birth of the sun. *Space Sci. Rev.* **92,** 133–152 (2000).
6. Arcones, A. & Martinez-Pinedo, G. Dynamical *r*-process studies within the neutrino-driven wind scenario and its sensitivity. *Phys. Rev. C.* **83,** 045809 (2011).
7. Goriely, S. *et al.* New fission fragment distributions and *r*-process origin of the rare-Earth elements. *Phys. Rev. Lett.* **111,** 25402 (2013).
8. Tanvir, K. *et al.* A 'kilonova' associated with the short-duration γ-ray burst GRB130603B. *Nature* **500,** 547–549 (2013).
9. Berger, E., Fong, W. & Chornock, R. An *r*-process kilonova associated with the short-hard GRB 130603B. *Astrophys. J. Lett.* **774,** L23 (2013).
10. Argast, D., Samland, M., Thielemann, F.-K. & Qian, Y.-Z. Neutron star mergers versus core-collapse supernovae as dominant *r*-process sites in the early Galaxy. *Astron. Astrophys.* **416,** 997–1011 (2004).
11. Goriely, S. & Arnould, M. Actinides: how well do we know their stellar production? *Astron. Astrophys.* **379,** 1113–1122 (2001).
12. Jacobson, H. R. & Frebel, A. Observational nuclear astrophysics: neutron-capture element abundances in old, metal-poor stars. *J. Phys. G* **41,** 044001 (2014).
13. Thielemann, F. *et al.* Heavy elements and age determinations. *Space Sci. Rev.* **100,** 277–296 (2002).
14. Cowan, J. J. & Sneden, C. h. Heavy element synthesis in the oldest stars and the early Universe. *Nature* **440,** 1151–1156 (2006).
15. Fields, B. D., Truran, J. W. & Cowan, J. J. A simple model for *r* process scatter and halo evolution. *Astrophys. J.* **575,** 845–854 (2002).
16. Huss, G. R., Meyer, B. S., Srinivasan, G., Goswami, J. N. & Sahijpal, S. Stellar sources of the short-lived radionuclides in the early solar system. *Geochim. Cosmochim. Acta* **73,** 4922–4945 (2009).
17. Diehl, R. *et al.* Radioactive 26Al from massive stars in the Galaxy. *Nature* **439,** 45–47 (2006).
18. Wasserburg, G. J., Busso, M., Gallino, R. & Nollett, K. M. Short-lived nuclei in the early Solar System: possible AGB sources. *Nucl. Phys. A* **777,** 5–69 (2006).
19. Turner, G., Harrison, T. M., Holland, G., Mojzsis, S. J. & Gilmour, J. Extinct 244Pu in ancient Zircons. *Science* **306,** 89–91 (2004).
20. Turner, G. *et al.* Pu–Xe, U–Xe, U–Pb chronology and isotope systematics of ancient zircons from Western Australia. *Earth Planet. Sci. Lett.* **261,** 491–499 (2007).
21. Kuroda, P. K. Nuclear fission in the Early history of the Earth. *Nature* **187,** 36–38 (1960).
22. Lachner, J. *et al.* Attempt to detect primordial 244Pu on Earth. *Phys. Rev.* **C85,** 015801 (2012).
23. Hoffman, D. C., Lawrence, F. O., Mewherter, J. L. & Rourke, F. M. Detection of plutonium-244 in nature. *Nature* **234,** 132–134 (1971).
24. Dauphas, N. The U/Th production ratio and the age of the Milky Way from meteorites and Galactic halo stars. *Nature* **435,** 1203–1205 (2005).
25. Hudson, G. B., Kennedy, B. M., Podosek, F. A. & Hohenberg, C. M. in *Proc. (A89-36486 15-91) Lunar and Planetary Science Conference, 19th,* Houston, TX, 14–18 March, 1988, 547–555 (Cambridge University Press/Lunar and Planetary Institute, Cambridge/Houston, TX, 1989).
26. Dwek, E. The evolution of the elemental abundances in the gas and dust phases of the galaxy. *Astrophys. J.* **501,** 643–665 (1998).
27. Mann, I. Interstellar dust in the Solar System. *Annu. Rev. Astron. Astrophys.* **48,** 173–203 (2010).
28. Knie, K. *et al.* 60Fe anomaly in a deep-sea manganese crust and implications for a nearby supernova source. *Phys. Rev. Lett.* **93,** 171103 (2004).
29. Fitoussi, C. *et al.* Search for supernova-produced 60Fe in a marine sediment. *Phys. Rev. Lett.* **101,** 121101 (2008).
30. Poutivtsev, M. *et al.* Highly sensitive AMS measurements of 53Mn. *Nucl. Instr. Meth. B* **268,** 756 (2010).
31. Altobelli, N. *et al.* Interstellar dust flux measurements by the Galileo dust instrument between the orbits of Venus and Mars. *J. Geophys.* **110,** A07102 (2005).
32. De Avillez, M. A. & Low, M.-M. M. Mixing Timescales in a Supernova-driven Interstellar Medium. *Astrophys. J.* **581,** 1047–1060 (2002).
33. Fuchs, B., Breitschwerdt, D., de Avillez, M. A., Dettbarn, C. & Flynn, C. The search for the origin of the Local Bubble redivivus. *Mon. Not. R. Astron. Soc* **373,** 993 (2006).
34. Baumgartner, V. & Breitschwerdt, D. Superbubble evolution in disk galaxies: I. Study of blow-out by analytical models. *Astron. Astrophys.* **557,** A140 (2014).
35. Ellis, J., Fields, B. D. & Schramm, D. N. Geological isotope anomalies as signatures of nearby supernovae. *Astrophys. J.* **470,** 1227 (1996).
36. Fields, B. D., Hochmuth, K. A. & Ellis, J. Deep-ocean crusts as telescopes: using live radioisotopes to probe supernova nucleosynthesis. *Astrophys. J.* **621,** 902–907 (2005).
37. Segl, M. *et al.* 10Be-dating of a mangenese crust from the Central North Pacific and implications for ocean paleocirculation. *Nature* **309,** 540–543 (1984).
38. Livingston, H. D. & Anderson, R. F. Large particle transport of plutonium and other fallout radionuclides to the deep ocean. *Nature* **303,** 228–231 (1983).
39. Bowen, V. T., Noshkin, V. E., Livingston, H. D. & Volchok, H. L. Fallout radionuclides in the pacific ocean: vertical and horizontal distributions, largely from GEOSECS stations. *Earth Planet. Sci. Lett.* **49,** 411–434 (1980).
40. Paul, M. *et al.* Experimental limit to interstellar 244Pu abundance. *Astrophys. J. Lett.* **558,** L133–L135 (2001).
41. Wallner, C. *et al.* Supernova produced and anthropogenic 244Pu in deep sea manganese encrustations. *N. Astron. Rev.* **48,** 145–150 (2004).
42. Baggaley, W. J. Advanced meteor orbit radar observations of interstellar meteoroids. *J. Geophys. Res.* **105,** 10353–10361 (2000).
43. Siebert, C. h., Nägler, T. F., von Blanckenburg, F. & Kramers, J. D. Molybdenum isotope records asa potential new proxy for paleoceanography. *Earth Planet. Sci. Lett.* **211,** 159–171 (2003).
44. Frank, M., O'Nions, R. K., Hein, J. R. & Banakar, V. K. 60 Myr records of major elements and Pb–Nd isotopes from hydrogenous ferromanganese crusts: Reconstruction of seawater paleochemistry. *Geochim. Cosmochim. Acta* **63,** 1689–1708 (1999).
45. Poutivtsev, M. Extraterrestrisches 53Mn in hydrogenetischen Mangankrusten. PhD thesis, 2007 (Tech. Univ. Munich, 2007).
46. Steier, P. *et al.* AMS of the Minor Plutonium Isotopes. *Nucl. Instr. Meth. B* **294,** 160–164 (2013).
47. Synal, H.-A. Developments in accelerator mass spectrometry. *Int. J. Mass Spectrom.* **349-350,** 192–202 (2013).
48. W. Kutschera, W. Applications of accelerator mass spectrometry. *Int. J. Mass Spectrom.* **349-350,** 203–218 (2013).
49. Feldman, G. J. & Cousins, R. D. Unified approach to the classical statistical analysis of small signals. *Phys. Rev. D* **57,** 3873–3889 (1998).
50. Clayton, D. D. The role of radioactive isotopes in astrophysics. *Lect. Notes Phys.* **812,** 25–79 (2011).
51. Lodders, K. Solar system abundances and condensation temperatures of the elements. *Astrophys. J.* **591,** 1220–1247 (2003).
52. Altobelli, N. *et al.* Cassini between Venus and Earth: detection of interstellar dust. *J. Geophys. Res.* **108,** 8032 (2003).
53. Westphal, A. J. *et al.* Evidence for interstellar origin of seven dust particles collected by the Stardust spacecraft. *Science* **345,** 786–791 (2014).
54. Donelly, J. *et al.* Actinide and ultra-heavy abundances in the local galactic cosmic rays: an analysis of the results from the *LDEF* ultra-heavy cosmic ray experiment. *Astrophys. J.* **747,** 40 (2012).
55. Wallner, C. *et al.* Development of a very sensitive AMS method for the detection of supernova-produced long-lived actinide nuclei in terrestrial archives. *Nucl. Instr. Meth. B* **172,** 333–337 (2000).
56. Paul, M. *et al.* An upper limit to interstellar 244Pu abundance as deduced from radiochemical search in deep-sea sediment: an account. *J. Radioanal. Nucl. Chem.* **272,** 243–245 (2007).
57. Ferrier, K. The interstellar environment of our galaxy. *Rev. Mod. Phys.* **73,** 1031 (2001).
58. Breitschwerdt, D., de Avillez, M. A., Feige, J. & Dettbarn, C. Interstellar medium simulations. *Astron. Nachr.* **333,** 486–496 (2012).
59. Cox, D. P. & Anderson, P. R. Extended adiabatic blast waves and a model of the soft x-ray background. *Astrophys. J.* **253,** 268–289 (1982).
60. Fry, B. J., Fields, B. D. & Ellis, J. R. Astrophysical shrapnel: discriminating among extra-solar sources of live radioactive isotopes. Preprint at http://arxiv.org/abs/arXiv:1405.4310 (2014).
61. Stuart, F. M. & Lee, M. R. Micrometeorites and extraterrestrial He in a ferromanganese crust from the Pacific Ocean. *Chem. Geol.* **322-323,** 209–214 (2012).
62. Basu, S., Stuart, F. M., Schnabel, C. & Klemm, V. Galactic-cosmic-ray-produced 3He in a ferromanganese crust: any supernova 60Fe excess on Earth? *Phys. Rev. Lett.* **98,** 141103 (2007).
63. Wallner, A. *et al.* 60Fe at the ANU – —search for a live supernova signature and a new half-life measurement. In: *Presentation at the 13th Intern. Conf. on Accelerator Mass Spectrometry,* 24 – 29 August 2014, 36 (Aix en Provence, France) (2014).
64. Ludwig, P. *et al.* Search for supernova produced 60Fe in Earth's microfossil record. In: *Presentation at the 13th International Conference on Accelerator Mass Spectrometry,* 24 – 29 August 2014, 37 (Aix en Provence, France) (2014).
65. Fimiani, L. *et al.* In Lunar and Planetary Science Conference, 18-22 March 2013, Vol. 45, 1778 (Woodlands, TX, USA, 2014).
66. Feige, J. *et al.* The search for supernova-produced radionuclides in terrestrial deep-sea archives. *Publ. Astron. Soc. Aust.* **29,** 109–111 (2012).
67. Wallner, A. Nuclear astrophysics and AMS—probing nucleosynthesis in the lab. *Nucl. Instr. Meth. B* **268,** 1277–1282 (2010).

68. Wallner, A. *et al.* A novel method to study neutron capture of ^{235}U and ^{238}U simultaneously at keV energies. *Phys. Rev. Lett.* **112**, 192501 (2014).

69. Steier, P. *et al.* Analysis and application of heavy isotopes in the environment'. *Nucl. Instr. Meth. B* **268**, 1045 (2010).

70. Dauphas, N. Multiple sources or late injection of short-lived *r*-nuclides in the early solar system? *Nucl. Phys. A.* **758**, 757c–760c (2005).

Acknowledgements

We thank the following organizations for supporting this work: part of this work was funded by the Austrian Science Fund (FWF): project No. P20434-N20 and I428-N16 (FWF & CoDustMas, Eurogenesis via ESF) and the Israel Science Foundation (ISF) under grant 43/01. We thank A. Lueckge and M. Wiedicke, Bundesanstalt für Geowissenschaften und Rohstoffe, Stilleweg 2, D-30655 Hannover, Germany for providing us the VA13 crust sample; D. Lal and W. Smith (Scripps Oceanographic Collections, USA) for locating and providing us with the deep-sea sediment samples (TRIP core) and N. Trubnikov (Hebrew U.) for the chemical processing of the sediment sample. Part of this work was also supported by the DFG Cluster of Excellence 'Origin and Structure of the Universe' (www.universe-cluster.de).

Author contributions

A.W. performed the data analysis and wrote the main paper together with M.P., and all authors discussed the results and commented on the manuscript. K.K., T.F. and G.K. organized the crust sample; M.P. provided the sediment sample. K.K. and F.Q. were primarily responsible for sample preparation of the crust; A.O and C.F. for the preparation of the sediment. P.S., A.W. and K.K. performed the AMS measurements.

Additional information

7

A low pre-infall mass for the Carina dwarf galaxy from disequilibrium modelling

Uğur Ural[1], Mark I. Wilkinson[2], Justin I. Read[3] & Matthew G. Walker[4]

Dark matter-only simulations of galaxy formation predict many more subhalos around a Milky Way-like galaxy than the number of observed satellites. Proposed solutions require the satellites to inhabit dark matter halos with masses 10^9–10^{10} Msun at the time they fell into the Milky Way. Here we use a modelling approach, independent of cosmological simulations, to obtain a pre-infall mass of $3.6^{+3.8}_{-2.3} \times 10^8$ Msun for one of the Milky Way's satellites: Carina. This determination of a low halo mass for Carina can be accommodated within the standard model only if galaxy formation becomes stochastic in halos below $\sim 10^{10}$ Msun. Otherwise Carina, the eighth most luminous Milky Way dwarf, would be expected to inhabit a significantly more massive halo. The implication of this is that a population of 'dark dwarfs' should orbit the Milky Way: halos devoid of stars and yet more massive than many of their visible counterparts.

[1] Leibniz Institute für Astrophysik Potsdam, An der Sternwarte 16, Potsdam 14482, Germany. [2] Department of Physics and Astronomy, University of Leicester, University Road, Leicester LE1 7RH, UK. [3] Astrophysics Research Group, Faculty of Engineering and Physical Sciences, University of Surrey, Guildford GU2 7XH, UK. [4] Department of Physics, McWilliams Center for Cosmology, Carnegie Mellon University, 5000 Forbes Avenue, Pittsburgh, Pennsylvania 15213, USA. Correspondence and requests for materials should be addressed to U.U. (email: uural@aip.de).

While the Cold Dark Matter paradigm for structure formation in the Universe has been very successful in reproducing observations on scales larger than ∼1 Mpc (refs 1–3), on galactic scales there have been long-standing puzzles. The discrepancy between the predictions of the Cold Dark Matter paradigm and the observed properties of the dwarf spheroidal galaxy satellites (dSphs) of the Milky Way has persisted for over a decade. Numerical models predict that thousands of dark matter subhalos should be found orbiting the Milky Way and Andromeda, yet only a few tens have been found to date[4,5]. This has become known as the 'Missing Satellites' problem. A popular resolution of this issue is to place stars only in the most massive satellite halos, implying a total or 'virial' mass for the Milky Way dSphs of ∼10^{10} Msun (see Fig. 1). However, in such halos, the required central stellar velocity dispersions of the dSphs would be too high to be consistent with the Milky Way dSphs[6]. More recent work has shown further that more refined mappings between luminous and dark matter result in a population of massive satellites, which have inexplicably failed to form stars (the 'Too Big To Fail' problem)[7–9]. Several solutions have been proposed, including lowering the mass of the Milky Way halo[10], or the central stellar velocity dispersion of the dSphs through the action of stellar feedback[11]. However, these still require the Milky Way dSphs to inhabit dark matter halos with pre-infall masses greater than ∼10^9 Msun[4].

Here we use a new 'disequilibrium' model fitting algorithm to constrain the mass of a dwarf galaxy—Carina—that appears to be in the process of tidal disruption by the Milky Way[12]. Previous

simulations of Carina, which assumed identical spatial distributions for the dark matter and stars, suggested that tides have played a role in the evolution of this system[13]. The extensive observed data for Carina, combined with our larger parameter space of non-equilibrium models, allows us to measure the mass of Carina over a far greater radial range than has been possible to date. Most significantly, we are able to 'wind the clock' back to estimate its mass before it fell into the Milky Way, without recourse to comparisons with cosmological simulations.

Results

Simulation procedure. Our method works by simulating the disruption of almost 19,000 N-body Carina models in a static Milky Way potential. To marginalize over the unknown model parameters, the N-body models are wrapped up inside a Markov Chain Monte Carlo (MCMC) pipeline (see Methods section). For the dark matter halo, we allowed both cusped and cored central density profiles. The former are found in cold dark matter simulations, while the latter give a better match to observations[14]. The main advantage of the method over previous studies is that by using full N-body simulations, we can model 'disequilibrium' systems like the tidally disrupting Carina dSph. A demonstration of the performance of our methodology on artificial data is shown in Fig. 2: we recover both the pre-infall and present-day mass of a mock dwarf.

Mass of the Carina dsph. In Fig. 3, as well as Table 1, we show our results for Carina. Figure 3 shows our best fit surface brightness (top) and projected velocity dispersion (bottom) profiles. Table 1 reports the median and 68% confidence intervals for all of our fitted parameters. To facilitate comparison with the predictions of cosmological simulations, we calculate the distribution of M_{200} (the mass within the radius where the mean density of the dSph reaches 200 times the critical density of the

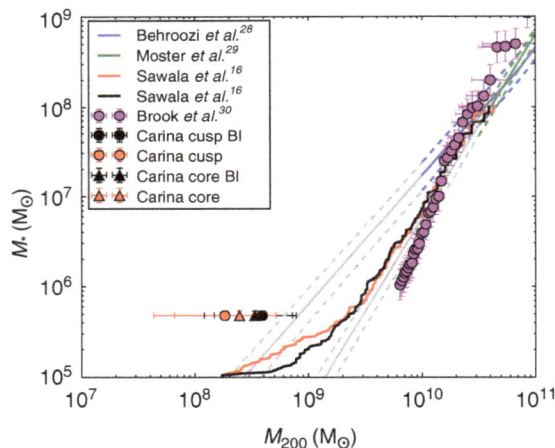

Figure 1 | The estimated pre-infall mass of the Carina dwarf compared with predictions from cosmological simulations. The estimated pre-infall mass of the carina dwarf compared with predictions from cosmological simulations. The blue[28] and green[29] lines show abundance matching estimates based on data from the Sloan Digital Sky Survey; below B1010 Msun (grey lines), they both become extrapolations. The red and black lines show the stellar mass halo mass relation taken from a recent cosmological hydrodynamical simulation[16], using the halo mass before infall (BI: black) and at the present time (red). Unlike the other lines and data points, these two curves are not based on abundance-matching[16]. The purple points show the abundance-matching results between Local Group dwarfs and a 'constrained' simulation of our local volume[30]. The black and red circles are our pre-infall mass estimates (M_{200}) for Carina for cusped and cored dark matter halos, respectively. They are lower than any of the curves, only marginally consistent with the extrapolation to low luminosities of the relation found in ref. 28. Finally, the black (cusped) and red triangles (cored) are our present-day mass estimates within 1.5 kpc. We calculate the 1σ error bars by including 68% of the good models around the median value as given in Table 1.

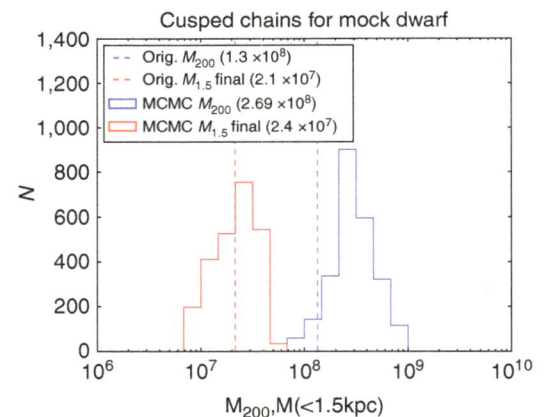

Figure 2 | The results of MCMC chains run with artificial input data. The present-day mass is very well constrained (red histogram)—the red dashed line shows the actual value from the target model. The pre-infall mass (blue histogram) has more uncertainty but is still close to the actual value (solid blue line), despite the large range of masses explored by the chains. The values quoted in the legends of the histograms are the median values that the best models in the MCMC chains found. The likely reason for the larger uncertainty in the pre-infall mass is that the target model in this test was chosen to be on a very eccentric orbit and has undergone stronger tidal disturbance than Carina (see Table 1). In addition, the noise we add to its 'observations' (in order to make the uncertainties similar to those in Carina) admits models with a larger range of orbital eccentricities. Both histograms show models with $\chi^2 < 9$.

Universe) values at the start of our simulations. We find pre-infall M_{200} values of 3.9×10^8 ($-2.4; +3.9$) and 3.37×10^8 ($-2.1; +3.8$) Msun for Carina models with cusped and cored halos, respectively. Interestingly, this low mass estimate agrees with an earlier study, which found that the halos in the Aquarius cosmological simulations that reproduced the mass of Carina at the present time were those with pre-infall masses of $<4 \times 10^8$ Msun[4]. After infall, M_{200} is no longer a meaningful quantity for a

satellite galaxy, and we therefore use the mass within 1.5 kpc (the radial extent of our surface brightness and velocity dispersion data) as our present-day mass estimate. This quantity is very well constrained, with $M(r<1.5\,\mathrm{kpc}) = 7.1 \times 10^7$ ($-3.5; +2.8$) and 9.7×10^7 ($-4.8; +4.9$) Msun for the cusped and cored halos, respectively. We find that both cusped and cored models fit the data very well, suggesting that there is little power in the binned data to distinguish between the two. In general, protected by their higher central densities, cusped halos are able to withstand the stronger tidal forces experienced during closer perigalactic passages.

Our favoured models include both cases in which Carina inhabits a rather low-mass halo, showing significant tidal disruption in its outer parts and cases where it is protected from external tides by a massive halo. These latter models may be favoured by a recent study that found no evidence for tidal tails at a radius of ~ 1 degree from the main body of Carina[15]. We explicitly tested which of our models show visible tidal tails when analysed similarly to that study and found that only the most tidally disrupted are inconsistent with their results. Table 1 presents results from the three sets of model chains, a cusped and a cored chain that use surface brightness, velocity dispersion and velocity gradient and an additional set of model chains with cusped haloes but which excluded the velocity gradient data. It is seen that ignoring the velocity gradient favours models with an even lower pre-infall mass (nevertheless, consistent with our other results within the errors). We expect our results for Carina to be relatively insensitive to the detailed properties of the Milky Way disk as the majority of successful models are not on disk-crossing orbits.

Discussion

The most striking aspect of our results is our upper bound on the mass of Carina both today and pre-infall. In Fig. 1, we compare this upper bound to predictions derived from 'abundance matching' schemes, where satellite luminosity is assumed to depend monotonically on the mass of the dark matter subhalos at infall, as well as to cosmological hydrodynamical simulations. Intriguingly, our data point for Carina lies to the low-mass side of all but one of the extrapolated relations. Even in this case, there is a tension at the $\sim 1\sigma$ level. Our pre-infall mass estimate, being based on N-body simulations constrained by the observed data for the Carina dSph, does not rely on any assumed cosmological model and, unlike previous studies, is entirely independent of

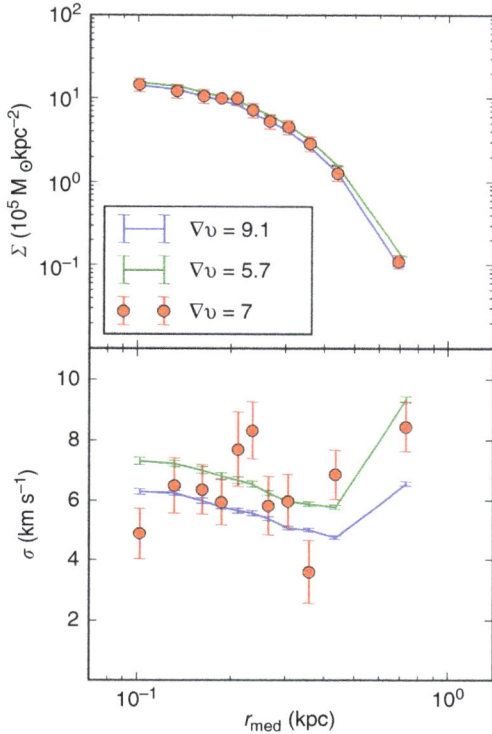

Figure 3 | A comparison between the best Carina models with cusped and cored dark matter halos and the observed data. The figure shows the surface brightness and projected velocity dispersion profiles (and the velocity gradient given in km s^{-1} in the legend) from two of our best models. The red data points are the observations with associated 1σ errors. The blue and green curves come from our best fit cusped and cored models, respectively, with their 1σ Poisson errors calculated in each bin.

Table 1 | Parameter constraints for Carina.

	Range	Cusp (Σ, σ, ∇v)	Core (Σ, σ, ∇v)	Cusp (Σ, σ)
M_s (10^5 Msun; pre-infall)	4.3; 21.5	4.8 ($-0.35; +0.6$)	4.8 ($-0.35; +0.49$)	4.9 ($-0.4; +0.55$)
r_s (kpc)	0.074; 0.74	0.2 ($-0.06; +0.04$)	0.19 ($-0.07; +0.06$)	0.22 ($-0.07; +0.06$)
M_h (10^8 Msun)	0.01; 100	4.0 ($-2.4; +4.1$)	3.48 ($-2.25; +3.87$)	2.57 ($-1.42; +3.7$)
r_h (kpc)	r_s; 5	2.06 ($-1.11; +1.96$)	0.88 ($-0.46; +0.48$)	1.55 ($-0.67; +1.78$)
$\mu_\alpha \cos(\delta)$ (mas cent^{-1})	-0.17; 0.61	0.15 ($-0.15; +0.36$)	0.15 ($-0.15; +0.33$)	0.23 ($-0.21; +0.25$)
μ_δ (mas cent^{-1})	-0.09; 0.57	0.11 ($-0.11; +0.36$)	0.30 ($-0.3; +0.17$)	0.26 ($-0.17; +0.18$)
R_{peri} (kpc)	—	91.3 ($-71.8; +10.7$)	98.6 ($-61.9; +3.3$)	98.2 ($-62.5; +3.6$)
$M200$ (10^8 Msun; pre-infall)	—	3.9 ($-2.4; +3.9$)	3.37 ($-2.1; +3.8$)	2.44 ($-1.3; +3.56$)
$M(r<1.5$ kpc; 10^7 Msun)	—	7.1 ($-3.4; +2.8$)	9.7 ($-4.8; +4.9$)	6.1 ($-2.4; +3.5$)
$v_{max,in}$ (pre-infall; km s^{-1})	—	17.4 ($-3.6; +3.2$)	18.3 ($-3.6; +5.4$)	15.0 ($-2.8; +4.2$)
v_{max} (present; km s^{-1})	—	15.0 ($-3.1; +2.9$)	17.0 ($-3.9; +4.4$)	13.7 ($-2.7; +4.0$)

MCMC, Markov Chain Monte Carlo.

The parameters listed in the first column are as follows: stellar mass and scale length (M_s, r_s); total halo mass we use to generate the model and scale length (M_h, r_h); the proper motions ($\mu_\alpha \cos(\delta)$, μ_δ); perigalactic distance (R_{peri}); the pre-infall and present epoch masses as explained in the text (M_{200}, $M(r<1.5\,\mathrm{kpc})$); and the maximum halo circular velocity before infall and today ($v_{max,in}$, v_{max}). The second column gives the range of values explored by our MCMC chains. Columns three and four list the constraints we obtain for all parameters for chains that used cored or cusped models and included the velocity gradient in the model likelihood. The fifth column presents results for chains with cusped models in which the velocity gradient was ignored. The estimated pre-infall and present-day masses for Carina are calculated as the average values given by our cored and cusped chains.

any cosmological simulations. The low mass that we find thus provides a new and complementary insight into the mapping between low-mass haloes and low-luminosity galaxies and suggests that a simple monotonic mapping between light and dark in our standard cosmological model fails. One solution is to posit that for halo masses below $\sim 10^{10}$ Msun, the physics of galaxy formation leads to a halo occupation function that is effectively stochastic, as suggested by some cosmological hydrodynamic simulations[16,17].

Our findings can be extended in the near future: dwarf galaxy proper motions from the Gaia satellite will significantly decrease the orbital uncertainties for Carina and the other Local Group dwarfs[18]. Combined with deep photometric observations of dwarf outskirts, we will be able to obtain a pre-infall mass distribution for the whole population. This will address the important question of whether Carina is an outlier in the mass distribution. However, even taken in isolation, our determination of the mass of Carina before interacting with the Milky Way precludes any models that associate the most luminous dSphs with the most massive subhalos.

Methods

Markov chain Monte Carlo. We performed a large suite of N-body simulations that compared the final state of the disrupting dwarf spheroidal galaxies in the tidal field of the Milky Way to a host of observational data for the Carina dwarf galaxy. The simulations were performed within a MCMC framework[19] in order to sample the parameter space effectively and constrain the properties of the dSph. Our method is reminiscent of ref. 20 that uses a genetic algorithm combined with restricted N-body simulations to explore the tidal disruption of NGC 205 around the Andromeda galaxy. However, our method is more general, using full rather than restricted N-body simulations and using an MCMC algorithm that allows us to fully explore parameter degeneracies.

First, a two-component, non-rotating and spherical N-body model of the dSph was built with 2×10^5 particles split equally between the dark matter and stellar components. The starting coordinates and velocities of the dSph were calculated by integrating the orbit of a point mass backwards in a static Milky Way potential using the present-day position and velocity. We then replaced the point mass with the live model and performed a full N-body simulation of its evolution around the Milky Way for 6 Gyr. At the end of the simulation, the radial profiles for the velocity dispersion (σ) and the surface brightness (Σ), as well as the velocity gradient (∇v) between the outermost bins along the major axis of the model dwarf were calculated, and a (reduced) χ^2-statistic was used to compare it with Carina data[12,21]. The likelihood ratio between the consecutive models was used to determine the more favourable region of the parameter space at each step, as the Markov Chain accepted either the new model or re-accepted the older one with the better likelihood. This is the learning process of the MCMC algorithm that chooses new initial conditions for the simulations at each step on the basis of the initial conditions of the last model that is accepted. In our pipeline, the proper motions are used as a prior on the basis of observational data and running N-body simulations that fit both the observational profiles and the proper motions, we are able to marginalize over orbit and halo properties simultaneously with the minimum number of assumptions for the dark matter mass and distribution of the satellite.

Supplementary Fig. 1 shows a schematic representation of the algorithm and the codes used at each step. MCMC chains that used models with cusped and cored halos were run separately, and each N-body simulation took ~ 1 h on 32 processors on average. As a very large number of simulations needed to be performed ($\sim 19,000$ in total for Carina), several chains were run in parallel. Before adding the chains together for the final analysis, the first 50–100 simulations of each of them were dismissed to account for the burn-in period of the algorithm.

Parameter space. The parameter space of initial conditions consisted of six free parameters ($\mu_\alpha \cos(\delta)$, μ_δ, M_s, r_s, M_h and r_h) for which the allowed ranges are given in Table 1. Given the potential for systematic errors in the observed proper motions, the allowed range for the present-day proper motions ($\mu_\alpha \cos(\delta)$, μ_δ) used for the orbit integration was the 3σ range of the observations (the values of the revised proper motions were obtained through private communication from the authors of ref. 22). The N-body models were generated with falcON[23] according to the split power profiles given in Equation 1, where α, β and γ were fixed: 1,4,1 for the cusped halo; 0,4,1 for the cored halo; and 0.515, 4.45 and 0.287 for the stellar component.

$$\rho = \frac{\rho_0}{\left(\frac{r}{r_s}\right)^\alpha \left(1 + \left(\frac{r}{r_s}\right)\right)^{(\beta-\delta)/\alpha}} \quad (1)$$

The latter was chosen to have a functional form providing a good fit to the present-day surface density in the central regions, albeit with variable amplitude

and scale radius that can take values between 0.2 and 2.5 times the original fit ($r_s = 0.237$ kpc). The initial stellar mass M_s was allowed to be up to five times larger than the current one.

The initial mass of the dark matter halo M_h could be as small as 10^6 Msun, making it only twice as massive as the stellar component, while the upper limit is high enough (10^{10} Msun) to allow the large mass models predicted by the cosmological simulations to be tested. Similarly, while the scale radius of the dark matter halo, r_h can be as large as 5 kpc, the lower limit is determined by that of the stellar component for each model.

Technical details of the N-body simulations and the Milky Way potential. The external Milky Way potential (pot 4a[24]) implemented for the point mass orbit integrator[25] and the N-body code PkdGRAV-1 (ref. 26) was provided by the GalPot programme provided in the NEMO Stellar Dynamics Toolbox[27]. The mass of the Milky Way reaches 5×10^{11} Msun at 100 kpc, which is Carina's current distance. The simulation time of 6 Gyr was chosen to allow a large enough timescale for the dwarf's evolution while avoiding the complications because of the evolution of the Milky Way potential, and hence Carina's orbit[25].

We keep the parameters describing the potential of the Milky Way fixed in our analysis. This is an additional systematic error that we do not currently marginalize over. However, much of the associated uncertainty is accounted for by the wide range of proper motions admitted by the large observed errors bars and so marginalizing over uncertainties in the Milky Way potential is not yet required. However, as Carina's proper motion errors shrink, this may become the leading error term. Such a situation would open up the possibility of actually constraining the Milky Way potential alongside the mass and orbit of Carina.

The parameters used in PkdGRAV were chosen on the basis of analysis of simulations with different number of particles, opening angle, time step and the softening length, taking into account the trade-off between CPU time and accuracy. Supplementary Table 1 shows a subset of these simulations where it is seen that the main factor that can affect the χ^2 is the number of particles. Therefore, we chose $N = 2 \times 10^5$, the largest number of particles that were computationally affordable—the slow down of the simulation when run with 2×10^6 particles instead of 2×10^5 was a factor of 10. We underline that even for the very tidally disrupting model we present in Supplementary Fig. 2, the effects of changing these numerical parameters is not large enough to change the results of the MCMC.

Tests with mock data. To test the MCMC pipeline described above, we ran the chains first for a mock dSph instead of Carina. The mock dwarf had a cusped dark matter halo profile to start with and its final Σ, σ and ∇v were similar to those of Carina. We modified the Σ and σ profiles, adding Gaussian noise to each data bin as well as increasing the error bars to match the s.d. of the data that resulted in a Mock dwarf with similar uncertainties to Carina. The noise in the velocity gradient in Carina was calculated for ~ 10 stars in the outermost data bin on the major axis. The error bars were obtained by using 100,000 random samples of 10 stars. Both the ideal N-body model and the noisy data are presented in Supplementary Fig. 3, where an increase in the velocity dispersion in the outer bins because of the eccentric orbit of the dwarf is observed.

Three sets of Markov Chains were run separately for the mock dwarf. Two chains where the likelihood comparisons between models were made using (Σ, σ, ∇v) were run separately testing either cored or cusped dark matter halos, while the third chain with a cusped halo used only Σ and σ in its 'observations'. Supplementary Table 2 shows the parameter values used in the mock data as well as those found by the three chains. The χ^2 cuts were chosen to include as many as possible 'good' models for our statistical calculations to be meaningful. The numbers presented in the Supplementary Table 2 are derived from 2,469 (50 unique) models with χ^2 (Σ, σ, ∇v) < 9 and 3,119 (335 unique) models with χ^2 (Σ, σ) < 6 for the cusped chains.

It is seen from Supplementary Table 2 that the current mass $M(r < 1.5$ kpc) is very well constrained by both the cored and the cusped chains within the radius probed by the data. It is also seen that when the velocity gradient was excluded from the χ^2 calculations, the chains found models with larger and more massive dark matter halos and hence overestimated the both the initial and the final mass. Interestingly, although the Mock dwarf was cusped to start with, the cored chains recovered the initial mass better than those with cusps that found models with M200 twice as massive. Nevertheless, the results for these chains are very promising considering the wide ranges of masses the algorithm explored. Supplementary Fig. 4 and Supplementary Fig. 5 show the distribution of the good models in the parameter space that was explored by the cusped (cored) chains, for all of the models as well as those within the $\chi^2 < 9$ cut. The figure shows that the orbit is the worst constrained property of the models; hence, any improvement of the observed uncertainties in the proper motions will improve the results.

As seen in Fig. 2, the final mass is very well constrained by the chains within the radius probed by the data.

MCMC for Carina. The same three sets of MCMC chains were then run for Carina, which is ideally suited for our study because of the quality of available photometric and spectroscopic data extending to large radii, as well as the evidence of a velocity gradient along the major axis[12]. These new chains found more models with small χ^2 than those for the mock dwarf. In order to test the validity of the χ^2 cut used above, we re-analysed the Carina chains for different cuts. Supplementary Figs 6 and 7 show the distribution of the models for a $\chi^2 < 6$ cut, which included 445 models (22 unique) and the $\chi^2 < 9$ cut that we use throughout this paper with 2,686 models (232 unique). As can be seen from the figures, these two options result in mass estimates consistent with each other. Therefore, as the overall power of the method we use is to provide a distribution of good models, we choose to use the same χ^2 cut for Carina and the mock dwarf. As a final test, in Supplementary Fig. 8, we compare two models with different χ^2 for both cusped and cored halos and demonstrate that even at the highest end of the χ^2 range the models used still provide a reasonable match for Carina.

Code availability. The simulation data, MCMC algorithm, pytipsy package necessary to read the binary data files and the codes analysing the simulation data are available on the project website: http://vo.aip.de/dwarfedmasses/.

The new version of PkdGRAV that was used for the N-body simulation can be obtained from https://hpcforge.org/projects/pkdgrav2/.

The falcON code used to generate the two-component N-body models is part of the NEMO package available at https://github.com/Milkyway-at-home/nemo/tree/master/nemo_cvs/usr/dehnen/falcON.

References

1. Planck CollaborationAde, P. A. R. *et al.* Planck 2013 results, XVI. Cosmological parameters.. *Astron. Astrophys.* **571**, 16–82 (2013).
2. Coc, A., Vangioni-Flam, E., Descouvemont, P., Adachour, A. & Angoulo, C. Updated Big Bang nucleosynthesis compared with wilkinson microwave anisotropy probe observations and the abundance of light elements. *Astrophys. J.* **600**, 544–552 (2004).
3. Seljak, U., Makarov, A. & McDonald, P. Cosmological parameter analysis including SDSS Lyα forest and galaxy bias: Constraints on the primordial spectrum of fluctuations, neutrino mass, and dark energy. *Phys. Rev.* **D71**, 103515.
4. Moore, B. *et al.* Dark matter substructure within galactic halos. *Astrophys. J. Lett.* **524**, 19–22 (1999).
5. Klypin, A., Kravtsov, A., Valenzuela, O. & Prada, F. Where are the missing galactic satellites? *Astrophys. J.* **522**, 82–92 (1999).
6. Read, J. I., Wilkinson, M. I., Evans, N. W., Gilmore, G. & Kleyna, J. T. The importance of tides for the Local Group dwarf spheroidals.. *Mon. Not. R Astron. Soc.* **367**, 387–399 (2006).
7. Boylan-Kolchin, M. Bullock, J. S. & Kaplinghat, M.Too big to fail?. The puzzling darkness of massive Milky Way subhaloes.. *Mon. Not. R Astron. Soc.* **415**, 40–44 (2011).
8. Boylan-Kolchin, M. Bullock, J. S. & Kaplinghat, M. The Milky Way's bright satellites as an apparent failure of ΛCDM. *Mon. Not. R Astron. Soc.* **422**, 1203–1218 (2012).
9. Wang, J., Carlos, S. F., Navarro, J. F., Gao, L. & Sawala, T. The missing satellites of the Milky Way. *Mon. Not. R Astron. Soc* **424**, 2715–2721 (2012).
10. Vera-Ciro, C. A., Helmi, A., Starkenburg, E. & Breddels, M. A. Not too big, not too small: the dark haloes of the dwarf spheroidals in the Milky Way. *Mon. Not. R. Astron. Soc* **428**, 1696–1703 (2013).
11. Brooks, A. M., Kuhlen, M., Zolotov, A. & Hooper, D. A baryonic solution to the missing satellites problem. *Astrophys. J.* **765**, 22–29 (2013).
12. Munoz, R. R. *et al.* Exploring halo substructure with giant stars. XI. The tidal tails of the Carina dwarf spheroidal galaxy and the discovery of magellanic cloud stars in the Carina foreground. *Astrophys. J.* **649**, 201–223 (2006).
13. Munoz, R. R., Majewski, S. R. & Johnston, K. V. Modeling the structure and dynamics of dwarf spheroidal galaxies with dark matter and tides. *Astrophys. J.* **679**, 346–372 (2008).
14. Walker, M. ,G. & Penarrubia, J. A method for measuring (slopes of) mass profiles of dwarf spheroidal galaxies. *Astrophys. J.* **742**, 20–38 (2011).
15. McMonigal, B., Bate, N. F. & Lewis, G. F. Sailing under the magellanic clouds: A DECam view of Carina dwarf. *Mon. Not. R Astron. Soc.* **444**, 3139–3149 (2014).
16. Sawala, T. *et al.* Bent by baryons: the low mass galaxy-halo relation. *Mon. R. Astron. Soc.* **448**, 2941–2947 (2015).
17. Kuhlen, Michael, Madau, Piero, Krumholz & Mark, R. Dwarf galaxy formation with H2-regulated Star Formation. II. Gas-rich Dark Galaxies at Redshift 2.5. *Astrophys. J.* **776**, 34–42 (2013).
18. Wilkinson, M. I. & Evans, N. W. The present and future mass of the Milky Way halo. *Mon. Not. R Astron. Soc.* **310**, 645–662 (1999).
19. Metropolos, N., Rosenbluth, A. W., Teller, A. H. & Teller, E. Equations of state calculations by fast computing machines. *J. Chem. Phys.* **21**(): 1087–1092 (1953).
20. Howley, K. M., Geha, M. & Guhathakurta, P. Darwin tames an andromeda dwarf: unraveling the orbit of NGC 205 using a genetic algorithm. *Astrophys. J.* **683**, 722–749 (2008).
21. Walker, M. G., Mateo, M. & Olszewski, E. W. Stellar velocities in the Carina, Fornax, Sculptor, and Sextans dSph Galaxies: data From the Magellan/MMFS survey. *Astron. J.* **137**, 3100–3138 (2009).
22. Piatek, S., Pryor, C. & Olsewski, E. W. 2004, Proper motions of dwarf spheroidal galaxies from hubble space telescope imaging. II. measurement for Carina. *Astron. J.* **126**, 2346–2361 (2003).
23. Dehnen, W. A Hierarchical O (N) force calculation algorithm. *J. Comp. Phys.* **179**, 27–42 (2002).
24. Dehnen, W. & Binney, J. Mass models of the Milky Way. *Mon. Not. R. Astron. Soc.* **294**, 429–438 (1998).
25. Lux, H., Read, J. I. & Lake, G. Determining orbits for the Milky Way's dwarfs. *Mon. Not. R. Astron. Soc.* **406**, 2312–2324 (2010).
26. Stadel, J. *Cosmological N-body Simulations and their Analysis.* PhD Thesis, Univ. Washington, Source DAI-B 62/08, p. 3657 (2002).
27. Teuben, P. J. The Stellar Dynamics Toolbox NEMO. in *Astronomical Data Analysis Software and Systems IV* (eds R. Shaw, H. E. Payne & J. J. E. Haye) PASP Conference Series 77, 398–401 (1995).
28. Behroozi, P. S., Conroy, C. & Wechsler, R. H. A comprehensive analysis of uncertainties affecting the stellar mass-halo mass relation for $0 < z < 4$. *Astrophys. J.* **717**, 379–403 (2010).
29. Moster, B. P., Sommerville, R. S. & Maulbetsch, C. Constraints on the relationship between stellar mass and halo mass at low and high redshift. *Astrophys. J.* **710**, 903–923 (2010).
30. Brook, C. B. *et al.* The stellar-to-halo mass relation for local group galaxies. *Astrophys. J.* **784**, 14–18 (2014).

Acknowledgements

We thank the developers of the codes we used in the pipeline: Hanni Lux for her orbit integration code; Walter Dehnen for the GalPot and falcON codes that we used to generate our two-component galaxies; Joachim Stadel for PkdGRAV; Jonathan Coles for his tipsy routines. We also thank Kristin Riebe who prepared the website of the project as well as the cover art. U.U. acknowledges funding for a large part of this project from the European Commission under the Marie Curie Host Fellowship for Early Stage Research Training SPARTAN, Contract No MEST-CT-2004-007512, University of Leicester, UK. M.I.W. acknowledges the Royal Society for support via a University Research Fellowship. J.I.R. would like to acknowledge support from SNF grant PP00P2 128540/1. M.G.W. is supported by National Science Foundation grants AST-1313045 and AST-1412999. The simulations in this paper used the Complexity HPC cluster at the University of Leicester, which is part of the DiRAC2 national facility, jointly funded by STFC and the Large Facilities Capital Fund of BIS.

Author contributions

U.U. performed and analysed all the simulations described in this paper. M.G.W. calculated the velocity dispersion and mean velocity profiles on the basis of his latest observed data on the Carina dSph. M.I.W. and J.I.R. assisted with technical aspects of the numerical simulations and provided input on the cosmological implications of the work. All authors contributed to the writing of the paper.

Additional information

8

Strong coronal channelling and interplanetary evolution of a solar storm up to Earth and Mars

Christian Möstl[1,2], Tanja Rollett[1], Rudy A. Frahm[3], Ying D. Liu[4], David M. Long[5], Robin C. Colaninno[6], Martin A. Reiss[2], Manuela Temmer[2], Charles J. Farrugia[7], Arik Posner[8], Mateja Dumbović[9], Miho Janvier[10], Pascal Démoulin[11], Peter Boakes[2], Andy Devos[12], Emil Kraaikamp[12], Mona L. Mays[13,14] & Bojan Vršnak[9]

The severe geomagnetic effects of solar storms or coronal mass ejections (CMEs) are to a large degree determined by their propagation direction with respect to Earth. There is a lack of understanding of the processes that determine their non-radial propagation. Here we present a synthesis of data from seven different space missions of a fast CME, which originated in an active region near the disk centre and, hence, a significant geomagnetic impact was forecasted. However, the CME is demonstrated to be channelled during eruption into a direction $+37\pm10°$ (longitude) away from its source region, leading only to minimal geomagnetic effects. *In situ* observations near Earth and Mars confirm the channelled CME motion, and are consistent with an ellipse shape of the CME-driven shock provided by the new Ellipse Evolution model, presented here. The results enhance our understanding of CME propagation and shape, which can help to improve space weather forecasts.

[1] Space Research Institute, Austrian Academy of Sciences, A-8042 Graz, Austria. [2] IGAM-Kanzelhöhe Observatory, Institute of Physics, University of Graz, A-8010 Graz, Austria. [3] Southwest Research Institute, 6220 Culebra Road, San Antonio, Texas 78238, USA. [4] State Key Laboratory of Space Weather, National Space Science Center, Chinese Academy of Sciences, Beijing 100190, China. [5] Mullard Space Science Laboratory, University College London, Holmbury St Mary, Dorking RH5 6NT, UK. [6] Space Science Division, Naval Research Laboratory, Washington, District of Columbia 20375, USA. [7] Space Science Center, Department of Physics, University of New Hampshire, Durham, New Hampshire 03824, USA. [8] NASA Headquarters, Washington, District of Columbia 20546, USA. [9] Hvar Observatory, Faculty of Geodesy, University of Zagreb, 10 000 Zagreb, Croatia. [10] Department of Mathematics, University of Dundee, Dundee DD1 4HN, Scotland. [11] Observatoire de Paris, LESIA, UMR 8109 (CNRS), F-92195 Meudon Principal, France. [12] Solar-Terrestrial Center of Excellence - SIDC, Royal Observatory of Belgium, 1180 Brussels, Belgium. [13] Catholic University of America, Washington, District of Columbia 20064, USA. [14] Heliophysics Science Division, NASA Goddard Space Flight Center, Greenbelt, Maryland 20771, USA. Correspondence and requests for materials should be addressed to C.M. (email: christian.moestl@oeaw.ac.at).

Coronal mass ejections (CMEs) are the 'hurricanes' of space weather and lead to massive disturbances in the solar wind plasma and magnetic fields through the inner heliosphere up to the heliospheric boundary[1-3]. During a CME, a mass of the order of 10^{12} kg is expelled from the Sun's corona, its outermost atmospheric layer, with kinetic energies of around 10^{23} J (ref. 1). CMEs play a pivotal role in solar and space physics. They are responsible for the strongest disturbances in the geophysical environment, potentially leading to power blackouts and satellite failures at Earth[4]. Increasingly, policy makers recognize CMEs as a serious natural hazard, and counter-measures for the protection of space and ground-based assets are implemented.

A major requirement for producing reliable CME forecasts is to know their direction as accurately as possible as they propagate away from the Sun. Indeed, previous research[5-9] has found that various heliospheric structures may alter the CME trajectory to change its geomagnetic impact drastically. The strongest change in the propagation direction from its solar source position, which coincides with the flare for strong eruptions[10], has been mainly argued to occur within the first few solar radii where the magnetic forces acting on the CME are strongest[7,11-13], although possible changes of CME direction during interplanetary propagation have also been put forward[5]. Non-radial CME eruptions of up to 25° in longitude were predicted[8] due to the channelling of the CME to the location of the streamer belt, where the coronal magnetic field strength has a minimum. Coronal holes, which are regions of high magnetic field strength from where the fast solar wind emanates, are able to deflect CMEs away from their source regions, depending on their area and distance to the CME source[6,14,15]. However, precise quantifications and the maximum possible amount of deflection remain unclear. Deflected motions from high to low solar latitudes have been described for prominences, which upon eruption often form part of a CME[16,17].

A major obstacle for quantifying the CME deflection in heliospheric longitude is the lack of coronagraphs that can image CMEs from outside the ecliptic plane. Consequently, deflections in heliospheric longitude are much more difficult to analyse than those in latitude. To accurately quantify this process, it is thus necessary to study the complete chain of solar, coronagraphic, heliospheric and *in situ* observations. This is now possible with the multiple imaging and *in situ* spacecraft available, forming the Great Heliospheric Observatory.

In this work, we discuss an event that emphasizes the pressing need for improved real-time predictions of the geomagnetic effects of CMEs. On 7 January 2014, a very fast CME erupted from a solar active region facing Earth. Fast CMEs that erupt from source regions close to the solar disk centre are usually expected to impact Earth[10,18], so many observers around the world predicted that this CME would be strongly geo-effective and yet, no geomagnetic storm followed. We demonstrate that this CME was strongly channelled into a non-radial direction by the effects of its locally surrounding magnetic field rather than by coronal holes (CHs). We also show how the CME evolved in the heliosphere up to its arrival at Earth and Mars, confirming the inferences from solar imaging. To this end, we introduce the Ellipse Evolution (ElEvo) model for studying the CME propagation, which lets us derive constraints on the global shape of the CME-driven shock. To explain what happened in this event, we take a tour from the solar observations into interplanetary space, providing a synthesis of observations from 13 instruments on seven space missions.

Results

Solar on-disk observations. On 7 January 2014 19 universal time (UT), a CME erupted from a source region at 12° south and 8°

west (S12W08) of disk centre, accompanied by a plethora of phenomena such as a flare, coronal dimmings, a global coronal wave and post-eruption arcades. These are the classical on-disk signatures of an erupting CME[19]. The flare, peaking at 18:32 UT, was of the highest class (X1.2). A very fast CME (projected speed of ~2,400 km s^{-1}) was observed by the coronagraph instruments onboard the Solar and Heliospheric Observatory (SOHO)[20] in real time. Alerts for a G3 class geomagnetic storm or higher (on a scale from G1 to G5) were sent out by various space weather prediction centres around the globe[21] and were picked up by the media.

Figure 1a shows the state of the solar corona on 7 January 2014 19:30 UT, about 1 h after the flare peak, imaged by the Solar Dynamics Observatory Atmospheric Imaging Assembly (SDO/AIA[22]) at 193 Å. Several active regions can be seen, with the largest one at the centre of the solar disk. Post-eruption arcades, magnetic loops filled with hot plasma and counterpart signatures of an erupting CME magnetic flux rope, are visible at the flare site, located at the southwest corner of the large active region. Two large CHs are visible in the northeastern quadrant. In Fig. 1b, the results from the automatic Coronal Pulse Identification and Tracking Algorithm[23] visualize the location of the front of a global coronal wave, which is thought to be driven by the lateral expansion of the CME[24]. The algorithm could successfully track the wave almost exclusively in the southwest quadrant of the solar disk. This is confirmed by a visual identification of the wave in running difference movies (Supplementary Movie 1) showing an asymmetric wave propagation. A similar pattern can be seen in the coronal dimming regions in Fig. 1c (derived[25] from SDO/AIA at 211 Å), which emphasize the evacuation of the corona at the locations of the footpoints of the erupting flux rope[19] (Supplementary Movie 2). The dimming appears earlier and is more cohesive in the southwest quadrant than in the northeast. Consequently, both the coronal wave and the dimming provide early suggestive evidence of the non-radial motion of the erupting CME to the southwest as seen from Earth. What could be the physical cause of this non-radial propagation?

Influence of CHs and solar magnetic fields. In Fig. 1a, the locations of two large CHs in the northeastern hemisphere of the disk seem consistent with the hypothesis that the CHs acted as to deflect the CME into the opposite direction. We took a closer look at this hypothesis within the framework of the so-called Coronal Hole Influence Parameter (CHIP[14,15]). The CHIP depends on the distance from the CH centre to the CME source region, the area of the CH as well as on the average magnetic field inside the CH. It can be considered as a parameter describing how strongly the CME is pushed away from the CH. The CHIP value for the CME event under study (see Methods) is $F = 0.9 \pm 0.2$ G, which is a factor of 3–5 below that necessary to deflect a CME originating close to the central meridian almost completely away from Earth[14,15]. In particular, the distance of the CHs to the flare is comparably large[15], which lets us dismiss the hypothesis that the CHs are mainly responsible for deflecting this CME.

The explanation for the non-radial propagation may rather be found in the solar magnetic fields near the eruption. Figure 1d shows the line-of-sight component of the photospheric magnetic field (SDO/Helioseismic and Magnetic Imager (HMI)[26]). An active region (AR 11944) is right at disk centre, a few degrees east of the flare position. The positive polarity (white) sunspot had a particularly strong vertical magnetic field of ~3,000 G, which exceeds usual values[27] by a factor of about 1.5. The flare is located in between this strong sunspot and the small negative polarity at its southwest (Fig. 1d), and the erupting CME orientation (see next section) is consistent with the direction of the photospheric inversion line in between those two magnetic polarities. The negative polarity is almost surrounded by positive polarities so that a coronal magnetic null point and related separatrices are expected. The study of this magnetic topology is of primary importance to understand flare reconnection (for example, flare ribbon locations). A further study, outside the scope of the present paper, including magnetic field extrapolation and data-driven magnetohydrodynamic (MHD) simulations, would be needed to understand precisely the role of the AR magnetic field complexity on the early CME development. However, as this topology is local, it is unlikely to be important for the CME development on scales larger than the AR, which is the focus of the present paper.

Next, looking at Fig. 1e,f a potential field source surface[28] model shows that the streamer belt of closed field lines is highly inclined with respect to the solar equator (typical of solar maximum), and runs from north to south right above the strong active region. The CME source region is not under the streamer, but close to an area of open flux further west (green field lines in Fig. 1e,f). This provides some evidence that the CME has erupted in the direction of least resistance in the solar global field[17], consistent with results of numerical simulations[12,13]. The solar observations thus imply that the strongly non-radial motion of this CME is due to a combination of two effects: (i) the strong nearby active region magnetic fields to the northeast, and (ii) the open coronal field to the west of the source. Both processes acted to channel the CME to the southwest of the solar disk, which was reflected in the asymmetries of the global coronal wave and dimmings.

Coronal evolution. We now take a look at multi-viewpoint coronagraph observations of the CME in Fig. 2a. We used two methods to estimate the CME propagation direction up to 30 solar radii ($R_\odot$). The first is the Graduated Cylindrical Shell (GCS) model, by which a wire-grid of a tapered hollow tube is fitted onto coronagraph images[29,30] by manual variation of several parameters controlling its shape. The triple viewpoints from the STEREO-B (COR1/2 (ref. 31)), SOHO (C2/C3) and STEREO-A (COR1/2) imagers constrain the results very well[30]. At the time of the event, the two STEREO spacecraft were 151° ahead (A) and 153° behind (B) in heliospheric longitude with respect to Earth, at distances of 0.96 AU (A) and 1.08 AU (B). The CME propagates to the east in STEREO-B, where the event is seen as backsided, which shows that the CME longitude must be greater than $-153° + 180° = 27°$ west of the Sun–Earth line. The GCS model was applied between 7 January 18:15 and 19:30 UT, when the resulting model apex position was between 2.1 and 18.5 $R_\odot$. The average three-dimensional speed of the CME apex was $2,565 \pm 250$ km s^{-1}, derived from a linear fit to $R(t)$, not far from the fastest speeds ever observed[3] ($\sim 3,000$ km s^{-1}). A constant CME direction is consistent with the time evolution in the coronagraph images, which gives $32 \pm 10°$ (west) and $-25 \pm 5°$ (south, with quoted errors common for the method[30,32]). This means that already very close to the Sun, at 2.1 $R_\odot$, the final direction of the CME was attained.

A second method was used to find the speed of the CME segment that propagates in the ecliptic plane. We applied a triangulation technique[33] to the CME leading edge in SOHO/C2/C3 and STEREO-A/COR2/HI1 (ref. 34) observations. Our results are averages of two methods (Fixed-β and Harmonic Mean)

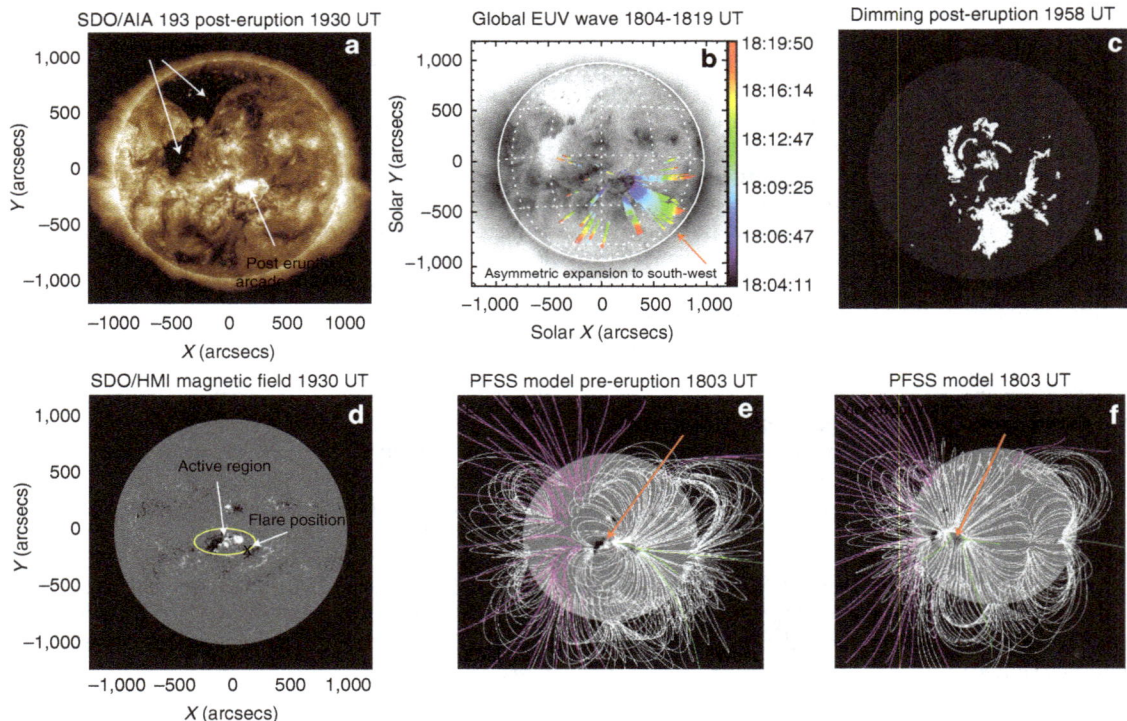

Figure 1 | Solar observations of the X1.2 flare and associated phenomena on 7 January 2014 18–20 UT. (a) Location of coronal holes and post eruption arcade in SDO/AIA 193 Å. **(b)** Extreme ultraviolet (EUV) wave evolution between 18:04 and 18:19 UT derived with the Coronal Pulse Identification and Tracking Algorithm (CorPITA) algorithm. Colours indicate the position of the wave front at different times. **(c)** Final extent of the coronal dimming (SDO/AIA 211 Å). **(d)** SDO/HMI line-of-sight magnetic field, showing the position of the large active region and the flare. White (black) colours indicate positive (negative) magnetic field polarities. **(e,f)** Pre-eruption potential field source surface model of the solar global magnetic field, as seen from Earth **(e)** and 40° west of Earth **(f)**. It depicts closed (white) and open field lines (pink negative polarity, green positive polarity). Solar east (west) is to the left (right) in all images.

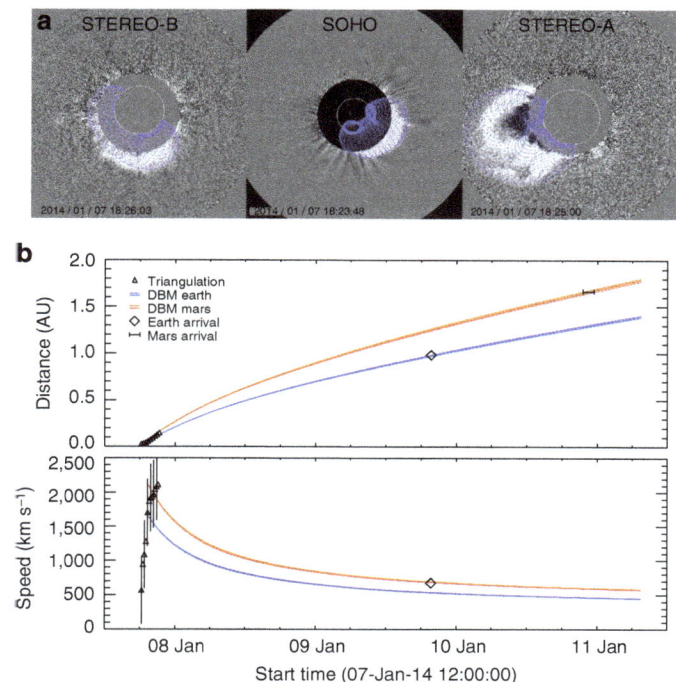

Figure 2 | Graduate Cylindrical Shell model of the CME and interplanetary shock kinematics. (**a**) Fit of the Graduated Cylindrical Shell (GCS) model (blue grid) overlaid on multipoint coronagraph observations, from left to right: STEREO-B, SOHO, STEREO-A. Shown are results for 7 January 2014 at 18:25 UT ±1 min, when the GCS apex was at 4.2 $R_\odot$. (**b**) Distance (top) and speed (bottom) of the CME shock in the ecliptic are shown as a function of time. Blue (orange) solid lines are the kinematics towards Earth (Mars) calculated with the ElEvo model, based on a DBM with $\gamma = 0.165 \times 10^{-7}$ km^{-1} and $w = 400$ km s^{-1}. Blue and orange dashed lines indicate errors from a variation of γ from 0.16 to 0.17 × 10^{-7} km^{-1}, which results from the uncertainty in t_{Mars} of ±1 h. Black triangles are the results of triangulation, with speeds and their errors deduced from a derivation of a spline fit on the distance measurements. The observed arrival times and speeds at Earth are indicated by black diamonds, and the arrival window at Mars with a black horizontal line.

assuming a small and wide CME extent in the ecliptic along the line of sight[33]. From a linear fit to $R(t)$ between 20 and 30 $R_\odot$, we find a speed of 2,124 ± 283 km s^{-1}, slightly lower than the apex speed from the GCS method. The direction of the CME front above 20 $R_\odot$ is W45 ± 10°. This is further west from Earth by about 13° compared with the GCS results, which can be expected since parts of CMEs seen at different latitudes may travel in slightly different directions, because of the CME three-dimensional tube shape. This is reasonable, as Fig. 2a shows the CME to be oriented with a moderate inclination angle to the ecliptic. Further from the Sun, we tracked the CME ecliptic leading edge to about 25° elongation with the STEREO-A Heliospheric Imager (HI), and applied the Fixed-Phi-Fitting method[32]. This results in a speed of 2,131 ± 210 km s^{-1}, consistent with triangulation. We use this as the initial speed for further modelling of the shock evolution in the ecliptic plane. We also assume that the CME leading edge is representative of the position of the CME-driven shock, which has been confirmed with imaging of CMEs at large elongations from the Sun and their in situ observations[35]. For the CME direction, we take 45 ± 10° west of Earth or 37 ± 10° away from the source region in heliospheric longitude, exceeding expected values for non-radial CME eruption in the corona[8].

Interplanetary evolution. Figure 2b extends the $R(t)$ and $V(t)$ functions of the CME shock up to Mars (1.66 AU). The interplanetary kinematics towards Earth and Mars are shown together with the arrival times at both planets, which will be discussed further below. We modelled the shock kinematics with the drag-based model (DBM[36]), which analytically describes the deceleration of CMEs by using equations of aerodynamic drag. It gives similar performances for arrival time predictions of CMEs at Earth as numerical simulations[37]. The DBM has two free, constant parameters: the drag parameter γ and the background solar wind speed w. Parameter γ contains information on the CME mass and size, the ambient solar wind density and the interaction between the CME and the solar wind[36]. For the background solar wind, we use $w = 400$ km s^{-1}, an average solar wind speed observed at Earth by the Wind spacecraft a few days around the time of the CME. This means that there are no high-speed solar wind streams west of Earth near the CME principal direction. A previous CME on 6 January 2014 was directed towards STEREO-A[38], and is not expected to influence the propagation of the 7 January CME towards Earth and Mars. Both inferences support the view that we can safely use a constant direction of motion and constant γ and w. Parameter γ may vary[36] from 0.1 to 2×10^{-7} km^{-1}. Because Mars is close to the apex of the ecliptic part of the CME shock (6° away), we can choose a value for γ that makes $R(t)$ consistent with the shock arrival time at Mars (t_{Mars}, see next section). This results in $\gamma = 0.165 \pm 0.005 \times 10^{-7}$ km^{-1}, an expected value[36] for a fast CME propagating into a slow, unstructured solar wind. The time t_{Mars} has an uncertainty of ±1 h (see below), providing the error margin in γ. We now constrain the interplanetary shock evolution further with a new model for the shape of the CME shock in combination with the available in situ observations.

Multipoint in situ observations of interplanetary CME (ICME) signatures, at longitude differences[39,40] from 10° to the size of the CME shock of ~100° can give constraints on how a CME evolves in the interplanetary medium[41–46]. Few such events have been described in the literature, and thus the global shape and extension of CMEs are poorly known. In a fortunate coincidence on 7 January 2014, Mars was at a heliospheric longitude of 51° west of Earth. Judging from the CME principal direction, Mars and Earth should see the apex and flank of the CME, respectively. Figure 3a visualizes the position of the planets and the STEREO spacecraft in the ecliptic plane, together with the shock evolution up to its arrival at Mars (see Supplementary Movie 3). We use the new ElEvo model for describing the global shape of the interplanetary shock (see Methods section), which is based on statistics of single-point shock observations[47]. Before we discuss the range of possible parameters and errors for this model, we take a look at the solar wind and planetary in situ data, serving as boundary conditions for ElEvo.

Arrival at Earth. Figure 3b–g shows the solar wind magnetic field and bulk plasma parameters observed by the Wind spacecraft at the L1 point near Earth. For a very fast CME from a source region near disk centre, such as the 7 January 2014 event, the expected in situ ICME signatures are a shock, followed by a sheath region of enhanced plasma density and temperature, and a magnetic flux rope with a size of the order of 0.1 AU in the radial direction[48]. However, in situ observations are limited to those acquired along a spacecraft trajectory through the global CME structure, and for a glancing encounter, the flux rope is likely to miss the spacecraft[49]. On 9 January 2014 at 19:40 UT, an interplanetary shock hit Wind, causing a sudden jump in the solar wind speed from 390 to 465 km s^{-1}. This time, which we label as t_{Earth}, defines the CME shock arrival at Earth.

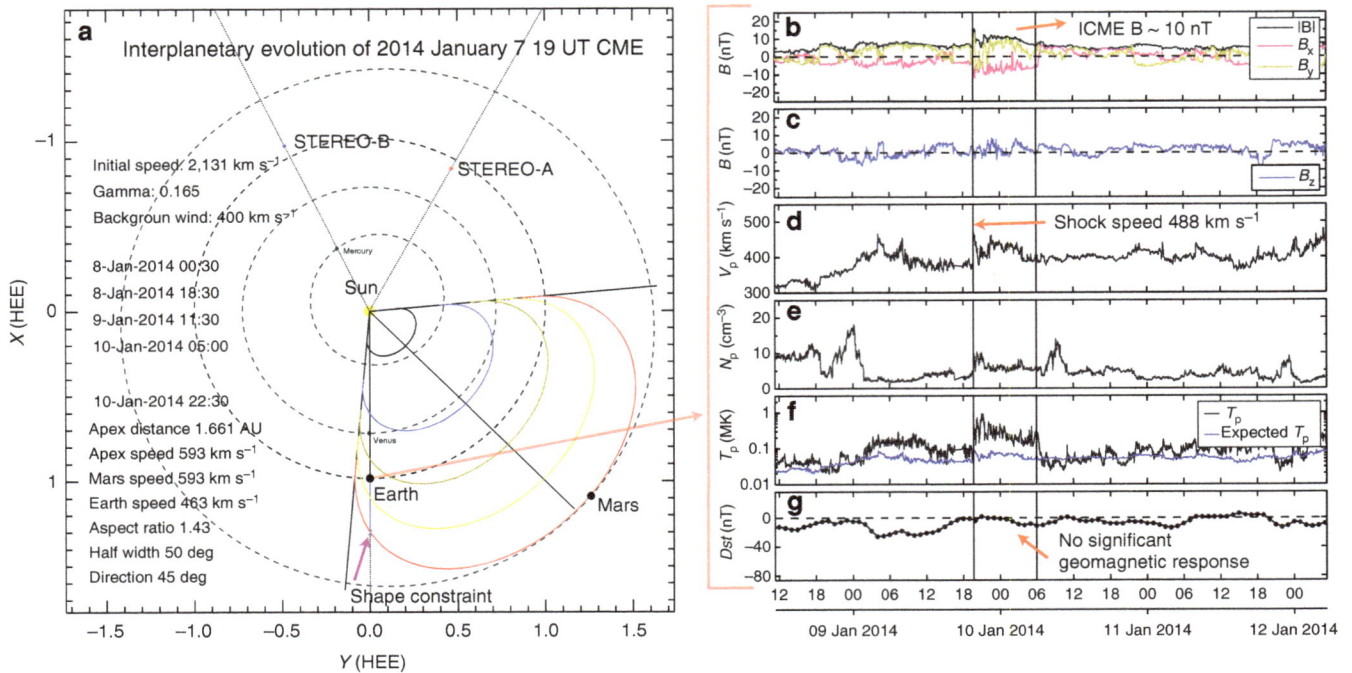

Figure 3 | Ellipse Evolution (ElEvo) model for the CME shock in the heliosphere and near-Earth solar wind. (a) Heliospheric positions of various planets and spacecraft on 7 January 2014 at 19:00 UT. The shape of the CME shock given by ElEvo is plotted for different timesteps as indicated by colours. The model parameters (bottom left) are stated for the last timestep $t_{Mars} = 10$ January 22:30 UT. For the same time, the 'shape constraint' gives a window for the heliocentric distance of the ellipse segment along the Sun–Earth line, making the ellipse shape consistent with t_{Earth}. **(b–g)** Solar wind magnetic field and bulk proton parameters in near-Earth space (Wind SWE/MFI) for 9–11 January 2014. **(b)** Total magnetic field (black) and components B_x (magenta) and B_y (yellow); **(c)** magnetic field B_z component (in Geocentric Solar Ecliptic coordinates); **(d)** proton bulk speed; **(e)** proton density; **(f)** proton temperature (black, expected temperature[40,63] from the solar wind speed in blue) and **(g)** geomagnetic Dst index. The first vertical solid line from the left indicates the arrival of the shock, and the second vertical line delimits the end of the ICME sheath region, which does not seem to be followed by a magnetic ejecta.

No other CME can explain the arrival of this shock. Figure 2b demonstrates that the arrival speed of the shock given by ElEvo at Earth indeed matches the *in situ* shock speed of $488 \, \mathrm{km \, s}^{-1}$, derived from the MHD Rankine–Hugoniot relations. The shock was weak, with a magnetosonic Mach number of 1.2. The orientation of the shock using the co-planarity theorem points in the radial direction away from the Sun and 26° northward, which is consistent with the main direction of the CME to the south of the ecliptic plane[45]. A sheath region followed, with an average speed of $417 \pm 20 \, \mathrm{km \, s}^{-1}$ and elevated proton temperatures, extending up to 10 January 2014 06:00 UT. This region has a radial size of 0.104 AU, with a magnetic field of $9.5 \pm 1.9 \, \mathrm{nT}$, which is enhanced compared with average solar wind values[48]. This magnetic field is relatively weak compared with the average field usually found in ICMEs, which contain flux ropes[48].

In summary, the above indicates that this CME almost entirely missed Earth, because a shock-sheath pair is seen but not any type of magnetic ejecta. Figure 3g shows the corresponding Disturbance storm time (*Dst*) index, which is derived from a combination of equatorial ground station magnetometers around the world. The sheath region has a magnetic field and radial size, which is comparable to non-cloud ejecta[48] at 1 AU, but due to the predominantly northward B_z (Fig. 3c), *Dst* does not even reach levels typical of a minor geomagnetic storm ($Dst < -50 \, \mathrm{nT}$). The maximum of the *Kp* index was only 3, which is below the NOAA threshold for a G1 category geomagnetic storm. However, the CME lead to a major solar energetic particle event near Earth, which resulted in an S3 solar radiation storm on the NOAA scale from S1 to S5.

It is also interesting to note that the particular orientation of the CME favours a miss of the flux rope at Earth too, because the east 'leg' is far below the ecliptic, as demonstrated by the SOHO image in Fig. 2a. Thus, in addition to the non-radial eruption, this orientation should have contributed to the false forecasts, because CMEs with a moderate to high inclination of their axes to the ecliptic plane have a small angular extent in the ecliptic[39], making it likely that the flux rope inside this CME has crossed the ecliptic to the west of Earth. As a consequence, the possibly strong southward magnetic fields of the CME flux rope have not impacted Earth's magnetosphere at all.

Arrival at Mars. Figure 4a shows an electron spectrogram by Mars Express (MEX) Electron Spectrometer instrument[50]. Enhanced fluxes of electrons, originating from both the planet's ionosphere and the solar wind and indicated by red colours in Fig. 4a, are seen starting on late 10 January. They show a drastic change on 11 January, when the electrons became both more intense and reached higher energies. These data suggest that a CME is arriving on this day at Mars, because the CME sheath contains both denser and faster plasma than the surrounding slow solar wind and additional energy is added into the induced magnetosphere of Mars by the CME. The MEX data do not allow a more precise definition of the arrival time than 11 January ± 1 day, so we turn to data from the Martian surface by the Radiation experiment (RAD[51]) onboard Mars Science Laboratory's Curiosity rover. RAD is able to observe Forbush decreases (FDs), a temporary decline in galactic cosmic ray intensity when an ICME passes a detector[52,53]. Figure 4b shows RAD count rates

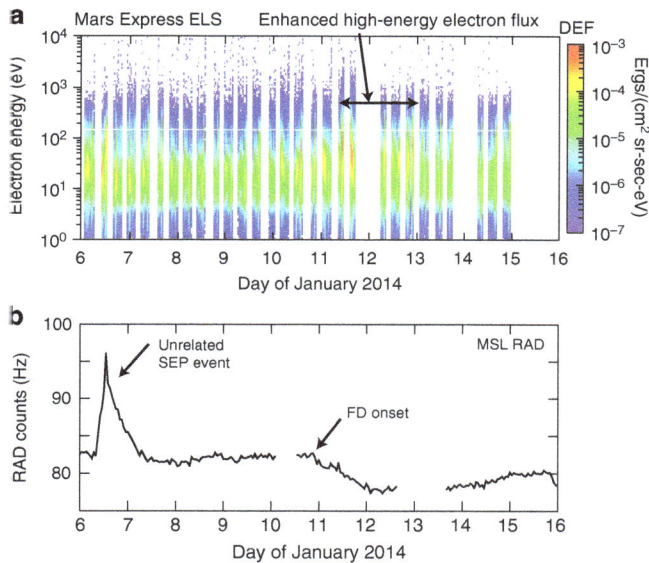

Figure 4 | Observations indicating the CME arrival at Mars.
(a) At Mars Express, the CME is observed by the Electron Spectrometer (ELS) as an increase in the electron magnetosheath and solar wind differential energy flux (colour coded) starting late on 10 January, with clear enhancements on early 11 January to late 12 January. The horizontal arrow bar delimits the interval of enhanced high-energy electrons (30–400 eV) on 11–12 January. (b) Counts of energetic particles per second by the RAD experiment on the surface of Mars onboard Mars Science Laboratory's Curiosity Rover. The high-energy solar energetic particle event stems from an eruption on 6 January. The CME of our study was launched from the Sun on 7 January and its shock hit Mars on 10 January 22:30 UT ±1h, as indicated by the onset of a Forbush decrease of the cosmic ray flux.

per second of energetic particles, which includes primary particles and secondary ionizing radiation created by solar and galactic cosmic rays in the Martian atmosphere and regolith. After a solar energetic particle event, related to a different CME[38], a return to normal counts is seen followed by a FD onset at $t_{FD} = 10$ January 2014 22:30 UT ± 1 h. What appears to be a single-step behaviour of the FD in Fig. 4b could indicate an arrival of only the ICME shock, and not the flux rope[53]. How is the time t_{FD} related to the ICME shock arrival at Mars? At Earth, there is in general a relatively tight correspondence in timing[54] between ICME arrivals at L1 and ground-based FD onsets, with the shock mostly arriving a few hours before the FD onset. Given that the corresponding FD onset at Earth (not shown) is within <2 h of the shock arrival at the Wind spacecraft, the start of the FD at Mars can be reasonably assumed to coincide, within a 2 h window, with the arrival at Mars of the presumably strong shock driven by this fast CME: $t_{Mars} = t_{FD}$.

Global shape of the shock. As we now know the arrival times of the CME at Earth and Mars, we can go back to Fig. 3a, which shows the shape for the CME shock in the ecliptic plane given by ElEvo for equidistant timesteps. The model assumes that the CME shock propagates along a constant direction with a constant ellipse aspect ratio (a_r), constant angular half width (λ) and one main axis of the ellipse oriented along the radial direction from the Sun. It is also possible to calculate the speeds and arrival times at *in situ* locations along the ellipse front analytically (see Methods section). The evolution of the ellipse apex, which is the point of the ellipse farthest from the Sun along the CME central direction, is modelled with the DBM. From an

optimization analysis (see Methods section) using the multipoint *in situ* arrival times, we find the ellipse aspect ratio to vary as $a_r = 1.4 \pm 0.4$, for half widths of the ellipse ranging from 35 to 60° (under the condition that the ellipse always hits Earth) and for the central ellipse direction at W45 ± 10°. Plotted in Fig. 3a is an ellipse with parameters $a_r = 1.43$, $\lambda = 50°$ and direction W45. The multipoint modelling of the CME shock thus implies an elliptical shape elongated perpendicular to the propagation direction[47].

Discussion
We presented for the first time the full evidence for a very strong non-radial motion of a CME, with complete observations from the Sun and two planetary impacts. The CME from 7 January 2014 19:00 UT almost entirely missed Earth despite its source being close to the centre of the solar disk. We attributed this to a non-radial CME propagation direction, which was attained very close to the Sun ($<2.1 \ R_\odot$), rather than to a deflection in interplanetary space[5]. The observations do not show a 'deflection', which implies a change in direction, but rather a 'channelled' CME motion, which is non-radial starting already with its inception on the Sun. We found a surprisingly large magnitude of this channelling with respect to the source region on the Sun ($+37 \pm 10°$ in heliospheric longitude), and a so far largely unrecognized process causing the channeling by nearby active region magnetic fields[49] rather than CHs[6]. The observations emphasize the need to understand the interplay between the active region and global magnetic fields in order to better predict the direction of CMEs, and support previous studies, which derive altered CME trajectories from modelling the background coronal magnetic field[7,8,11].

We also showed suggestive evidence that the non-radial CME motion can be seen in extreme ultraviolet observations of the corona, which showed asymmetries in the global coronal waves and dimming regions with respect to the flare position. Because such asymmetries can also arise from the structure of the solar corona[55], further research is needed on the possibility of diagnosing non-radial CME motions within extreme ultraviolet images. Finally, the arrivals of the CME-driven shock have been observed *in situ* by the Wind spacecraft near Earth and by the RAD experiment on Mars Science Laboratory's Curiosity rover on the Martian surface. These observations, together with results provided by the new ElEvo model, show that these arrivals are consistent with the CME direction given by solar and coronagraph imaging. The *in situ* arrival times allowed to directly constrain the global shape of the CME-driven shock to an ellipse with aspect ratio of 1.4 ± 0.4, with the ellipse elongated perpendicular to the direction of CME motion[47].

The enhanced understanding of non-radial CME propagation presented in our paper will be helpful for real-time CME forecasting[18] in order to avoid false positive predictions, as it was the case for the event studied, that is, a CME that was predicted to impact Earth actually missed. However, this means that a false negative forecast is also possible: a CME that is launched from a source region as much as 40° away in longitude from the Sun–Earth line may impact Earth centrally. It needs to be emphasized that non-radial CME motion in heliospheric longitude is very difficult to study because the images showing the CME radial distance from the Sun result from integrations along the line of sight. The upcoming Solar Orbiter mission[56], imaging for the first time the Sun and the heliosphere from outside of the ecliptic plane, will provide better insights into the dynamical processes responsible for non-radial CME eruptions.

In summary, the presented observations demonstrate the high value of many different instruments in spatially distant locations to study solar storms. We were able to draw a consistent picture

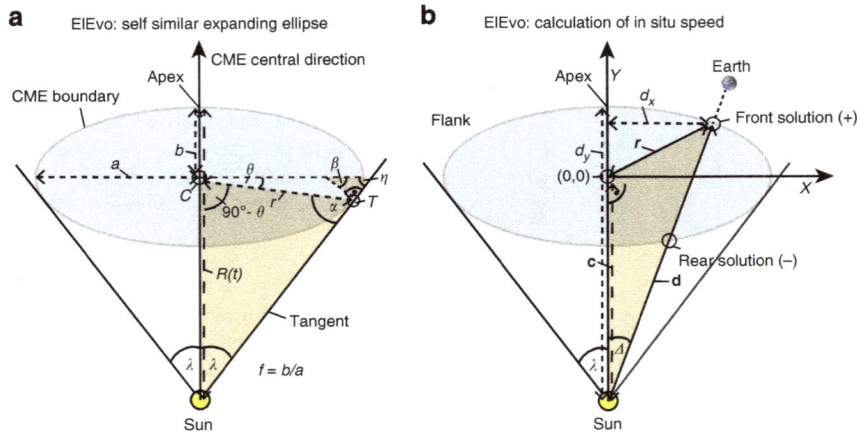

Figure 5 | Derivation of the geometry of a self-similar expanding ellipse in the ElEvo model. (**a**) A CME leading edge, a shock or the front of a flux rope, described as an ellipse, propagates away from the Sun in the ecliptic or solar equatorial plane with constant angular width and aspect ratio. (**b**) Geometry for deriving the speed of any point along the front of the CME leading edge.

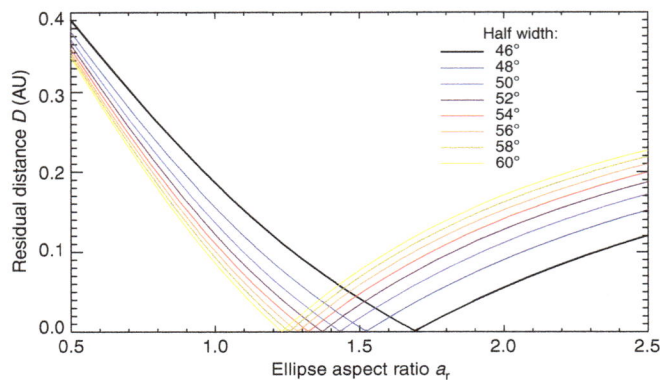

Figure 6 | Optimization of the ElEvo model shape with multipoint *in situ* observations of a CME shock arrival. The average residual distances (D) between the ellipse and the heliocentric distances of Earth and Mars, at the observed *in situ* shock arrival times, are plotted as function of the ellipse aspect ratio a_r, for half widths of 46–60° heliospheric longitude as indicated by the colours. The CME direction is set to W45, the mean value derived from triangulation.

of the evolution of a CME from its inception on the Sun, during which the event under study experienced a strongly non-radial motion, to the impacts at planets and spacecraft. These fundamental results should help to improve the reliability of real-time forecasts of space weather.

Methods
Calculating the CH influence parameter.
The CHIP is given as[14]

$$\mathbf{F} = \left(\frac{\langle B \rangle A}{d^2}\right)\mathbf{e_F}, \qquad (1)$$

with $\langle B \rangle$ the average magnetic field inside the CH, A the CH area corrected for projection, d the distance from the CH centre to the eruption site and $\mathbf{e_F}$ the unit vector along this direction. This vector defines the direction in which an erupting CME will be deflected due to the presence of the coronal hole, under the condition that $\mathbf{F}$ is sufficiently large. The values for A and the barycenter position (the centre of the CH region weighted by pixel intensity) were provided by the Heliophysics Event Knowledge base[57], obtained with the SPoCA algorithm[58] used on SDO/AIA 193 Å images[22]. We calculated the distance from the CH barycentre to the eruption site (S12W08) with the great circle distance on the solar sphere. The areas of the northern (southern) CHs are $1.2 \pm 0.04 \times 10^{11}$ km² ($0.3 \pm 0.01 \times 10^{11}$ km²), the distances to the source are 7.0×10^5 km (4.5×10^5 km) and the average magnetic fields are 1.38 (3.76) G (measured from SDO/HMI[26]). As there are two CHs present, the CHIP is treated as a vector, which is summed up for all CHs present on the disk[14].

The resulting CHIP from both CHs at the location of the source region at S12W08 is F = 0.9 ± 0.2 G, with the uncertainties arising from the calculation of the area with different thresholding techniques and an uncertainty in the determination of the magnetic field. The direction $\mathbf{e_F}$ arising from the CHIP analysis is towards a position angle (PA) of about 230° (PA is measured from solar north at 0° to east—left side in a solar image—at 90°, south at 180° and west at 270°), which also seems at first glance consistent with the CME propagation direction to the southwest, but the low CHIP[14,15] means that the combined coronal holes are not sufficiently close, have a large enough area or strong magnetic field to explain the non-radial propagation of the CME.

Derivation of the ElEvo model for the evolution of CMEs. To model the shape of the shock driven by the CME in the interplanetary medium, we introduce a new method that describes the shock as an ellipse in the ecliptic plane, which we call the ElEvo model. It allows the extension of one-dimensional models for heliospheric CME propagation[36,59], which provide the distance–time $R(t)$ and speed–time $V(t)$ functions of the CME front, into two-dimensional models of the evolution of CME boundaries in the ecliptic plane with only a few lines of code. ElEvo can thus be used to visualize the shape of a CME shock between the Sun to a given planet or *in situ* observing spacecraft using fully analytical formulas. Further, the speed and arrival time of any point along the ellipse can be calculated analytically.

ElEvo is an extension to a model describing CME boundaries as self-similar expanding (SSE) circles[60,61]. Whereas SSE has been designed to derive CME parameters from observations of CMEs at large angles from the Sun with a heliospheric imager[32,61], it can also be used to propagate a CME into the heliosphere and calculate expected planetary arrivals and speeds[60]. The initial speed, direction and width of a CME, which are known from coronagraph observations, can be used as initial conditions. The advantage of ElEvo is that the shape of an ellipse is more flexible than an SSE circle, and thus, better suited for consistent modelling with multipoint *in situ* observations because the aspect ratio of the ellipse is a free parameter.

The assumptions of the ElEvo model are: (i) the angular width in heliospheric longitude of the CME boundary remains the same for all times, (ii) one principal axis of the ellipse is oriented along to the propagation direction, and (iii) the ellipse aspect ratio and (iv) direction are both constants. However, it is possible to change the aspect ratio and direction in a code as a function of time, but this is not implemented in the current study. For describing the interplanetary deceleration of the CME, we use the DBM[36], which describes the kinematics for the distance $R(t)$ and the speed $V(t)$ of the CME as function of time (t). It has two free parameters: the drag parameter γ (on the order of 0.1 to 2×10^{-7} km^{-1}), which describes the amount of drag exerted by the solar wind on—strictly speaking—the CME flux rope, and w, which is an average of the background solar wind speed. By choosing low values of γ it is possible to describe the shock propagation up to 1 AU with DBM, so it can be used to calculate CME shock arrival times and speeds. In summary, the ElEvo model in combination with the DBM has four free parameters: (i) the inverse aspect ratio f, (ii) the ellipse half angular width λ, (iii) the drag parameter γ and (iv) the background solar wind speed w. We now derive the equations necessary to code the geometry of the ElEvo model to visualize the SSE ellipse as it propagates away from the Sun as well as the speed for each point along the ellipse front.

Visualizing a self-similar propagating ellipse. Figure 5a shows the geometry of an ellipse under the assumptions described above. The $R(t)$ of the ellipse apex (the point of the ellipse farthest from the Sun along the ellipse central direction) is given by DBM[36]. In this section, we derive equations for the ellipse semi-major axis

a and semi-minor axis b as a function of $R(t)$, the inverse aspect ratio f and the half width λ. We use $f = b/a$ rather than $a_r = a/b$ because it simplifies the following calculation. The equations

$$
\begin{aligned}
f &= b/a \\
\beta &= \lambda \\
\theta &= \arctan\left(b^2/a^2 \tan\beta\right),
\end{aligned} \tag{2}
$$

follow from the definition of f and the definition of angle β, which is the angle between the semi-major axis a and the normal to the tangent at point T (Fig. 5a). The location of T is the point of tangency on the ellipse for a line originating at the Sun. It can be easily seen that $\beta = \lambda$ by checking the sum of the angles of the small orange triangle in Fig. 5a in relation to a larger triangle (not highlighted) containing the angles λ and η. The polar angle of the ellipse θ is given by a relationship from general ellipse geometry between β and θ. It is important to emphasize that we further construct the ellipse based on this particular value of θ, for which a line with distance r connects the ellipse center C to point T. Combining equation (2) gives a relationship for the polar angle θ based on known parameters,

$$
\theta = \arctan\left(f^2 \tan\lambda\right). \tag{3}
$$

From the law of sines on the large orange-shaded triangle in Fig. 5a we derive:

$$
\frac{\sin\lambda}{r} = \frac{\sin\alpha}{R(t) - b}. \tag{4}
$$

Angle α follows from the angles of the orange-shaded triangle, and distance r from the definition of an ellipse in polar coordinates:

$$
\begin{aligned}
\alpha &= 90\deg + \theta - \lambda, \\
r &= ab/\sqrt{(b\cos\theta)^2 + (\alpha\sin\theta)^2}.
\end{aligned} \tag{5}
$$

The last equation can be rewritten with the definition of f as

$$
\begin{aligned}
\omega &= \sqrt{(f^2-1)\cos^2\theta + 1}, \\
r &= b/\omega.
\end{aligned} \tag{6}
$$

Introducing α from equation (5) and the last equation for r into equation (4) then eliminates the unknowns (α, r) and expresses b in function of known variables:

$$
\begin{aligned}
b &= \frac{R(t)\omega \sin\lambda}{\cos(\lambda - \theta) + \omega \sin\lambda}, \\
a &= b/f, \\
c &= R(t) - b.
\end{aligned} \tag{7}
$$

Equations (7) are the final description of the ellipse parameters. The minor axis b of the ellipse depends on all known variables $(R(t), f, \lambda)$ through θ and ω, from equations (3) and (6). The major axis a then simply follows from the definition of f in equations (2). The heliocentric distance of the centre of the ellipse is parameter c (Fig. 5b), closing the model equations necessary for visualizing the ellipse.

Calculation of speeds along the ellipse front. For comparison to *in situ* observations, which give parameters such as the speed and arrival time of the CME shock with very good accuracy[48], one needs to know the speed of any point along the ellipse front as a function of the ellipse parameters. This problem has been solved analytically for the circular SSE geometry[60], and we introduce here the corresponding analytic solution for ellipses.

Figure 5b demonstrates the geometry, with Δ being the known angle between the CME central direction and for example Earth, which could also be any other planet or spacecraft in the solar wind. The direction of the apex with respect to a coordinate system including the Sun and Earth depends on different methods for CMEs observed with coronagraphs[29,33] or heliospheric imagers[61,62].

We introduce a coordinate system centred on the ellipse (Fig. 5b), with coordinate X being perpendicular to the CME propagation direction and Y orthogonal to X. Here, $\mathbf{c}$ is the vector from the Sun to the ellipse centre, $\mathbf{d}$ connects the Sun to the front edge of the ellipse in the direction of Earth (that is, $\mathbf{d}$ stops at the ellipse boundary and does not connect the Sun to the planet), and $\mathbf{r}$ connects the centre to the end point of $\mathbf{d}$ on the ellipse:

$$
\begin{aligned}
\mathbf{c} &= (0, c), \\
\mathbf{d} &= \mathbf{c} + \mathbf{r}.
\end{aligned} \tag{8}
$$

The problem consists in finding the norm of $\mathbf{d}$ as a function of Δ. There are two crossings of $\mathbf{d}$ with the ellipse, one at the rear and one at the front (Fig. 5b), which will form the two solutions of the problem. The coordinates of $\mathbf{r}$ can be expressed with the projections of the vector $\mathbf{d}-\mathbf{c}$ in the X/Y coordinates:

$$
\begin{aligned}
\mathbf{r} &= \mathbf{d} - \mathbf{c}, \\
d_x &= d\sin(\Delta), \\
d_y &= d\cos(\Delta), \\
r_x &= d_x, \\
r_y &= d_y - c.
\end{aligned} \tag{9}
$$

Then, r_x and r_y can be introduced into the definition of an ellipse in cartesian coordinates,

$$
\left(\frac{r_x}{a}\right)^2 + \left(\frac{r_y}{b}\right)^2 = 1. \tag{10}
$$

This results in the following expression, which was simplified with the definition of f from equations (2):

$$
d^2 f^2 \sin^2\Delta + (d\cos\Delta - c)^2 = b^2. \tag{11}
$$

This quadratic equation gives two analytic solutions of the front and rear crossings of d with the ellipse as a function of the parameters b, c from equation (7), f from equation (2) and Δ:

$$
d_{1,2} = \frac{c\cos\Delta \pm \sqrt{(b^2 - c^2)f^2 \sin^2\Delta + b^2 \cos^2\Delta}}{f^2 \sin^2\Delta + \cos^2\Delta}. \tag{12}
$$

The solution with the positive sign in front of the root is the 'front' solution (d_1) and the one with the negative sign the 'rear' solution (d_2). The speed $V_\Delta(t)$ of the ellipse at the position defined by the angle Δ is derived from the self-similar expansion of the ellipse, which implies a constant half width λ. The assumption of self-similar expansion means that the shape of the ejection must not change in time, so the ratio between speeds and distances for all points along the ellipse must be constant[60]:

$$
V_\Delta(t) = \frac{d_1(t)}{R(t)} V(t). \tag{13}
$$

Further, the time when $d_1(t)$ is equal to the heliocentric distance of the planet or the *in situ* observer gives the arrival time of the ellipse at the *in situ* location, and the speed $V_\Delta(t)$ at this time the arrival speed.

Analysis of the aspect ratio and width of CME shocks. In this section, we discuss how to find optimal solutions for the ElEvo shape when multipoint observations of the ICME arrival are available. We first create a shock apex kinematic $R(t)$ with the DBM with parameters $\gamma = 0.165 \times 10^{-7}\,\mathrm{km}^{-1}$ and $w = 400\,\mathrm{km\,s}^{-1}$, which yields a DBM arrival time at Mars consistent with the observed arrival time t_{Mars}. From the $R(t)$ apex, we calculated with equation (12) for a range of half widths from $45° < \lambda < 60°$ the parameter $d_{1,\mathrm{Earth}}$ for the longitude of Earth, at Earth arrival time t_{Earth}. This range for values of λ is chosen because the half shock extension is thought to be around $50°$ in heliospheric longitude[40]. From the *in situ* observations, we know that the shock has impacted Earth. Thus, for an ellipse apex direction of W45 $\pm 10°$, λ must be larger than $45° \pm 10°$, or the shock would not reach Earth. This also means that the half width (λ) of the shock is $> 35°$ in the ecliptic, consistent with the value of $\sim 50°$.

Because the ellipse shape needs to be consistent with both Earth and Mars arrival times, we repeated the same procedure for Mars and calculated the average of the residual distances D between the ellipse front (d_1 for the corresponding values of Δ for Earth and Mars) and the heliocentric distances of both planets (d_{Earth} and d_{Mars}) at the respective observed arrival times t_{Earth} and t_{Mars}:

$$
\begin{aligned}
\Delta d_{\mathrm{Earth}} &= \left| d_{\mathrm{Earth}} - d_{1,\mathrm{Earth}} \right|, \\
\Delta d_{\mathrm{Mars}} &= \left| d_{\mathrm{Mars}} - d_{1,\mathrm{Mars}} \right|, \\
D &= (\Delta d_{\mathrm{Earth}} + \Delta d_{\mathrm{Mars}})/2.
\end{aligned} \tag{14}
$$

Figure 6 shows parameter D as function of a_r varied from 0.5 to 2.5 and for different λ indicated by colours. This plot was made with the average CME direction of W45. For each λ, an optimal solution for a_r exists where D has a minimum. At the minimum, the ellipse impacts Earth and Mars at the observed arrival times. For two *in situ* spacecraft that observe an ICME arrival, there are optimal solutions for pairs of λ and a_r that match the two given *in situ* arrival times for the CME shock if either λ or a_r is known or assumed. We note in passing that for CME events where three or more *in situ* arrival times would be available, there would be higher residuals, and a stronger constraint might be derived. The important new result is that λ and a_r are not independent of each other, and the result of the optimization procedure is a range of $1.69 > a_r > 1.23$ for half widths of $45° < \lambda < 60°$ when keeping the direction constant at W45. For a half width of $50°$, the optimal $a_r = 1.43$ (used in Fig. 3a). In general, as the width increases the aspect ratio must become smaller to be consistent with the *in situ* arrival times. To fully include the errors from the direction determination, we also experimented with varying the shock apex direction by $\pm 10°$ around W45, a typical error for CME directions by the triangulation method[33]. For the W55 case, $55° < \lambda < 60°$, the optimal aspect ratio is in the range $1.84 > a_r > 1.58$. For the W35 case, $35° < \lambda < 60°$, the aspect ratio is $1.51 > a_r > 0.99$. Thus, including the errors from both the direction and varying the half width within reasonable values leads to a considerable possible variation in $< a_r > = 1.4 \pm 0.4$. However, $< a_r >$ indicates the global shock shape to be a slightly elongated ellipse, being only close to a circular shape for a very extreme choice of parameters.

To better visualize this optimization process, we illustrated in Fig. 3a a distance window named 'shape constraint' along the Sun–Earth line, where the ellipse at Mars arrival time has to pass through in order to be consistent with the shock arrival time at Earth. After impacting the Wind spacecraft, the shock travelled 26 h 50 min ± 1 h until its apex hit Mars. We assume that during this time the shock travelled with a speed of $488\,\mathrm{km\,s}^{-1}$, which is close to the slow solar wind speed so we do not expect much deceleration. With such a speed, the shock was at t_{Mars} at a distance of 0.315 ± 0.015 AU further away from Earth, along the Sun–Earth line. This distance has an error (indicated on the figure by small horizontal lines on Fig. 3a) due to the uncertainty of ± 1 h in t_{Mars}. One can see in Fig. 3a that the outermost (red) ellipse indeed crosses the 'shape constraint' window, which means that the implementation of ElEvo is consistent with the observed multipoint arrival times.

References

1. Webb, D. F. & Howard, T. A. Coronal mass ejections: observations. *Living Rev. Solar Phys* **9**, 3 (2012).
2. Liu, Y. D., Richardson, J. D., Wang, C. & Luhmann, J. G. Propagation of the 2012 March coronal mass ejection from the Sun to the heliopause. *Astrophys. J.* **788**, L28 (2014).
3. Liu, Y. D. *et al.* Observations of an extreme storm in interplanetary space caused by successive coronal mass ejections. *Nat. Commun* **5**, 3481 (2014).
4. National Research Council. *Severe Space Weather Events—Understanding Societal and Economic Impacts: A Workshop Report* (National Academies, 2008).
5. Wang, Y. *et al.* Impact of major coronal mass ejections on geospace during 2005 September 7-13. *Astrophys. J.* **646**, 625–633 (2006).
6. Gopalswamy, N., Mäkelä, P., Xie, H., Akiyama, S. & Yashiro, S. CME interaction with coronal holes and their interplanetary consequences. *J. Geophys. Res.* **114**, A00A22 (2009).
7. Shen, C., Wang, Y., Gui, B., Ye, P. & Wang, S. Kinematic evolution of a slow CME in corona viewed by STEREO-B on 8 October 2007. *Sol. Phys.* **269**, 389–400 (2011).
8. Kay, C., Opher, M. & Evans, R. M. Forecasting a coronal mass ejections's altered trajectory: ForeCAT. *Astrophys. J.* **775**, 5 (2013).
9. Isavnin, A., Vourlidas, A. & Kilpua, E. K. J. Three-dimensional evolution of flux-rope CMEs, and its relation to the local orientation of the heliospheric current sheet. *Sol. Phys.* **289**, 2141–2156 (2014).
10. Yashiro, S., Michalek, G., Akiyama, S., Gopalswamy, N. & Howard, R. A. Spatial relationship between solar flares and coronal mass ejections. *Astrophys. J.* **673**, 1174–1180 (2008).
11. Gui, B. *et al.* Quantitative analysis of CME deflections in the Corona. *Sol. Phys.* **271**, 111–139 (2011).
12. Zuccarello, F. P. *et al.* The role of streamers in the deflection of coronal mass ejections: comparison between STEREO three-dimensional reconstructions and numerical simulations. *Astrophys. J.* **744**, 66 (2012).
13. Lynch, B. J. & Edmondson, J. K. Sympathetic magnetic breakout coronal mass ejections from pseudostreamers. *Astrophys. J.* **764**, 87 (2013).
14. Mohamed, A. A. *et al.* The relation between coronal holes and coronal mass ejections during the rise, maximum, and declining phases of solar cycle 23. *J. Geophys. Res.* **117**, A01103 (2012).
15. Mäkelä, P. *et al.* Coronal hole influence on the observed structure of interplanetary CMEs. *Sol. Phys.* **284**, 59–75 (2013).
16. Kilpua, E. K. J. *et al.* STEREO observations of interplanetary coronal mass ejections and prominence deflection during solar minimum period. *Ann. Geophys.* **27**, 4491–4503 (2009).
17. Panasenco, O., Martin, S. F., Velli, M. & Vourlidas, A. Origins of rolling, twisting, and non-radial propagation of eruptive solar events. *Sol. Phys.* **287**, 391–413 (2013).
18. Tobiska, W. K. *et al.* The Anemomilos prediction methodology for Dst. *Space Weather* **11**, 490–508 (2013).
19. Webb, D. F. *et al.* The origin and development of the May 1997 magnetic cloud. *J. Geophys. Res.* **105**, 27251–27259 (2000).
20. Brueckner, G. E. *et al.* The large angle spectroscopic coronagraph (LASCO). *Sol. Phys.* **162**, 357–402 (1995).
21. NASA/GSFC Community-Coordinated Modeling Center, CME ScoreBoard http://kauai.ccmc.gsfc.nasa.gov/SWScoreBoard/ (2014).
22. Lemen, J. R. *et al.* The atmospheric imaging assembly (AIA) on the Solar Dynamics Observatory (SDO). *Sol. Phys.* **275**, 17 (2012).
23. Long, D. M., Bloomfield, D. S., Gallagher, P. T. & Pérez-Suárez, D. CorPITA: An automated algorithm for the identification and analysis of coronal "EIT Waves". *Sol. Phys.* **289**, 3279–3295 (2014).
24. Veronig, A. M., Muhr, N., Kienreich, I. W., Temmer, M. & Vršnak, B. First observations of a dome-shaped large-scale coronal extreme-ultraviolet wave. *Astrophys. J.* **716**, L57 (2010).
25. Kraaikamp, E. & Verbeeck, C., The AFFECTS team. Solar Demon - Flares, Dimmings and EUV waves detection http://www.solardemon.oma.be/ (2013).
26. Scherrer, P. H. *et al.* The helioseismic and magnetic imager (HMI) investigation for the Solar Dynamics Observatory (SDO). *Sol. Phys.* **275**, 207–227 (2012).
27. Schad, T. A. On the collective magnetic field strength and vector structure of dark umbral cores measured by the Hinode spectropolarimeter. *Sol. Phys.* **289**, 1477–1498 (2014).
28. Schrijver, C. J. & De Rosa, M. L. Photospheric and heliospheric magnetic fields. *Sol. Phys.* **212**, 165–200 (2003).

29. Thernisien, A., Vourlidas, A. & Howard, R. A. Forward modeling of coronal mass ejections using STEREO/SECCHI data. *Sol. Phys.* **256**, 111–130 (2009).
30. Colaninno, R. C., Vourlidas, A. & Wu, C. C. Quantitative comparison of methods for predicting the arrival of coronal mass ejections at Earth based on multiview imaging. *J. Geophys. Res.* **118**, 6866–6879 (2013).
31. Howard, R. A. *et al.* Sun earth connection coronal and heliospheric investigation (SECCHI). *Space Sci. Rev.* **136**, 67–115 (2008).
32. Möstl, C. *et al.* Connecting speeds, directions and arrival times of 22 coronal mass ejections from the Sun to 1 AU. *Astrophys. J.* **787**, 119 (2014).
33. Liu, Y. *et al.* Reconstructing coronal mass ejections with coordinated imaging and in situ observations: global structure, kinematics, and implications for space weather forecasting. *Astrophys. J.* **722**, 1762 (2010).
34. Eyles, C. J. *et al.* The heliospheric imagers onboard the STEREO mission. *Sol. Phys.* **254**, 387–445 (2009).
35. Möstl, C. *et al.* STEREO and Wind observations of a fast ICME flank triggering a prolonged geomagnetic storm on 5-7 April 2010. *Geophys. Res. Lett.* **37**, L24103 (2010).
36. Vršnak, B. *et al.* Propagation of interplanetary coronal mass ejections: the drag-based model. *Sol. Phys.* **285**, 295–315 (2013).
37. Vršnak, B. *et al.* Heliospheric propagation of coronal mass ejections: comparison of numerical WSA-ENLIL + Cone model and analytical drag-based model. *Astrophys. J. Suppl. Ser.* **213**, 21 (2014).
38. Thakur, N. *et al.* Ground level enhancement in the 2014 January 6 solar energetic particle event. *Astrophys. J.* **790**, L13 (2014).
39. Kilpua, E. K. J. *et al.* Multispacecraft observations of magnetic clouds and their solar origins between 19 and 23 May 2007. *Sol. Phys.* **254**, 325–344 (2009).
40. Richardson, I. G. & Cane, H. V. Signatures of shock drivers in the solar wind and their dependence on the solar source location. *J. Geophys. Res.* **98**, 15295–15304 (1993).
41. Burlaga, L., Sittler, E., Mariani, F. & Schwenn, R. Magnetic loop behind an interplanetary shock - Voyager, Helios, and IMP 8 observations. *J. Geophys. Res.* **86**, 6673–6684 (1981).
42. Bothmer, V. & Schwenn, R. The structure and origin of magnetic clouds in the solar wind. *Ann. Geophys.* **16**, 1–24 (1998).
43. Mulligan, T. & Russell, C. T. Multispacecraft modeling of the flux rope structure of interplanetary coronal mass ejections: cylindrically symmetric versus nonsymmetric topologies. *J. Geophys. Res.* **106**, 10581–10596 (2001).
44. Farrugia, C. J. *et al.* Multiple, distant (40°) in situ observations of a magnetic cloud and a corotating interaction region complex. *J. Atmos. Solar Terr. Phys* **73**, 1254–1269 (2011).
45. Möstl, C. *et al.* Multi-point shock and flux rope analysis of multiple interplanetary coronal mass ejections around 2010 August 1 in the inner heliosphere. *Astrophys. J.* **758**, 10 (2012).
46. Rollett, T. *et al.* Combined multipoint remote and in situ observations of the asymmetric evolution of a fast solar coronal mass ejection. *Astrophys. J.* **790**, L6 (2014).
47. Janvier, M., Démoulin, P. & Dasso, S. Mean shape of interplanetary shocks deduced from in situ observations and its relation with interplanetary CMEs. *Astron. Astrophys.* **565**, A99 (2014).
48. Richardson, I. G. & Cane, H. V. Near-Earth interplanetary coronal mass ejections during Solar Cycle 23 (1996–2009): catalog and summary of properties. *Sol. Phys.* **264**, 189–237 (2010).
49. Xie, H., Gopalswamy, N. & St. Cyr, O. C. Near-Sun flux-rope structure of CMEs. *Sol. Phys.* **284**, 47–58 (2013).
50. Barabash, S. *et al.* The analyzer of space plasmas and energetic atoms (ASPERA-3) for Mars Express mission. *Space Sci. Rev.* **126**, 113–164 (2006).
51. Hassler, D. M. *et al.* Mars' surface radiation environment measured with Mars Science Laboratory's Curiosity rover. *Science* **343**, 1244797 (2014).
52. Forbush, S. E. On world-wide changes in cosmic-ray intensity. *Phys. Rev* **54**, 975–988 (1938).
53. Richardson, I. G. & Cane, H. V. Galactic cosmic ray intensity response to interplanetary coronal mass ejections/magnetic clouds in 1995–2009. *Sol. Phys.* **270**, 609–627 (2011).
54. Dumbović, M., Vršnak, B., Čalogović, J. & Karlica, M. Cosmic ray modulation by solar wind disturbances. *Astron. Astrophys.* **531**, A91 (2011).
55. Gopalswamy, N. *et al.* EUV wave reflection from a coronal hole. *Astrophys. J. Lett.* **691**, L123–L127 (2009).
56. Müller, D., Marsden, R. G., St. Cyr, O. C. & Gilbert, H. R.The Solar Orbiter Team. Solar Orbiter. *Sol. Phys.* **285**, 25–70 (2012).
57. Hurlburt, N. *et al.* Heliophysics event knowledgebase for the Solar Dynamics Observatory (SDO) and beyond. *Sol. Phys.* **275**, 67–78 (2012).
58. Verbeeck, C., Delouille, V., Mampaey, B. & De Visscher, R. The SPoCA-suite: Software for extraction, characterization, and tracking of active regions and coronal holes on EUV images. *Astron. Astrophys.* **561**, A29 (2014).
59. Gopalswamy, N., Lara, A., Yashiro, S., Kaiser, M. L. & Howard, R. A. Predicting the 1-AU arrival times of coronal mass ejections. *J. Geophys. Res.* **106**, 29207–29217 (2001).

60. Möstl, C. & Davies, J. A. Speeds and arrival times of solar transients approximated by self-similar expanding circular fronts. *Sol. Phys.* **285**, 411–423 (2013).

61. Davies, J. A. *et al.* A Self-similar expansion model for use in solar wind transient propagation studies. *Astrophys. J.* **750**, 23 (2012).

62. Lugaz, N. Accuracy and limitations of fitting and stereoscopic methods to determine the direction of coronal mass ejections from heliospheric imagers observations. *Sol. Phys.* **267**, 411–429 (2010).

63. Lopez, R. E. Solar cycle invariance in solar wind proton temperature relationships. *J. Geophys. Res.* **92**, 11189–11194 (1987).

Acknowledgements

This study was supported by the Austrian Science Fund (FWF): [P26174-N27, V195-N16]. T.R. gratefully acknowledges the JungforscherInnenfonds of the Council of the University Graz. D.M.L. is a Leverhulme Early-Career Fellow funded by the Leverhulme Trust. M.D. and B.V. acknowledge financial support by the Croatian Science Foundation under the project 6212 SOLSTEL. Y.D.L. was supported by the Recruitment Program of Global Experts of China, NSFC under grant 41374173 and the Specialized Research Fund for State Key Laboratories of China. The presented work has received funding from the European Union Seventh Framework Programme (FP7/2007–2013) under grant agreement No. 606692 [HELCATS] and No. 284461 [eHEROES]. Part of this work was supported by NASA grants NNX13AP39G, NNX10AQ29G and STEREO Farside Grant to UNH. MEX/ASPERA-3 is supported in the United States of America by NASA contract NASW-00003. RAD is supported by the National Aeronautics and Space Administration (NASA, HEOMD) under Jet Propulsion Laboratory (JPL) subcontract #1273039 to the Southwest Research Institute and in Germany by DLR and DLR's Space Administration grant numbers 50QM0501 and 50QM1201 to the Christian Albrechts University, Kiel. A.D. acknowledges support from the Belgian Federal Science Policy Office through the ESA-PRODEX program, grant No. 4000103240. E.K. acknowledges support from the European Commission's Seventh Framework Programme (FP7/2007-2014) under the grant agreement nr. 263506 (AFFECTS project), and grant agreement nr. 263252 (COMESEP project). This research has made use of the Heliophysics Event Knowledge database and the ESA JHelioviewer software. We thank Janet G. Luhmann and Julia K. Thalmann for discussions, and the Center for Geomagnetism in Kyoto for providing the Dst indices.

Author contributions

C.M. has designed and coordinated the study. All authors have contributed to the analysis of data, the creation of the figures and to writing the manuscript.

Additional information

9

Observations of discrete harmonics emerging from equatorial noise

Michael A. Balikhin[1,*], Yuri Y. Shprits[2,3,*], Simon N. Walker[1], Lunjin Chen[4], Nicole Cornilleau-Wehrlin[5,6], Iannis Dandouras[7,8], Ondrej Santolik[9,10], Christopher Carr[11], Keith H. Yearby[1] & Benjamin Weiss[3]

A number of modes of oscillations of particles and fields can exist in space plasmas. Since the early 1970s, space missions have observed noise-like plasma waves near the geomagnetic equator known as 'equatorial noise'. Several theories were suggested, but clear observational evidence supported by realistic modelling has not been provided. Here we report on observations by the Cluster mission that clearly show the highly structured and periodic pattern of these waves. Very narrow-banded emissions at frequencies corresponding to exact multiples of the proton gyrofrequency (frequency of gyration around the field line) from the 17th up to the 30th harmonic are observed, indicating that these waves are generated by the proton distributions. Simultaneously with these coherent periodic structures in waves, the Cluster spacecraft observes 'ring' distributions of protons in velocity space that provide the free energy for the waves. Calculated wave growth based on ion distributions shows a very similar pattern to the observations.

[1] Department of Automatic Control and Systems Engineering, University of Sheffield, Mappin Street, Sheffield S1 3JD, UK. [2] Department of Earth Planetary and Space Sciences, UCLA, 595 Charles Young Drive East, Box 951567, Los Angeles, California 90095-1567, USA. [3] Department of Earth Atmospheric and Planetary Sciences, MIT, 77 Massachusetts Avenue, Cambridge, Massachusetts 02139-4307, USA. [4] W.B. Hanson Center for Space Sciences, Department of Physics, The University of Texas at Dallas, 800 West Campbell Road, Richardson, Texas 75080-3021, USA. [5] LPP, CNRS, École Polytechnique, Palaiseau 91128, France. [6] LESIA, Observatoire de Paris, Section de Meudon, 5, Place Jules Janssen, Meudon 92195, France. [7] CNRS, IRAP, 9, Avenue du Colonel Roche, Toulouse BP 44346-31028, France. [8] UPS-OMP, IRAP, 14, Avenue Edouard Belin, Toulouse 31400, France. [9] Department of Space Physics, Institute of Atmospheric Physics ASCR, Bocni II/1401, 14131 Praha 4, Czech Republic. [10] Faculty of Mathematics and Physics, Charles University in Prague, V Holesovickach 2, 18000 Praha 8, Czech Republic. [11] Blackett Laboratory, Imperial College London, South Kensington Campus, London SW7 2AZ, UK. * These authors contributed equally to this work. Correspondence and requests for materials should be addressed to Y.Y.S. (email: yshprits@igpp.ucla.edu).

Oscillations of the electric and magnetic field in plasmas, usually referred to as plasma waves, have been observed in the Earth's magnetosphere, interplanetary space, and most recently, outside of the heliosphere. In space, plasma waves exhibit a wide variety of modes and are classified according to their frequency, polarization characteristics, types of oscillation (longitudinal or transverse) and their dispersion relation, which is the relation between the frequency of the wave and its vector of propagation.

The OGO (Orbiting Geophysical Observatory) 3 space mission detected plasma waves that were very closely confined to the terrestrial magnetic equatorial region[1,2]. These emissions were observed above the proton gyrofrequency—the frequency at which a proton gyrates around the field line. Due to their close confinement to the equator, they were named 'equatorial noise', but are also referred to as fast magnetosonic waves or magnetosonic noise due to their properties. These emissions are one of the most common waves observed in space. While these waves are observed only very close to the geomagnetic equator, they are seen on around 60% of equatorial satellite traversals in the inner magnetosphere[3]. When these waves were discovered[1], it was also noted that they may also be in resonance with harmonics of electron bounce motion (periodic motion of trapped electrons along the field line between the mirror points) and thus may be potentially generated by electrons in the plasma.

Observations by Interplanetary Monitoring Platform 6 Satellite, Hawkeye 1 (Explorer 52) Satellite and the Geostationary Operational Environmental Satellite[2,4] showed cursory evidence for discrete frequency bands, suggesting that these waves may interact with protons, alpha particles and heavy ions trapped near the equator. However, the width and spacing of these bands in frequency appeared to be non-uniform and could not be accurately measured, except at the frequencies of the lowest harmonics. The spectral frequencies of these bands were in some cases approximately at harmonics of the proton gyrofrequency but did not match them exactly (see Supplementary Note 2). Clear observational and analytical evidence for this type of frequency spectrum has so far remained elusive. A number of very detailed follow-up studies, including a recent detailed statistical study using measurements from the Polar mission[5], either failed to find the discrete waves, or found spectral structures at frequencies different from harmonics of the local proton gyrofrequency[6]. The suggested explanation for the discrepancy between theory and observations was that the waves may be generated at different locations (near the equator) and propagate to the point of observation. However, since observations showing a clear harmonic structure were not available, the theory remained unverified by observations.

Multi-point Cluster observations presented in this study show remarkable observations of very distinct harmonic emissions coinciding with multiples of gyrofrequency on two Cluster spacecraft. The waves are observed exactly in the source region. Using the observed distributions of rings, we calculated the growth rates of magnetosonic waves and show that the results of the calculations are consistent with the observed harmonics between the 17th and 30th harmonic resonances. The presented observations of distinct periodic emissions exactly at the harmonics of the gyrofrequency together with the simulations of wave growth that are based on the observed ion distributions, definitively show that magnetosonic emissions are generated by unstable ion ring distributions.

Results

Cluster mission. The European Space Agency (ESA) Cluster mission[7] consists of four identically instrumented spacecraft in a polar, eccentric orbit (apogee 18.6, perigee 3 Earth radii) with a period of 57 h. Launched in August 2000, the mission has been operating since February 2001. During its lifetime, the inter-satellite separation has varied from less than a few hundred kilometres to over 20,000 km, to explore processes occurring within the magnetosphere at different spatial scales (see Supplementary note 3).

To resolve the long-standing scientific question of the generation and propagation of the equatorial noise, ESA's Cluster mission conducted a special Inner Magnetosphere Campaign aimed at studying the structure of these waves in their source region. Figure 1 shows that on 6 July 2013, all four spacecraft were close to the geomagnetic equator. Clusters 3 and 4 were very close to each other, within 60 km, while Cluster 1 was ~800 km from Clusters 3 and 4, and Cluster 2 was around 4,400 km in the earthward direction from the trio. Supplementary Fig. 6 and Supplementary Note 8 show the expected wavelength of the magnetosonic waves. Satellites 3 and 4 are separated by a maximum of 3–5 wavelengths, depending upon the propagation direction with respect to the separation vector.

The observations of waves made by the Cluster Spatio-Temporal Analysis of Field Fluctuations (STAFF) instrument on 6 July 2013, between 18:40 and 18:55 UT, not only present observational evidence for their generation, but also show the most remarkable example of their banded structure ever observed in space. Despite being commonly referred to as magnetosonic noise, the emissions observed by the Clusters 3 and 4 spacecraft separated by 60 km have a remarkably clear discrete structure between the 17th and 30th harmonics of the proton gyrofrequency (Fig. 2) in the frequency range in which equatorial noise is usually observed. This previously unobserved, well organized and periodic structure provides definitive evidence that these waves are generated by protons. The exact match between the harmonics and observed emissions lines shows that these observations are made right in the wave source region.

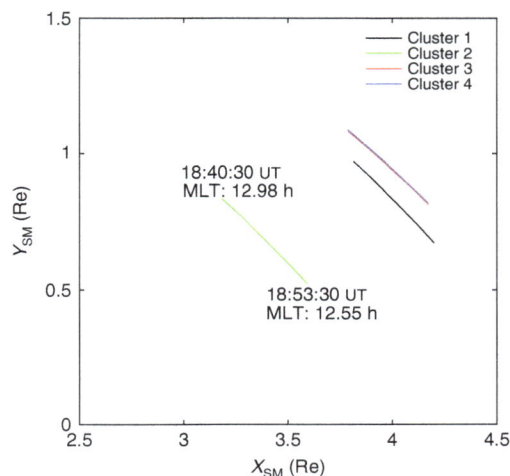

Figure 1 | Location of Cluster spacecraft. The location and motion of the four Cluster spacecraft during the period 18:40:30 to 18:53:30 UT on 6 July 2013 in which the emissions were observed (Clusters 1, 2, 3 and 4 are shown in black, green, red and blue respectively). The coordinate system used (known as Solar Magnetic (SM)) is aligned with the Earth's magnetic field, (the z direction is aligned with the magnetic dipole, and the Sun direction lies in the xz plane) in units of Earth radii (Re). Since spacecraft Cluster 3 and Cluster 4 (C3,4) are separated by only 60 km, their traces lie virtually on top of each other. Cluster 1 is around 1,000 km from C3,4, while C2 is around 4,000 km distant.

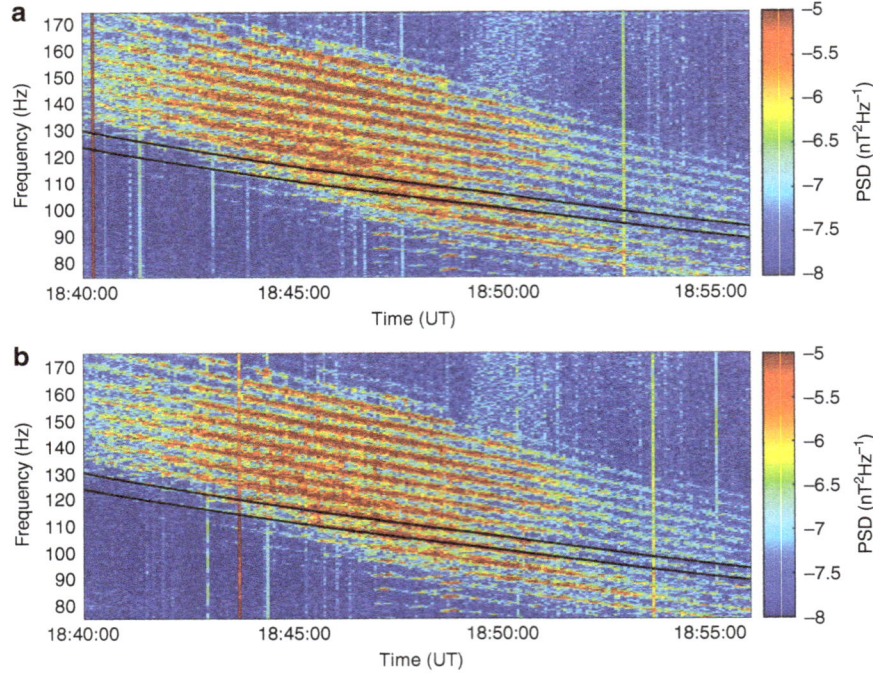

Figure 2 | Dynamical spectrograms. Observations by the STAFF instrument on (**a**) Cluster 3 and (**b**) Cluster 4 of the harmonic structure of magnetosonic waves near the equator. The figure shows the colour coded magnetic field Power Spectral Density (PSD) as a function of time and frequency. The 20th and 21st harmonics of the proton gyrofrequency are marked by solid black lines. Harmonics up to 30th are clearly seen, and one can outline the traces of the 31st and higher harmonics.

The Cluster measurements enabled not only the observation of the fine structure of the wave spectrum but also provided multi-satellite measurements of this emission at very short separation distances. The periodic pattern of emissions between the 17th and 30th harmonics observed on Cluster 4 is almost an exact replication of that observed by Cluster 3. The similarity of the signals has been analysed with the use of the coherency function (Supplementary Fig. 1 and Supplementary Note 1). The high coherence (>0.8) between the signals at harmonic frequencies of the gyrofrequency show that their separation is less than the wave coherency length and that this remarkably organized periodic structure is at least 60 km in scale which encompasses several wavelength (Supplementary Fig. 1, Supplementary Note 6).

Supplementary Fig. 2 and Supplementary Note 4 shows a comparison of the wave observations made by all four Cluster spacecrafts. While Cluster 1 observes similar discrete pattern of waves, the waves are not coherent with the Cluster 3 and Cluster 4 observations. Cluster 2 is more distant from Clusters 3 and 4, and did not observe similar type emissions.

Cluster measurements also allow to determine the polarization properties of waves to confirm that the observed emissions are the same type as the usually observed magnetosonic noise waves (Fig. 3). The fluctuating wave magnetic field on Cluster 4 is orientated parallel/antiparallel to the background magnetic field, the wave propagates at highly oblique wave normal angles, and shows linear polarization, confirming that these are typical equatorial magnetosonic waves[8]. For comparison, the polarization properties resulting from spacecrafts 1 and 3 are shown in Supplementary Figs 3 and 4.

The Cluster spacecraft also provided an opportunity to observe the source of free energy for this wave. It has been suggested[4,9] that ring-like particle distributions in velocity space may lead to wave generation through the development of instabilities. Figure 4 shows the momentum space distribution of protons near Alfvén speed (the characteristic speed at which low

frequency waves propagate within a plasma) observed by the Cluster Ion Spectrometry Composition Distribution function (CIS CODIF) analyser instrument. Particle distributions at all CIS measured energies are shown in Supplementary Figs 5 and 6 and Supplementary Note 5. The observed 'ring' distribution is unstable and results in the generation of waves[9].

The unique observations by the multiple Cluster spacecraft in the vicinity of the geomagnetic equator clearly show the fine periodic structure of magnetosonic waves generated in their source region and the simultaneous occurrence with 'ring-type' ion distributions.

Excitation of waves and growth rates. The linear growth rate can be expressed as a sum of different harmonics of an integral over perpendicular velocity that depends on the gradients of ion phase space density in the velocity space, and can be expressed as[10]

$$\sum_{n=-\infty}^{+\infty} \int_0^\infty dv_\perp \left(W_{n\perp} \frac{\partial f}{\partial v_\perp} + W_{n\parallel} \frac{\partial f}{\partial v_\parallel} \right)_{v_\parallel = v_{\mathrm{res}\parallel}} \quad (1)$$

where n is the harmonic number, $v_\perp$ and $v_\parallel$ are the perpendicular and parallel velocities with respect to background magnetic field, and $W_{n\perp}$ and $W_{n\parallel}$ are weighting functions. $v_\parallel$ in equation (1) is taken at resonance velocities corresponding to different order resonances of harmonic number n. Since the waves are highly oblique, the resonance occurs with the ions of $v_\parallel \sim 0$ only when the wave frequency is approximately equal to multiples of the ion gyrofrequency, while there are few resonant ions when the wave frequency is not in the vicinity of multiples of the ion gyrofrequency. The injection of protons will create a ring-type distribution, where phase space density has a positive df/dv along the v direction. This ion distribution may be unstable and provide the free energy for the wave excitation with growth rate maximizing at multiples of the ion gyrofrequency.

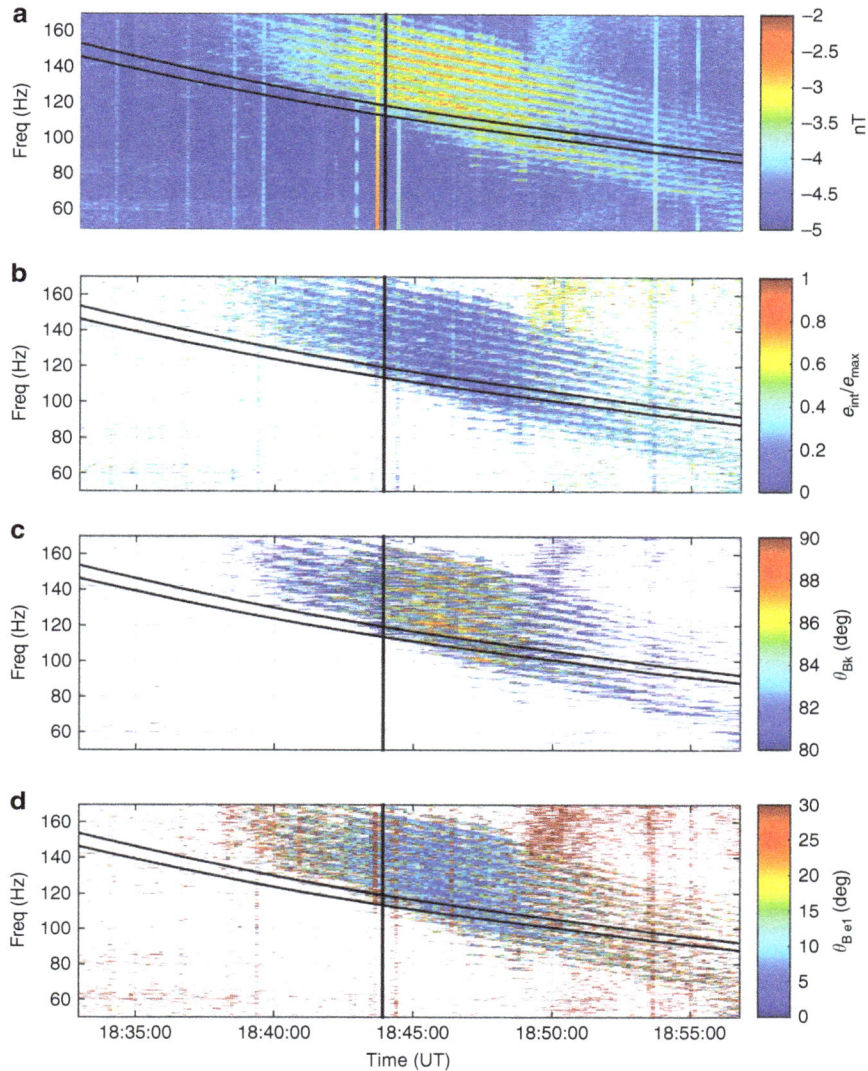

Figure 3 | Polarization properties of the magnetosonic waves observed by Cluster 4 on 6 July 2013. (**a**) The spectrum of the waveform STAFF-SC Bz component. (**b**) The ellipticity of the waves representing the polarization of the emissions. Values close to unity indicate circular polarization while those in the region of zero are indicative of linear polarization. (**c**) The wave normal angle with respect to the external magnetic field. (**d**) The angle between the external magnetic field and the oscillating magnetic field of the wave. The horizontal black lines represent the 20th and 21st harmonics of the proton gyrofrequency. Cluster 4 crossed the geomagnetic equator at the time marked by the vertical black line.

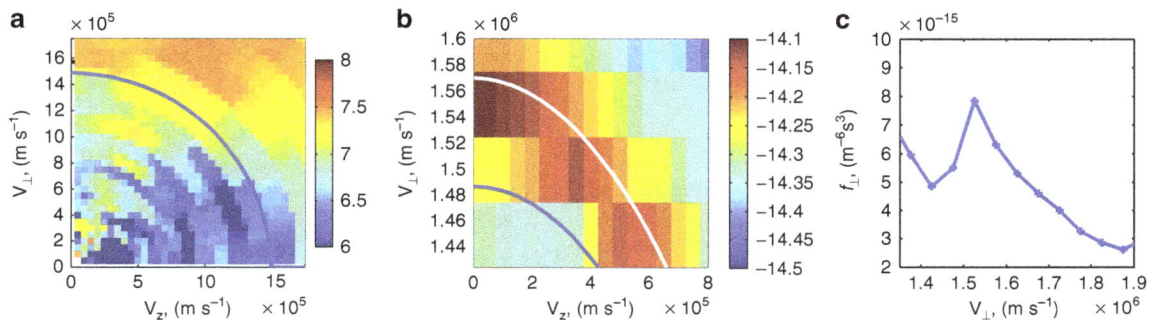

Figure 4 | Observations of the ion distribution in velocity space. (**a**) Distributions of proton fluxes in velocity space at 18:54:33 UT. (**b**) Distribution of phase space density for quasi-perpendicular ions at 18:54:33 UT. The white line denotes velocity contour of $1.57 \times 10^6\,\mathrm{m}^{-3}$ while the blue line denotes a velocity contour of value equal to Alfven speed $1.487 \times 10^6\,\mathrm{m}^{-3}$. (**c**) A plot of the ring distribution for the phase space density of protons gyrating near the equatorial plane (particles bouncing very near the equator). The blue line connecting the data points outlines the shape of the distribution. y axis is the density in phase space, and x axis is the velocity of particles.

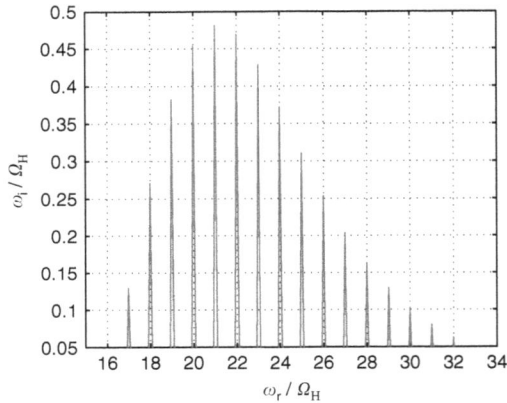

Figure 5 | Theoretical linear growth rates based on the measured ion distributions. Growth rates ω_i normalized to the proton gyrofrequency are given as a function of wave frequency ω_r which is also normalized to the proton gyrofrequency (Ω_H). Growth rate of magnetosonic waves is calculated for nearly perpendicularly directed wave vector (89.5° angle between wave vector direction and the background magnetic field).

Observations of ion distributions enable the calculation of the wave growth rates. Linear growth rates[11] are calculated using the measured ion distributions, magnetic field and plasma density measurements inferred from wave observations. For the growth rate calculation, we use a background magnetic field $B_0 = 305$ nT, electron number density of 20×10^6 m^{-3} and corresponding Alfvén speed 1.487×10^6 m s^{-1}. To represent the observed ring, we assume a Gaussian ring distribution with number density 0.008×10^6 m^{-3}, peak velocity 1.57×10^6 m s^{-1} and a width of 0.2×10^6 m s^{-1} (as shown in Supplementary Fig. 7) and evaluate the gradient of the proton distribution in velocity space to calculate the growth rate.

The growth rates show the frequencies at which waves should be theoretically observed. Figure 5 shows the linear growth rate as a function of frequency normalized to the proton gyrofrequency. The general structure of peaks is very similar to those observed (as shown in Fig. 2), with maximum growth rates occurring between the 17th and 30th harmonics. Traces of the higher harmonics can be also seen in Fig. 2

Discussion

Fast magnetosonic waves have recently attracted much attention because they are capable of accelerating particles to high energies or providing a mechanism that results in the loss of these particles into the atmosphere[5,12,13] and may be important for space weather. The observed discrete structure of magnetosonic waves may play an important role in the acceleration and scattering of electrons and ions by these waves. The discrete nature of these waves may change how these waves interact with electrons during both gyro and bounce resonance interactions and may determine acceleration and loss rates for electrons in the radiation belts. The presence of such highly structured waves may be also used in the future as a tell-tale sign of ion ring distributions.

Similar wave generation mechanisms may also operate in the magnetospheres of the outer planets, close to the Sun and in distant corners of the universe. Understanding the mechanisms behind the generation of waves is most important for laboratory plasma and for finding new ways to remotely heat plasma.

Methods

Wave propagation analysis. The data sets obtained by the STAFF-SC instrument consist of a time series of vector measurements of the magnetic field. During the period of observation, the data are sampled at 450 Hz. The spectral information, shown in Fig. 2, is obtained in this particular case with the use of the fast Fourier transform technique. This results in a frequency representation of the time series data. The polarization parameters are obtained as follows. For each frequency resulting from the fast Fourier transform process, the three spectral components corresponding to the three component measurements are combined to form the spectral matrix. By analysing the complete spectral matrix using singular value decomposition (SVD) techniques[14], it is possible to obtain the wave vector direction (without distinguishing between parallel and antiparallel directions), the orientation and the size of the polarization ellipse, and the planarity of the polarization (not shown).

SVD is a general method used to factorize a real or complex matrix, and a corresponding detailed implementation is discussed in a study[14]. This factorization process is similar to performing a least-squares fit to the data, but without actually solving the minimization problem.

Wave growth calculation. Wave growth is calculated by solving the kinetic dispersion relation in a uniformly magnetized plasma[11,15] consisting of three components: a cold electron component, a cold proton component and a hot proton ring. Several assumptions are made to facilitate the calculation using a non-Maxwellian ion distribution. First, cold plasma is dominant over the hot proton component, which allows the cold dispersion relation to be used to approximate the real part of the kinetic dispersion relation. This assumption is valid because the measured proton ring density is much smaller than the cold plasma density. Second, the growth rate is small compared with the wave frequency, which is also verified by our calculation results. With these two assumptions, we can obtain the temporal growth rate in terms of proton phase space density gradients in velocity space (as shown in the equation (1)), evaluated at resonant protons satisfying $\omega - k_{\parallel} v_{\parallel} = n\Omega$.

Instrumentation. The data presented in this paper were collected by the Fluxgate Magnetometer (FGM)[16], the STAFF-SC search coil magnetometer[17] and the CIS CODIF mass-resolving ion spectrometer[18]. During the period of observation presented in this paper, the satellites were operating in science burst mode 1. In this mode, FGM sampled the DC magnetic field at 67 Hz whilst STAFF-SC sampled the AC magnetic field at 450 Hz through a 180 Hz filter.

References

1. Russell, C. T., Holzer, R. E. & Smith, E. J. OGO 3 observations of ELF noise in the magnetosphere 2. The nature of the equatorial noise. *J. Geophys. Res.* **75**, 755–768 (1970).
2. Gurnett, D. A. Plasma wave interactions with energetic ions near the magnetic equator. *J. Geophys. Res.* **81**, 2765–2770 (1976).
3. Santolik, O. *et al.* Systematic analysis of equatorial noise below the lower hybrid frequency. *Ann. Geophys.* **22**, 2587–2595 (2004).
4. Perraut, S. *et al.* A systematic study of ULF waves above FH + from GEOS 1 and 2 measurements and their relationship with proton ring distributions. *J. Geophys. Res.* **A87**, 6219–6236 (1982).
5. Tsurutani, B. T. *et al.* Extremely intense ELF magnetosonic waves: a survey of polar observations. *J. Geophys. Res.* **119**, 964–977 (2014).
6. Santolik, O., Pickett, J. S., Gurnett, D. A., Maksimovic, M. & Cornilleau-Wehrlin, N. Spatiotemporal variability and propagation of equatorial noise observed by Cluster. *J. Geophys. Res.* **107**, 1495 (2002).
7. Escoubet, C. P., Schmidt, R. & Goldstein, M. L. Cluster—science and mission overview. *Space Sci. Rev.* **79**, 11–32 (1997).
8. Boardsen, S. A. *et al.* Van Allen Probe observations of periodic rising frequencies of the fast magnetosonic mode. *Geophys. Res. Lett.* **41**, 8161–8168 (2014).
9. Horne, R. B., Wheeler, G. V. & Alleyne, H. S. C. K. Proton and electron heating by radially propagating fast magnetosonic waves. *J. Geophys. Res.* **105**, 27597–27610 (2000).
10. Kennel, C. F. Low Frequency Whistler Mode. *Phys. Fluids* **9**, 2190 (1966).
11. Chen, L., Thorne, R. M., Jordanova, V. K. & Horne, R. B. Global simulation of magnetosonic wave instability in the storm time magnetosphere. *J. Geophys. Res.* **115**, A11222 (2010).
12. Horne, R. B. *et al.* Electron acceleration by fast magnetosonic waves. *Geophys. Res. Lett.* **34**, L17107 (2007).
13. Shprits, Y. Y. Potential waves for pitch-angle scattering of near-equatorially mirroring energetic electrons due to the violation of the second adiabatic invariant. *Geophys. Res. Lett.* **36**, L12106 (2009).
14. Santolik, O., Parrot, M. & Lefeuvre, F. Singular value decomposition methods for wave propagation analysis. *Radio Sci.* **38**, 10–11 (2003).
15. Boardsen, S. A., Gallagher, D. L., Gurnett, D. A., Peterson, W. R. & Green, S. L. Funnel-shaped, low-frequency equatorial waves. *J. Geophys. Res.* **97**, 967–14,976 (1992).

16. Balogh, A. *et al*. The Cluster magnetic field investigation. *Space Sci. Rev.* **79**, 65–91 (1997).
17. Cornilleau-Wehrlin, N. *et al*. The Cluster Spatio-Temporal Analysis of Field Fluctuations (STAFF) experiment. *Space Sci. Rev.* **79**, 107–136 (1997).
18. Rème, H. *et al*. First multispacecraft ion measurements in and near the Earth's magnetosphere with the identical Cluster ion spectrometry (CIS) experiment. *Ann. Geophys.* **19**, 1303–1354 (2001).

Acknowledgements

This research was supported by the Cluster mission. Y.Y.S. would like to acknowledge the support of The Presidential Early Career Award for Scientists and Engineers through NASA grant NNX10AK99G, NSF GEM AGS-1203747 and UC Lab Fee grant 12-LR-235337. O.S. acknowledges support from grants Praemium Academiae, LH12231, and LH14010. M.A.B. acknowledges a Royal Society International Collaboration Grant and EPSRC grant EP/H00453X/1, project PROGRESS funded from the European Union's Horizon 2020 research and an innovation programme under the grant agreement No 637302, and International Space Science Institute (Bern). L.C. acknowledges the support of NSF grant AGS-1405041.

Author contributions

Y.Y.S. and M.A.B. led the study and coordinated the efforts among different institutions. The manuscript was largely written by Y.Y.S. with contributions from M.A.B. and all other coauthors. N.C.W., C.C., I.D., B.W. and K.H.Y. provided the data sets and advice on their usage. S.N.W., M.A.B. and O.S. performed the wave analysis, Y.Y.S. and L.C. performed the particle analysis, and L.C. performed growth rate calculations. All authors commented on the manuscript and participated in the analysis of the results.

Additional information

10

Neutrino and cosmic-ray emission from multiple internal shocks in gamma-ray bursts

Mauricio Bustamante[1,2,3], Philipp Baerwald[4,5], Kohta Murase[4,5,6] & Walter Winter[2]

Gamma-ray bursts (GRBs) are short-lived, luminous explosions at cosmological distances, thought to originate from relativistic jets launched at the deaths of massive stars. They are among the prime candidates to produce the observed cosmic rays at the highest energies. Recent neutrino data have, however, started to constrain this possibility in the simplest models with only one emission zone. In the classical theory of GRBs, it is expected that particles are accelerated at mildly relativistic shocks generated by the collisions of material ejected from a central engine. Here we consider neutrino and cosmic-ray emission from multiple emission regions since these internal collisions must occur at very different radii, from below the photosphere all the way out to the circumburst medium, as a consequence of the efficient dissipation of kinetic energy. We demonstrate that the different messengers originate from different collision radii, which means that multi-messenger observations open windows for revealing the evolving GRB outflows.

[1] Center for Cosmology and AstroParticle Physics (CCAPP), The Ohio State University, 191 W. Woodruff Avenue, Columbus, Ohio 43210, USA. [2] DESY, Platanenallee 6, D-15738 Zeuthen, Germany. [3] Institut für Theoretische Physik und Astrophysik, Universität Würzburg, Am Hubland, D-97074 Würzburg, Germany. [4] Department of Astronomy and Astrophysics, Center for Particle and Gravitational Astrophysics, Institute for Gravitation and the Cosmos, Pennsylvania State University, 525 Davey Lab, University Park, Pennsylvania 16802, USA. [5] Department of Physics, Pennsylvania State University, 525 Davey Lab, University Park, Pennsylvania 16802, USA. [6] Institute for Advanced Study, Princeton, New Jersey 08540, USA. Correspondence and requests for materials should be addressed to M.B. (email: bustamanteramirez.1@osu.edu).

Gamma-ray bursts (GRBs) are violent outbreaks of energy distributed over cosmological distances. Most of the energy is detected as gamma rays during the so-called prompt phase, lasting from a few seconds to several hundred seconds (see refs 1–3 for reviews). The common view is that relativistic jets are ejected from a central engine, triggered by a collapsing star or a neutron star merger, in the direction of the observer. The inhomogeneity in the jets naturally leads to internal shocks, at which charged particles can be accelerated. In the classical GRB scenario, the observed gamma-ray emission is attributed to synchrotron radiation from non-thermal electrons. It is natural to expect that protons are accelerated as well, and GRBs have also been considered as a possible candidate class for the origin of the ultra-high-energy cosmic rays (UHECRs)[4–6]. Whereas the charged cosmic rays cannot be traced back to their origin because of their deflection on magnetic fields during propagation, neutrinos from GRBs, which would be generated via proton–gas or proton–radiation interactions, point back to the sources and could provide crucial clues to the UHECR mystery[7].

Neutrinos up to PeV energies from presumably extragalactic sources have now been detected in the IceCube neutrino telescope[8]. While even the signal shape seems compatible with a GRB origin[9–12], stacking searches for prompt GRB neutrinos using the timing and directional information coming from gamma-ray observations have been so far unsuccessful[13,14]. Because some of the early analytical predictions of the GRB neutrino fluxes[14,15] have shortcomings that are independent of astrophysical uncertainty (although these do not exist in some numerical works such as ref. 9), the model used by IceCube in ref. 14 has been revised by about one order of magnitude[16–18]. The current data are even pushing into the expected regime of the latest predictions, enabling us to address whether GRBs can be the sources of the UHECRs, and what the neutrinos can tell us about that.

In most of the earlier discussions, a simple one-zone model is assumed: this approach considers one representative collision between two relativistic plasma blobs representing the inhomogeneity in the jet, calculates the emission from this collision and scales the result for the whole burst by assuming many such identical collisions within the jet. In this simple model, the GRB parameters are fixed during its duration. In particular, the internal shock radius R_C, where the representative collision occurs and which is crucial for neutrino and UHECR production[9], is often estimated from geometric arguments[19]. Taking the blobs as spherical shells, and using the representative value of the Lorentz factor $\Gamma \equiv (1 - (v/c)^2)^{-1/2}$ of the plasma blobs, with the average velocity of the blobs, the variability timescale t_v and the burst redshift z, the collision radius can be estimated as

$$R_C \simeq 2\Gamma^2 c t_v / (1 + z). \qquad (1)$$

The variability timescale can be obtained by inspection of the pulse rising time of the burst's light curve; the Lorentz factor can only be estimated using various approaches[20,21]; and the redshift can be estimated via the observation of the host galaxy of the GRB. In the internal shock model, using the typical variability timescale (which is about three times shorter than the pulse width cf $\sim 1\,\mathrm{s}$[22,23]), $R_C \sim 10^8$–$10^{10.5}$ km is expected[24] and neutrino predictions correspondingly vary[9,18,25]. Specifically, in dissipative photospheric scenarios[10,26–29], internal shocks may occur under or around the radius known as the photosphere, at which the Thomson optical depth for $e\gamma$ scattering[30] is unity. Gamma rays can directly escape above the photosphere, where the optical depth is low. Even beyond it, high-energy gamma rays are attenuated by $\gamma\gamma$ interactions. Since the photospheric radius $R_{ph} \sim 10^8$–$10^{8.5}$ km is small, neutrino production is expected to be highly efficient around the photosphere.

UHECR production is also sensitive to R_C; UHECR escape also depends on GRB parameters[31]. Although it is often assumed that UHECRs can escape after the dynamical timescale (that is, the shock crosses the shell), this is not the case if magnetic fields do not decay. Especially, strong constraints on the UHECR–neutrino connection can be obtained if cosmic rays escape only as neutrons, which are produced in the same interactions as the neutrinos ('neutron escape model')[32,33]. While this specific model is essentially excluded[34,35], a hard flux of protons leaking from the sources (hereafter called 'direct escape' around the maximum energy and/or 'diffusion escape' at lower energies) can dominate the UHECR emission, which is largely allowed by neutrino observations[31,34]. As demonstrated in ref. 31, the dominant UHECR escape mechanism is in fact a function of the shell parameters.

Since the one-zone model is not realistic in the internal shock picture, R_C and Γ should evolve even within one GRB. The R_C dependence of neutrino production efficiency has been discussed in the internal shock model[12,18,25,36], but its integrated effects have not been studied in detail. Hence, it is conceivable that not all collisions occur at the same radius, which has significant consequences for the neutrino and cosmic-ray production, as we show in this work. For example, different UHECR escape mechanisms will dominate in different phases of the evolving GRB.

In this work, we demonstrate that the different messengers originate from different collision radii. Even in the internal shock model, the neutrino production can be dominated by emission from around the photosphere, that is, the radius where the ejecta become transparent to gamma-ray emission. Possible subphotospheric contributions enhance the detectability. We predict a minimal neutrino flux per flavour at the level of $E^2 J \sim 10^{-11}$ GeV cm^{-2} sr^{-1} s^{-1} for the contribution from beyond the photosphere, with a spectral shape similar to the original theoretical prediction. However, in striking contrast to earlier approaches, this prediction turns out to hardly depend on model parameters such as the Lorentz boost, the baryonic loading or the variability timescale.

Results

Dynamical burst model. To demonstrate neutrino and cosmic-ray emission from various R_C, we follow the internal collision model of ref. 37; see Fig. 1 for illustration. We set out a number N_{sh} of shells from a central engine with equal initial kinetic energies but a spread in the bulk Lorentz factor, around Γ_0, of

$$\ln\left(\frac{\Gamma_{k,0} - 1}{\Gamma_0 - 1}\right) = A_\Gamma x, \qquad (2)$$

where $\Gamma_{k,0}$ is the initial Lorentz factor of the k-th shell and x follows a Gaussian distribution $P(x)\mathrm{d}x = e^{-x^2/2}/\sqrt{2\pi}\,\mathrm{d}x$. Note that a large dispersion $A_\Gamma \gg 0.1$ is required to achieve high efficiencies[38], as we have explicitly tested, since the energy dissipation is proportional to the difference between the Lorentz factors of the colliding shells. The shells are assumed to be emitted with an uptime of the emitter δt_{eng}, followed by an equally long downtime, which is an input of the simulation. The variability timescale t_v will be obtained after running the simulation from the light curve as an output, with a value that is typically similar to δt_{eng}. For simplicity, we have assumed constant uptime and downtime, but ref. 24 explored a scenario where δt_{eng} is different for each emitted shell and follows a log-normal distribution; post simulation, it is possible also in this case to infer a variability timescale for the whole burst. While the shells evolve, their widths, masses and speeds (that is, their Lorentz factors) are assumed to be constant, and their mass

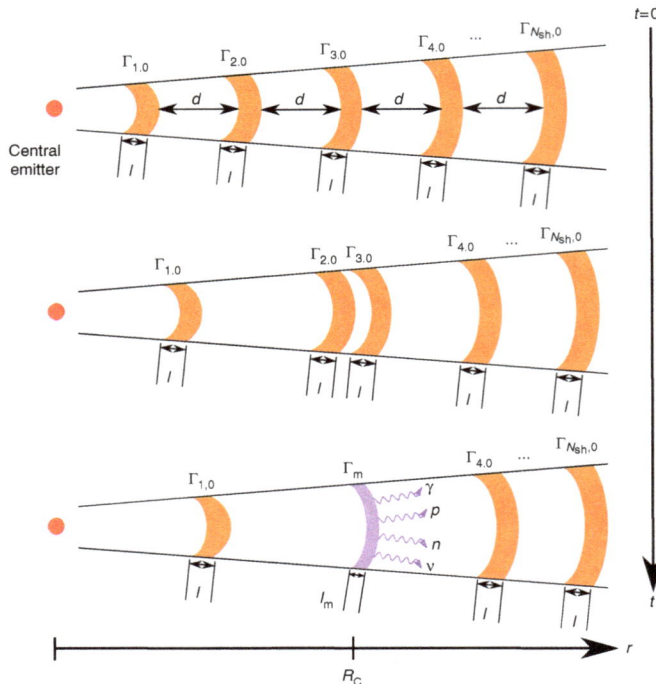

Figure 1 | Illustration of the internal collision model of gamma-ray bursts used in this study. A set of N_{sh} shells with equal energies, widths and separations $l = d = c\delta t_{eng}$ are emitted from a central engine, where δt_{eng} is the uptime of the central emitter. The shells have a spread in the bulk Lorentz factor, but initially equal bulk kinetic energies. The shells propagate, collide and merge (marked by the shell coloured purple) as soon as they meet other shells (multiple collisions are allowed), whereupon their masses, widths and speeds change. The energy dissipated in the collision is assumed to be radiated away immediately.

density decreases $\propto r^{-2}$, with r the radial distance to the emitter. Because of the different speeds of the shells, a shell will collide with another and merge into a new one; see Fig. 1. During the burst evolution, shells may collide several times. We assume that after a collision the new shell immediately cools by prompt radiation of the internal energy into gamma rays, cosmic rays and neutrinos. Derivations of the properties of the newly formed shell are given in refs 37,39 and are maintained in the simulations presented here. Our results match the analytical predictions for the dissipation of modest-amplitude fluctuations from refs 40,41. Note that we simplify the evolution of the internal shocks in several points, although our approach is enough for the purpose of this work. First, since we focus on the classical internal shock scenario where optically thin synchrotron emission is the most relevant mechanism, we assume situations where most of the dissipation occurs in the optically thin regime. If significant dissipation occurs in the optically thick regime, the internal energy scales adiabatically $\propto r^{-2/3}$, which is spent to accelerate the outflow. Second, since we do not consider cases where only a fraction of the internal energy made available after a collision is released as radiation[38], this means the efficiency issue of the internal shock model may remain unresolved[42]. Third, observed light curves from real GRBs may have slow variability components as well as fast variability components[43], which are not easily explained by a discrete number of shells from a continuous emitter, whereas continuous outflow models give better agreement[44–46].

In this study, we choose for our base model the parameter values $\Gamma_0 = 500$, $N_{sh} = 1{,}000$, $\delta t_{eng} = 0.01$ s and $A_\Gamma = 1$, as well as

a perfect acceleration efficiency of $\eta = 1$ (defined by $t'^{-1}_{acc} \equiv \eta c^2 eB'/E'_p$, with E'_p the proton energy; see ref. 31). The simulation yields 990 collisions, $t_v \simeq 0.06$ s from the average obtained rise time of the light curve pulses (see ref. 37), a burst duration $T \simeq N_{coll}t_v \approx 59$ s and an average $\langle\Gamma\rangle \approx 370$ (average Lorentz factor of the merged shells, corresponding to the observable Γ), that is, the GRB is sufficiently close to conventional assumptions in neutrino production models. Our study focuses on long-duration GRBs, which typically last tens of seconds, and our chosen parameter sets indeed yield burst durations of that order. We normalize the total isotropic photon energy of all collisions in the source frame to $E_{iso} = 10^{53}$ erg, consistent with GRB observations. Note that the fraction of photon energy dissipated in subphotospheric collisions is only about 9%, which means that a renormalization of the gamma-ray energy output to only collisions above the photosphere would hardly affects our result. For the cosmic-ray and neutrino production, we follow refs 31,47 to compute the spectra for each collision individually, choosing equal energies in electrons (that is, photons) and magnetic field, and a baryonic loading of ten (that is, ten times more dissipated energy in protons than in photons). Neutrinos are produced in $p\gamma$ interactions. The target photon spectrum is assumed to be a broken power law with spectral indices $\alpha_\gamma = 1$ and $\beta_\gamma = 2$, respectively, with a fixed break energy of $\epsilon'_{\gamma,break} = 1$ keV in the merged-shell rest frame (primed quantities are in the merged-shell rest frame). That is, it is implied that the target photon spectrum corresponds to conventional GRB observations regardless of the underlying radiation processes leading to this spectral shape.

Simulation results. The light curve of the simulated burst is shown in Fig. 2a as a black curve. Although we show the light curves for only two representative simulations in this study (the aforementioned one and another one with $N_{sh} = 100$ and $\delta t_{eng} = 0.1$ s, in Fig. 2b), we will present a more detailed parameter space study in a future work (Bustamante et al., manuscript in preparation). We do not investigate effects of the spectral evolution during the dynamical time for one collision[48], as we imply that taking into account contributions from multiple shells is more relevant, like in the case of gamma rays[39]. Note that, although we do not calculate hadronic cascades, their feedback on neutrino spectra is unimportant, given the value of the baryonic loading factor used in this work.

We show in Fig. 3 the neutrino fluence (a), maximal proton energy (b) and maximal gamma-ray escape energy (c) for each collision (dot) as a function of R_C. The maximal proton energies are obtained from comparing acceleration, dynamical, synchrotron loss and photohadronic (for protons) timescales. As a result, we find that the collisions are spread between about 10^6 km and our choice of 5.5×10^{11} km for the deceleration radius[49], where outflows terminate by the external shock into the circumburst medium. Most collisions occur around 10^{10} km—slightly above the estimate from the geometry equation (1), $R_C \approx 1.6 \times 10^9$ km. Red dots mark collisions in the neutron escape model regime (optically thick to $p\gamma$ interactions) and blue empty circles, collisions in the direct proton escape regime.

Black squares mark subphotospheric collisions, that is, those for which the Thomson optical depth is larger than unity. The optical depth is obtained by calculating the proton number density from the masses of the shells and assuming that the electron number density is as high as the proton density, which is expected for an electrically neutral plasma. In reality, however, the electron and positron densities may be somewhat higher if there is a significant non-thermal contribution from electron–positron pair production. The obtained photospheric radius

$R_{ph} \approx 2 \times 10^8$ km is somewhat larger than the conventional expectation calculated using the dissipated energy in gamma rays ($R_{ph} \approx 3 \times 10^7$ km). This estimate is affected by the efficiency

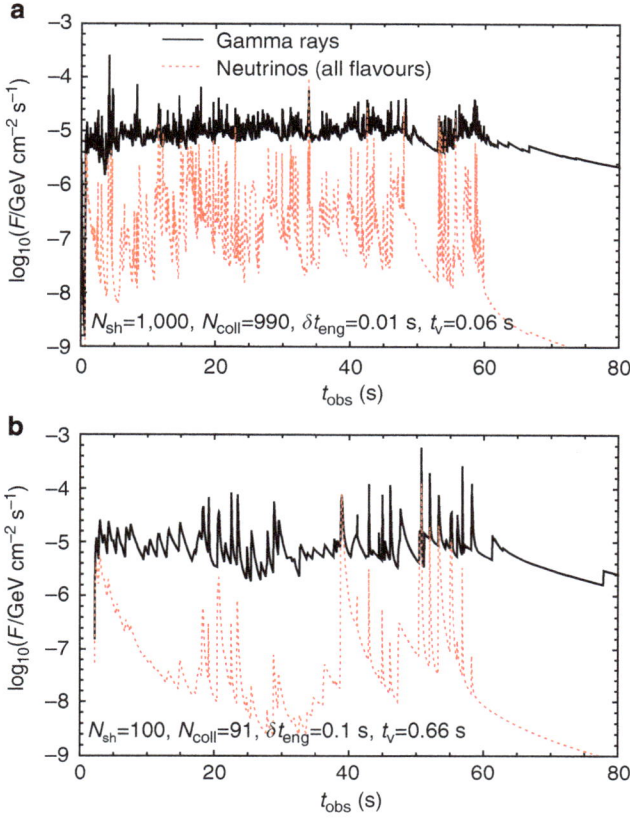

Figure 2 | Two simulated light curves. The curves for the energy flux of gamma rays (solid, black) and neutrinos (dotted, red) are built from the collisions of shells output by an engine emitting shells with equal kinetic energies, with t_{obs} the time in the observer's frame. The light curves in **a** (**b**) correspond to a simulation with $N_{sh} = 1{,}000$ (100), $N_{coll} = 990$ (91), $\delta t_{eng} = 0.01$ s (0.1s) and $t_\nu = 0.06$ s (0.66 s). A redshift of $z = 2$ was assumed to produce these light curves.

of the conversion from kinetic energy into dissipated energy, which is roughly 25% in our cases. However, the more important reason is the significant baryonic loading: since most of the energy is dissipated into protons, the masses of the shells have to be upscaled to match the required energy output in gamma rays (10^{53} erg), which leads to larger radii of the photosphere because of higher electron densities. It can therefore be expected that the large baryonic loadings that are needed to describe the UHECR observations[25,34] will lead to larger fractions of subphotospheric collisions.

We find that the obtained range of collision radii is large, from under the photosphere out into the deceleration radius, since dissipation occurs for a wide range of R_C especially when the spread of the Lorentz factor A_Γ is large[40,41]. Note that $\lesssim 12\%$ of collisions occurs under the photosphere for the chosen parameter set, altogether 118 out of the total 990 collisions, but most of the energy dissipation occurs at large radii $> 10^{10}$ km. In the internal shock model, gamma-ray emission should be produced beyond the photosphere, so we only consider collisions beyond the photosphere in the following, unless noted otherwise. This is conservative, since the baryonic loading may be smaller under the photosphere[10] and particle acceleration becomes inefficient for radiation-mediated shocks[50]. The ratio of total energy emitted as neutrinos via optically thin internal shocks to the total energy emitted by these collisions as gamma rays is 4.8% for this representative parameter set.

Most importantly, Fig. 3a demonstrates that neutrinos are dominantly produced at small collision radii $R_C \lesssim 10^9$ km, close to the photospheric radius $R_{ph} \approx 2 \times 10^8$ km. This result can be understood as follows. In each collision, the emitted gamma-ray energy, $E_{\gamma-sh}^{iso}$, is a fraction of the total dissipated energy. The pion production efficiency (fraction of proton energy going into produced pions) at the photon spectral break $\epsilon'_{\gamma,break}$, which is neglecting spectral effects, can be approximated as[9,10]

$$f_{p\gamma} \propto \frac{\kappa_p \sigma_{p\gamma} E_{\gamma-sh}^{iso}}{4\pi R_C^2 \Gamma_m \epsilon'_{\gamma,break}}. \tag{3}$$

Here Γ_m is the Lorentz factor of the merged shell, $\sigma_{p\gamma}$ is the photohadronic cross section, and $\kappa_p \simeq 0.2$ is the fraction of proton energy going into the pion per interaction. Since the internal shock model predicts[40] $E_{\gamma-sh}^{iso} \propto R_C^{-q}$ for $0 \lesssim q \lesssim 2/3$, we expect $f_{p\gamma} \propto R_C^{-2-q} \epsilon'^{-1}_{\gamma,break}$. Hence, since A_Γ has to be sufficiently large

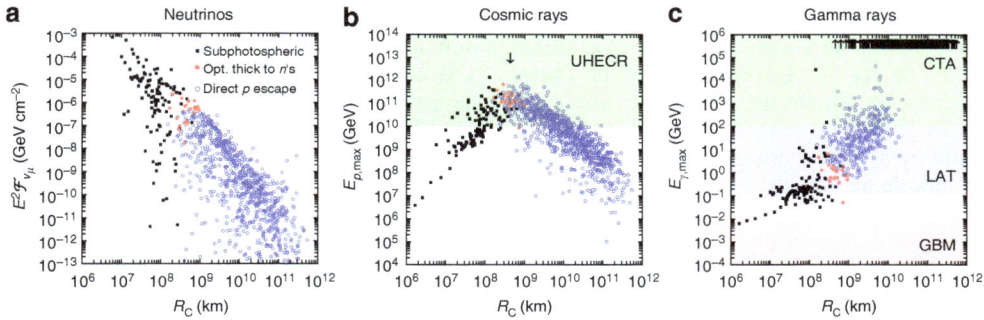

Figure 3 | Neutrino fluence and maximum proton and gamma-ray energies for each internal collision. (**a**) Muon-neutrino fluence ($\nu_\mu + \bar{\nu}_\mu$, in the observer's frame), (**b**) maximal proton energy (in the source frame, for ideal ($\eta = 1$) acceleration) and (**c**) maximal allowed gamma-ray energy (in the source frame, where $\tau_{\gamma\gamma}(E_{\gamma,max}) = 1$) as a function of the collision radius. Each dot represents one collision: red filled dots represent collisions where cosmic rays mainly escape as neutrons (optical thickness to $p\gamma$ interactions larger than unity), blue empty circles represent collisions where cosmic-ray leakage dominates over the neutron escape model, and black squares denote subphotospheric collisions or collisions where this picture cannot be maintained (that is, where the Thomson optical depth is large). In **b**, the ultra-high energy range for cosmic rays, above 10^{10} GeV, is shown as a green band; the downward-pointing arrow marks the approximate energy above which adiabatic energy losses dominate. In **c**, the energy ranges that can be reached by the Fermi-GBM (pink), Fermi-LAT (blue) and Cherenkov Telescope Array (CTA) (green) instruments are illustrated as coloured bands. Collisions in which photons with energies above 10^6 GeV are able to escape are marked as upward-pointing arrows.

for efficient energy dissipation ($A_\Gamma = 1$ in the simulations in the present study), neutrino production is typically dominated by collisions at radii around the photosphere. UHECR protons come from collisions in the range $10^{8.5}\,\text{km} \lesssim R_C \lesssim 10^{10}\,\text{km}$; see Fig. 3b. First the maximal proton energy increases with collision radius as (close to the peak) synchrotron losses limit the maximal proton energy, and the magnetic field drops with R_C. The peak occurs where adiabatic losses take over, and the decline comes from a decrease of the acceleration timescale for dropping magnetic fields; the expressions for the different energy-gain and energy-loss timescales can be found, for example, in refs 9,31. Note that the UHECRs come from two different components dominating at different collision radii: for $R_C \lesssim 10^{8.5}\,\text{km}$, neutron escape dominates and for $R_C \gtrsim 10^{8.5}\,\text{km}$, protons directly escaping from the source dominate—which are obviously not related to strong neutrino production; see Fig. 3a. In the chosen example, the main contribution to cosmic rays actually comes from direct escape. Finally, Fig. 3c illustrates that high gamma-ray energies, which can only be observed in Cherenkov Telescope Array (CTA) or other next-generation imaging atmospheric Cherenkov telescopes, come from large collision radii $R_C \gtrsim 10^9\,\text{km}$, since for lower radii the optical depth for $\gamma\gamma$ interactions is too high. As a consequence, neutrinos, cosmic rays and Fermi Large Area Telescope (LAT)/CTA gamma rays probe different emission radii. Neutrinos are useful to probe dissipation at small radii, including subphotospheric dissipation. For dissipation at large radii, where heavy nuclei survive, the TeV gamma-ray diagnostics of a GRB would be useful[25].

There has been some evidence that the composition of UHECRs is heavy[51]. Initial studies such as refs 52–54 concluded that heavy nuclei cannot survive inside GRBs: photodisintegration on fireball photons would break them up into lighter nuclei and protons. Anchordoqui et al.[55] calculated the neutrino emission from the injection of both protons and nuclei and found that the latter cannot survive in internal shocks; however, only collisions at very small radii, around $10^8\,\text{km}$, were considered. It has been argued that the typical collision radius is much larger (see, for example, ref. 43 and references therein) and that heavy nuclei can be largely loaded in GRB jets[56,57]. Therefore, acceleration of nuclei to ultra-high energies and their survival against photodisintegration are possible, provided R_C is large enough[25,58,59]. In Fig. 4, we show that this is indeed the case for our simulations. The figure shows the maximum energy to which iron nuclei ($A = 56$, $Z = 26$) can be accelerated at each of the collisions. The energy is a factor of 26 higher than for protons (compare with Fig. 3b), where its absolute magnitude is a consequence of the assumed acceleration efficiency. Here the photodisintegration timescale has been calculated using the approximation in ref. 25. Triangles (blue) and circles (red) represent collisions in which the maximum energy is limited, respectively, by the break-up of the nucleus due to photodisintegration and by adiabatic losses. Even though photodisintegration losses dominate up to $\sim 10^9\,\text{km}$, after which adiabatic losses take over, maximum energies well within the UHE band can be achieved at the turning point, where most of the UHECR emission would come from. Note that this turning point is about a factor of five higher in R_C than for protons (compare arrows in Figs 3b and 4), which means that UHECR nuclei on average reach their peak energy at higher R_C than UHECR protons. UHECR nuclei may also escape directly at the highest energies, but there is no such thing as neutron escape. It is therefore expected that nuclei come from somewhat larger collision radii than protons at the highest energies, where the radiation densities are too low to break up the nuclei. Since the actual energy output of heavy nuclei depends on the nuclear loading (that is, an additional assumption is required), we do not show their energy output explicitly in the following.

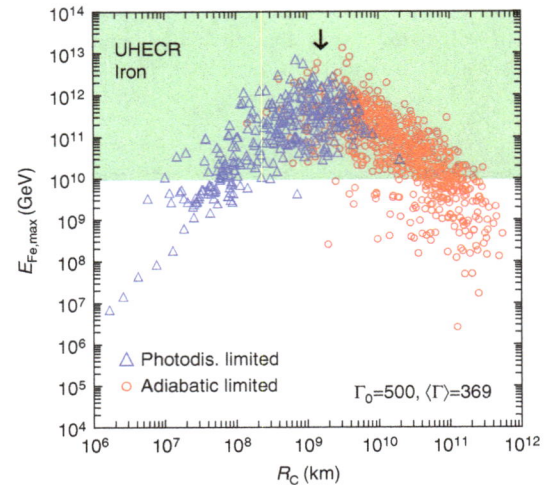

Figure 4 | Maximum energy to which iron nuclei can be accelerated in each collision of a simulated GRB. For the iron nuclei, $A = 56$, $Z = 26$. Our standard parameter set is assumed for the simulation. Energy is shown in the source frame and is calculated for ideal ($\eta = 1$) acceleration. Triangles (blue) and circles (red) represent collisions where the energy is limited by break-up due to photodisintegration and by adiabatic losses, respectively. The photodisintegration timescale has been computed using the approximation in ref. 25. The ultra-high energy range for cosmic rays, above 10^{10} GeV, is shown as a green band. The downward-pointing arrow marks the approximate energy above which adiabatic energy losses dominate.

To obtain an even more quantitative statement of how much energy is released as a function of collision radius, we show in Fig. 5 binned distributions for the prompt gamma rays, neutrinos and cosmic-ray protons, which are all directly calculable within our model. Figure 5a shows the energy output per bin, while Fig. 5b shows the fraction of energy in each bin compared with the total, for each particle species. We note that the energy per messenger per bin is obtained as a product of energy released per collision, and the number of collisions occurring per R_C-bin; especially the latter number is important to get the proper weighing of R_C. The result confirms the above observations: the neutrino production is dominated by small values of R_C just beyond the photosphere from within a relatively narrow region $R_C \approx 10^{8.5}$–$10^9\,\text{km}$, the cosmic-ray production by intermediate $R_C \approx 10^9$–$10^{10}\,\text{km}$ and the prompt gamma-ray emission is, in fact, dominated by large R_C, at around 10^{10}–$10^{11.5}\,\text{km}$—compatible with what is typically expected in the literature[24]. These results have significant implications: our knowledge of the prompt phase of GRBs is obtained from gamma rays, of course, and, consequently, R_C is derived from gamma-ray observations. This collision radius is, however, not the one to be used for neutrino or cosmic-ray calculations. It is therefore conceivable that multi-zone predictions are different from the naive one-zone expectation based on the gamma-ray emission radius. One can also read off from Fig. 5 that a significant amount of energy in UHECRs is transported away by direct escape, unrelated to neutrino production, which may affect the predicted neutrino flux if normalized to the observed UHECRs, as in, for example, ref. 34.

We show in Fig. 6 the predicted quasi-diffuse neutrino spectra from collisions beyond the photosphere as thick orange curves for three different values of Γ_0, where Fig. 6b corresponds to our standard assumptions. Note that the neutrino fluence per burst has been rescaled to a quasi-diffuse flux prediction by assuming 667 (identical) bursts per year and is significantly below the current diffuse neutrino signal reported by IceCube at the level of

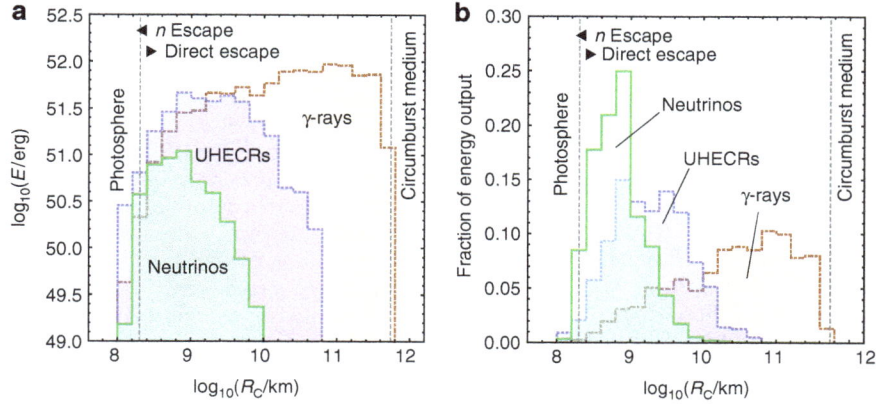

Figure 5 | Energy dissipated beyond the photosphere. We consider energy dissipated in (prompt) gamma rays, neutrinos (all flavours) and CR protons (UHECRs from 10^{10} to 10^{12} GeV). Energies are binned as a function of the collision radius. (**a**) Absolute energy values; (**b**) the fraction of energy output normalized to one for each messenger. Neutron escape dominates the cosmic-ray emission below $R_C \approx 10^{8.5}$ km, while proton escape dominates above this radius. The rough value of the photospheric radius and the assumed radius of the circumburst medium are indicated as dashed lines.

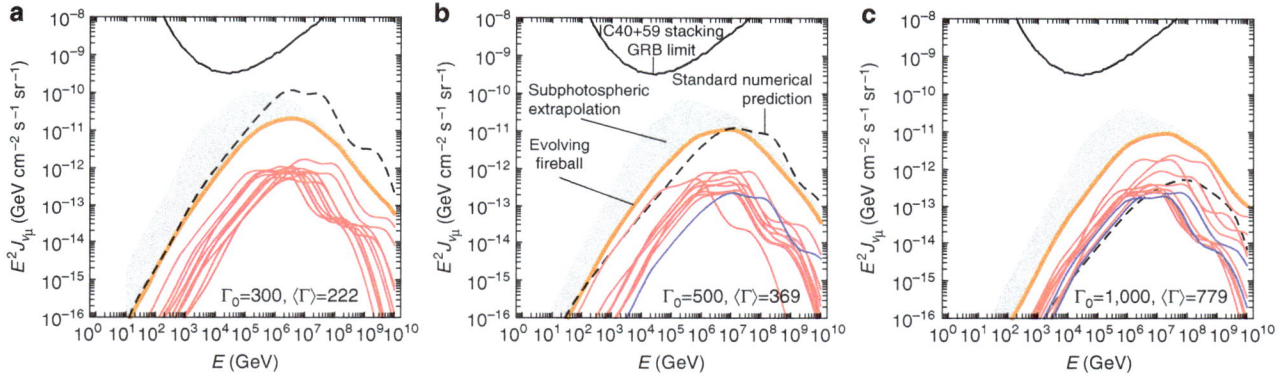

Figure 6 | Quasi-diffuse neutrino spectra from simulations of multiple internal shocks. Muon-neutrino spectra $(v_\mu + \bar{v}_\mu)$ from collisions beyond the photosphere (thick orange curves), reference spectra computed from averaged burst parameters in the conventional approach (dashed curves) and maximal subphotospheric extrapolations (shaded regions) for three different values of Γ_0 in the different panels: 300 (**a**), 500 (**b**) and 1,000 (**c**). The individual (dominant) collisions (contributing to the thick orange curves) are shown also as thin red and blue curves corresponding, respectively, to the optically thick to $p\gamma$ interactions regime, with the neutron escape dominating at the maximum energy, and to the regime dominated by direct proton escape instead. The (thick orange) spectra ('evolving fireball') are estimates of the diffuse flux obtained from the single-burst fluence $\mathcal{F}$ (one GRB at $z = 2$) by assuming $\dot{N} = 667$ per year long bursts per year over the whole sky ($J = \mathcal{F} \times \dot{N} \times (4\pi)^{-1}$). The diffuse GRB flux limit from the IC40 + 59 analysis[14] is shown as a thin black curve. The obtained average values $\langle \Gamma \rangle$ from the simulation, corresponding to the observable Γ, are depicted as well.

10^{-8} GeV cm^{-2} s^{-1} sr^{-1} flux[60]. The dashed curves correspond to the standard assumption that all collisions occur at the same radius, derived from gamma-ray observations. To generate these curves, we use the parameters N_{coll}, t_v, $\langle \Gamma \rangle$ and T obtained from the simulation assuming identical shells with a collision radius obtained from equation (1) ($R_C \approx 10^{9.2}$ km in Fig. 6b). The reference flux in Fig. 6b is significantly lower than the prediction in ref. 16. In that reference, the same parameters as in the IceCube analysis[61] were used for comparison, implying that $R_C \approx 1.9 \times 10^8$ km. That is about one order of magnitude smaller than the R_C used here; cf. equation (3) for its impact on the neutrino flux. The reference flux in Fig. 6a is comparable to ref. 16.

We first of all find that the neutrino spectra from collisions beyond the photosphere (thick orange curves) all exhibit the same flux level quite independently of Γ_0 (and even of A_Γ, as we have explicitly tested). The expected neutrino flux per flavour is at the level of $E^2 J \sim 10^{-11}$ GeV cm^{-2} sr^{-1} s^{-1}, peaking between 10^5 and 10^7 GeV. This contribution can be regarded as a minimal

prediction for the neutrino flux, as it can be inferred from gamma-ray observations and hardly depends on the parameters. Note that this flux is probably outside the sensitivity of the existing IceCube experiment, but it will provide a target for the optimization of the planned high-energy volume upgrade. There is a significant qualitative difference to conventional models such as refs 7,15, for which the pion production efficiency contains a factor Γ^{-4} coming from the collision radius estimate in equation (1) applied to equation (3). However, the optical thicknesses to Thomson scattering and photohadronic interactions both scale $\propto R_C^{-2}$, which leads to the following estimate for the pion production efficiency at the photosphere independent of Γ (ref. 10):

$$ f_{p\gamma}^{ph} \sim 5 \times \frac{\varepsilon}{0.25} \times \frac{\epsilon_e}{0.1} \times \frac{1 \text{keV}}{\epsilon_{\gamma,\text{break}}'}. \tag{4} $$

Here ϵ_e is the fraction of the dissipated energy going into photons and ε is the dissipation efficiency (ratio between dissipated and kinetic energies). Notably, Γ drops out of $f_{p\gamma}^{ph}$—unless the break energy is fixed in the observer's frame, in which case there is a single factor of Γ retained.

When the innermost collisions give the dominant contributions, the time-integrated neutrino fluence roughly scales as

$$\mathcal{F}_\nu \propto \frac{N_{coll}\left(f_{p\gamma}\gtrsim 1\right)}{N_{coll}} \times \min\left[1, f_{p\gamma}^{ph}\right] \times \frac{\epsilon_p}{\epsilon_e} \times E_{iso}, \qquad (5)$$

where $N_{coll}(f_{p\gamma}\gtrsim 1)$ is the number of collisions with efficient neutrino production close to the photosphere, N_{coll} is the total number of collisions and ϵ_p is the fraction of energy going into protons. Since the number of dominant collisions contributing to the neutrino flux is of order ten almost independently of the model parameters (see thin solid curves in Fig. 6), the neutrino flux prediction is relatively robust. The neutrino prediction above the photosphere hardly depends on the baryonic loading (ϵ_p/ϵ_e) as well, as long as most of the energy is dissipated into protons. Increasing the baryonic loading in equation (5) is compensated by a correspondingly smaller ϵ_e in equation (4). As a result, the neutrino flux is roughly independent of ϵ_p/ϵ_e—which we have explicitly tested numerically.

We have also tested that this prediction does not depend on the variability timescale of the burst: Fig. 7 shows predictions for two different values of the emitter uptime $\delta t_{eng} = 0.1$ s (a) and 1 s (b), where the fixed $N_{sh} = 1,000$ leads to a longer burst duration T. Clearly, the quasi-diffuse flux coming from the simulations is independent of the value of t_v, as expected from equation (4). This markedly contrasts with the standard numerical prediction, in which larger variability timescales unavoidably imply lower particle densities at the source and, therefore, a reduced neutrino production. In Fig. 7c, $\delta t_{eng} = 0.1$ s with a reduced number of shells $N_{sh} = 100$ is chosen, corresponding to the light curve in Fig. 2b, that has a similar duration as the light curve for our standard simulation, but fewer pulses. In this case, the obtained result depends on the actual realization of the Γ-distribution, as only a few collisions dominate the neutrino flux and lead to strong variations—six different realizations of the Γ-distribution are shown in Fig. 7c. As expected, our conclusions hold for a sufficiently large sample of GRBs centred around our average prediction. We expect that these fluctuations become more severe for even fewer pulses from fewer collisions, as it has been studied for the contributions from different bursts in section 4 of ref. 47. The independence on the model parameters implies that the predicted flux $E^2 J \sim 10^{-11}$ GeV cm^{-2} sr^{-1} s^{-1} is very robust.

The only exception may be increasing E_{iso} (see equation (5)) or the baryonic loading, which may in fact be required to match the injected energy needed to describe UHECR observations; see section 2 in ref. 34.

The photon spectra can still be approximated by the Band function up to a Thomson optical thickness of 10 or so[62,63], which occurs under the photosphere. This means that we can extrapolate our assumptions to below the photosphere to some degree. In the most extreme case, all energy may be dissipated into neutrinos, whereas the energy of neutrons, protons and gamma rays is reconverted into kinetic energy—this is, however, very speculative, as nonthermal particle acceleration may not occur efficiently[10]. We show the corresponding subphotospheric extrapolations for the neutrino spectra as highly uncertain shaded regions in Fig. 6, corresponding to the contribution to the black squares in Fig. 3. Since the photospheric radius increases with decreasing Γ, the number of subphotospheric collisions increases with it, and their contribution in Fig. 6a can be much higher than in Fig. 6b (and in Fig. 6c much lower). As a consequence, the subphotospheric extrapolation may even reach the current sensitivity limit, and can be already constrained with current data. However, note again that this extrapolation is highly uncertain, as gamma-ray data cannot be used to obtain information about below the photosphere.

Finally, we show the 'neutrino light curve' for our standard parameter set in Fig. 2a as a dotted (red) curve; Fig. 2b shows it for a simulation with fewer collisions and longer emitter uptime, corresponding to the neutrino spectra in Fig. 7c. It can be clearly seen that the neutrino flux is typically much lower than the gamma-ray flux, except in some rare cases where the collision occurs close to the photosphere. Furthermore, the variation of the neutrino flux is larger due to the strong dependence of the pion production efficiency on R_C. One qualitative prediction that could help neutrino searches is that neutrinos are more likely to be associated with gamma-ray spikes that are pulses with very short variability timescales.

Discussion

In summary, we have studied neutrino, gamma-ray (at different energies) and cosmic-ray production in an evolving GRB outflow. We have demonstrated that they are produced at different collision radii. Consequently, the typical emission radius derived from prompt gamma rays cannot be directly applied to neutrino and UHECR production, and the GRB will look very different

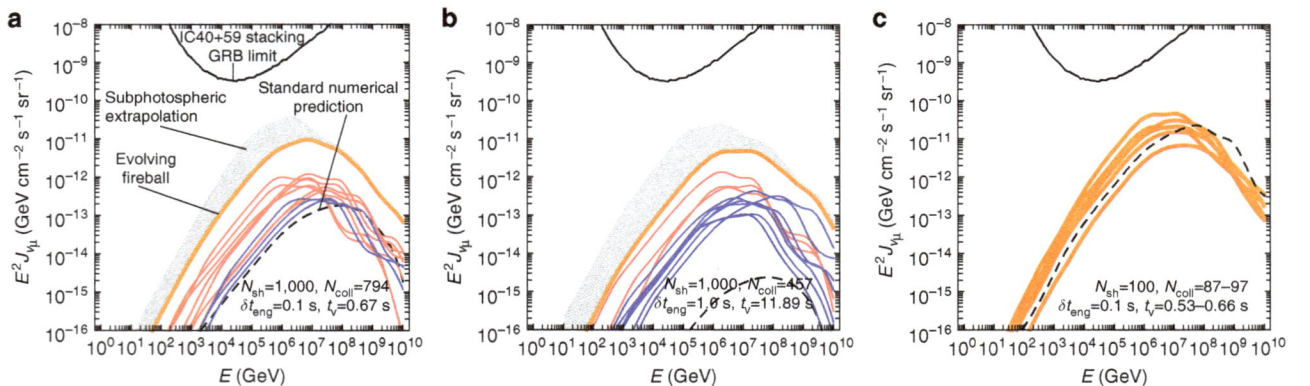

Figure 7 | Quasi-diffuse muon-neutrino spectra from simulations with alternative parameter sets. The $\nu_\mu + \bar{\nu}_\mu$ neutrino spectra in this figure should be compared with the ones obtained using our standard parameter set, Fig. 6b ($\delta t_{eng} = 0.01$ s, $N_{sh} = 1,000$ and $\Gamma_0 = 500$). (**a,b**) Two larger values of δt_{eng} for $N_{sh} = 1,000$, which leads to a longer burst duration T. (**c**) Using $\delta t_{eng} = 0.1$ s with a reduced number of shells $N_{sh} = 100$ (corresponding to the light curves in Fig. 2b), that is, T is similar to the original example, but the light curve is less spiky because of fewer collisions. Since in that case the statistical fluctuations from burst to burst increase, we show six different realizations of the predicted flux.

from the point of view of different messengers. This concept is well known from conventional astronomical observations, where astrophysical objects look very different in different wavelength bands.

The neutrino spectra derived from gamma-ray observations are dominated by the emission close to the photosphere at $R_C \approx 10^{8.5}$–10^9 km, as the pion production efficiency depends on the collision radius in a nonlinear way. UHECR protons have been shown to be produced at intermediate collision radii $R_C \approx 10^{8.5}$–10^{10} km, where the magnetic fields are high enough for efficient acceleration, but not so high that synchrotron losses limit the maximal proton energies. We have taken into account two possibilities for UHECR escape: emission as neutrons, which are not magnetically confined, and emission as protons from the edges of the shells—the dominant mechanism in each collision depends on the parameters of the colliding shells. Since the neutrons come from photohadronic interactions, their production dominates at smaller collision radii, where the $p\gamma$ optical depth is higher, whereas protons tend to be directly emitted at large radii. Heavier nuclei can also survive for sufficiently large collision radii; their actual contribution depends on the nuclear loading. The main energy in gamma rays is deposited between around 10^{10} and $10^{11.5}$ km, compatible with earlier estimates. In particular, gamma rays at the highest energies, such as in the energy range only accessible to CTA, cannot come from collision radii $\lesssim 10^9$ km as the photon densities are too high there to let them escape.

For the quasi-diffuse neutrino flux prediction, we have identified two distinctive contributions. Above the photosphere, gamma-ray observations can be used to infer the pion production efficiency, which leads to a neutrino flux per flavour $E^2 J \sim 10^{-11}$ GeV cm^{-2} sr^{-1} s^{-1} for an assumed isotropic energy of 10^{53} erg emitted in gamma rays. Especially, there is no strong dependence on the Lorentz boost Γ, in contrast to conventional one-zone models, as both the photosphere and the pion production efficiency scale with the collision radius in the same way. This is the minimal neutrino flux that one would expect in stacking analyses based on the actual gamma-ray observations, such as ref. 14. There is also a significantly milder dependence on the baryonic loading, as this parameter changes the photosphere of the model at the same time that it rescales the neutrino flux. The prediction hardly depends on the time variability or number of pulses in the GRB light curve within a certain time window either. However, if the overall number of pulses is low, these will only come from a very small number of collisions, which means that large statistical fluctuations of the neutrino flux from burst to burst are expected even for the same parameter values. In that case, our observations have to be instead interpreted for a large enough ensemble of bursts. Note that the chosen isotropic energy and baryonic loading may not be sufficient to describe UHECR observations, see section 2 of ref. 34 for a detailed discussion, which will need to be addressed in a future study.

The neutrino flux is significantly lower than earlier predictions[16] because (a) we have explicitly excluded subphotospheric contributions, (b) large photospheric radii have been obtained as a consequence of significant baryonic loadings (10) and the moderate energy dissipation efficiency of the fireball (25%) and (c) only a small number of collisions beyond the photosphere occurs at radii where the neutrino production efficiency is high. This expected 'minimal' flux is beyond the sensitivity of the current IceCube experiment, but could be reached in future high-energy extensions[64]. No gamma-ray information from deep below the photosphere can be directly obtained, and the neutrino production in that regime is more speculative[50]. In principle, however, a high-energy extension of the detector could also constrain the subphotospheric neutrino production.

Our results imply that model-dependent studies of the multi-messenger connection, such as a GRB stacking analysis of neutrino fluences, can be improved and give a stronger case for testing the hypothesis that UHECRs originate from GRBs. Compared with the one-zone model, some additional assumptions need to be made for the distribution of the collision radii. In particular, the width of the initial distribution of the bulk Lorentz factor A_Γ, with which the shells are set out by the central engine, turns out to be the key additional parameter. It can in principle be obtained from comparing the light curves between simulation and observation. On the other hand, we have the advantage that the uncertainty in R_C, which is the key issue in the standard model, disappears, as a collision radius distribution is now predicted by the theory. While we expect that the bulk Lorentz factor distribution has to be broad in some way to maintain a high dissipation efficiency, it remains to be studied how the results change for qualitatively different distributions. There should also be new opportunities stemming from our results: different messengers can be used to study different regions of an evolving GRB outflow. For instance, direct neutrino and gamma-ray observations, in CTA, of a single GRB would open windows to very different regions of the GRB.

During completion of this work, ref. 59 appeared, which shares some common aspects.

References

1. Piran, T. The physics of gamma-ray bursts. *Rev. Mod. Phys.* **76**, 1143–1210 (2004).
2. Mészáros, P. Gamma-Ray Bursts. *Rep. Prog. Phys.* **69**, 2259–2322 (2006).
3. Zhang, B. Gamma-ray bursts in the Swift era. *Chin. J. Astron. Astrophys.* **7**, 1–50 (2007).
4. Milgrom, M. & Usov, V. Possible association of ultrahigh-energy cosmic ray events with strong gamma-ray bursts. *Astrophys. J. Lett.* **449**, L37–L40 (1995).
5. Waxman, E. Cosmological gamma-ray bursts and the highest energy cosmic rays. *Phys. Rev. Lett.* **75**, 386–389 (1995).
6. Vietri, M. On the acceleration of ultrahigh-energy cosmic rays in gamma-ray bursts. *Astrophys. J.* **453**, 883–889 (1995).
7. Waxman, E. & Bahcall, J. N. High-energy neutrinos from cosmological gamma-ray burst fireballs. *Phys. Rev. Lett.* **78**, 2292–2295 (1997).
8. Aartsen, M. G. et al. (IceCube Collaboration). Evidence for high-energy extraterrestrial neutrinos at the IceCube detector. *Science* **342**, 1242856 (2013).
9. Murase, K. & Nagataki, S. High energy neutrino emission and neutrino background from gamma-ray bursts in the internal shock model. *Phys. Rev. D* **73**, 063002 (2006).
10. Murase, K. Prompt high-energy neutrinos from gamma-ray bursts in the photo-spheric and synchrotron self-compton scenarios. *Phys. Rev. D* **78**, 101302 (2008).
11. Baerwald, P., Hümmer, S. & Winter, W. Magnetic field and flavor effects on the gamma-ray burst neutrino flux. *Phys. Rev. D* **83**, 067303 (2011).
12. Zhang, B. & Kumar, P. Model-dependent high-energy neutrino flux from gamma-ray bursts. *Phys. Rev. Lett.* **110**, 121101 (2013).
13. Aartsen, M. G. et al. (IceCube Collaboration). First observation of PeV-energy neutrinos with IceCube. *Phys. Rev. Lett.* **111**, 021103 (2013).
14. Abbasi, R. et al. (IceCube Collaboration). An absence of neutrinos associated with cosmic-ray acceleration in γ-ray bursts. *Nature* **484**, 351–353 (2012).
15. Guetta, D., Hooper, D., Álvarez-Muñiz, J., Halzen, F. & Reuveni, E. Neutrinos from individual gamma-ray bursts in the BATSE catalog. *Astropart. Phys.* **20**, 429–455 (2004).
16. Hümmer, S., Baerwald, P. & Winter, W. Neutrino emission from gamma-ray burst fireballs, revised. *Phys. Rev. Lett.* **108**, 231101 (2012).
17. Li, Z. Note on the normalization of predicted GRB neutrino flux. *Phys. Rev. D* **85**, 027301 (2012).
18. He, H.-N. et al. Icecube non-detection of GRBs: constraints on the fireball properties. *Astrophys. J.* **752**, 29 (2012).
19. Halzen, F. & Hooper, D. High-energy neutrino astronomy: the cosmic ray connection. *Rep. Prog. Phys.* **65**, 1025–1078 (2002).
20. Piran, T. Gamma-ray bursts and the fireball model. *Phys. Rep.* **314**, 575–667 (1999).
21. Zou, Y.-C. & Piran, T. Lorentz factor constraint from the very early external shock of the gamma-ray burst ejecta. *Mon. Not. R. Astron. Soc.* **402**, 1854–1862 (2010).
22. Nakar, E. & Piran, T. Time scales in long GRBs. *Mon. Not. R. Astron. Soc.* **331**, 40–44 (2002).
23. Bhat, P. N. Variability time scales of long and short GRBs, Proceedings of the 7th Huntsville Gamma-Ray Burst Symposium (GRB 2013), 14–18 April, 2013, Nashville, Tennessee. Preprint at http://arxiv.org/abs/1307.7618 (2013).

24. Nakar, E. & Piran, T. Gamma-ray burst light curves—another clue on the inner engine. *Astrophys. J. Lett.* **572**, L139–L142 (2002).
25. Murase, K., Ioka, K., Nagataki, S. & Nakamura, T. High-energy cosmic-ray nuclei from high- and low-luminosity gamma-ray bursts and implications for multi-messenger astronomy. *Phys. Rev. D* **78**, 023005 (2008).
26. Wang, X.-Y. & Dai, Z.-G. Prompt TeV neutrinos from dissipative photospheres of gamma-ray bursts. *Astrophys. J. Lett.* **691**, L67–L71 (2009).
27. Murase, K., Kashiyama, K. & Mészáros, P. Subphotospheric neutrinos from gamma-ray bursts: the role of neutrons. *Phys. Rev. Lett.* **111**, 131102 (2013).
28. Bartos, I., Beloborodov, A. M., Hurley, K. & Márka, S. Detection prospects for GeV neutrinos from collisionally heated gamma-ray bursts with IceCube/DeepCore. *Phys. Rev. Lett.* **110**, 241101 (2013).
29. Gao, S., Asano, K. & Mészáros, P. High energy neutrinos from dissipative photospheric models of gamma ray bursts. *J. Cosmol. Astropart. Phys.* **1211**, 058 (2012).
30. Rees, M. J. & Mészáros, P. Dissipative photosphere models of gamma-ray bursts and x-ray flashes. *Astrophys. J.* **628**, 847–852 (2005).
31. Baerwald, P., Bustamante, M. & Winter, W. UHECR escape mechanisms for protons and neutrons from GRBs, and the cosmic ray-neutrino connection. *Astrophys. J.* **768**, 186 (2013).
32. Rachen, J. P. & Mészáros, P. Photohadronic neutrinos from transients in astrophysical sources. *Phys. Rev. D* **58**, 123005 (1998).
33. Mannheim, K., Protheroe, R. J. & Rachen, J. P. On the cosmic ray bound for models of extragalactic neutrino production. *Phys. Rev. D* **63**, 023003 (2001).
34. Baerwald, P., Bustamante, M. & Winter, W. Are gamma-ray bursts the sources of ultra-high energy cosmic rays? *Astropart. Phys.* **62**, 66–91 (2015).
35. Ahlers, M., González-García, M. C. & Halzen, F. GRBs on probation: testing the UHE CR paradigm with IceCube. *Astropart. Phys.* **35**, 87–94 (2011).
36. Guetta, D., Spada, M. & Waxman, E. On the neutrino flux from gamma-ray bursts. *Astrophys. J.* **559**, 101–109 (2001).
37. Kobayashi, S., Piran, T. & Sari, R. Can internal shocks produce the variability in GRBs? *Astrophys. J.* **490**, 92–98 (1997).
38. Kobayashi, S. & Sari, R. Ultra efficient internal shocks. *Astrophys. J.* **551**, 934–939 (2001).
39. Aoi, J., Murase, K., Takahashi, K., Ioka, K. & Na-gataki, S. Can we probe the Lorentz factor of gamma-ray bursts from GeV-TeV spectra integrated over internal shocks? *Astrophys. J.* **722**, 440–451 (2010).
40. Beloborodov, A. M. On the efficiency of internal shocks in gamma-ray bursts. *Astrophys. J. Lett.* **539**, L25–L28 (2000).
41. Li, Z. Prompt GeV emission from residual collisions in GRB outflows. *Astrophys. J.* **709**, 525–534 (2010).
42. Kumar, P. Gamma-ray burst energetics. *Astrophys. J. Lett.* **523**, L113–L116 (1999).
43. Zhang, B. & Yan, H. The internal-collision-induced magnetic reconnection and turbulence (ICMART) model of gamma-ray bursts. *Astrophys. J.* **726**, 90 (2011).
44. Daigne, F. & Mochkovitch, R. Gamma-ray bursts from internal shocks in a relativistic wind: temporal and spectral properties. *Mon. Not. R. Astron. Soc.* **296**, 275–286 (1998).
45. Daigne, F. & Mochkovitch, R. The physics of pulses in gamma-ray bursts: emission processes, temporal profiles and time lags. *Mon. Not. R. Astron. Soc.* **342**, 587–592 (2003).
46. Bosnjak, Z., Daigne, F. & Dubus, G. Prompt high-energy emission from gamma-ray bursts in the internal shock model. *Astron. Astrophys.* **498**, 677–703 (2008).
47. Baerwald, P., Hümmer, S. & Winter, W. Systematics in the interpretation of aggregated neutrino flux limits and flavor ratios from gamma-ray bursts. *Astropart. Phys.* **35**, 508–529 (2012).
48. Asano, K. & Mészáros, P. Neutrino and cosmic-ray release from gamma-ray bursts: time-dependent simulations. *Astrophys. J.* **785**, 54 (2014).
49. Rees, M. J. & Mészáros, P. Relativistic fireballs—energy conversion and time-scales. *Mon. Not. R. Astron. Soc.* **258**, 41–43 (1992).
50. Murase, K. & Ioka, K. TeV-PeV neutrinos from low-power gamma-ray burst jets inside stars. *Phys. Rev. Lett.* **111**, 121102 (2013).
51. Abraham, J. *et al.*(Pierre Auger Observatory Collaboration). Measurement of the depth of maximum of extensive air showers above 10^{18} eV. *Phys. Rev. Lett.* **104**, 091101 (2010).
52. Lemoine, M. Nucleosynthesis in gamma-ray bursts outflows. *Astron. Astrophys.* **390**, L31–L34 (2002).
53. Pruet, J., Guiles, S. & Fuller, G. M. Light element synthesis in high entropy relativistic flows associated with gamma-ray bursts. *Astrophys. J.* **580**, 368–373 (2002).
54. Beloborodov, A. M. Nuclear composition of gamma-ray burst fireballs. *Astrophys. J.* **588**, 931–944 (2003).
55. Anchordoqui, L. A., Hooper, D., Sarkar, S. & Taylor, A. M. High-energy neutrinos from astrophysical accelerators of cosmic ray nuclei. *Astropart. Phys.* **29**, 1–13 (2008).
56. Metzger, B. D., Giannios, D. & Horiuchi, S. Heavy nuclei synthesized in gamma-ray burst outflows as the source of UHECRs. *Mon. Not. R. Astron. Soc.* **415**, 2495–2504 (2011).
57. Horiuchi, S., Murase, K., Ioka, K. & Mészáros, P. The survival of nuclei in jets associated with core-collapse supernovae and gamma-ray bursts. *Astrophys. J.* **753**, 69 (2012).
58. Wang, X.-Y., Razzaque, S. & Mészáros, P. On the origin and survival of UHE cosmic-ray nuclei in GRBs and hypernovae. *Astrophys. J.* **677**, 432–440 (2008).
59. Globus, N., Allard, D., Mochkovitch, R. & Parizot, E. UHECR acceleration at GRB internal shocks, Preprint at http://arxiv.org/abs/1409.1271 (2014).
60. Aartsen, M. G. *et al.* (IceCube Collaboration). Observation of high-energy astrophysical neutrinos in three years of IceCube data. *Phys. Rev. Lett.* **113**, 101101 (2014).
61. Abbasi, R. *et al.* (IceCube Collaboration). Limits on neutrino emission from gamma-ray bursts with the 40 string IceCube detector. *Phys. Rev. Lett.* **106**, 141101 (2011).
62. Pe'er, A. Temporal evolution of thermal emission from relativistically expanding plasma. *Astrophys. J.* **682**, 463–473 (2008).
63. Beloborodov, A. M. Radiative transfer in ultra-relativistic outflows. *Astrophys. J.* **737**, 68 (2011).
64. Aartsen, M. G. *et al.* (IceCube Collaboration). IceCube-Gen2: a vision for the future of neutrino astronomy in Antarctica, Preprint at http://arxiv.org/abs/1412.5106 (2014).

Acknowledgements
We thank V. Mangano, E. Waxman and B. Zhang for discussion and comments. This work is supported by NASA through Hubble Fellowship Grant No. 51310.01 awarded by the STScI, which is operated by the Association of Universities for Research in Astronomy Inc., for NASA, under Contract No. NAS 5-26555 (K.M.). M.B. and W.W. would also like to acknowledge support from DFG grants WI 2639/3-1 and WI 2639/4-1, and the 'Helmholtz Alliance for Astroparticle Physics HAP', funded by the Initiative and Networking Fund of the Helmholtz Association. P.B. acknowledges support from NASA grant NNX13AH50G. M.B., K.M. and W.W. would like to thank the Kavli Institute for Theoretical Physics at UCSB for its hospitality during the development of part of this work. This research was supported in part by the National Science Foundation under Grant No. NSF PHY11-25915.

Author contributions
All authors contributed to all aspects of this work, discussed the results and commented on the manuscript.

Additional information
Competing financial interests: The authors declare no competing financial interests.

11

Full-Sun observations for identifying the source of the slow solar wind

David H. Brooks[1,†], Ignacio Ugarte-Urra[1] & Harry P. Warren[2]

Fast ($>700\,\mathrm{km\,s^{-1}}$) and slow ($\sim400\,\mathrm{km\,s^{-1}}$) winds stream from the Sun, permeate the heliosphere and influence the near-Earth environment. While the fast wind is known to emanate primarily from polar coronal holes, the source of the slow wind remains unknown. Here we identify possible sites of origin using a slow solar wind source map of the entire Sun, which we construct from specially designed, full-disk observations from the Hinode satellite, and a magnetic field model. Our map provides a full-Sun observation that combines three key ingredients for identifying the sources: velocity, plasma composition and magnetic topology and shows them as solar wind composition plasma outflowing on open magnetic field lines. The area coverage of the identified sources is large enough that the sum of their mass contributions can explain a significant fraction of the mass loss rate of the solar wind.

[1] College of Science, George Mason University, 4400 University Drive, Fairfax, Virginia 22030, USA. [2] Space Science Division, Naval Research Laboratory, 4555 Overlook Avenue SW, Washington, District Of Columbia 20375, USA. † Present address: Hinode Team, ISAS/JAXA, 3-1-1 Yoshinodai, Chuo-ku, Sagamihara, Kanagawa 252-5210, Japan. Correspondence and requests for materials should be addressed to D.H.B. (email: dhbrooks@ssd5.nrl.navy.mil).

Understanding the flow of energy and matter throughout the solar system is a fundamental goal of heliophysics, and identifying the solar sources of this flow would be a major step forward in achieving that objective. It would allow us to determine the physical properties of the plasma in the source regions, a significant constraint for theoretical models. Models of the solar wind, for example, are very sensitive to boundary conditions at the site of origin[1].

The solar wind is comprised of a fast and a slow component[2–4], both of which interact with and affect Earth's magnetic environment[5]. The origin of the fast wind is generally well established[6,7], but there is still no consensus on the source of the slow wind. Many sources have been suggested, from helmet streamers[8] and 'blobs' disconnecting from their cusps[9], to equatorial coronal holes[10], active regions[10,11] and their boundaries[12] or chromospheric jets[13,14]. Narrow open-field corridors that connect coronal holes of the same polarity have also been proposed theoretically[15]. Recently, observations from the Hinode satellite have also identified specific outflow sites at the edges of active regions[16–18].

Unfortunately, the two-component velocity structure of the solar wind cannot be used to distinguish sources low down in the solar atmosphere. The theoretically predicted[3], and observationally confirmed[2], high velocities of the wind at Earth are not observed in the low corona[19,20], and the wind does not reach supersonic velocities until more than a solar radius above the Sun[21], with acceleration not complete until at least 10 solar radii[22]. So, the acceleration to high velocities must take place at larger heights[23]. An important clue to the origin of the slow solar wind, however, is that the plasma composition (elemental abundance) is similar to that of the solar corona[24], rather than to the solar photosphere, and this is a difference that can be exploited. The composition of the corona is enhanced with low first ionization potential (FIP) elements relative to the photosphere, and the degree of enhancement (fractionation) can be measured using the intensities of spectral lines from elements with different FIP, for example, Si (low FIP) and S (high FIP). Spectroscopic measurements of plasma composition from the Hinode EUV imaging spectrometer (EIS) have therefore become a valuable tool for attempting to establish a link between these candidate source regions on the Sun and in situ measurements at Earth, but this has only been achieved for the active region outflows and only for one active region observed in December 2007 (refs 25,26).

A significant problem with these measurements, however, is that the slow scanning time of spectrometers permits only limited field-of-view coverage, and so the presence of other, unobserved sources on the Sun at the same time cannot be ruled out. Estimates of the mass loss rate associated with the active region outflows, for example, suggest that they could account for 1/4 of the mass loss rate of the solar wind (10^{12} g s^{-1}; ref. 16), but studies of other regions have shown that some portion of the mass may flow along large-scale closed loops and return to the surface in the vicinity of distant active regions[27], suggesting that the estimates may be too large. More recent measurements of velocities and densities in fact suggest that the mass loss rate may be overestimated by as much as an order of magnitude[25,28]. So, even if individual outflow regions do contribute to the solar wind, the mass loss deficit needs to be made up from elsewhere on the Sun, implying that other sources are likely.

During 16–18 January 2013, we overcame this shortcoming using a new EIS-observing programme that scanned the entire Sun over a 48-h period, thus allowing us to map the whole disk to look for candidate sources. Despite being near solar maximum, the Sun was relatively quiescent during the scan, with only two flares that reached higher than Geostationary Operational

Environmental Satellite (GOES) C-class according to the Hinode flare catalogue[29]. One of these was associated with a (partial) halo coronal mass ejection (CME) that caused a temporary increase in the GOES proton flux, but both of the events occurred near the solar limb and were located far from where EIS was scanning at that time. The scan is based on a full-disk mosaic programme that we run every 3–4 weeks as part of efforts to monitor the instrument sensitivity in direct comparison with the Extreme Ultraviolet Variability Experiment on the Solar Dynamics Observatory (SDO). We execute a specially designed observing sequence at 26 positions on the solar disk by re-pointing the spacecraft to 15 positions and performing the scan using the top and bottom of the EIS CCD (charge-coupled device) as needed to cover the whole Sun. The programme takes a few hours to complete using the 40″ (arcsecond) slit. It is ordinarily not practical to run a similar sequence using the full spectral resolution EIS slits, however, because the scan would take several days and consume a large amount of telemetry.

For the purpose of this study, however, we designed a new EIS-observing sequence that matches the field of view of the regular scan (492″ by 512″) using the 2″ slit and coarse 4″ steps. The programme includes a series of Fe lines that we can use to measure the density and emission measure (EM—the distribution of plasma as a function of temperature). The specific Fe lines used are: Fe VIII 185.213 Å, Fe IX 188.497 Å, Fe X 184.536 Å, Fe XI 188.216 Å, Fe XI 188.299 Å, Fe XII 195.119 Å, Fe XII 203.72 Å, Fe XIII 202.044 Å, Fe XIII 203.826 Å, Fe XIV 264.787 Å, Fe XV 284.16 Å and Fe XVI 262.984 Å. The ratio of the Fe XIII lines at 202.044 Å and 203.826 Å is sensitive to electron density. The line list also includes Si X 258.37 Å and S X 264.22 Å that we can use to make abundance measurements. Their ratio is sensitive to the degree of fractionation of the plasma, when convolved with the EM distribution derived from the Fe VIII-XVI lines listed above. We and others have assessed the reliability of the line list extensively in studies of several different solar features[30–33].

From our observations, we derive pure temperature images of the full Sun at the highest spatial resolution yet achieved, and Doppler velocity maps of the corona extending to higher temperatures than previously possible. We also compute the first plasma composition map of the entire Sun. By combining these observations with a magnetic field extrapolation model, we construct a unique slow-wind source map.

Results

Full-Sun images and plasma composition. Figure 1 shows pure temperature images derived from spectral fits to the full-Sun data for a selection of the lines acquired by our observing programme. These images cover a range of temperatures from 450,000 K (0.45 MK) to 2,800,000 K (2.8 MK). Although similar images to Fig. 1 are routinely produced by the SDO Atmospheric Imaging Assembly (AIA), and instruments on other spacecraft, for a subset of these wavelengths, they are broad-band images with a spectral width of at least a few Å, which is significantly worse than the 0.0223 Å spectroscopic resolution of the images we show in Fig. 1, and thus contain multiple contributions from many different temperature spectral lines[34]. Spectrally pure temperature images with a broad temperature coverage are necessary to reduce the uncertainties associated with the integral inversion techniques that allow us to convert the line intensities into properties of the emitting plasma, such as electron density, EM or the degree to which the plasma composition is enhanced. We prepared coaligned full-Sun intensity images like that in Fig. 1 for all the spectral lines used in our analysis. Figure 2 shows an expanded

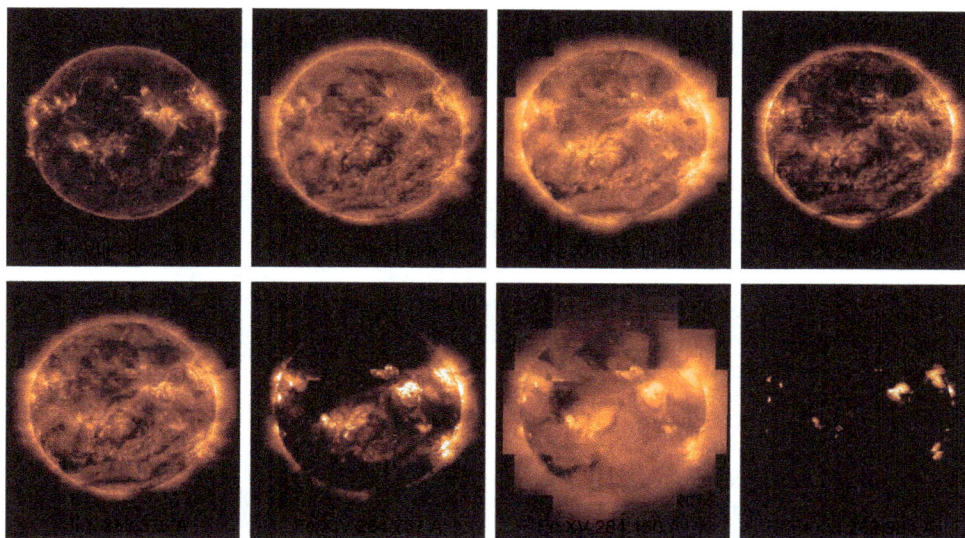

Figure 1 | EIS images of the solar corona. These pure temperature images are constructed from EIS spectral line intensities and cover a range of temperatures from 0.45 to 2.8 MK. They are used to construct EM distributions at every pixel. The mosaic is constructed by recording the top and bottom readings of the EIS CCD at 26 positions on the solar disk. The observing sequence scans across the central disk, around the North limb and finally around the South limb. It is sheared East-West in some locations because of solar rotation during the 2-day spectrometer scan. Details of the observing sequence are given in the main text.

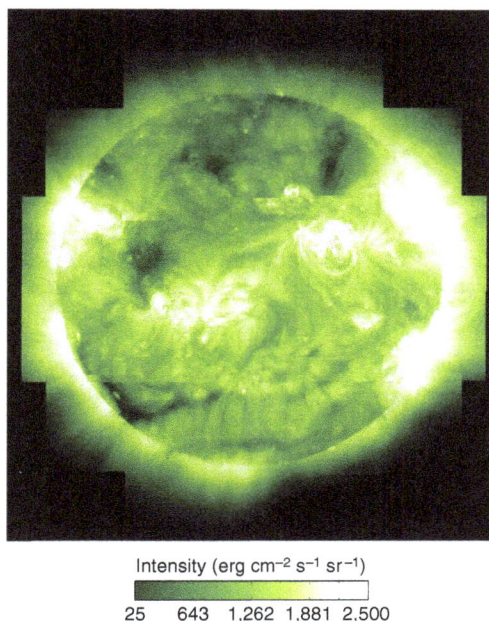

Intensity (erg cm^{-2} s^{-1} sr^{-1})

25   643   1,262 1,881 2,500

Figure 2 | EIS image of the solar corona at 2 MK. An expanded image of the solar corona at 2 MK constructed from Fe XIII 202.044 Å spectral line intensities. This line was used to determine Doppler velocities, and its ratio with Fe XIII 203.83 Å was used to compute electron densities. It is formed close in temperature to the Si X 258.37 Å/S X 264.22 Å ratio used to measure elemental abundances.

image for the Fe XIII 202.044 Å line formed at 2 MK. We draw attention to this image because it is central to our analysis.

The ratio of the intensities of spectral lines from low FIP and high FIP elements can be used to calculate their relative abundances and thus the plasma composition for our full-Sun map. Previous studies[25,35] have examined the abundance diagnostics in the EIS spectra and concluded that the Si X

258.37 Å/S X 264.22 Å ratio is one of the best. Compared with the other available ratios, it is relatively insensitive to the electron temperature and density. The variation is 30–40% in the temperature region where the lines are formed (around 1.4 MK). There is, however, a strong variation at high temperatures, and a factor of 2.3 sensitivity to density in the log $n = 8$–10 range. We therefore need to measure the density to account for that sensitivity, and convolve the ratio with the EM distribution to account for any significant high-temperature emission. Following our previous work[25], we performed these calculations for every pixel in the full-Sun data set by first deriving the electron density using the Fe XIII 202.04/203.83 ratio and then using that density to compute contribution functions (the equivalent of an imager filter's temperature response) for all the spectral lines in our observing programme. We used the CHIANTI database v.7 (refs 36,37) assuming a photospheric composition for the plasma[38]. We then fit the observed intensities by convolving them with an EM distribution derived from a Monte Carlo simulation. The Monte Carlo code is available in the PINTofALE software package[39,40] and uses a Markov-Chain algorithm to find the best-fit solution. Only the Fe lines were used for the EM calculation to minimize any uncertainties due to elemental abundances. Since Fe and Si are low FIP elements, their abundances (and hence intensities) are expected to be enhanced in the corona to a similar degree. Most of the Fe lines used for the EM analysis, however, lie on the short-wavelength (SW) EIS detector, whereas the Si X 258.37 Å and S X 264.22 Å lines lie on the long-wavelength detector. So, uncertainties in the cross-calibration of the detectors, and their evolution with time[41,42], could lead to a mismatch between the Fe EM and Si X 258.37 Å absolute intensity. We therefore scaled the derived EM distributions to ensure that the Si X 258.37 Å line is reproduced. This procedure also accounts for any uncertainties in the Fe/Si abundance. In our previous work[25], we found that this scaling was always < 20%, but that study used observations from early in the mission. So, we checked whether the method accounts for any sensitivity evolution in these more recent data by examining an area of one of the rasters where many of the pixels require larger scaling and re-calibrating the line intensities using two different methods that attempt to account for the sensitivity changes[41,42]. Our experiment verified that the scaling was reduced to under 30%, which is comparable to the accuracy of the method (see Methods section). Using the derived distribution, we then calculated the degree to which the plasma is fractionated by computing the expected intensity of the S X 264.22 Å line. The ratio of the predicted to observed intensity for this line gives us our level of fractionation compared with photospheric values.

Given the size of the full-Sun data set, 16 million calculations were needed to produce a plasma composition map including every pixel over the full Sun. We show a display version of this map in Fig. 3, created from the ratio of Si X 258.37 Å and S X 264.22 Å lines. This image captures the main features of the composition map, such as whether a structure has photospheric or coronal abundances and the relative level of enhancement, but was not used for any of our analysis. We show this map for presentation because it is only affected by bad pixels or missing data in the Si X and S X lines, and this is relatively easier to filter out. The full composition map, however, is affected by bad pixels and missing data in all of the spectral lines from Fe VIII to Fe XVI and is therefore much noisier and more difficult to interpret visually. We stress that we used the full composition map created using all the spectral lines for all of the quantitative analysis.

We then filtered the full composition map to define areas with an enhanced (slow wind) composition. We used an enhancement threshold of 60% to include the entire range of fractionation values, which accounts for the fact that the slow wind has a variable composition[24]. It also attempts to account for the fact that the Si/S ratio does not always show a clear fractionation pattern. Although the EIS observations show that the Si/S ratio can detect variations in composition between, for example, polar coronal holes (photospheric composition) and active region outflows (coronal composition)[25], the two elements lie close to the traditionally defined boundary between low and high

FIP elements, and some models[43] suggest that the ratio may underestimate the enhancement factor due to possible under-fractionation of Si and overfractionation of S. Here we only use the measurement to determine whether the plasma in a pixel is fractionated; the actual enhancement factor itself is not used in any computations. We stress, however, that the EIS composition measurements are in agreement with the general trends seen in the *in situ* data.

Full-Sun velocity map. Regions of enhanced composition could be possible sources of the slow speed wind, but our full-Sun composition map alone cannot show us whether the plasma from these regions is actually upflowing from the solar atmosphere. We obtained this information from the Doppler shift of the spectral line centroids, and derived radial velocity maps for several of the lines in our observations.

The EIS spectrometer has several peculiar characteristics that make Doppler velocity measurements difficult. For example, EIS does not observe any photospheric spectral lines, so it is not possible to obtain an absolutely calibrated wavelength scale. The Doppler velocities we use are therefore relative velocity measurements. Furthermore, the EIS slit is not perfectly aligned to the vertical axis of the CCD and there is a drift of the spectrum on the CCD due to thermal variations in the instrument around the satellite orbit. These effects have been extensively investigated and we accounted for them using our best current knowledge of the instrument[28,44]. Most of the instrumental effects are corrected using the recommended neural network model[44]. This model uses the Fe XII 195.12 Å line as calibration standard and assumes that velocities in this line are 0 when averaged over the entire mission. There is some evidence that this assumption is not accurate[45] and that coronal lines may exhibit blue shifts of a few km s^{-1}. Therefore, some care needs to be taken when choosing a reference wavelength to calibrate the Doppler velocities. Here we refined the velocity measurements by correcting to an off-limb reference wavelength for the Fe XII 195.12 Å line[28]. This reference wavelength was obtained by averaging the line profiles measured in two large quiet regions above the East and West limbs where the spectral line is expected to be close to its rest (or slightly blue shifted) wavelength. The wavelength scale was then shifted to the reference wavelength, and the correction was then applied to the strong Fe XIII 202.04 Å line that is within the same wavelength band. The final velocity measurements have uncertainties of ~4.5 km s^{-1}, and they are converted to radial velocities using a simple cos θ expression. Here we only use radial Doppler velocities calculated from the Fe XIII 202.04 Å data.

We show an example velocity map in Fig. 4. It shows regions of plasma upflow and we can compare their locations with features in the intensity image in Fig. 2. We see, for example, that the bright active region in the North West hemisphere (AR11654) has large areas of upflow, but mostly on the solar Eastern side, suggesting that we can rule out the red-shifted downflow areas in and around the active region (AR) core as a solar wind source (in a direct sense).

Open/closed magnetic field model. The velocity and composition maps reveal the locations of slow-wind composition plasma that is upflowing, but the magnetic field topology is key to determining whether these upflows become outflows, really escape into interplanetary space and are directed towards the ecliptic plane where they can be measured *in situ*. No direct open magnetic field channel, for example, has been established for the December 2007 region[46], and other cases where the magnetic topology has been inferred show that not all of the upflows can escape on open field[47]. Therefore, we also generated a full-Sun

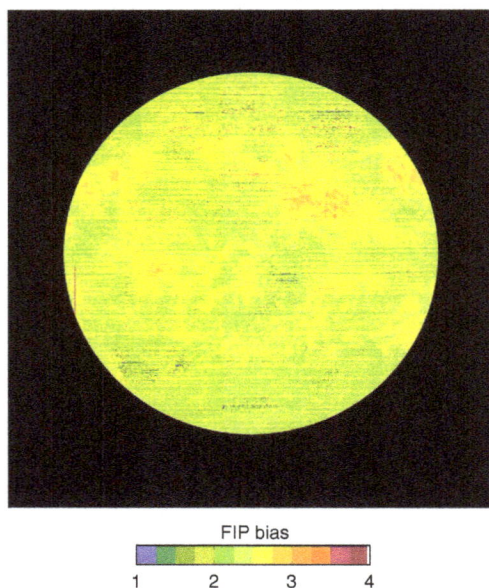

FIP bias

1   2   3   4

Figure 3 | EIS plasma composition map. Display version of the full-sun plasma composition map created from the ratio of the Si X 258.37 Å and S X 264.22 Å spectral lines. Darker areas correspond to regions with photospheric abundances. Lighter areas correspond to regions with enhanced (coronal) abundances. To reduce noise, we treated the map using a Fast Fourier Transform filtered by a Hanning mask, and excluded bad pixels and regions outside the solar limb. All of the analysis was performed on the untreated data[60].

potential field source surface (PFSS) extrapolation[48,49] to add this final piece to the puzzle.

The PFSS approximation has significant limitations. In particular, the corona is unlikely to be free of electric currents, and these alter both the strength and connectivity of the magnetic field compared with a potential configuration. Conversely, the model has been relatively successful in capturing the large-scale coronal field[10,49,50], which is the objective here, and we only use it to determine whether a field line is open and whether it extends

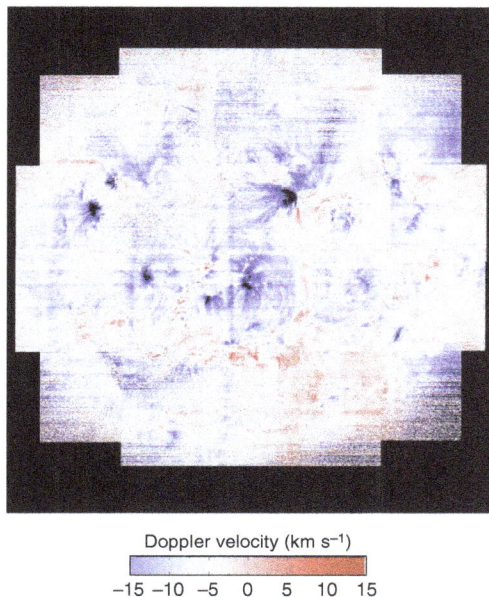

Doppler velocity (km s⁻¹)

-15 -10 -5 0 5 10 15

Figure 4 | EIS Doppler velocity map of the corona at 2 MK. Full-Sun coronal Doppler velocity map derived from single Gaussian fits to the Fe XIII 202.044 Å spectral line. Blue areas highlight plasma that is flowing towards the observer. Red areas highlight plasma that is flowing away from the observer. Vertical artefacts result from the thermal orbital variation of the spectra and should be ignored. The image is scaled to within ±15 km s⁻¹. We discuss the details of the velocity derivation method in the main text.

down towards the ecliptic plane. Our view is that the most likely shortcoming of our use of the PFSS model is that we may miss field lines that bend more dramatically towards the ecliptic. Definitive confirmation of our results will come from applying more sophisticated magnetic field models in the future.

We used the PFSS package available in SolarSoft[49]. This package allows access to a database of samples of potential field models (at 6 h cadence), constructed from Helioseismic and Magnetic Imager magnetogram observations, for any heliographic latitude and Carrington longitude. Field lines are then traced out from these locations until they either close back onto the Sun or open out to reach the source surface where they are forced to be radial. We extrapolated magnetic field lines from each of the EIS coordinates, corresponding to every pixel, using the nearest sample to the time of the centre of each EIS raster scan. This ensures that the extrapolation is always made from a magnetic field model sampled within 3.5 h (often much less), since the PFSS model generally does not evolve significantly during that time-frame. We then converted the EIS solar coordinates in arcseconds to heliographic coordinates, corrected for the solar B angle, and finally converted to Carrington angles.

We computed a total of ∼1.6 million potential field lines from this model to cover the full Sun. We show a subset of these field lines in Fig. 5, overlaid on the intensity images in Fig. 1, and again in Fig. 6, overlaid on the velocity map in Fig. 4. Figure 5 clearly shows that many of the open-field lines are associated with active regions, and Fig. 6 shows, for example, that the Eastern outflow from AR11654 does indeed lie on open magnetic field lines, some of which extend down into the ecliptic plane.

From the PFSS extrapolation, we determined which magnetic field lines reach the source surface, and are therefore open, for every EIS pixel in the data set. These data can then be mapped to our Doppler velocity and plasma composition maps to find potential solar wind sources.

Slow solar wind sources and mass loss rate. We combined all this information to produce our solar wind source map, adopting a number of criteria to decide whether a pixel should be counted as a candidate slow-wind source and ultimately included in our

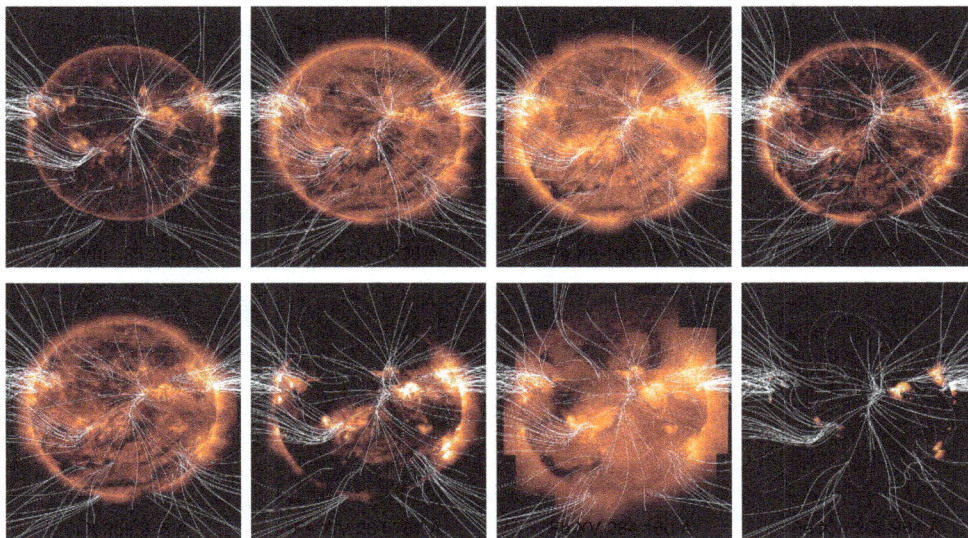

Figure 5 | EIS images of the solar corona with magnetic field lines overlaid. We have overlaid magnetic field lines from our PFSS calculation on the full-Sun intensity images in Fig. 1 that cover a broad range of temperatures from 0.45 to 2.8 MK. The solid field lines are open and the dotted field lines are closed. Only a small subset of the total number of field lines we computed are shown. The extrapolations appear different because the subset was chosen randomly for each intensity image. We have drawn relatively more open-field lines for emphasis.

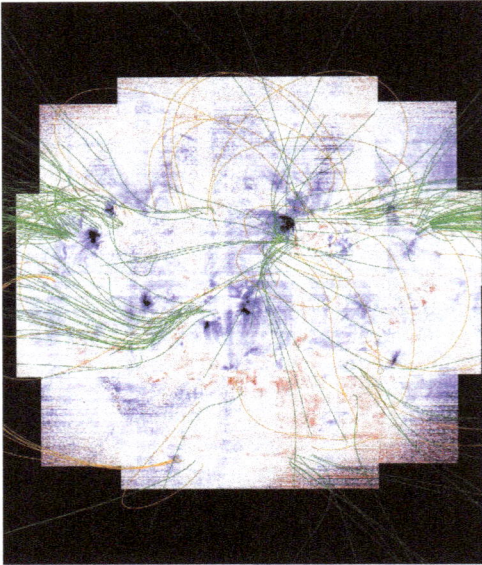

Figure 6 | EIS Doppler velocity map of the solar corona at 2 MK with magnetic field lines overlaid. Overlay of magnetic field lines from our PFSS calculation on the Fe XIII 202.044 Å Doppler velocity map in Fig. 4. This time, the green field lines are open and the orange field lines are closed. Again, only a small subset of the total number of field lines we computed are shown: 287 in this case. As with Fig. 5, the field lines are selected randomly and we have drawn relatively more open-field lines for emphasis.

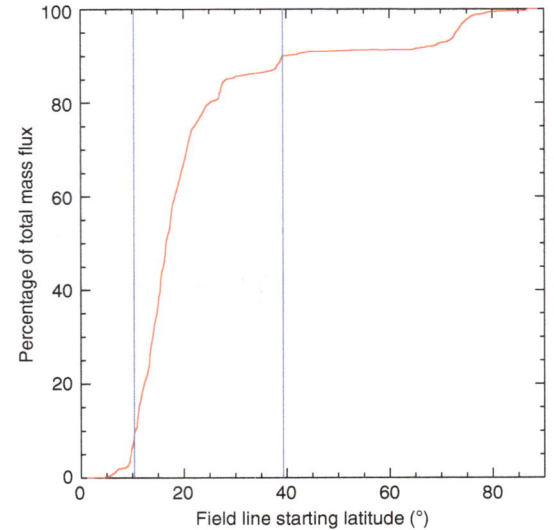

Figure 7 | Relationship between total mass flux and magnetic field line-starting latitude. Percentage of total mass flux as a function of magnetic field line-starting latitude (red line). Most of the mass flux (90%) comes from field lines that originate from below 40° latitude (marked by the vertical blue line on the right hand side). The vertical line on the left hand side indicates that most of this flux comes from field lines that originate above a starting latitude of ~11°.

mass loss rate calculation (see below). Some of these are purely technical. For example, we only included pixels if the numerical calculation of the EM distribution was well constrained. We set the condition that the χ^2 should be no larger than the number of lines in the integral inversion. This ensures that the difference between the calculated and observed intensities is generally within the calibration uncertainty. Since the EM calculation depends on the density, we excluded values well outside the range of sensitivity of the ratio ($8 < \log n < 11$). We also excluded pixels with a poor spectral fit to the Fe XIII 202.04 Å line.

A number of other criteria were also used. First, we only included pixels where the PFSS field line trace reaches the source surface, that is, the magnetic field is open. Second, as we show in Fig. 7, 90% of the mass flux comes from pixels below ~40°, so we only included pixels whose traced field line originated from below this latitude. Practically speaking, none of the field lines from above this latitude influence our study because they do not extend down close to the ecliptic plane, but it is unclear how close the other field lines must reach to be able to deliver mass flux to the ecliptic plane that can later be observed *in situ*. Most of the mass flux originates from above 11°, however (Fig. 7), which implies that some field lines from these latitudes should not be excluded. So we set this as the threshold, but it is clearly dependent on the model, which is of course simplistic. Third, we only included pixels within the solar radius on January 17 (midway through the scan). Pixels high above the limb do correspond to locations in the magnetic field models and a field line can be traced from them. But those locations rotate over the limb, not radially off-limb, so the back projection and radial velocity correction have increasing uncertainty close to and above the limb. Fourth, we assumed that the plasma was fractionated if the enhancement above photospheric levels was >60%; this is well above the radiometric calibration uncertainty (~23%; ref. 51).

The fractionation measurements also depend on updates to the radiometric calibration. Using an alternative re-calibration of the sensitivity evolution of the instrument since 2006 (ref. 41), we

calculated that <5% of the pixels would change by more than this amount. Finally, given the uncertainties in the radial Doppler velocities (4.5 km s^{-1}), we have to make a careful choice of velocity threshold to decide whether the plasma in a pixel is upflowing or not. There are two possible approaches: assume (1) that any motions along the line of sight average to 0 so that the coronal lines are at rest or (2) that they have a blue shift[45] of a few km s^{-1}. In case (1), the mass flux would be underestimated if they actually have a small blue shift because fewer pixels would be included, while case (2) would overestimate the mass flux if they are at rest because more pixels would be included and their velocities would also be larger. Both cases will underestimate the mass flux if we exclude pixels with velocities below 4.5 km s^{-1}. Fortunately, the identified source regions are not particularly sensitive to this choice: the main effect is that they become more extended and/or denser due to more pixels being included, but the mass flux calculation itself is significantly affected, as we discuss below.

We show the final map in Fig. 8. The red and green areas show regions where enhanced composition plasma is outflowing on open-field lines that extend down close to the ecliptic plane, and these areas meet all of the criteria that, in our view, make them possible sources of the slow-speed solar wind. Their area coverage is at least 50 times greater than the outflow area estimate by Sakao *et al.*[16], which is more than enough to overturn the lower density and velocity measurements found recently for active region outflows when calculating the mass loss rate contribution to the slow wind. We calculated the total mass loss rate for our candidate sources using the formula

$$ M = \sum_{i=1}^{N} m_{\mathrm{p}} n_i v_i l^2 $$

where m_{p} is the proton mass, n_i is the electron density of pixel i, v_i is the radial velocity of pixel i, l^2 is the area of an EIS pixel and N is the total number of pixels that meet all of our selection criteria. Using the Fe XIII densities and velocities, we measured the mass

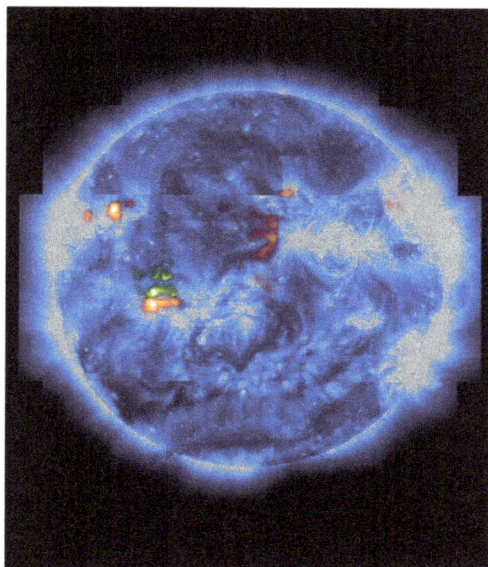

Figure 8 | Slow solar wind source map. The sources are overlaid on an AIA 193 Å composite intensity image (blue), which we used to correctly coalign and place the EIS raster data in the mosaic. It shows all regions where coronal plasma is outflowing on open-field lines that reach close to the ecliptic plane. These are smoothed with a Gaussian filter to emphasize areas where there is a larger concentration of sources (red). The map is then filtered to identify weaker concentrations, and these are merged on to the image in green. The AIA images have been treated with an unsharp mask to bring out the details.

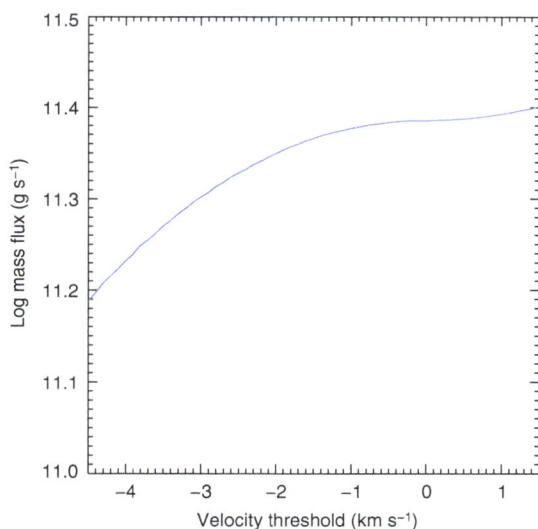

Figure 9 | Relationship between total mass flux and velocity threshold. Logarithm of the total mass flux as a function of the chosen velocity threshold.

loss rate at every pixel, to include every possible type of source, and calculated the mass flux. As discussed, the mass flux depends on the choice of velocity threshold, which is illustrated in Fig. 9. In one extreme, we assume the coronal lines are at rest and only pixels with velocities above the uncertainty are included. In the other extreme, we assume the coronal lines have a small upflow of $1.5\,km\,s^{-1}$ and include all blue-shifted pixels. This leads to calculated mass loss rates of 1.5–$2.5 \times 10^{11}\,g\,s^{-1}$. Assuming an Earth-directed isotropic distribution, these

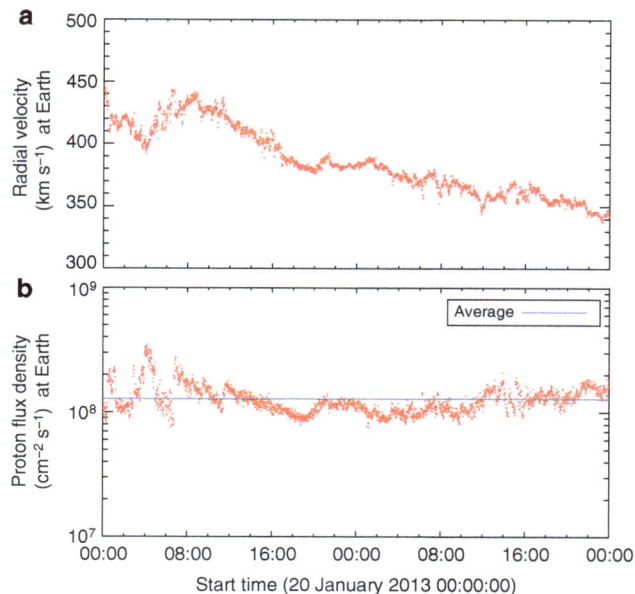

Figure 10 | ACE measurements of the solar wind near Earth. Near-Earth *in situ* radial velocity profile for 20–22 January measured by ACE/SWEPAM (**a**). Proton flux density for the same period (**b**). The average proton flux density is shown as the horizontal blue line.

measurements translate to proton flux densities at Earth of 6.6×10^7–$1.1 \times 10^8\,cm^{-2}\,s^{-1}$, which can be compared with the *in situ* measurements made in the days following our scan, when the plasma has had sufficient time to travel to Earth, by the Advanced Composition Explorer (ACE) Solar Wind Electron, Proton and Alpha Monitor (SWEPAM)[52]. The ACE data (Fig. 10) show that the radial solar wind velocity at Earth was $\sim 400\,km\,s^{-1}$ during the 20–22 January period, which is typical of the slow wind. The proton flux density is quite stable, with an average of $1.3 \times 10^8\,cm^{-2}\,s^{-1}$, and 50–80% of this can be accounted for by the EIS measurements. The mass flux comparison and ACE velocity data also imply that there is no significant contribution from the fast solar wind during the observation period.

Discussion

The comparison between Hinode and ACE observations obviously has large uncertainties because ACE makes measurements corresponding to the features that it actually sees along the Sun-Earth line and projected back to the surface, but since 90% of the outflow mass flux comes from below 40° latitude (Fig. 7), the comparison at least shows that the observed sources in the low corona can potentially supply enough mass flux into the heliosphere, and towards the ecliptic, to explain most of the actual *in situ* particle measurement, rather than a generic value for the solar wind mass loss rate.

We stress that other candidate sources in the low corona, such as magnetically confined plasma, un-fractionated photospheric plasma or downflowing plasma, cannot contribute directly to the wind: the plasma must first be fractionated to FIP bias levels measured *in situ* in the slow solar wind[24] and then expelled on open magnetic field lines. At that point, they would appear in a solar wind source map similar to ours, with exactly the same signature that we are showing. The remaining mass flux may be more likely to come from other sources that are only visible in the higher corona.

On the basis of our analysis, the majority of the mass flux from the low corona, however, appears to come from the edges of active regions (red areas in Fig. 8)[10,11,17,49,53,54]. Like coronal hole boundaries, active regions can be bisected by the heliospheric current sheet[46], so outflows from either side could explain why the heliospheric current sheet is always surrounded by slow wind[55]. A minority flux component also flows from a few coronal hole-like regions, whose shapes follow the boundary between quiet and active areas, but these sources are less concentrated. As such, they are less visible in our Figure, so we have highlighted them in green in Fig. 8. Our results support the view that the slow wind flows from several contributing sources: the red and green areas in Fig. 8 and an unknown source, possibly in the higher corona, that contributes the rest of the mass flux. The figure clearly shows many more red areas than green areas in the low corona, however, so active region outflows appear to be the primary source, at least for the time interval of our observations.

There are of course uncertainties in our results because of the methodology of the analysis. For example, we implicitly assume ionization equilibrium to perform the EM analysis, but this might be violated by the high-speed wind motion at the outflow sites. We do not know the driver of the outflows, but since the ionization relaxation time is less than $\sim 200\,\mathrm{s}$ in the low corona[56], equilibrium is reached fairly rapidly even if they are generated by some impulsive heating mechanism such as chromospheric jets or spicules. Departures from equilibrium could lead to observable effects, such as anomalously high intensities of Li-like spectral lines that have relatively longer relaxation timescales than spectral lines of other iso-electronic sequences[57]. As evidenced by the low values of χ^2 in our calculations, we did not, however, detect any non-equilibrium-based discrepancy in our EM analysis. These effects would be interesting to investigate more systematically in the future by attempting to reproduce the observed intensities with coupled time-dependent ionization and hydrodynamic models of the outflows[58]. The longevity of the outflows (they can persist for several days[25]), however, and the fact that the FIP bias has had time to evolve to coronal values, possibly suggests that the outflows have reached a quasi-equilibrium state by the time they are detected in the corona, and are less likely to be the result of material being ejected rapidly from below the photosphere[26].

In an ideal scenario, we would further constrain the sources by examining the *in situ* composition data from ACE/SWICS (Solar Wind Ion Composition Spectrometer), as we did in our previous study[25]. By filtering out all the regions in our map where the composition measured by EIS does not match that of SWICS, we could make a conclusive link between the wind sources on the Sun and the *in situ* plasma measurements. This comparison, however, was not possible for the time period of our observations, so we should remain cautious. A definitive analysis awaits future observations and will be a focus of the upcoming Solar Orbiter mission.

Methods

Data reduction. We treated the EIS data for cosmic rays, dusty, hot and warm pixels and dark current using the standard routine eis_prep, which is available in SolarSoftware. We corrected other instrumental effects such as grating tilt, orbital spectral drift and the spatial offsets between detectors using a neural network model[44] and refined the coalignment between different wavelengths using the routine eis_ccd_offset[59]. We then calibrated the data to physical units $(\mathrm{erg\,cm^{-2}\,s^{-1}\,sr^{-1}})$.

We fit all the spectral lines in the EIS-calibrated scans using single Gaussian functions, except for a few that are blended (the Fe XI and XII lines and Fe XIII 203.826 Å) and so required multiple Gaussian fits to cleanly separate them. We then extracted the intensities for each spectral line at every coaligned pixel.

As discussed in the main text, we improved the placement of the EIS scans within the mosaic using the coaligned full-Sun AIA 193 Å images taken closest in time to the EIS scans. We downloaded these from the Virtual Solar Observatory. They are level 1 data sets and have been corrected for flat fields, cosmic rays and bad pixels and have been converted to DN per pixel s^{-1}.

We downloaded the SWEPAM data from the ACE science center at http://www.srl.caltech.edu/ACE/ASC/level2/lvl2DATA_SWEPAM.html. The data are calibrated level 2 and have been verified by the ACE team.

Test of the FIP bias measurement method. The simplest method for computing the FIP bias would be to take the S X 264.22 Å to Si X 258.37 Å intensity ratio. These lines are formed at similar temperatures and are very close in wavelength so that calibration issues are minimized. This approach would not, however, account for the temperature and density sensitivity of the line ratio. A more rigorous method is to use density sensitive line ratios to infer the electron density and to use a series of emission lines to compute a temperature distribution. The ratio of the S X 264.22 Å intensity computed from the Differential Emission Measure (DEM) to the observed intensity would yield the FIP bias.

Our analysis method combines these two ideas. We rely primarily on the DEM but introduce a simple scaling using the Si X 258.37 Å line to account for any residual calibration issues because most of the lines used to compute the DEM are on the SW detector, whereas Si and S are on the long-wavelength detector. Using a generative model, we have tested whether the FIP bias determined by our method is sensitive to cross-detector calibration problems, a significant difference in fractionation level between Fe and Si, and uncertainties in the atomic data.

We construct a Gaussian DEM distribution with the peak temperature (T) randomly assigned from the range, $\log T = 5.8$–6.4, and the DEM width (w) randomly assigned from the range, $\log w = 4.5$–5.9. The peak EM is calculated from a randomly assigned density (n) in the range $\log n = 8.5$–9.5. This Gaussian DEM is then used to calculate the intensities of all the spectral lines in our analysis, and they are then randomly perturbed within the calibration uncertainty. At this point, we can use the generated intensities to calculate the DEM and FIP bias using our analysis method as outlined in the main text.

We performed three sensitivity tests. First, we calculated the FIP bias for 100 simulations, and then, to mimic a significant cross-detector calibration problem, we reduced the SW intensities by a factor of 2 and re-computed the FIP bias factors. We show the results in the left hand panel of Supplementary Fig. 1. The FIP bias factors calculated from the calibration-error model remain within 10% of the original values.

Second, we calculated the FIP bias for 100 simulations, and then, to mimic the effects of a significant difference in fractionation level between Fe and Si, we reduced the Fe abundance by a factor of 2 and re-computed the FIP bias factors. We show the results in the centre panel of Supplementary Fig. 1. The FIP bias factors calculated from the fractionation-difference model remain within 5% of the original values.

Third, to mimic the effects of atomic data uncertainties, we checked the dispersion in FIP bias measurements from 100 simulations that results from increasing the line intensity errors. We show the results of this final test in the right hand panel of Supplementary Fig. 1. Here we found that the FIP bias remains within 30% of the original value until the intensity errors become as large as 40–50%.

In summary, the method we use to determine the FIP bias is robust even if the calibration is significantly in error, Fe and Si are fractionated significantly differently and/or the uncertainties in the atomic data are as large as 40–50%.

References

1. Wang, Y.-M., Ko, Y.-K. & Grappin, R. Slow solar wind from open regions with strong low-coronal heating. *Astrophys. J.* **691**, 760–769 (2009).
2. Neugebauer, M. & Snyder, C. W. Solar plasma experiment. *Science* **138**, 1095–1097 (1962).
3. Parker, E. N. Dynamics of the interplanetary gas and magnetic fields. *Astrophys. J.* **128**, 664–676 (1958).
4. Neugebauer, M. & Snyder, C. W. Mariner 2 observations of the solar wind, 1, average properties. *J. Geophys. Res.* **71**, 4469–4484 (1966).
5. Neupert, W. M. & Pizzo, V. Solar coronal holes as sources of recurrent geomagnetic disturbances. *J. Geophys. Res.* **79**, 3701–3709 (1974).
6. Krieger, A. S., Timothy, A. F. & Roelof, E. C. A coronal hole and its identification as the source of a high velocity solar wind stream. *Sol. Phys.* **29**, 505–525 (1973).
7. Zirker, J. B. Coronal holes and high-speed wind streams. *Rev. Geophys. Space Phys.* **15**, 257–269 (1977).
8. Crooker, N. U. et al. Multiple heliospheric current sheets and coronal streamer belt dynamics. *J. Geophys. Res.* **98**, 9371–9381 (1993).
9. Sheeley, N. R. et al. Measurements of flow speeds in the corona between 2 and 30R$_{SUN}$. *Astrophys. J.* **484**, 472–478 (1997).
10. Neugebauer, M., Liewer, P., Smith, E. J., Skoug, R. M. & Zurbuchen, T. H. Sources of the solar wind at solar activity maximum. *J. Geophys. Res.* **107**, SSH 13-1-SSH 13-15 (2002).
11. Liewer, P. C., Neugebauer, M. & Zurbuchen, T. H. Characteristics of active-region sources of solar wind near solar maximum. *Sol. Phys.* **223**, 209–229 (2004).
12. Ko, Y.-K. et al. Abundance variation at the vicinity of an active region and the coronal origin of the slow solar wind. *Astrophys. J.* **646**, 1275–1287 (2006).
13. Pneuman, G. W. & Kopp, R. A. Downflow in the supergranulation network and its implications for transition region models. *Sol. Phys.* **57**, 49–64 (1978).
14. de Pontieu, B., McIntosh, S. W., Hansteen, V. H. & Schrijver, C. J. Observing the roots of solar coronal heating - in the chromosphere. *Astrophys. J. Lett.* **701**, L1–L6 (2009).

15. Antiochos, S. K., Mikic, Z., Titov, V. S., Lionello, R. & Linker, J. A. A m-odel for the sources of the slow solar wind. *Astrophys. J.* **731**, 112–122 (2011).
16. Sakao, T. *et al.* Continuous plasma outflows from the edge of a solar active region as a possible source of solar wind. *Science* **318**, 1585–1588 (2007).
17. Doschek, G. A. *et al.* Flows and nonthermal velocities in solar active regions observed with the EUV imaging spectrometer on Hinode: a tracer of active region sources of heliospheric magnetic fields? *Astrophys. J.* **686**, 1362–1371 (2008).
18. Harra, L. K. *et al.* Outflows at the edges of active regions: contribution to solar wind formation? *Astrophys. J.* **676**, L147–L150 (2008).
19. Doschek, G. A., Bohlin, J. D. & Feldman, U. Doppler wavelength shifts of transition zone lines measured in Skylab solar spectra. *Astrophys. J.* **205**, L177–L180 (1976).
20. Sandlin, G. D., Brueckner, G. E. & Tousey, R. Forbidden lines of the solar corona and transition zone: 975-3000Å. *Astrophys. J.* **214**, 898–904 (1977).
21. Kohl, J. L. *et al.* First results from the SOHO Ultraviolet Coronagraph Spectrometer. *Sol. Phys.* **175**, 613–644 (1997).
22. Grail, R. R. *et al.* Rapid acceleration of the polar solar wind. *Nature* **379**, 429–432 (1996).
23. Feldman, U., Landi, E. & Schwadron, N. A. On the sources of fast and slow solar wind. *J. Geophys. Res.* **110**, A07109 (2005).
24. Von Steiger, R. *et al.* Composition of quasi-stationary solar wind flows from Ulysses/Solar Wind Ion Composition Spectrometer. *J. Geophys. Res.* **105**, 27217–27238 (2000).
25. Brooks, D. H. & Warren, H. P. Establishing a connection between active region outflows and the solar wind: abundance measurements with EIS/Hinode. *Astrophys. J. Lett.* **727**, L13–L17 (2011).
26. Brooks, D. H. & Warren, H. P. The coronal source of extreme-ultraviolet line profile asymmetries in solar active region outflows. *Astrophys. J. Lett.* **760**, L5–L10 (2012).
27. Boutry, C., Buchlin, E., Vial, J.-C. & Regnier, S. Flows at the edge of an active region: observation and interpretation. *Astrophys. J.* **752**, 13–23 (2012).
28. Warren, H. P., Ugarte-Urra, I., Young, P. R. & Stenborg, G. The temperature dependence of solar active region outflows. *Astrophys. J.* **727**, 58–62 (2011).
29. Watanabe, K., Masuda, S. & Segawa, T. Hinode flare catalogue. *Sol. Phys.* **279**, 317–322 (2012).
30. Brooks, D. H., Warren, H. P., Williams, D. R. & Watanabe, T. Hinode/EUV imaging spectrometer observations of the temperature structure of the quiet corona. *Astrophys. J.* **705**, 1522–1532 (2009).
31. Warren, H. P. & Brooks, D. H. The temperature and density structure of the solar corona. 1. observations of the quiet Sun with the EUV imaging spectrometer on Hinode. *Astrophys. J.* **700**, 762–773 (2009).
32. Testa, P., Reale, F., Landi, E., de Luca, E. E. & Kashyap, V. L. Temperature distribution of a non-flaring active region from simultaneous XRT and EIS observations. *Astrophys. J.* **728** p 30–41 (2011).
33. Winebarger, A. R., Schmelz, J. T., Warren, H. P., Saar, S. H. & Kashyap, V. L. Using a differential emission measure and density measurements in an active region core to test a steady heating model. *Astrophys. J.* **740**, 2–13 (2011).
34. O'Dwyer, B., Del Zanna, G., Mason, H. E., Weber, M. A. & Tripathi, D. SDO/ AIA response to coronal hole, quiet Sun, active region, and flare plasma. *Astron. Astrophys.* **521**, 21–25 (2010).
35. Feldman, U., Warren, H. P., Brown, C. M. & Doschek, G. A. Can the composition of the solar corona be derived from Hinode/extreme-ultraviolet imaging spectrometer spectra? *Astrophys. J.* **695**, 36–45 (2009).
36. Dere, K., Landi, E., Mason, H. E., Monsignori Fossi, B. C. & Young, P. R. CHIANTI - an atomic database for emission lines. *Astron. Astrophys. Sup.* **125**, 149–173 (1997).
37. Landi, E., Del Zanna, G., Young, P. R., Dere, K. P. & Mason, H. E. CHIANTI - an atomic database for emission lines. XII. version 7 of the database. *Astrophys. J.* **744**, 99–107 (2012).
38. Grevesse, N., Asplund, M. & Sauval, A. J. The solar chemical composition. *Space Sci. Rev.* **130**, 105–114 (2007).
39. Kashyap, V., Drake, J. J. & Markov-chain Monte, J. J. Carlo reconstruction of emission measure distributions: application to solar extreme-ultraviolet spectra. *Astrophys. J.* **503**, 450–466 (1998).
40. Kashyap, V. & Drake, J. J. PINTofALE: package for the interactive analysis of line emission. *Bull. Astron. Soc. India* **28**, 475–476 (2000).
41. Warren, H. P., Ugarte-Urra, I. & Landi, E. The absolute calibration of the EUV imaging spectrometer on Hinode. *Astrophys. J. Suppl. Ser.* **213**, 11–22 (2014).
42. Del Zanna, G. A revised radiometric calibration for the Hinode/EIS instrument. *Astron. Astrophys.* **555**, 47–66 (2013).
43. Laming, J. M. Non-WKB models of the first ionization potential effect: the role of slow mode waves. *Astrophys. J.* **744**, 115–127 (2012).
44. Kamio, S., Hara, H., Watanabe, T., Fredvik, T. & Hansteen, V. H. Modeling of EIS spectrum drift from instrumental temperatures. *Sol. Phys.* **266**, 209–223 (2010).
45. Peter, H. & Judge, P. G. On the Doppler shifts of solar ultraviolet emission lines. *Astrophys. J.* **522**, 1148–1166 (1999).
46. Culhane, J. L. *et al.* Tracking solar active region outflow plasma form its source to the near-Earth environment. *Sol. Phys.* **289**, 3799–3816 (2014).
47. Del Zanna, G., Aulanier, G., Klein, K.-L. & Torok, T. A single picture for solar coronal outflows and radio noise storms. *Astron. Astrophys.* **526**, A137–A148 (2011).
48. Schrijver, C. J. Simulations of the photospheric magnetic activity and outer atmospheric radiative losses of cool stars based on characteristics of the solar magnetic field. *Astrophys. J.* **547**, 475–490 (2001).
49. Schrijver, C. J. & De Rosa, M. L. Photospheric and heliospheric magnetic fields. *Sol. Phys.* **212**, 165–200 (2003).
50. Schrijver, C. J., Sandman, A. W., Aschwanden, M. J. & De Rosa, M. L. The coronal heating mechanism as identified by full Sun visualizations. *Astrophys. J.* **615**, 512–525 (2004).
51. Lang, J. *et al.* Laboratory calibration of the Extreme-Ultraviolet Imaging Spectrometer for the Solar-B satellite. *Appl. Opt.* **45**, 8689–8705 (2006).
52. McComas, D. J. *et al.* Solar Wind Electron Proton Alpha Monitor (SWEPAM) for the Advanced Composition Explorer. *Space Sci. Rev.* **86**, 563–612 (1998).
53. Luhmann, J. G., Li, Y., Arge, C. N., Gazis, P. R. & Ulrich, R. Solar cycle changes in coronal holes and space weather cycles. *J. Geophys. Res.* **107**, SMP 3-1–SMP 3-12 (2002).
54. Wang, Y.-M. & Sheeley, N. R. Sunspot activity and the long-term variation of the Sun's open magnetic flux. *J. Geophys. Res.* **107**, SSH 10-1–SSH 10-15 (2002).
55. Burlaga, L. F., Ness, N. F., Wang, Y.-M. & Sheeley, N. R. Heliospheric magnetic field strength and polarity from 1 to 81 AU during the ascending phase of solar cycle 23. *J. Geophys. Res.* **107**, SSH 20-1–SSH 20-11 (2002).
56. Lanzafame, A. C. *et al.* ADAS analysis of the differential emission measure structure of the inner solar corona: application of the data adaptive smoothing approach to the SERTS-89 active region spectrum. *Astron. Astrophys.* **384**, 242–272 (2002).
57. Judge, P. G., Woods, T. N., Brekke, P. & Rottman, G. J. On the failure of standard emission measure analysis for solar extreme-ultraviolet and ultraviolet irradiance spectra. *Astrophys. J. Lett.* **455**, L85–L88 (1995).
58. Bradshaw, S. J. & Mason, H. E. A self-consistent treatment of radiation in coronal loop modelling. *Astron. Astrophys.* **401**, 699–709 (2003).
59. Young, P. R., Watanabe, T., Hara, H. & Mariska, J. T. High-precision density measurements in the solar corona. I. analysis methods and results for Fe XII and Fe XIII. *Astron. Astrophys.* **495**, 587–606 (2009).
60. Culhane, J. L. *et al.* The EUV Imaging Spectrometer for Hinode. *Sol. Phys.* **243**, 19–61 (2007).

Acknowledgements

D.H.B. thanks M.L. DeRosa for guidance on the use of the PFSS software. This work was performed under contract with the Naval Research Laboratory and was funded by the NASA Hinode program. Hinode is a Japanese mission developed and launched by ISAS/ JAXA, with NAOJ as domestic partner and NASA and STFC (UK) as international partners. It is operated by these agencies in co-operation with ESA and NSC (Norway). CHIANTI is a collaborative project involving George Mason University, the University of Michigan (USA) and the University of Cambridge (UK).

Author contributions

D.H.B. initiated the study, obtained the EIS observations, processed and analysed the data and wrote the paper. All authors participated in designing the EIS-observing sequence and satellite pointing scheme, writing analysis software and discussing the results and manuscript.

Additional information

12

Investigating Alfvénic wave propagation in coronal open-field regions

R.J. Morton[1,2], S. Tomczyk[2] & R. Pinto[3,4]

The physical mechanisms behind accelerating solar and stellar winds are a long-standing astrophysical mystery, although recent breakthroughs have come from models invoking the turbulent dissipation of Alfvén waves. The existence of Alfvén waves far from the Sun has been known since the 1970s, and recently the presence of ubiquitous Alfvénic waves throughout the solar atmosphere has been confirmed. However, the presence of atmospheric Alfvénic waves does not, alone, provide sufficient support for wave-based models; the existence of counter-propagating Alfvénic waves is crucial for the development of turbulence. Here, we demonstrate that counter-propagating Alfvénic waves exist in open coronal magnetic fields and reveal key observational insights into the details of their generation, reflection in the upper atmosphere and outward propagation into the solar wind. The results enhance our knowledge of Alfvénic wave propagation in the solar atmosphere, providing support and constraints for some of the recent Alfvén wave turbulence models.

[1] Department of Mathematics and Information Sciences, Northumbria University, Newcastle Upon Tyne NE1 8ST, UK. [2] High Altitude Observatory, National Center for Atmospheric Research, Boulder, Colorado 80307-3000, USA. [3] UPS-OMP, IRAP, Université de Toulouse, 14 Avenue Edouard Belin -314000 Toulouse, France. [4] CNRS, IRAP, 9 Avenue colonel Roche, BP 44346, F-31028 Toulouse, France. Correspondence and requests for materials should be addressed to R.J.M. (email: richard.morton@northumbria.ac.uk).

It is now well known that the fast solar wind originates from within open magnetic field regions in the Sun's atmosphere, for example, coronal holes, with wind acceleration beginning in the transition region and low corona[1-5]. However, the mechanism(s) responsible for heating the coronal plasma and accelerating the solar wind within the open fields is still unclear. A host of different physical models exist to explain this phenomena[6] and one broad category of models relies on the propagation, reflection and dissipation of Alfvén waves[7-11]. Alfvén waves are perfect for the transport of energy as they are incompressible, making it difficult for their energy to be dissipated without the presence of large gradients in the Alfvén speed. This property should allow them to propagate out into the extended solar atmosphere and deposit their energy and momentum beyond the critical point and the point of temperature maximum, an important feature needed to explain the acceleration of fast solar winds[9,12,13]. These Alfvén waves would likely be generated towards the foot points of magnetic flux tubes, which are rooted in the photosphere and extend out into interplanetary space. In theory, after traversing the lower solar atmosphere, the waves reach the corona where they propagate and are partially reflected by the gradual variation in Alfvén speed, enabling the development of magneto-hydrodynamic (MHD) wave turbulence[7-11,14] and the subsequent dissipation of the energy transported by the waves.

The textbook Alfvén wave, first described by Hannes Alfvén, requires the existence of an infinite, homogenous plasma. In this scenario, other perturbations of the plasma can also be categorized neatly as slow and fast MHD waves. The distinct characteristics of Alfvén waves are that they are the sole propagator of vorticity perturbations and magnetic tension is the dominant restoring force, in addition to incompressibility. Conversely, slow and fast MHD waves are characterized by compressibility and do not perturb vorticity[15]. However, these characteristics are textbook ideas of MHD waves. In highly inhomogeneous plasmas, such as the solar atmosphere, MHD wave modes have mixed properties and cannot be so precisely categorized. The fine-scale structure in the solar atmosphere is frequently modelled as a collection of over-dense, magnetic waveguides. In such a setting, the wave energy is confined predominantly to the regions of enhanced density. Two wave modes in the system share characteristics reminiscent of the textbook Alfvén wave, namely the fundamental modes of the torsional Alfvén wave and the fast kink wave[15]. The torsional mode causes an incompressible rotation of magnetic surfaces. The fast kink wave generally has mixed properties. It is identified by a non-axisymmetric displacement of the waveguide (or a swaying motion) where magnetic tension is the dominant restoring force. The fast kink wave is also highly incompressible, especially in the limit where the wavelength is much larger than the transverse scale of the density inhomogeneity that defines the waveguide[16]. The kink mode also transports vorticity and shares a remarkable similarity with the surface Alfvén wave[15]. From theses considerations, we follow Goossens et al. (ref. 15, 16) and use the adjective Alfvénic to describe wave modes with characteristics similar to the textbook Alfvén waves. At present, it is unknown whether the models of Alfvén wave turbulence remain valid if the Alfvénic waves of inhomogeneous plasmas were used instead. The common characteristics shared by the waves could suggest that central features of the wave turbulence models may survive.

The existence of Alfvénic fluctuations far into the solar wind (at 1 AU) has been known for a number of years[17] and their presence in the corona have been anticipated from the measurements of non-thermal line widths[18-25]. Recently, imaging and spectroscopic observations confirmed the presence of Alfvénic waves throughout the solar atmosphere[26-31], with fast

kink modes appearing omnipresent along fine-scale magnetic structures. The exact relationship between the motions encapsulated by the non-thermal width measurements and the wave motions seen in images is still unclear, typified by the apparent disparate measured amplitudes (and, hence, energetics). This is likely someway due to the complex dependence of non-thermal line widths on the resolution of the instrument, line of sight integration and the addition of different sources of motion, that is, flows and the multitude of MHD wave modes[32-34].

Nevertheless, the associated flux of energy estimated to be carried by the Alfvénic waves at the coronal base appears to be sufficient to meet the requirements for heating the coronal plasma and/or solar wind acceleration. Although, again, there are still some remaining issues that need to be clarified with respect to the energy flux[31,35-37] (also, see the Results section).

The presence of Alfvénic waves in the chromosphere and corona does not, however, provide sufficient support for acceleration of the solar wind via Alfvénic waves. For Alfvénic waves to be a viable option for wind acceleration, the presence of counter-propagating Alfvénic waves, over an extended frequency range, is crucial for MHD turbulence to occur in the open-field regions. Counter-propagation allows for the nonlinear interaction of the waves, which leads to a turbulent cascade. To date, only measurements of wave amplitudes and periods exist in such regions[18-25,30,31]. Further, current coronal observations of Alfvénic waves place only relatively simple constraints on the models, that is, non-thermal widths have provided frequency-integrated velocity amplitudes. More rigorous constraints can be placed on the models if insights are obtained into the evolution of the Alfvénic waves as they propagate from the corona out into the solar wind, along with measurements of key plasma parameters, for example, density, magnetic field strength.

In the following, these issues are explored through the analysis of both spectroscopic and imaging data from ground- and space-based observatories. We investigate whether the conditions in an open magnetic field region are suitable for MHD turbulence to develop, search for indications of the waves origin and whether the observed coronal Alfvénic waves are connected to those found far out in the solar wind. It is envisioned that the results will help constrain the variety of Alfvénic wave models. This is supported by an exploitation of the observed waves to probe some of the key plasma parameters of the open magnetic field region.

Results

The coronal multi-channel polarimeter. The Coronal Multi-Channel Polarimeter (CoMP)[38] is used to analyse Doppler velocity fluctuations and their propagation in an open-field region located at the northern pole of the Sun. CoMP was the first instrument to confirm the ubiquity of Alfvénic waves in the corona[29] and, still, has an unequalled ability to observe coronal velocity fluctuations, over a large field of view at high temporal cadence.

The Sun, on 27 March 2012, had two dominant open-field regions with the southern one more readily identifiable. The southern region is marked by the presence of a coronal hole that has extremely low emission in Fe XIII images of the corona obtained with CoMP (Fig. 1). This lack of emission means that the data preparation technique (see Methods) fails low in the coronal hole and as such there is no line centre intensity measurements (or Doppler velocities/widths). The region for which velocity measurements could be made is highlighted by the contour on the CoMP image in Fig. 1a. The northern open-field region is less obvious but there is reduced emission at the pole compared with the rest of the corona. The lack of an obvious coronal hole is also apparent in coronal Fe IX and XII emission

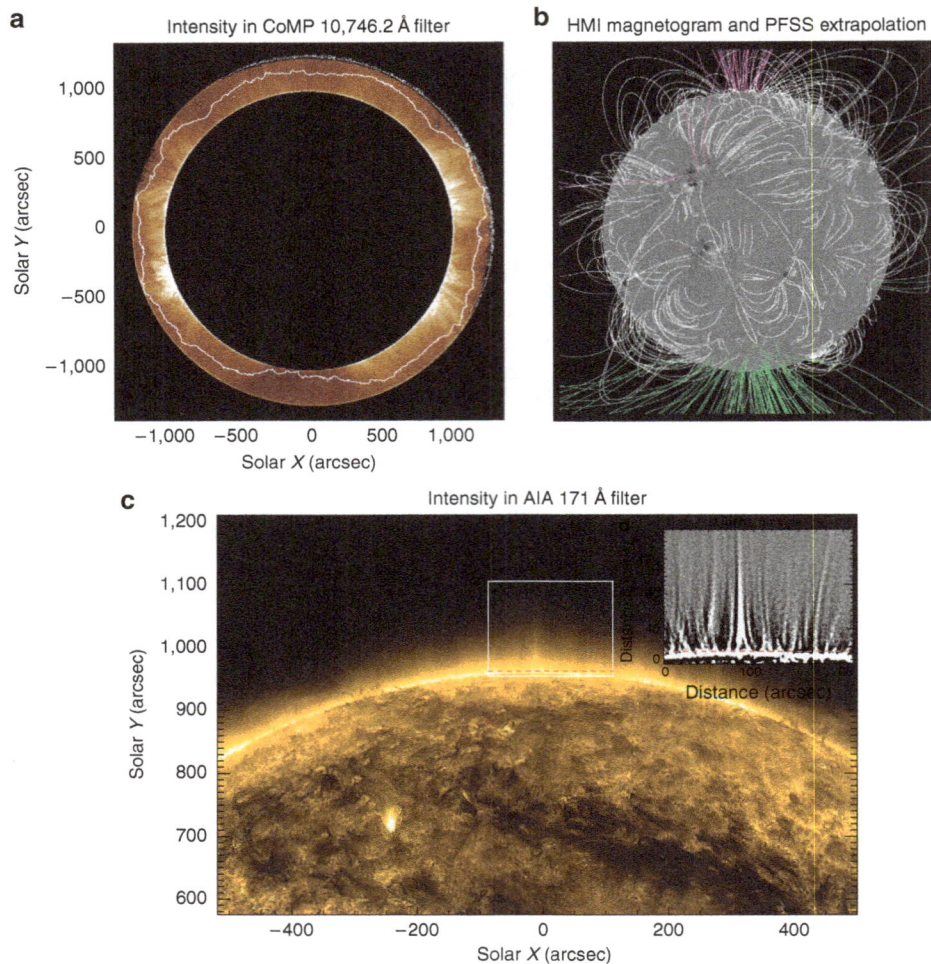

Figure 1 | The solar corona. (a) An intensity image of the corona as seen in the infrared Fe XIII emission line, where the disk of the Sun has been occulted. The image is enhanced with a radial gradient filter for visual purposes only. The white contour in the corona demarcates the region for which Doppler velocities could be measured. **(b)** A magnetogram, based on Helioseismic and Magnetic Imager (HMI) data and evolved with the surface-flux assimilation model, demonstrates the photospheric magnetic field and is over-plotted with magnetic field lines from the PFSS potential field extrapolation. The purple and green lines show open-field regions of negative and positive polarity, while the white lines show closed field regions. **(c)** An enlarged image of the polar region as seen with SDO/AIA in the Fe IX coronal emission line. Fine-scale structures exist at the limb and are visible as elongated intensity enhancements. **(d)** An unsharp masked cut-out of a portion of the open-field region (box in **c**) clearly reveals this fine-scale magnetic structure extending radially away from the surface. The red dashed lines in **c** and **d** shows the position of the slit used to generate the time–distance diagram in Fig. 5a.

from higher resolution, full disk images provided by the Solar Dynamics Observatory (SDO) Atmospheric Imaging Assembly (AIA), although radially orientated, field-aligned fine-scale magnetic structures are visible (Fig. 1c,d). The relationship of this fine-scale magnetic structure to plumes and inter-plumes is unclear from the current data set. We are able to say that they are much narrower than the classic plumes, which are reported to span ~ 30 Mm (40 arcseconds) on average.

To confirm the presence of open fields in this region, we use the evolving surface-flux assimilation model and potential field source surface (PFSS) extrapolation tool[39] (the extrapolation is shown in Fig. 1b). The southern coronal hole is seen in the extrapolation as a large open-field region that has some presence on the solar disk, while the northern open-field region is restricted to high latitudes. There is a direct correspondence between the locations of weak emission at low coronal heights in the CoMP data and the open-field lines in the PFSS extrapolation. We note that the visible structure in the region of interest remains practically unchanged throughout the observations, indicating little or no large-scale changes in the magnetic topology of the region.

The existence of counter-propagating Alfvénic waves. We focus on the northern open-field region and examine the Doppler velocities in this part of the corona. It is evident from the data that there is a predominantly outward propagating Doppler signal in this region. To analyse the fluctuations in the Doppler velocities, we use two different methods to extract the data in the form of velocity time–distance diagrams (Fig. 2). The two independent methods are used to validate the analysis techniques and results of each other. For the first method, the crude assumption is made that the direction of potential wave propagation is strictly in the North–South direction. Individual Doppler velocity time–distance diagrams are generated from vertical strips of pixels in the boxed region (shown in Fig. 2a). The presence of the outward propagating Doppler velocity signals is visible in these diagrams and so is the periodic nature of these features, implying that the observed signals correspond to MHD waves.

The power spectrum for the Doppler velocity time–distance diagrams is determined for a range of frequencies (f) and wavenumbers (k) in the vertical direction, and the outward and inward propagating signals are decomposed. To analyse the waves, both an f–k diagram (Fig. 2) and wavenumber-integrated

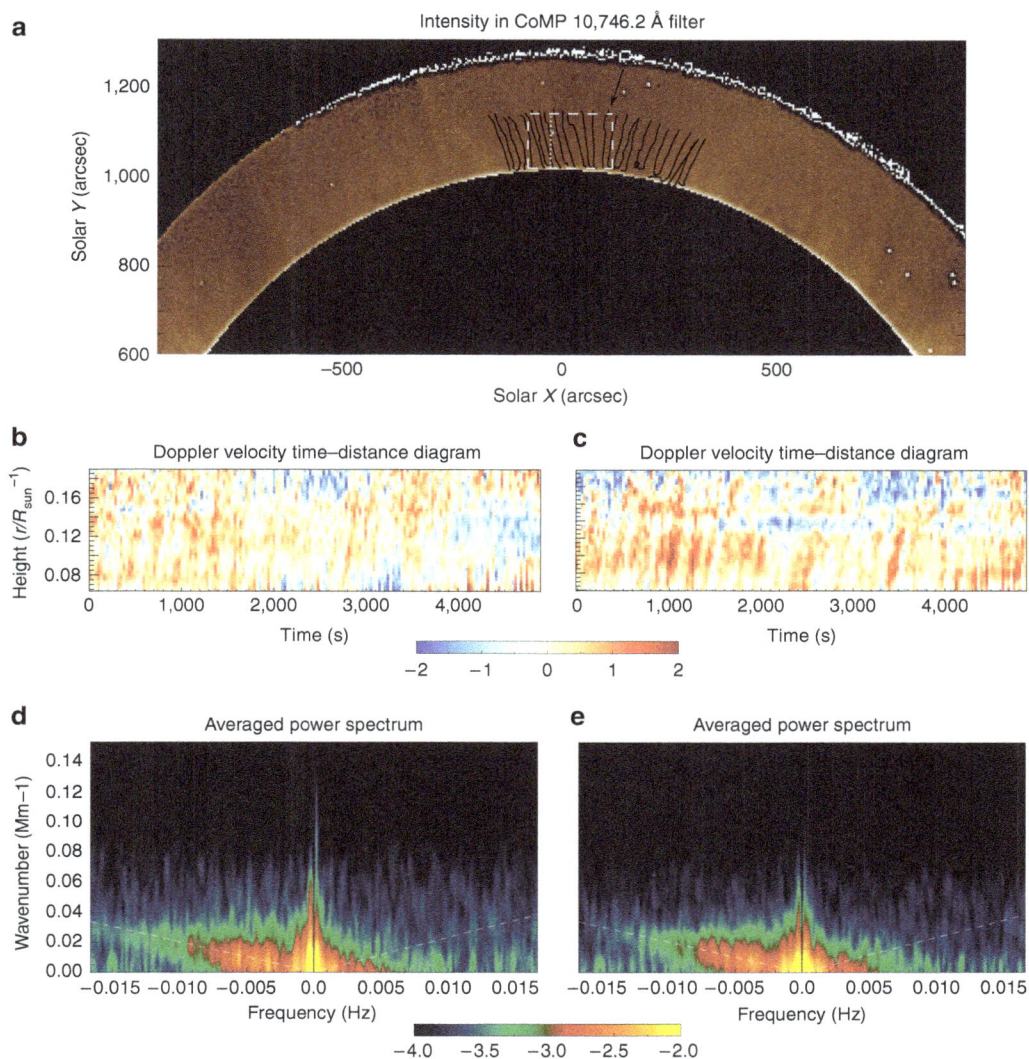

Figure 2 | Propagating Alfvénic waves in the open-field region. (a) A close-up CoMP intensity image of the northern polar region that demonstrates the locations of the box (dashed white lines) and the tracks (black lines) used in the two methods for isolating propagating waves in the open-field region. The predominantly upwardly propagating waves can be seen in the Doppler velocity time–distance diagrams from the box (**b**) and from the track (**c**) methods, where the scale bar indicates the velocity (km s^{-1}). The locations of these Doppler velocity time–distance diagrams are highlighted in **a** by the white dotted line (**b**) and black arrow (**c**). The time–distance diagrams are used to derive the corresponding frequency-wavenumber power plots for the box (**d**) and track (**e**) methods. A pronounced ridge of power in the negative frequency domain is clearly visible corresponding to the dominance of outward propagating waves. Crucially, the power plots also reveal the existence of inwardly propagating waves, but with reduced power. The scale bar indicates the power (km^2 s^{-2}) to the log base 10. The over-plotted white dashed lines show the calculated average inward and outward wave propagation velocities from the Doppler velocity time–distance diagrams.

power spectra are calculated (Fig. 3). A prominent ridge in the left half of the f–k diagram corresponding to negative frequencies confirms the dominance of outward propagating waves. No clear ridge exists for the inward propagating waves although there is significant power in the low frequency part of the spectrum. Such ridges have also been found in f–k diagrams for large quiescent coronal loops[40].

The second method is a more refined version of the first, providing the f–k diagram, power spectra and, in addition, the propagation velocities of the waves (see Methods). The method explicitly determines the direction of propagation of the waves and assuming that these propagating waves are Alfvénic in nature, the direction of propagation defines the orientation of the magnetic field in the plane of the sky. Building up this picture of orientations over the entire atmosphere, the local direction of the magnetic field is determined.

Starting at a particular pixel, the direction of the magnetic field in the plane of sky is followed and tracks are built that are aligned with the direction of wave propagation. The tracks are almost vertical with respect to the solar disk, highlighting the radial structure of the magnetic field (black lines in Fig. 2). The Doppler velocity time series at each point along the track is then used to build Doppler velocity time–distance maps and, as with the preceding method, the f–k diagram (Fig. 2) and the k-integrated power spectra (Fig. 3) are obtained. The two methods show qualitatively similar results suggesting that the wave propagation direction is predominantly in the North–South direction, that is, radial for this small open-field region.

An estimate of the propagation speed of the waves is obtained from these Doppler velocity time–distance diagrams with a weighted average giving the typical speeds and standard deviation for outward propagation as 444 ± 2 km s^{-1} and inward propagation as 365 ± 76 km s^{-1}. The propagation speed of Alfvénic

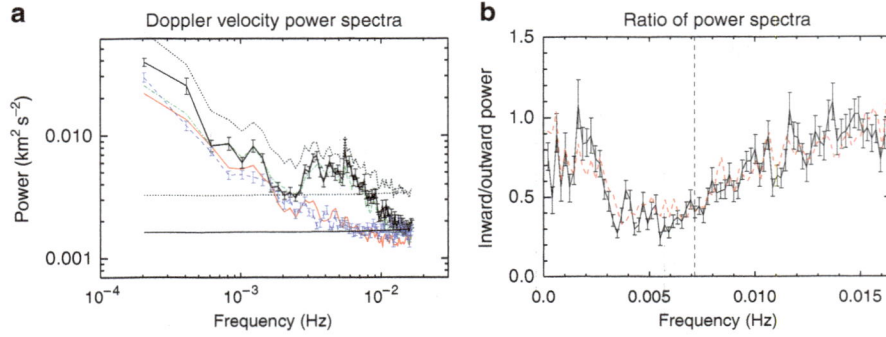

Figure 3 | Average frequency power spectra and ratios of counter-propagating waves. The predominance of power in outward propagating waves (box method—black solid and track method—green dash-dot) is clear in the wavenumber-integrated power spectra (**a**). The spectra also provide insight into the power of the inward propagating waves (box method—blue dash and track method—red dash-triple dot). The total power spectrum is also shown for reference (dots). The horizontal dotted line is the estimate for the data noise associated with the total power spectra, while the solid horizontal line is the data noise split between inward and outward components. (**b**) The ratio of inward to outward power is displayed as a function of frequency and reveals a frequency-dependent trend suggestive of partial reflection of the waves in the corona. The vertical dashed line highlights the frequency after which noise begins to dominate the inward power spectra. The error bars in both plots show the s.e.m.

waves is given by $c_p \approx \omega/k$ and this relationship is used to overplot the outward and inward velocities on the f–k diagrams in Fig. 2. Good agreement between the location of the ridge and the estimated outward propagation speed is clear, giving confidence in the values. These speeds are significantly greater than the sound speed for a 1.6 MK plasma $c_s \sim 150$ km s^{-1} and, hence, the observed waves are likely Alfvénic.

As mentioned, counter-propagating waves are a crucial component of Alfvén wave heating/acceleration models if MHD wave turbulence is to develop[14,41], generating the small scales necessary for the dissipation of Alfvénic wave energy through their interaction. The power spectra obtained here provide evidence for the presence of both outward and inward propagating Alfvénic waves in open-field regions, suggesting that the Alfvénic waves are at least partially reflected in the open magnetic field regions of corona, or further out.

The inwardly propagating waves can be seen to have significantly less power, both in the f–k diagrams (Fig. 2c,d) and the wavenumber-integrated power spectra (Fig. 3a). In Fig. 3b, the ratio of inward to outward wave power is given as a function of frequency. The ratio reveals a frequency dependence of the power ratio, potentially indicative of wave reflection in the corona. The results suggest (for f < 6 mHz) that the low frequency waves are more readily reflected than high frequency waves, which is to be expected for coronal Alfvén waves[7,9]. However, the ratio begins to increase after f ∼ 6–7 mHz. The increase in the ratio for these higher frequencies is likely related to the inward power being dominated by noise for f > 6 mHz. The noise in the data has a flat power spectrum and the continued decrease in the outward wave power with frequency > 6 mHz leads to the increase in the ratio. The ratio flattens off for f > 10 mHz, a sign that the outward power spectra also becomes dominated by noise. The influence of noise is confirmed upon estimating the contribution of the data noise to the power spectra.

Connecting the coronal waves to the solar wind. The power spectra of the outward and inward propagating waves are found to have a slope of ∼1/f (see Table 1 for exact values) and an enhancement of power at ∼3–5 mHz. The functional 1/f form of the coronal power spectra is also found in magnetic and velocity fluctuation power spectra observed in solar wind streams in the heliosphere at 0.3 AU (64 R$_\odot$)[42,43]. The similarity between the slopes of the coronal and wind power spectra lends support to

Table 1 | Properties of the $10^a f^b$ fit to the outward (Out) and inward (In) power spectra shown in Fig. 3, for the box (M1) and track (M2) methods.

	Frequency range (mHz)	a	b	χ^2_ν
Out M1	0.2–2	−5.0 ± 0.1	−0.92 ± 0.03	4.3
	4–10	−4.8 ± 0.1	−1.04 ± 0.05	3.8
In M1	0.2–2	−5.3 ± 0.1	−0.98 ± 0.03	3.8
Out M2	0.2–2	−4.4 ± 0.1	−0.7 ± 0.04	0.9
	4–10	−5.0 ± 0.1	−1.1 ± 0.05	1.4
In M2	0.2–2	−5.1 ± 0.1	−0.89 ± 0.04	2.7

the ideas presented in ref. 43, which suggest that the shape of the wind spectrum is already set in the low solar atmosphere, at least in the low corona and transported outwards.

Interestingly, the presence of an enhancement in power is observed at ∼1–2 mHz in some solar wind spectra[42], similar to the enhancement in the coronal power spectra at frequencies of 3–5 mHz. This may imply that the features seen in solar wind spectra are potential remnants of the functional form of the power spectra set in the low solar atmosphere (< 1.3 R$_\odot$). The frequency shift of this feature between the two spectra, that is, coronal and at 0.3 AU, may be explained by the stretching of the perpendicular wavenumber scale with height owing to the radial expansion of the magnetic field combined with solar wind acceleration and magnetic field line rotation[44].

Evidence p-modes play a role in Alfvénic wave generation. The enhancement of power seen in the coronal power spectra is of interest in its own right. The frequency range occupied by the power enhancement is coincidental with the frequency range of peak power of p-modes and has also been observed in CoMP power spectra for large quiescent coronal loops[40]. This leads us to suspect that p-modes play an important role in the generation of Alfvénic waves in open-field regions, as well as in the quiescent Sun.

Theoretical considerations support the idea that p-modes can be guided into the corona by the magnetic field. In particular, mode conversion would play a dominant role in generating coronal Alfvénic waves[45]. At some height in the lower solar atmosphere, dependent upon the local pressure and magnetic field, there will exist a canopy where the gas pressure and magnetic pressure are approximately equal in magnitude ($\beta = 8\pi p/B^2 \approx 1$). At the canopy, a transformation from slow to fast magnetoacoustic waves can occur, with the efficiency of this transformation dependent upon the angle between the magnetic field and the direction of wave propagation. This phenomenon has been well studied in a variety of magnetic field configurations and plasma conditions[46–49]. Although slow magnetoacoustic modes above this canopy are expected to steepen and shock owing to the rise in temperature in the upper chromosphere, the fast magnetoacoustic waves are reflected due to the steep gradient in Alfvén speed at the transition region. In the case when the gravity, magnetic field and wave vector of the fast magnetoacoustic wave are not co-planar, then there is a coupling of the fast wave to the Alfvén wave[45,50–52]. This coupling enables wave energy to cross the transition region and propagate into the corona as an Alfvén wave. Finally, regions of strong magnetic field provide the necessary conditions for p-modes to leak out from the interior into the atmosphere[53–55], where they can then be converted via various methods of mode coupling. This is of particular importance as the open coronal fields are generally thought to be rooted in the kilo-Gauss faculae that form the magnetic network in the photosphere[56,57]. It has been estimated that a sufficient amount of energy can be converted from p-modes to coronal Alfvén waves to explain the approximate energy content of observed Alfvénic motions in the corona[45].

We note that the discussed theoretical work[45,50–52] has only been undertaken for a homogenous atmosphere, that is, the excitation of classical Alfvén waves. It is still unknown whether

p-modes could excite Alfvénic waves in a corona with a cross-field density structuring. However, it is clear from the theory that vorticity perturbations must be excited in the corona by the mode conversion at the transition region; otherwise Alfvén waves would not be produced. In the case of a structured corona, any generated vorticity at the transition region must also be transported and the Alfvénic modes would seemingly be the best choice for this. The observed enhanced power of the coronal Alfvénic waves coincident with the frequency range of the dominant p-mode power may support this but further investigation is required for the confirmation of this hypothesis.

Probing open-field regions with solar magneto-seismology. Moving away from the power spectra, another interesting aspect observed is that the propagation velocities of the outward waves is greater than the inward. Under the assumption that there exists a background outflow, in this case most likely the solar wind, then a simplified relationship describing the Alfvénic waves can be obtained (see Methods),

$$\omega = (U \pm c_{ph})k_z. \qquad (1)$$

Here, c_{ph} is the phase speed of the wave, U is the flow speed and z is the direction parallel to the magnetic field. The propagation speed of the outward (plus) and inward (minus) waves is modified by the presence of the flow. Hence, outward and inward wave speed estimates obtained earlier from the Doppler velocity time–distance diagrams suggest the presence of an outward moving flow with an average speed $U = 31 \pm 7$ km s^{-1}. This value is comparable to the spectroscopic

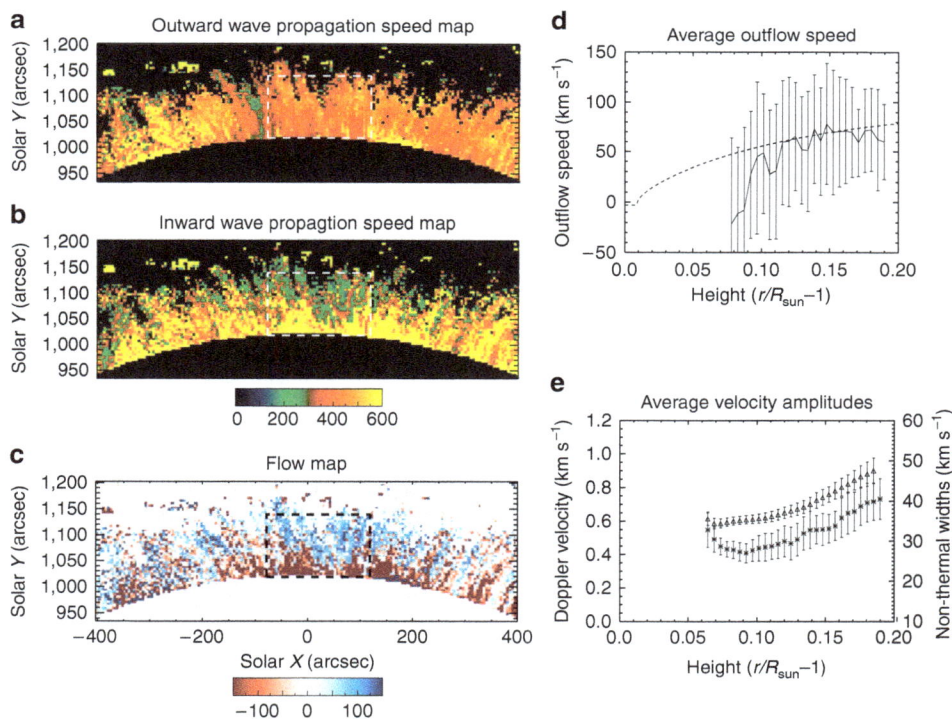

Figure 4 | Measurements of the flow speed along the open fields. The measured propagation speed for the outward (**a**) and inward (**b**) Alfvénic waves can be used to determine the flow speed of plasma along the magnetic field (**c**). The black and white boxes in the propagation and flow maps mark the region that is used for determining the average values in the following panels. The scale bars indicate propagation speed (km s^{-1}) and flow speed (km s^{-1}). The open-field region is dominated by outflowing plasma (**d**), which is likely the beginnings of the solar wind and is comparable to outflows obtained from hydrodynamic wind models (dashed line). The error bars show the standard deviation of the average flow. (**e**) The RMS Doppler velocity (stars) and non-thermal widths (triangles) are shown as a function of height revealing that the Alfvénic wave amplitude increases as they propagate away from the limb. The error bars show the s.d. of the respective quantities.

estimate of outflows in open-field regions obtained from the transition region and low coronal emission lines[58].

This basic process is taken further. Following the same principles outlined for creating Doppler velocity time–distance diagrams, estimates for the outward and inward propagation velocities throughout the corona are obtained (Fig. 4a,b, see Methods for details). Subsequently, a flow map for the northern polar region is derived (Fig. 4c), which reveals a scene that is dominated by outflows, as should be expected if the open magnetic fields are a source region for the solar wind. The flow is seen to accelerate as a function of height (Fig. 4d), tending towards a value of 60–70 km s^{-1} at 1.2 $R_\odot$. The observed acceleration of the wind above $\sim 1.09\ R_\odot$ is consistent with predictions of wind acceleration in the low corona found from solutions of one-dimensional hydrodynamic solar wind models (ref. 59—wind speed from the model is shown in Fig. 4d). Interestingly, the wind speed values from the model at $\sim 1.4\ R_\odot$ (not shown in the figure) are comparable to previously measured outflow speeds in open-field regions at the corresponding height[4]. Note, the measured variation of average flow speed from inflow to outflow below 1.09 $R_\odot$ is thought to be an artefact of erroneously high-estimated inward propagation speeds, hence inflow speeds, near the occulting disk (see Methods).

The measured properties of the Alfvénic waves, that is, amplitude and propagation speed, are further exploited to gain estimates for the gradients of magnetic field strength and density in the open-field region (see Methods). These gradients are shown in combination with electron density measurements available from CoMP, an analytic model of electron densities in an open magnetic field region from measurements in white light[7] and the magnetic field estimated from the PFSS extrapolation model (Fig. 5a–c).

The estimated electron density from CoMP is found to follow the expected trend from the analytic model and this is supported by the density gradient obtained from the magneto-seismological inversion. In addition, the magnetic field gradients from the magneto-seismology show good agreement with that predicted by the PFSS. The quality of the agreement between the different measures for the plasma parameters suggests that the physical properties of this particular open magnetic field region are well constrained by this set of observations.

Energetics of the waves. Now, combining CoMP and SDO observations, we are able to provide insight and constraints on the energetics of the observed Alfvénic waves. Using SDO, we measure the swaying motions of the fine-scale structures (Fig. 1d) at a height of $\sim 1.01\ R_\odot$ (Fig. 6), which is interpreted in terms of the MHD kink wave mode[28,31,60]. It is likely to be these swaying motions that correspond to the periodic fluctuations of the CoMP Doppler velocities[33]. The observational signature of torsional motions is oppositely directed Doppler velocities on either side of the fine structure. Owing to large spatial resolution of CoMP and its inability to resolve the finest structure in the corona, these oppositely directed velocities will average to zero and have a negligible contribution to the measured Doppler velocities. The kink motions are observed to be periodic, with the distribution of periods peaking at ~ 300 s (3.3 mHz)—coinciding with the frequency range of the enhanced power in the CoMP observations (Fig. 3a). The measured velocity amplitudes have an average value of 14 km s^{-1}. This value is significantly larger than CoMP Doppler velocity amplitudes (< 1 km s^{-1}—Fig. 4e), although the poor spatial resolution of CoMP is known to lead to a significant underestimation of the Doppler velocities due to bulk plasma motions[33,34]. Conversely, the measurements from SDO data are almost a third of the value of the non-thermal velocities (Fig. 4e) that CoMP measures (and other instruments[20,21,23–25])—although

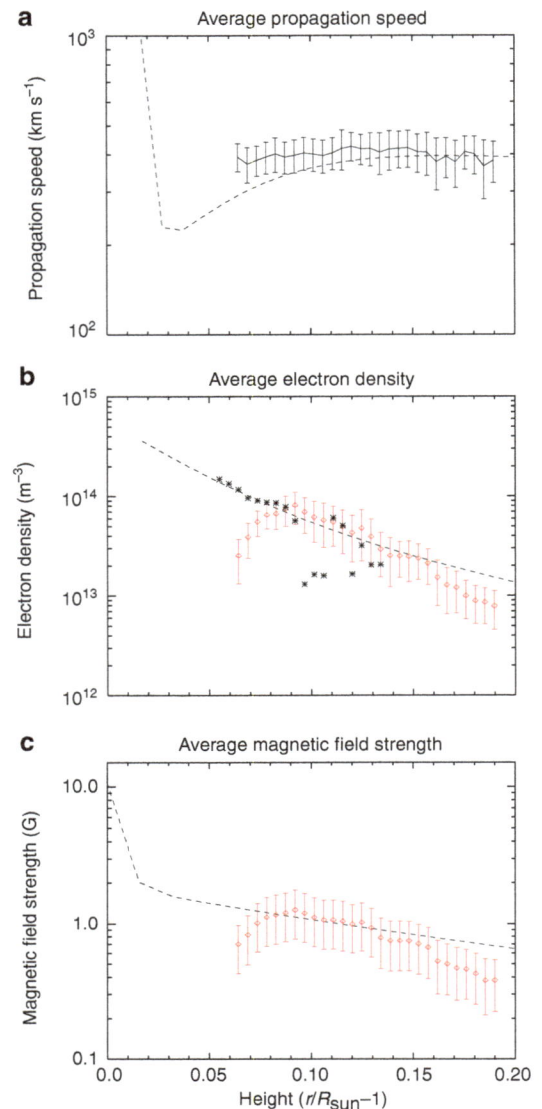

Figure 5 | Physical properties of the open-field region. Multiple complementary techniques are used to estimate the physical properties of the magnetized plasma. (**a**) The measured outward propagation speed (solid line), averaged over the boxed region in Fig. 4a, is compared with the estimated propagation speed calculated from the analytic density profile (dashed line in **b**) and PFSS magnetic field (dashed line in **c**). (**b**) The electron density estimates from CoMP (stars) is shown and is compared with the density profiles obtained from magneto-seismology of solar atmosphere (red diamonds) and an analytic density profile (dashed line—ref. 7). (**c**) The magnetic field strength profile from the magneto-seismology (red diamonds) is compared with the magnetic field strength estimated from the PFSS potential field extrapolation (dashed line). The error bars in all plots are the s.d. of the respective quantities.

velocity amplitudes up to 40 km s^{-1} are found in the SDO measurements (Fig. 6). The reasons for the larger amplitudes inferred from the non-thermal widths are still unclear. Should unresolved torsional Alfvénic motions be present along fine structure, it would then be expected that they would contribute to non-thermal line broadening[61]. In addition, unresolved kink motions with small displacement amplitudes (so as to be also unobservable with SDO) would also contribute.

The upper limit on the energy flux of propagating Alfvénic waves[37] is given by $F \approx 2 < \rho > v_{rms}^2 c_{ph} \delta$, we estimate the

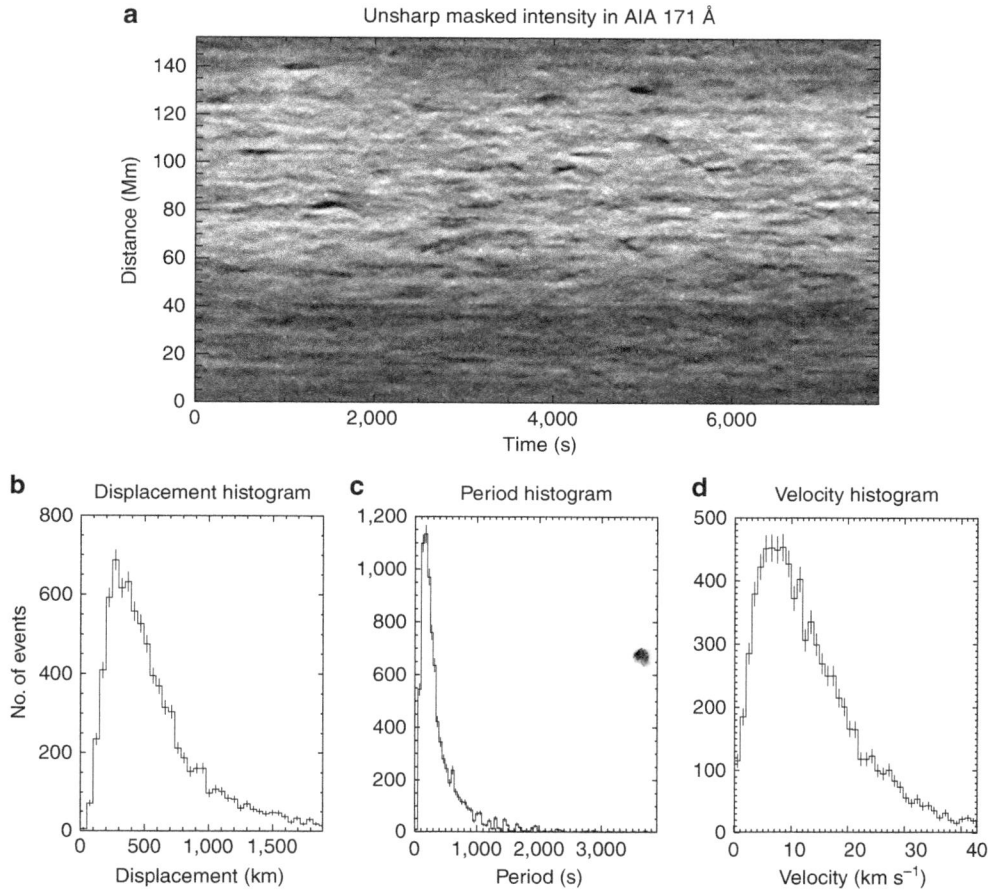

Figure 6 | Direct observations of Alfvénic waves with SDO. An example time–distance diagram (location shown in Fig. 1c,d) reveals that the fine-scale magnetic structure in the open-field region sways periodically (**a**), and this motion is interpreted from MHD wave theory as the kink mode. These Alfvénic waves observed with SDO contribute to both the CoMP Doppler velocities and non-thermal widths. The wave motion is measured directly (**b–d**) and the mean values with s.e.m. are 590 ± 5 km for displacement amplitude, 470 ± 6 s for the period and 14.7 ± 0.2 km s^{-1} for the velocity amplitude. Error bars show the sample s.d. for each bin in the histogram.

contribution of the kink MHD waves is on the order of $\sim 50\delta$ W m^{-2}, where v_{rms} is the root mean square velocity amplitude and $<\rho> = 9.5 \times 10^{-13}$ kg m^{-3} is the average density at 1.01 $R_\odot$ extrapolated from CoMP electron density measurements. A phase speed of 250 km s^{-1} is also used[30]. The factor δ corresponds to the filling factor of the waveguides in the observed region. The value of δ is unclear although it likely <50% (see Fig. 1). The total wave energy including both kink and torsional motions may be best calculated by the non-thermal widths, giving a total value of $\sim 300\delta$ W m^{-2} at 1.06 $R_\odot$ ($<\rho> = 2.7 \times 10^{-13}$ kg m^{-3}, $c_{ph} = 450$ km s^{-1})— which can meet the energy flux requirements for accelerating the solar wind assuming $\delta \sim 50\%$ (ref. 62). However, this number should be treated with some caution as it is uncertain exactly what physical phenomena are contributing to the non-thermal broadening[25,33].

Discussion

The initial study presented here demonstrates the potential for using Doppler velocity time series of the corona to understand Alfvénic wave propagation through the solar atmosphere and connecting the dynamics of the corona to the solar wind. The results reveal the existence of both outward and inward propagating Alfvénic waves in open magnetic field regions—a crucial element required if MHD wave turbulence is to develop.

We remind the reader that it is presently unknown whether the details of the nonlinear interaction of fast kink waves are similar to those of Alfvén waves. Although the kink waves are Alfvénic, the turbulent cascade of vorticity to small scales through their nonlinear interaction seems entirely plausible. Alternatively, the kink waves are subject to mode conversion via resonant absorption[15] and the energy could be transferred to torsional Alfvén waves before it is cascaded to higher frequencies.

Further, the ratio of outward to inward wave power demonstrates a frequency dependence that is suggestive of the waves being partially reflected in the corona, that is, a decreasing proportion of inward power to outward power with increasing frequency, which is also a feature in the models of Alfvénic wave propagation.

In addition to this, the power spectra reveal insights into the generation of the waves and their propagation. The slope of the power spectra for the Alfvénic waves shows a $1/f$ functional form that is comparable to previously measured solar wind spectra at 0.3 AU. This implies that the waves propagate outwards from the corona into the solar wind and at least partially maintain the shape of the power spectra[44]. Moreover, the observations reveal the presence of enhanced power centred on the frequencies routinely associated with peak p-mode power. Current theoretical models suggest that the acoustic waves generated in the solar interior can generate coronal Alfvénic waves through mode conversion[45,50]. We again highlight that these models use a homogenous corona, which raises questions about whether these results are directly applicable to the observations presented here. However, the current models suggest that vorticity is generated in the corona and the fast kink modes are one of the key transporters of vorticity in an inhomogeneous plasma[15]. This may advocate that p-modes could play a dominant role in generating coronal Alfvénic waves in the frequency range neighbouring 3 mHz. The role of acoustic waves is excluded from certain surface to heliosphere solar wind acceleration models dependent on Alfvén waves due to a simplifying assumption of incompressibility throughout the system (note that not all Alfvén wave models assume incompressibility). The indication that p-modes could play an important role in wave generation suggests that this assumption would neglect a fundamental physical process.

Currently, the estimates of energy flux from observable Alfvénic (that is, kink) waves in open-field regions advocate that there is insufficient energy to meet the demands for solar wind acceleration[31]. Although, CoMP measurements of the 10,747 Å line profile (and a host of other coronal emission line profile measurements from different instruments) reveal that the velocity amplitudes from non-thermal widths imply otherwise. This would indicate that only by including contributions from both resolved and unresolved (and both kink and torsional) Alfvénic waves can the energy flux requirement for solar wind acceleration be potentially met exclusively by waves. Further work is clearly needed to deconstruct the motions that comprise the non-thermal widths and provide a clear estimate for the energy flux of all Alfvénic waves.

This study also demonstrates the power of magneto-seismology of the solar atmosphere to probe the corona. Using the measured properties of the waves, evidence for the origins of the solar wind are seen in flow maps and significant constraints can be placed on other physical properties of the open magnetic field region, that is, magnetic field strength, density and the gradients of the two quantities. It is hoped that the measurements of the plasma parameters, along with the measured wave properties, for example, amplitude and power spectra, should provide a significant test scenario for models of solar wind acceleration, especially those which focus on MHD Alfvénic wave turbulence.

Methods

Observations and data reduction. The data used here were obtained with the CoMP[38] instrument on 27 March 2012 at 18:51:02 UT to 20:13:02 UT. The details of the acquisition and reduction of the data are fully described in refs 29,38,40 and we make use of the final data product that has a cadence of 30 s and a pixel size of 4″.46 (3,234 km). The final data product is intensities at three infrared wavelengths (10,745.0 Å—I_1, 10,746.2 Å—I_2 and 10,747.4 Å—I_3), which are positions centred on the Fe XIII emission line at 10,747 Å. The filter has a bandpass of 1.3 Å. Following ref. 63, for each pixel in the CoMP field of view in each time frame, we calculate the central intensity, Doppler shift and Doppler width of the line profile using an analytic fit of a Gaussian to the intensity values at each wavelength position.

We also make use of SDO AIA[64] and the Helioseismic and Magnetic Imager[65] data, which are taken from 18:00 UT to 20:10 UT and processed following standard procedures.

Velocity time–distance maps. Here, we describe the two complementary methods for analysing the wave power spectra.

For the first method, a box is defined over the northern region, which is 200″ wide and 125″ tall (shown in Fig. 2). We make the crude assumption that, in the boxed region, the direction of potential wave propagation is strictly in the North–South direction. Displayed in Fig. 2 is an example of Doppler velocity time–distance diagram from a strip of pixels within the box, oriented in the vertical direction and one pixel wide. The presence of the outward propagating Doppler velocity signals is visible and so is the periodic nature of these features.

For each pixel in the box, the time-averaged value of the velocity is subtracted from the time series. Then, for each vertical strip in the box, the time–distance diagram is subjected to a two-dimensional Hamming window before a two-dimensional Fast Fourier Transform (FFT) is performed and the power for a range of frequencies and wavenumbers in the vertical direction is calculated. The

outward propagating motions are decomposed from the inward propagating motions. Averaging the outward and inward power over the different strips, the power is integrated over wavenumber (k) and the outward and inward power spectra are determined (Fig. 3). The spectra are fit with a power law of the form $10^a f^b$. The outward spectra are fit in two frequency ranges while, the inward wave power spectra are not fit for the high frequency ranges due to the power spectra being predominantly noise in this regime. The results for the fits are given in Table 1.

It is highly instructive to examine the averaged f–k diagram obtainable from the time–distance diagrams. To improve the frequency/wavenumber resolution in the f–k diagrams, the mean subtracted and windowed time–distance diagrams are padded with zeros before the FFT is performed. The results are shown in Fig. 2d.

The second method used is a more refined version of the first, providing the f–k power spectra and, in addition, the propagation velocities of the waves. This technique was demonstrated in ref. 40 and involves the tracking of wave packets via a coherence approach. In brief, for each pixel in the CoMP data, the time series is filtered at 3.5 mHz and the coherence of this filtered time series with the filtered time series in neighbouring pixels is calculated. This provides islands of high coherence, typically elongated in a particular direction that corresponds to the angle of wave propagation. Assuming that these propagating waves are Alfvénic, the direction of propagation defines the orientation of the magnetic field in the plane of the sky. It has been demonstrated that this technique provides magnetic field orientations that are in excellent agreement with the CoMP magnetic field measurements from the linear polarization of the 10,747 Å emission line[40], and, hence, supporting the assumption of Alfvénic waves.

Starting at a particular pixel, we follow the direction of the magnetic field in the plane of sky using the calculated wave angles. Doing so allows tracks to be built that are aligned with the direction of wave propagation. We select 36 of these tracks that lie in the northern open-field region and these tracks are over-plotted on an intensity image in Fig. 2. The velocity time series at each point along the track is then extracted via cubic interpolation. An example of one of the time–distance diagrams from a track is shown in Fig. 2c. As with the first method, these Doppler velocity time–distance diagrams are used to create an average f–k diagram (Fig. 2e) and the k-integrated power spectra (Fig. 3a). Note that the power spectra and ratios in Fig. 3 are calculated from the original Doppler velocity time–distance diagrams, zero padding was not used to create these figures.

Note that the power in the f–k plots in Fig. 2 has been clipped between the given ranges. This leads to no loss of useful information as the signal is progressively noisier for larger k values and is done purely for aesthetic reasons.

Propagation speed calculation. An estimate of the propagation speed of the waves is obtained from the Doppler velocity time–distance diagrams. The procedure is based on the technique described in ref. 40, so we give a brief overview of the technique and additions made to the process.

For each pixel, a time–distance diagram is created from tracks that are built using the wave propagation angles. Each time–distance diagram is subject to an FFT and then decomposed into inward and outward parts, both of which are then subject to an inverse FFT. The following is applied to both the inward and outward time series separately. The time series central pixel along the track is used as the 'master time series' and is cross-correlated with the time series of the other pixels along the track to find values for the phase lag between the series. The lag values between the different time-series are fit with a linear function, which is then used to align the time series to the 'master time series'. The aligned series are then averaged to create a higher signal-to-noise 'master time series'. This new 'master time series' is then cross-correlated with the time series of the other pixels along the track to get improved estimates on the lag values. A sub-resolution estimate of the phase lag is obtained by fitting the cross-correlation peak with a parabola. This process still leaves a number of high error or outlier lag data points that can negatively influence the result. To remove these points, two steps are applied. First, a null hypothesis test is performed. The value of the cross-correlation of any of the two series is tested to see whether the two series are uncorrelated and each series is not auto-correlated. For two such series with populations that are normally distributed, the sample cross-correlations would have zero mean and a variance of $1/N$, N being the number of samples. For a 99.7% confidence interval, this assumes that the null hypothesis is incorrect if the cross-correlation value is $> 2.58/\sqrt{N}$. Lag data points with cross-correlation values below the cut-off are discarded. Having removed these data points, a linear fit to the lag data points is performed. From the residuals to the fit, the residual sample standard deviation is calculated. Here, the second cut is applied. Assuming the residuals are sampled from a normal distribution, a 99.7% significance level is calculated using t-distribution statistics for a two-tailed test. Any lag data point that lies outside this significance range is deemed an outlier and also removed. For the number of lag data points, this condition has a much stricter rejection criterion than, for example, Chauvenet's criterion. These two steps appear to be a highly robust way to remove erroneous data points. A linear function is then fit to the remaining lag data points and the propagation speed of the wave can be obtained from the gradient, with the error on the gradient calculated using the sample standard deviation.

For the propagation speed maps (Fig. 4), the time–distance diagrams used have varying spatial lengths depending on the position of the pixel under consideration.

The standard length is 25 spatial pixels but can be shorter, for example, for pixels near the occulting disk.

Note that the inward propagation speeds suffer from great uncertainty in the lowest part of the corona. Typically this results in unphysically large values for the propagation speed that are well in excess of the outward propagation speed and the estimated Alfvén speed using the PFSS estimates for magnetic field and density measurements from CoMP (Fig. 5a). It can be seen in the average power spectra (Fig. 3) that the amplitude of the inward propagating waves is comparable to the noise level for the higher frequencies. This has the greatest influence in the lower corona where amplitudes of waves are smaller (Fig. 4a–e). The amplitude of both inward and outward waves grow with height above the limb (Fig. 4d) and at a certain point the inward wave amplitude becomes large enough to overcome the noise and provide reliable measurements. In addition, the time–distance diagrams for pixels of the lower corona will typically have less data points spatially, as mentioned, which leads to greater uncertainties in the results. For this reason, the propagation speeds are clipped at $600\,\mathrm{km\,s^{-1}}$ and this predominantly affects the inward wave speed (Fig. 4b).

Density measurements. CoMP is able to measure both the 10,798 Å and the 10,747 Å Fe XIII emission lines, the ratio of which is known to be sensitive to the electron number density, n_e, in the corona[66]. At 20:29 UT, after the main 10,747 Å sequence was taken, CoMP was set to take scans at both 10,798 Å and 10,747 Å. The scans are used to calculate the central intensity for each wavelength and the intensity ratio is calculated for a number of consecutive frames and averaged. Then, using the CHIANTI database v7.0 (ref. 67), electron density versus intensity ratio curves are calculated for a range of heights above the photosphere taking into account the strong influence of photo-excitation on the formation of the two lines. The curves are then used to calculate the electron number density in the open-field region, which can be converted to coronal mass density using $\rho = \mu m_p n_e$, where μ is the mean atomic weight (1.3 for coronal abundances) and m_p is the proton mass, $1.67 \times 10^{-27}\,\mathrm{kg}$. The electron density measurements as a function of height (stars) are shown in Fig. 5b and compared with a scaled analytic density profile (dashed line—ref. 7). The measurements only cover a relatively short section of the corona due to weak emission in the corresponding region in 10,798 Å. As the emission decreases in strength with height, the associated errors increase and the number of pixels with significant signal decreases. Both these factors are likely to be responsible for the increased deviation in measured density from the analytic profile with height.

Estimating quantities from magneto-seismology. In the main text, we discussed the process of estimating the flow speed in the corona via magneto-seismology of the solar atmosphere by exploiting the measured propagation speeds of the waves. This relationship can be obtained from the following equation:

$$\rho_i\left(\frac{\partial}{\partial t} + U_i\frac{\partial}{\partial z}\right)^2\eta + \rho_e\left(\frac{\partial}{\partial t} + U_e\frac{\partial}{\partial z}\right)^2\eta = \frac{2B^2}{\mu_0}\frac{\partial^2\eta}{\partial z^2}. \qquad (2)$$

Here, η is the perturbation of the displacement, μ_0 is the magnetic permeability of free space, and the subscript i and e correspond to quantities inside and external to the waveguide, respectively. The magnetic field inside and external to the waveguide is assumed equal. This equation describes the kink wave in a thin over-dense waveguide in a low-beta plasma[68]. The condition for the waveguide to be thin is satisfied if the tube radius is much less than the wavelength. This is satisfied for the current observations where $\lambda \sim 100$ Mm. Looking for plane wave solutions and under the assumption that $U_i \approx U_e$, it is straightforward to arrive at equation (1), with c_{ph} equal to the kink speed. The square of the kink speed is given by $c_k^2 = B^2/(\mu_0 <\rho>)$, where $<\rho> = (\rho_i + \rho_e)/2$ is the average density.

Further information about coronal plasma properties can also be obtained via magneto-seismology of the solar atmosphere, if we exploit measurements of the wave amplitude and propagation speed together. In particular, gradients in the density and magnetic field strength can be estimated[69,70]. The results can then be compared with estimates from complementary techniques, that is, the magnetic field from PFSS extrapolations; density from CoMP line ratios and the analytic model from ref. 7.

MHD wave theory reveals that the relationship between the measureable quantities and the quantities to be estimated is given by,

$$v = C\sqrt{\frac{c_k}{\omega}}\frac{1}{\sqrt{B}}, \qquad (3)$$

where v is the velocity perturbation amplitude of the wave, ω is the angular frequency, C is a constant and B is the magnetic field strength. Due to the presence of the unknown C, it is only possible to find the gradients of the density and magnetic field strength. The magnitude of the quantities can only be known if we have initial values for both the quantities. Hence, the gradient of each quantity is normalized and can be scaled for comparison with the magnetic field and density estimates measured by other means.

To exploit equation (3), the propagation speed is averaged horizontally over the boxed region shown in Figs 2 and 4 and plotted as a function of height in Fig. 5a.

In addition to the propagation speed, the Doppler velocity amplitude is also required as a function of height. Each velocity time series in the boxed regions is smoothed using boxcar function of length 3. The root mean square (RMS) value of

the removed signal due to smoothing is found to match the estimated RMS of the noise, implying that the smoothing isolates the real velocity signal and suppresses the noise. Taking the RMS value of the smoothed velocity signal and averaging, we obtain the Doppler velocity displayed in Fig. 4e (stars).

Now, the variation in velocity amplitude for Alfvénic waves is expected to be due to changes in density alone and is given by $v \propto n_e^{-1/4}$ (assuming electron and ion number densities are equal and wave damping is negligible). Using this expression, the density gradient inferred from the RMS Doppler velocity amplitude (red diamonds) shows good agreement with the analytic density profile (Fig. 5b). Note that the density gradient determined from the Doppler velocity has been scaled for comparison with the analytic density profile and the density values obtained from the CoMP line ratios.

Using the measured Doppler velocity amplitude and propagation speed, the gradient in magnetic field is obtained using equation (3) and shown in Fig. 5c (red diamonds). We can compare this to the average magnetic field strength in the open-field region estimated from the SDO/Helioseismic and Magnetic Imager magnetograms via the PFSS extrapolation (dashed line). As we are unable to provide an exact boundary for the open-field region, the PFSS magnetic field is examined when averaged over latitudes > 80 degrees and > 85 degrees, both results incorporate the open magnetic flux region seen in Fig. 1. The results from both latitude boundaries show minor differences, mainly near the solar surface and are practically identical in the corona. It can be seen that there is good agreement between the two measures for the gradients of the magnetic field. Note that, again the magneto-seismological profile has been scaled to allow comparison with the PFSS profile.

The PFSS and magneto-seismology results suggest that the magnetic field strength decreases only by ~ 0.5 Gauss from 1.05 to 1.2 $R_\odot$. However, the magnetic field strength in the corona is weak, which means that this corresponds to a substantial decrease of 30–50%.

Finally, to complete the comparison between the different measures, the PFSS magnetic field and analytic density profile are combined to provide an estimate for the averaged Alfvén speed (dashed line—Fig. 5a). As noted in the main text, the agreement between magneto-seismology results and the PFSS and line ratio measurements provides re-assurance that the plasma properties of this open magnetic field region are well constrained.

Non-thermal widths. The non-thermal widths of the spectral lines are often used as an indicator of Alfvénic wave activity[20–25], so we also derive them to provide a comparison with the Doppler shift velocities. The non-thermal widths are also calculated from the Fe XIII emission line profile. The peak formation temperature of Fe XIII is ~ 1.6 MK, so it is expected that the line profile has a thermal width $\sigma_T = \sqrt{(2k_b T_e/m_{ion})} = 21\,\mathrm{km\,s^{-1}}$, where k_b is the Boltzmann constant, T_e is the electron temperature (assumed equal to the ion formation temperature) and m_{ion} is the mass of the ion. In addition, the line will be broadened by an instrumental width that has a value of $21\,\mathrm{km\,s^{-1}}$. The thermal and instrumental widths can be removed from the measured Doppler width to provide the non-thermal width. As with Doppler velocity, the non-thermal width is averaged horizontally over the boxed region and plotted as a function of height in Fig. 4e. The gradients in the amplitude of both the Doppler velocities and non-thermal widths are found to be approximately equal, demonstrating that both are measures of Alfvénic wave activity in the corona.

Wave energy flux. SDO/AIA is used to measure the swaying motions of the magnetic field, which can be interpreted as MHD kink waves. The magnetic field is outlined by density enhancements in the corona seen in the open-field region (Fig. 1d). The measurements are performed at a height of ~ 1.01 $R_\odot$ and follow the methods in ref. 31. The results are shown as histograms in Fig. 6. A median value for the sample population of each quantity, that is, displacement amplitude, period and velocity amplitude, is calculated. The distributions for each of these quantities show approximately log-normal behaviour; hence, the mean value of the log-normal distribution is calculated to give the average values. These values are given in Table 2.

To estimate the wave energy flux, we require an estimate for the mass density in the corona. This is determined from extrapolating from the CoMP electron density measurements back to 1.01 $R_\odot$ using the scaled analytic function[7]. The value of the mass density used is then $\rho = 9.5 \times 10^{-13}\,\mathrm{kg\,m^{-3}}$. In addition, the phase speed at 1.01 $R_\odot$ is taken from the estimate of average Alfvén speed in ref. 30.

The wave energy flux from the non-thermal widths uses the density and value for non-thermal widths at a height of 1.06 $R_\odot$.

Table 2 | Measured wave properties from SDO/AIA data.

	Median	Mean from log-normal
Displacement (km)	470	590 ± 5
Period (s)	268	414 ± 5
Velocity (km s⁻¹)	11.2	14.7 ± 0.2

References

1. Krieger, A. S., Timothy, A. F. & Roelof, E. C. A coronal hole and its identification as the source of a high velocity solar wind stream. *Sol. Phys.* **29**, 505–525 (1973).
2. Hassler, D. M. *et al.* Solar wind outflow and the chromospheric magnetic network. *Science* **283**, 810–813 (1999).
3. Peter, H. & Judge, P. G. On the Doppler shifts of solar ultraviolet emission lines. *Astrophys. J.* **522**, 1148–1166 (1999).
4. Giordano, S. *et al.* Identification of the coronal sources of the fast solar wind. *Astrophys. J.* **531**, L79–L82 (2000).
5. Tu, C. Y. *et al.* Solar wind origin in coronal funnels. *Science* **308**, 519–523 (2005).
6. Cranmer, S. R. Self-consistent models of the solar wind. *Space Sci. Rev.* **172**, 145–156 (2012).
7. Cranmer, S. R. & van Ballegooijen, A. A. On the generation, propagation, and reflection of Alfvén waves from the solar photosphere to the distant heliosphere. *Astrophys. J. Suppl.* **156**, 265–293 (2005).
8. Suzuki, T. K. & Inutsuka, S.-i. Making the corona and the fast solar wind: a self-consistent simulation for the low-frequency Alfvén waves from the photosphere to 0.3 AU. *Astrophys. J.* **632**, L49–L52 (2005).
9. Verdini, A. *et al.* A turbulence-driven model for heating and acceleration of the fast Wind in coronal holes. *Astrophys. J.* **708**, L116–L120 (2010).
10. Perez, J. C. & Chandran, B. D. G. Direct numerical simulations of reflection-driven, reduced magnetohydrodynamic turbulence from the sun to the Alfvén critical point. *Astrophys. J.* **776**, 124 (2013).
11. van der Holst, B. *et al.* Alfvén wave solar model (AWSoM): coronal heating. *Astrophys. J.* **782**, 81 (2014).
12. Leer, E. & Holzer, T. E. Constraints on the solar coronal temperature in regions of open magnetic field. *Sol. Phys.* **63**, 143–156 (1979).
13. Leer, E. & Holzer, T. E. Energy addition in the solar wind. *J. Geophys. Res.* **85**, 4681–4688 (1980).
14. Matthaeus, W. H. *et al.* Coronal heating by magnetohydrodynamic turbulence driven by reflected low-frequency waves. *Astrophys. J.* **523**, L93–L96 (1999).
15. Goossens, M. *et al.* Surface Alfvén waves in solar flux tubes. *Astrophys. J.* **753**, 111 (2012).
16. Goossens, M. *et al.* On the nature of kink MHD waves in magnetic flux tubes. *Astron. Astrophys.* **503**, 213 (2009).
17. Belcher, J. W. & Davis, Jr. L. Large-amplitude Alfvén waves in the interplanetary medium-2. *J. Geophys. Res.* **76**, 3534–3563 (1971).
18. Hassler, D. M. *et al.* Line broadening of Mg X 609 and 625 A coronal emission lines observed above the solar limb. *Astrophys. J.* **348**, L77–L80 (1990).
19. Seely, J. F. *et al.* Turbulent velocities and ion temperatures in the solar corona obtained from SUMER line widths. *Astrophys. J.* **484**, L87–L90 (1997).
20. Doyle, J. G., Banerjee, D. & Perez, M. E. Coronal line-width variations. *Sol. Phys.* **181**, 91–101 (1998).
21. Banerjee, D., Teriaca, L., Doyle, J. G. & Wilhelm, K. Broadening of SI VIII lines observed in the solar polar coronal holes. *Astron. Astrophys.* **339**, 208–214 (1998).
22. Moran, T. G. Interpretation of coronal off-limb spectral line width measurements. *Astron. Astrophys.* **374**, L9–L11 (2001).
23. Banerjee, D., Pèrez-Suàrez, D. & Doyle, J. G. Signatures of Alfvén waves in the polar coronal holes as seen by EIS/Hinode. *Astron. Astrophys.* **501**, L15–L18 (2009).
24. Bemporad, A. & Abbo, L. Spectroscopic signature of Alfvén waves damping in a polar coronal hole up to 0.4 solar radii. *Astrophys. J.* **751**, 110 (2012).
25. Chae, J., Schühle, U. & Lemaire, P. SUMER measurements of non-thermal motions: constraints on coronal heating mechanisms. *Astrophys. J.* **505**, 957–973 (1998).
26. De Pontieu, B. *et al.* Chromospheric Alfvénic waves strong enough to power the solar wind. *Science* **318**, 1574–1577 (2007).
27. Morton, R. J. *et al.* Observations of ubiquitous compressive waves in the sun's chromosphere. *Nat. Commun.* **3**, 1315 (2012).
28. Morton, R. J. & McLaughlin, J. A. High-resolution observations of active region moss and its dynamics. *Astrophys. J.* **789**, 105 (2014).
29. Tomczyk, S. *et al.* Alfvén waves in the solar corona. *Science* **317**, 1192–1196 (2007).
30. McIntosh, S. W. *et al.* Alfvénic waves with sufficient energy to power the quiet solar corona and fast solar wind. *Nature* **475**, 477–480 (2011).
31. Thurgood, J. O., Morton, R. J. & McLaughlin, J. A. First direct measurements of transverse waves in solar polar plumes using SDO/AIA. *Astrophys. J.* **790**, L2 (2014).
32. Goossens, M. *et al.* The transverse and rotational motions of magneto-hydrodynamic kink waves in the solar atmosphere. *Astrophys. J.* **788**, 9 (2014).
33. McIntosh, S. W. & De Pontieu, B. Estimating the 'dark' energy content of the solar corona. *Astrophys. J.* **761**, 138 (2012).
34. De Moortel, I. & Pascoe, D. J. The effects of line-of-sight integration on multi-strand coronal loop oscillations. *Astrophys. J.* **746**, 31 (2012).
35. Morton, R. J. & McLaughlin, J. A. Hi-C and AIA observations of transverse magnetohydrodynamic waves in active regions. *Astron. Astrophys.* **553**, 10 (2013).
36. Goossens, M. *et al.* Energy content and propagation in transverse solar atmospheric waves. *Astrophys. J.* **768**, 191 (2013).
37. Van Doorsselaere, T. *et al.* Energy propagation by transverse waves in multiple flux tube systems using filling factors. *Astrophys. J.* **795**, 18 (2014).
38. Tomczyk, S. *et al.* An instrument to measure coronal emission line polarization. *Sol. Phys.* **247**, 411–428 (2008).
39. Schrijver, C. J. & De Rosa, M. L. Photospheric and heliospheric magnetic fields. *Sol. Phys.* **212**, 165–200 (2003).
40. Tomczyk, S. & McIntosh, S. W. Time-distance seismology of the solar corona with CoMP. *Astrophys. J.* **697**, 1384–1391 (2009).
41. Verdini, A. & Velli, M. Alfvén waves and turbulence in the solar atmosphere and solar wind. *Astrophys. J.* **662**, 669–676 (2007).
42. Roberts, D. A. Demonstrations that the solar wind is not accelerated by waves or turbulence. *Astrophys. J.* **711**, 1044–1050 (2010).
43. Roberts, D. A. Evolution of the spectrum of solar wind velocity fluctuations from 0.3 to 5 AU. *J. Geophys. Res.* **115**, A12101 (2010).
44. Verdini, A., Grappin, R., Pinto, R. & Velli, M. On the Origin of the 1/f spectrum in the solar wind magnetic field. *Astrophys. J.* **750**, L33 (2012).
45. Hansen, S. C. & Cally, P. S. Benchmarking fast-to-Alfvén mode conversion in a cold MHD plasma. II. how to get Alfvén waves through the solar transition region. *Astrophys. J.* **751**, 31 (2012).
46. Bogdan, T. J. *et al.* Waves in the magnetized solar atmosphere. II. waves from localized sources in magnetic flux concentrations. *Astrophys. J.* **599**, 626–660 (2003).
47. Khomenko, E., Collados, M. & Felipe, T. Nonlinear numerical simulations of magneto-acoustic wave propagation in small-scale flux tubes. *Sol. Phys.* **251**, 589–611 (2008).
48. Vigeesh, G., Hasan, S. S. & Steiner, O. Wave propagation and energy transport in the magnetic network of the sun. *Astron. Astrophys.* **508**, 951–962 (2009).
49. Fedun, V., Shelyag, S. & Erdélyi, R. Numerical modeling of footpoint-driven magneto-acoustic wave propagation in a localized solar flux tube. *Astrophys. J.* **727**, 17 (2011).
50. Cally, P. S. & Goossens, M. Three-dimensional MHD wave propagation and conversion to Alfvén waves near the solar surface. I. direct numerical solution. *Sol. Phys.* **251**, 251–265 (2008).
51. Cally, P. S. & Hansen, S. C. Benchmarking fast-to-Alfvén mode conversion in a cold magnetohydrodynamic plasma. *Astrophys. J.* **738**, 119 (2011).
52. Khomenko, E. & Cally, P. S. Numerical simulations of conversion to Alfvén waves in sunspots. *Astrophys. J.* **746**, 68 (2012).
53. Schunker, H. & Cally, P. S. Magnetic field inclination and atmospheric oscillations above solar active regions. *Mon. Not. R. Astron. Soc* **372**, 551–564 (2006).
54. Jain, R., Gascoyne, A. & Hindman, B. W. Interaction of p modes with a collection of thin magnetic tubes. *Mon. Not. R. Astron. Soc* **415**, 1276–1279 (2011).
55. Gascoyne, A., Jain, R. & Hindman, B. W. Energy loss of solar p modes due to the excitation of magnetic sausage tube waves: importance of coupling the upper atmosphere. *Astrophys. J.* **789**, 109 (2014).
56. Dowdy, Jr. J. F., Rabin, D. & Moore, R. L. On the magnetic structure of the quiet transition region. *Sol. Phys.* **105**, 35–45 (1986).
57. Peter, H. On the nature of the transition region from the chromosphere to the corona of the Sun. *Astron. Astrophys.* **374**, 1108–1120 (2001).
58. Tian, H. *et al.* The nascent fast solar wind observed by the EUV imaging spectrometer on board hinode. *Astrophys. J.* **709**, L88–L93 (2010).
59. Pinto, R. *et al.* Time-dependent hydrodynamical simulations of slow solar wind, coronal inflows, and polar plumes. *Astron. Astrophys.* **497**, 537–543 (2009).
60. Van Doorsselaere, T. *et al.* Seismological demonstration of perpendicular density structuring in the solar corona. *Astron. Astrophys.* **491**, L9–L12 (2008).
61. Zaqarashvili, T. V. Observation of coronal loop torsional oscillation. *Astron. Astrophys.* **399**, L15–L18 (2003).
62. Hansteen, V. H. & Leer, E. Coronal heating, densities, and temperatures and solar wind acceleration. *J. Geophys. Res.* **100**, 21577–21594 (1995).
63. Tian, H. *et al.* Observations of coronal mass ejections with the coronal multichannel polarimeter. *Sol. Phys.* **288**, 637–650 (2013).
64. Lemen, J. R. *et al.* The atmospheric imaging assembly (AIA) on the solar dynamics observatory (SDO). *Sol. Phys.* **275**, 17–40 (2012).
65. Scherrer, P. H. *et al.* The helioseismic and magnetic imager (HMI) investigation for the solar dynamics observatory (SDO). *Sol. Phys.* **275**, 207–227 (2012).
66. Flower, D. R. & Pineau des Forets, G. Excitation of the Fe XIII spectrum in the solar corona. *Astron. Astrophys.* **24**, 181 (1973).
67. Landi, E. *et al.* CHIANTI—An atomic database for emission lines. XII. version 7 of the database. *Astrophys. J.* **744**, 99 (2012).
68. Ruderman, M. S. The effects of flows on transverse oscillations of coronal loops. *Sol. Phys.* **267**, 377–391 (2010).

69. Verth, G., Goossens, M. & He, J.-S. Magnetoseismological determination of magnetic field and plasma density height variation in a solar spicule. *Astrophys. J.* **733**, L15 (2011).
70. Morton, R. J. Magneto-seismological insights into the penumbral chromosphere and evidence for wave damping in spicules. *Astron. Astrophys.* **566**, A90 (2014).

Acknowledgements

R.J.M. is grateful to Northumbria University for the award of the Anniversary Fellowship and the Leverhulme Trust for the award of an Early Career Fellowship. R.J.M. also thanks the Higher Education Funding Council for England and the High Altitude Observatory for financial assistance that enabled this work and acknowledges IDL support provided by STFC. The National Center for Atmospheric Research is sponsored by the National Science Foundation. R.P. was supported by the FP7 project #606692 (HELCATS).

Author contributions

R.J.M. and S.T. performed the analysis of the observations. R.J.M. and R.P. interpreted the observations. S.T. designed the instrument; the observing runs and handled the initial processing of the raw data. All the authors discussed the results and commented on the manuscript.

Additional information

Competing financial interests: The authors declare no competing financial interests.

13

Optical aperture synthesis with electronically connected telescopes

Dainis Dravins[1], Tiphaine Lagadec[1,†] & Paul D. Nuñez[2,3,†]

Highest resolution imaging in astronomy is achieved by interferometry, connecting telescopes over increasingly longer distances and at successively shorter wavelengths. Here, we present the first diffraction-limited images in visual light, produced by an array of independent optical telescopes, connected electronically only, with no optical links between them. With an array of small telescopes, second-order optical coherence of the sources is measured through intensity interferometry over 180 baselines between pairs of telescopes, and two-dimensional images reconstructed. The technique aims at diffraction-limited optical aperture synthesis over kilometre-long baselines to reach resolutions showing details on stellar surfaces and perhaps even the silhouettes of transiting exoplanets. Intensity interferometry circumvents problems of atmospheric turbulence that constrain ordinary interferometry. Since the electronic signal can be copied, many baselines can be built up between dispersed telescopes, and over long distances. Using arrays of air Cherenkov telescopes, this should enable the optical equivalent of interferometric arrays currently operating at radio wavelengths.

[1] Lund Observatory, Lund University, Box 43, Lund SE-22100, Sweden. [2] Collège de France, 11 Place Marcelin Berthelot, Paris FR-75005, France. [3] Laboratoire Lagrange, Observatoire de la Côte d'Azur, BP 4229, Nice FR-06304, France. † Present addresses: ESTEC, European Space Research and Technology Centre, Keplerlaan 1, NL-2200 AG Noordwijk, The Netherlands (T.L.); JPL, Jet Propulsion Laboratory, California Institute of Technology, 4800 Oak Grove Drive, Pasadena, California 91109-8099, USA (P.D.N.). Correspondence and requests for materials should be addressed to D.D. (email: dainis@astro.lu.se).

In optical astronomy, the highest angular resolution is presently offered by phase/amplitude interferometers combining light from telescopes separated by baselines up to a few hundred metres. Tantalizing results show how stellar disks start to become resolved, revealing stars as a diversity of individual objects, although until now feasible only for a small number of the largest ones. Concepts have been proposed to extend such facilities to scales of a kilometre or more, as required for surface imaging of bright stars with typical sizes of a few milliarcseconds. However, their realization remains challenging, both due to required optical and atmospheric stabilities within a fraction of an optical wavelength, and the need to span many interferometric baselines (given that optical light cannot be copied with retained phase, but must be divided and diluted by beamsplitters to achieve interference among multiple telescope pairs). While atmospheric issues could be avoided by telescope arrays in space, such are impeded by their complexity and their cost. However, atmospheric issues can be circumvented by measuring higher-order coherence of light through intensity interferometry.

In the following, we present a laboratory demonstration of a multi-telescope array to verify the end-to-end sequence of operation from observing star-like sources to reconstruction of their images.

Results

Origin and principles of intensity interferometry.
The technique of intensity interferometry[1], once pioneered by Hanbury Brown and Twiss[5] is effectively insensitive to atmospheric turbulence and permits very long baselines at short optical wavelengths also. What is observed is the second-order coherence of light (that of intensity I, not of amplitude or phase), by computing temporal correlations of arrival times between photons recorded in different telescopes observing the same target. When the telescopes are close together, the intensity fluctuations measured in both telescopes are correlated in time, but when further apart the degree of correlation diminishes.

Although this behaviour can be understood in terms of classical optical waves undergoing random phase shifts, it is fundamentally a two-photon process measuring the degree of photon 'bunching' in time, and its full explanation requires a quantum description. The concept is normally seen as the first experiment in quantum optics[2-4], with its first realization already in the 1950s, when Hanbury Brown and Twiss[5] succeeded in measuring the diameter of the bright and hot star Sirius. This laid the foundation for experiments of photon correlations including also states of light that do not have classical counterparts (such as photon antibunching), and has found numerous applications outside optics, particularly in particle physics since the same bunching and correlation properties apply to other bosons, that is, particles with integer quantum spin, whose statistics in an equilibrium maximum-entropy state follow the Bose–Einstein distribution. The method can also be seen as mainly exploiting the corpuscular rather than the wave nature of light, even if those properties cannot be strictly separated.

The quantity measured is $<I_1(t) \cdot I_2(t)> = <I_1(t)> <I_2(t)> (1 + |\gamma_{12}|^2)$, where $<>$ denotes temporal averaging and γ_{12} is the mutual coherence function of light between locations 1 and 2, the quantity commonly measured in phase/amplitude interferometers. Compared with randomly fluctuating intensities, the correlation between intensities I_1 and I_2 is 'enhanced' by the coherence parameter, and an intensity interferometer thus measures $|\gamma_{12}|^2$ with a certain electronic time resolution. Realistic values of a few nanoseconds correspond to light-travel distances around 1 m, and permitted error budgets relate to such a number rather than to the optical wavelength, making the method

practically immune to atmospheric turbulence and optical imperfections. The above coherence relation holds for ordinary thermal ('chaotic') light where the light wave undergoes 'random' phase jumps on timescales of its coherence time but not necessarily for light with different quantum statistics[2-4].

Intensity interferometry in the modern era. The method has not been pursued in astronomy since Hanbury Brown's[1] early two-telescope intensity interferometer in Narrabri, Australia, but new possibilities are opening up with the erection of large arrays of air Cherenkov telescopes. Although primarily devoted to measuring optical Cherenkov light in air induced by cosmic gamma rays, their optical performance, telescope sizes and distributions on the ground would be suitable for such applications[6].

Realizing that potential, various theoretical simulations of intensity interferometry observations with extended telescope arrays have been made[7-11], in particular examining the overall sensitivity and limiting stellar magnitudes. Different studies conclude that, with current electronic performance and current Cherenkov telescope designs, one night of observation in one single wavelength band permits measurements of hotter stars down to visual magnitude about $m_v = 8$, giving access to thousands of sources. Such values agree with extrapolations from the early pioneering observations, also when including observational issues such as background light from the night sky[10]. Fainter sources could be reached once simultaneous measurements in multiple wavelength bands can be realized, or if full two-dimensional (2D) data are not required.

Interferometer array in the laboratory. In this work, we report the first end-to-end practical experiments of such types of measurements. A set-up was prepared in a large optics laboratory with artificial sources ('stars'), observed by an array of small telescopes of 25 mm aperture, each equipped with a photon-counting solid-state detector (silicon avalanche photodiode operated in Geiger-mode). The continuous streams of photon-counts were fed into electronic firmware units in real time (with 5 ns time resolution) computing temporal cross correlations between the intensity fluctuations measured across baselines between many different pairs of telescopes. The degree of mutual correlation for any given baseline provides a measure of the second-order spatial coherence of the source at the corresponding spatial frequency. Numerous telescope pairs of different baseline lengths and orientations populate the interferometric Fourier-transform plane, and the measured data provide a 2D map of the second-order spatial coherence of the source, from which its image can be reconstructed.

Experiment set-up. Artificial 'stars' (single and double, round and elliptic) were prepared as tiny physical holes drilled in metal (using miniature drills otherwise intended for mechanical watches). Typical aperture sizes of 0.1 mm subtend an angle of ~1 arcsec as observed from the 23 metre source-to-telescope distance in the laboratory.

The technique of intensity interferometry, in principle, requires chaotic light for its operation[2-4] in order for 'random' intensity fluctuations to occur and, in practice, also sources of high surface brightness (such as stars hotter than the Sun). This brightness requirement of a high effective source temperature comes from the need for telescopes to be small enough to resolve the structures in the Fourier domain, while still receiving sufficiently high photon fluxes. Of course, the same limitations apply to laboratory sources as to celestial ones. There are numerous hot stars in the sky but correspondingly hot laboratory sources are less easy to find. Ordinary sources producing chaotic light have too low effective temperatures (for example, high-pressure arc lamps of ~ 5,000 K) to permit conveniently short integration times while laser light cannot be directly used since it is not chaotic (ideally, undergoing no intensity fluctuations whatsoever).

Following various tests, the required brilliant illumination was produced by scattering $\lambda = 532$ nm light from a 300 mW laser against microscopic particles in thermal (Brownian) motion. Those were monodispersive polystyrene spheres of 0.2 μm diameter, suspended in room-temperature distilled water, undergoing random motion due to collisions with the water molecules surrounding them. On scattering, the laser light is broadened through Doppler shifts, producing a spectral line with a Lorentzian wavelength shape, and with a chaotic (Gaussian) distribution of the electric field amplitudes[12,13]. Such light is thus equivalent in its photon statistics and intensity fluctuations to the thermal (white) light expected from any normal star, the difference being that the spectral passband is now very much narrower. The width of the scattered line was estimated from its measured temporal coherence to be ~ 10–100 kHz, with a brightness temperature of the order 10^5 K. The relatively long coherence times enabled convenient measurements on microsecond scales, with typical photon-count rates between 0.1 and 1 MHz, with integration times of a few minutes for each telescopic baseline. Such count rates assure negligible photon-noise contributions although there may still remain issues of systematics due to, for example, slightly inhomogeneous illumination of the artificial stars.

Here we note that the signal-to-noise ratio in intensity interferometry in principle is independent of optical bandpass, be it broadband white light or just a narrow spectral line[1]. For any broader bandpass, realistic electronic time resolutions are slower than optical coherence times, and the noisier measurement of a smaller photon flux from a narrower passband (with longer coherence time) is compensated by lesser averaging over fewer coherence times. Thus, also the present quasi-monochromatic light source can represent measurements over broader optical bandpasses.

Simulating a large telescope array. Angular resolution in optical systems is often quantified by the Airy disk diffraction radius $\theta \approx 1.22 \ \lambda/D$ rad. At $\lambda = 532$ nm, the baseline D required to resolve 1 arcsec is ~ 10 cm, dictating a rather compact set-up of small telescopes. Our five units offered baselines between 3 and 20.5 cm, the former constraining the largest field to nominally 4.5 arcsec and the latter limiting the resolution to 0.65 arcsec. The zero baseline (needed for calibration) was realized with two additional separate telescopes behind one beamsplitter, both telescopes observing the same signal.

To mimic a large telescope array, the effective number of telescopes and baselines was increased by measuring across different interferometer angles. The array was mounted on a horizontal optical table and thus covered horizontal baselines only. To cover also oblique and vertical baselines in the 2D Fourier-transform plane, the position angle of the artificial 'star' was successively rotated relative to the plane of the telescopes. Various details of the overall laboratory arrangements and of measuring techniques are described elsewhere[14]. With N telescopes, N ($N-1$)/2 different baselines can be constructed. Five telescopes give 10 baselines at any given angle, and a series of 18 measurements in steps of 10° over a total of 180° produces 180 baselines. The second-order coherence in a point (u,v) of the Fourier-transform plane equals that in ($-u, -v$), so each baseline provides two points; data for angles up to 360° are copies of the first ones. Figure 1 shows the coherence pattern for one thus measured asymmetric binary 'star', with one larger and

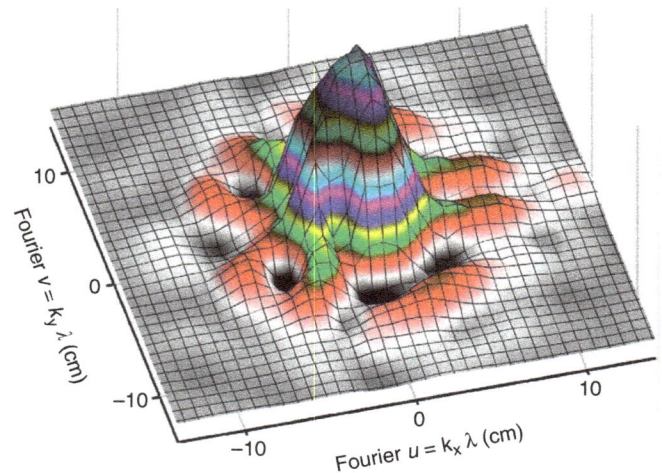

Figure 1 | Second-order optical coherence surface: measurements for an artificial asymmetric binary star built up from intensity-correlation measurements over 180 baselines between pairs of small optical telescopes in the laboratory. The coordinates refer to the plane of the Fourier transform of the source image and correspond to different telescope baseline lengths and orientations. The image reconstructed from these data is in Fig. 2, right.

one smaller component (flux ratio 4:1). No smoothing was applied to the measured coherence values, neither for this plot nor for the later image reconstructions. The overplotted surface merely visualizes the global measured pattern, whose individual points are too numerous to be readily displayed.

Image reconstruction principles. Image reconstruction from second-order coherence implies some challenges not present in first-order phase/amplitude techniques. While intensity interferometry has the advantage of not being sensitive to phase errors in the optical light paths, it also does not measure such phases but rather obtains the absolute magnitudes of the respective Fourier-transform components of the source image. Such data can by themselves fit model parameters such as stellar diameters or binary separations but images cannot be directly computed through merely an inverse Fourier transform.

Theoretical simulations have verified that, provided the Fourier-transform plane is well-filled with data, 2D image restoration becomes feasible[8,9,11]. The pattern of second-order coherence is equivalent to the intensity pattern produced by diffraction of light in a correspondingly sized aperture, and it is already intuitively clear that a thorough mapping of that pattern must put stringent constraints on the source geometry.

Image reconstruction procedures and results. Strategies in analysing intensity interferometry data involve estimating Fourier phases from Fourier magnitudes[8,15,16]. Since the Fourier transform of a finite object is an analytic function, the Cauchy–Riemann equations can be used to find derivatives of the phase from derivatives of the magnitude, producing images that are unique except for translation and reflection. Here one could start with a 1D phase estimate along a single slice through the (u,v) plane origin. 2D coverage requires combining multiple 1D reconstructions, while ensuring mutual consistency between adjacent ones. However, these algorithms appear to require a very dense coverage of the (u,v) plane and, despite our numerous baselines, did not appear efficient with current data.

Instead, an inverse-problem approach was taken, imposing very general constraints to interpolate between Fourier frequencies ('regularization'). The Multi-aperture Image Reconstruction Algorithm, MiRA (previously tested on simulated data[9]), produced more stable results. This iterative procedure maximizes the agreement between measurements and the squared modulus of the Fourier transform of the reconstructed image. Regularization techniques can be applied to favour solutions with certain properties, for example, if something is initially known about the source. To reconstruct the elliptical star in Fig. 2, a 'compact' solution (quadratic regularization) was favoured, and a 'smooth' solution for the binary (quadratic smoothness regularization)[17].

Figure 2 shows examples of such image reconstructions at $\lambda = 532$ nm from measurements with 100 and 180 optical baselines. As far as we are aware, these are the first diffraction-limited images in visual light that have been reconstructed from an array of dispersed optical telescopes, connected by electronic software only, without any optical links between them.

Similar to any phase/amplitude interferometry or aperture synthesis, the fidelity of the reconstructed images depends on measurement noise, the density of coverage across the Fourier (u,v) plane and the character of the 'cleaning' algorithms applied. In the present measurements, the random part of the noise is essentially negligible but some systematics are likely present, for example, due to a probably somewhat inhomogeneous illumination of the artificial stars that makes their brightness shapes not identical to their geometrical apertures.

Reconstructions were tested on different sources. For the single elliptic star (Fig. 2, left), its major and minor axes could be determined to a precision of 2%, a number estimated from several image reconstructions with different initial random images. Even slightly better values were obtained for a symmetric binary. However, the present asymmetric one was chosen to be a more difficult target, because of the contrast between its components, and since the Fourier-plane coverage is marginal for both the largest and smallest scales, contributing to some image artefacts seen in Fig. 2. Still, the reconstruction precision is consistent with what is found in previous detailed theoretical simulations[8], where the reconstructed radii gradually become less precise for greater brightness ratios between both binary members.

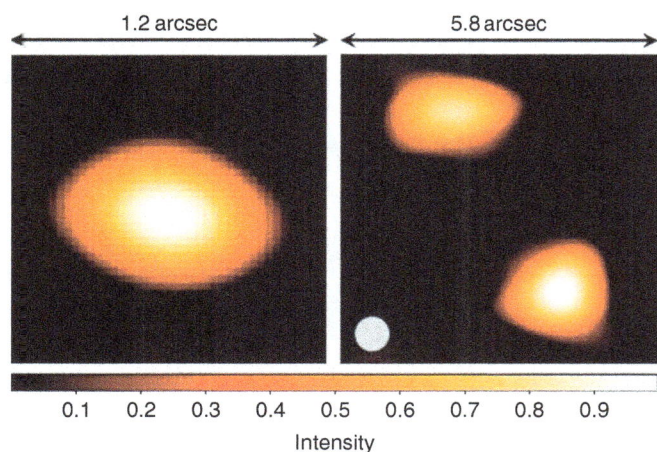

Figure 2 | Optical images reconstructed from electronic intensity interferometry; measurements with 100 and 180 baselines, respectively, of a single elliptical artificial 'star' and an asymmetric binary one (flux ratio 4:1). Sufficiently dense coverage of the interferometric Fourier-transform plane enables a full reconstruction of source images, despite the lack of Fourier phase information. In principle, the images are unique except for their possible mirrored reflections. The spatial scales are similar on both axes; colours show reconstructed linear intensities, the circle indicates the 0.65 arcsec resolution set by the longest baseline.

The computing effort is modest (and not comparable to the extensive numerical efforts involved in radio aperture synthesis that involve also spectral analysis). To obtain an image of 10,000 pixels from measurements across 100 baselines, starting from a random image, requires just some 10 s on a 3 GHz processor. The exact time can be shorter if one already has some initial guess (perhaps obtained from some 'raw' reconstruction from a Cauchy–Riemann analysis), or be longer for additional baselines or more numerous pixels.

The image quality was further explored as a function of the sampling or smoothing in the (u,v) plane, affecting the pixel sizes in the reconstructed image. The best performance was found when the number of samples within the Fourier plane roughly matches the number of baseline measurements. Possibly, there could be other dependences for noise-limited data, where some kind of smoothing may be required. In general, the optimization of reconstruction for different numbers of sample points in the presence of various noise levels and incomplete Fourier-plane coverages is a somewhat complex problem that will require further studies (analogous to what in the past has been made for aperture synthesis in the radio).

Long-baseline optical interferometry. The longer-term ambition is to connect widely separated telescopes in a large optical aperture synthesis array. Similar to current radio interferometry, the electronic signal from any telescope can be freely copied or even stored for possible later analysis. Realistic signal bandwidths for measured intensity fluctuations are in the order of 100 MHz, very comparable to those encountered in radio arrays. However, the requirements on the signal correlators are much more modest than for radio phase/amplitude interferometers. Optical intensity interferometry can in practice only detect spatial coherence while the wavelengths are selected by colour filters or through detector responses, not—as the case in radio—by temporal coherence measurements. Consequently, the correlator electronics can be simple table-top devices rather than massive computing facilities.

The signal-to-noise properties[1,6,7] of intensity interferometry require the telescopes to be large to reach adequately low photon noise, but their optical precision need not be higher than that of ordinary air Cherenkov telescopes. However, the limiting signal-to-noise ratios may still depend on their exact optical design. For example, how high electronic time resolutions that can be exploited may be constrained by the time spread of light paths inside the telescope. Details of such and other parameters are discussed elsewhere[6,7,10].

Obvious candidates for a kilometre-sized optical imager are the telescopes of the planned Cherenkov Telescope Array[18], foreseen to have in the order of 100 telescopes, distributed over a few square kilometres. Assuming it becomes available for also interferometric observations, the spatial resolution at the shortest optical wavelengths may approach $\sim 30\,\mu$as. Such resolutions have hitherto been reached only in the radio, and it is awkward to predict what features on, or around, stellar surfaces will appear in the optical[6]. However, to appreciate the meaning of such resolutions, we note that a hypothetical Jupiter-size exoplanet in transit across some nearby bright star such as Sirius would subtend an easily resolvable angle of some 350 μas (ref. 14). While spatially resolving the disk of an exoplanet in its reflected light may remain unrealistic for the time being, the imaging of its dark silhouette on a stellar disk—while certainly very challenging[19]—could perhaps be not quite impossible.

References

1. Hanbury Brown, R. *The Intensity Interferometer. Its Applications to Astronomy* (Taylor & Francis, London, 1974).

2. Labeyrie, A., Lipson, S. G. & Nisenson, P. *An Introduction to Optical Stellar Interferometry* (Cambridge Univ. Press, Cambridge, 2006).

3. Saha, S. K. *Aperture Synthesis. Methods and Applications to Optical Astronomy* (Springer, 2011).

4. Shih, Y. *An Introduction to Quantum Optics. Photon and Biphoton Physics* (CRC Press, Boca Raton, 2011).

5. Hanbury Brown, R. & Twiss, R. Q. A test of a new type of stellar interferometer on Sirius. *Nature* **178**, 1046–1048 (1956).

6. Dravins, D., LeBohec, S., Jensen, H. & Nuñez, P. D. for the CTA consortium, Optical intensity interferometry with the Cherenkov Telescope Array. *Astropart. Phys.* **43**, 331–347 (2013).

7. Dravins, D., LeBohec, S., Jensen, H. & Nuñez, P. D. Stellar intensity interferometry: Prospects for sub-milliarcsecond optical imaging. *New Astron. Rev.* **56**, 143–167 (2012).

8. Nuñez, P. D., Holmes, R., Kieda, D. & LeBohec, S. High angular resolution imaging with stellar intensity interferometry using air Cherenkov telescope arrays. *Mon. Not. R. Astron. Soc.* **419**, 172–183 (2012).

9. Nuñez, P. D., Holmes, R., Kieda, D., Rou, J. & LeBohec, S. Imaging submilliarcsecond stellar features with intensity interferometry using air Cherenkov telescope arrays. *Mon. Not. R. Astron. Soc.* **424**, 1006–1011 (2012).

10. Rou, J., Nuñez, P. D., Kieda, D. & LeBohec, S. Monte Carlo simulation of stellar intensity interferometry. *Mon. Not. R. Astron. Soc.* **430**, 3187–3195 (2013).

11. Dolne, J. J., Gerwe, D. R. & Crabtree, P. N. Cramer-Rao lower bound and object reconstruction performance evaluation for intensity interferometry. Proc. SPIE 9146, Optical and Infrared Interferometry IV, 914636 (2014).

12. Berne, B. J. & Pecora, R. *Dynamic Light Scattering. With Applications to Chemistry, Biology and Physics.* (Dover, 2000).

13. Crosignani, B., Di Porto, P. & Bertolotti, M. *Statistical Properties of Scattered Light* (Academic Press, 1975).

14. Dravins, D. & Lagadec, T. Stellar intensity interferometry over kilometer baselines: Laboratory simulation of observations with the Cherenkov Telescope Array. Proc. SPIE 9146, Optical and Infrared Interferometry IV, 91460Z (2014).

15. Holmes, R. B. & Belen'kii, M. S. Investigation of the Cauchy-Riemann equations for one-dimensional image recovery in intensity interferometry. *J. Opt. Soc. Am. A* **21**, 697–706 (2004).

16. Holmes, R., Calef, B., Gerwe, D. & Crabtree, P. Cramer-Rao bounds for intensity interferometry measurements. *Appl. Opt.* **52**, 5235–5246 (2013).

17. Thiébaut, E. Image reconstruction with optical interferometers. *New Astron. Rev.* **53**, 312–328 (2009).

18. Actis, M. *et al.* Design concepts for the Cherenkov Telescope Array CTA: An advanced facility for ground-based high-energy gamma-ray astronomy. *Exp. Astron.* **32**, 193–316 (2011).

19. Strekalov, D. V., Erkmen, B. I. & Yu, N. Intensity interferometry for observation of dark objects. *Phys. Rev. A* **88**, 053837 (2013).

Acknowledgements

This work was supported by the Swedish Research Council and The Royal Physiographic Society in Lund. Early experiments in intensity interferometry and photon correlation techniques at the Lund Observatory involved also Toktam Calvén Aghajani, Daniel Faria, Johan Ingjald, Hannes Jensen, Lennart Lindegren, Eva Mezey, Ricky Nilsson and Helena Uthas.

Author contributions

D.D. conceived this project; D.D. and T.L. together set up the laboratory experiment at the Lund Observatory, performed all measurements and their reductions; P.D.N. carried out the image reconstructions. D.D. wrote the draft manuscript but all authors contributed to editing its final version.

Additional information

14

Formation of lunar swirls by magnetic field standoff of the solar wind

Timothy D. Glotch[1], Joshua L. Bandfield[2], Paul G. Lucey[3], Paul O. Hayne[4], Benjamin T. Greenhagen[4], Jessica A. Arnold[1], Rebecca R. Ghent[5] & David A. Paige[6]

Lunar swirls are high-albedo markings on the Moon that occur in both mare and highland terrains; their origin remains a point of contention. Here, we use data from the Lunar Reconnaissance Orbiter Diviner Lunar Radiometer to support the hypothesis that the swirls are formed as a result of deflection of the solar wind by local magnetic fields. Thermal infrared data from this instrument display an anomaly in the position of the silicate Christiansen Feature consistent with reduced space weathering. These data also show that swirl regions are not thermophysically anomalous, which strongly constrains their formation mechanism. The results of this study indicate that either solar wind sputtering and implantation are more important than micrometeoroid bombardment in the space-weathering process, or that micrometeoroid bombardment is a necessary but not sufficient process in space weathering, which occurs on airless bodies throughout the solar system.

[1] Department of Geosciences, Stony Brook University, Stony Brook, New York 11794-2100, USA. [2] Space Science Institute, 4750 Walnut St #205, Boulder, Colorado 80301, USA. [3] Hawaii Institute of Geophysics and Planetology, University of Hawaii, Honolulu, Hawaii 96822, USA. [4] Jet Propulsion Laboratory, M/S 183-301, 4800 Oak Grove Drive, Pasadena, California 91109, USA. [5] Department of Earth Sciences, University of Toronto, Toronto, Ontario, Canada M5S 3B1. [6] University of California Los Angeles, Box 951567, Los Angeles, California 90095-1567, USA. Correspondence and requests for materials should be addressed to T.D.G. (email: timothy.glotch@stonybrook.edu).

Lunar swirls have been documented since the Apollo era[1], and since that time, a variety of swirl morphologies has been identified, ranging from the complex structures present in the Reiner Gamma formation, Mare Ingenii and Mare Marginis swirls, to simple bright patches with diffuse edges, as at Descartes. Intermediate morphologies occur near Airy and Gerasimovich craters[2]. Several swirl formation mechanisms have been proposed, including (1) solar wind standoff due to the presence of local magnetic fields, preventing solar wind sputtering and implantation, nanophase iron formation, and the resulting surface darkening associated with space weathering[3], (2) recent comet impacts or micrometeoroid swarms that scoured the lunar surface, leaving a fine-gained, unweathered material and possibly imparting a remnant magnetization[4–6], and (3) electrostatic levitation and deposition of high-albedo, fine-grained, feldspar-enriched dust[7]. The association of all swirls with magnetic field anomalies of varying strength[8–10] has driven the development of each of these hypotheses. In this work, we use the unique mid-infrared data set from the Diviner Lunar Radiometer to distinguish between these hypotheses. We show that Diviner data support the solar wind standoff model for lunar swirl formation and disqualify the impact swarm and dust transport hypotheses.

Previous near-infrared observations of lunar swirls show them to be optically immature compared with the surrounding terrains[10]. Space weathering leads to optical maturation of the surfaces of airless bodies and is thought to be caused by two main processes: (1) solar wind sputtering and/or implantation of hydrogen atoms, leading to the formation of nanophase metallic iron blebs and (2) micrometeoroid bombardment that leads to the formation of agglutinitic glass and a reduced vapour-deposited coating[11]. Under the solar wind standoff model, horizontal magnetic fields at the lunar swirl sites deflect the solar wind, preventing most sputtering and implantation of solar wind ions[12,13]. If this is the case, micrometeoroid bombardment should be the major relevant space-weathering process[14] at swirl sites. A relative lack of solar wind interaction with the lunar surface at swirl sites is supported by recent observations from the Moon Mineralogy Mapper (M^3) instrument that confirm the Clementine observations of a substantial reduction in optical maturity and also show a reduction of the presence of surface hydroxyls at the swirl sites[15]. Recent observations in the ultraviolet by the Lunar Reconnaissance Orbiter Camera Wide Angle Camera subsystem also suggest that micrometeoroid bombardment, and not solar wind interaction, plays the dominant role in space weathering at the lunar swirls[16].

Diviner has three narrow band infrared channels near 8 μm, from which the silicate Christiansen Feature (CF) can be calculated[17]. The position of the CF is related to the degree of mineral silicate polymerization and provides an indicator of composition[18]. However, analysis of the global observations by Diviner has revealed that the CF position is also influenced by optical maturity. Optically mature surfaces are shifted to longer wavelengths by ~0.1 μm (or about ~20% of the full range of lunar CF values that have been measured) compared with immature surfaces with the same composition[19]. A global CF map[19] shows substantial anomalies related to young, fresh craters such as Tycho and Jackson, which have CF positions at shorter wavelengths than the surrounding mature terrain.

Here we find that Diviner data show the CF anomalies are due to abnormal space weathering at both the mare and highland swirl sites. This result, in addition to the relative lack of thermophysical anomalies at the swirl sites, strongly supports the solar wind standoff model and disqualifies the micrometeoroid/comet swarm and dust levitation models for swirl formation. Diviner data can be used to determine the composition and degree of space weathering of the lunar surface[17,19,20] and to determine thermophysical properties using daytime and night-time thermal infrared measurements[21,22]. The ability to characterize both the compositional and thermophysical properties of the lunar regolith make Diviner well suited to examine the swirls and differentiate between the three proposed formation mechanisms.

Results

Diviner CF anomaly at swirl sites. We analysed Diviner data covering 12 swirl regions including both the mare and highlands sites (Figs 1 and 2; Table 1, Supplementary Figs 1–10). At both the Reiner Gamma and Van de Graaff crater sites, the distributions of CF values on and off the swirls are clearly separated. Garrick-Bethell et al.[7] demonstrated that only a small portion of the Reiner Gamma 'dark lanes' (small curvilinear low-albedo regions that occur between some high-albedo swirls) are truly darker than the surrounding terrain at visible wavelengths. We found that the CF distribution of this portion of the dark lanes within Reiner Gamma is nearly identical to the off-swirl distributions (Fig. 1). The off-swirl locations were defined to be well outside the visible light albedo boundaries of the swirls identified in Lunar Reconnaissance Orbiter Camera wide angle camera or Clementine data. The average off-swirl CF values at Reiner Gamma (8.31 μm) and Van de Graaff (8.20 μm) are close to the previously determined global average mare and highland CF values[17]. The average on-swirl values for Reiner Gamma and Van de Graaff are shifted to shorter wavelengths by 0.08 and 0.09 μm, respectively. In addition, three-point Diviner spectra acquired both on and off the swirls at both the sites show clear differences in the spectral shape, which is reflected in the differences in modelled CF values for these sites. Swirl spectra were acquired from the regions outlined by boxes in Fig. 1 and the average spectra and resulting modelled CF values are shown in Fig. 2.

The low CF position values of each of the swirls are clearly associated with lower optical maturity compared with the surrounding terrains (Table 1). Figure 3 shows the relationship between CF position and the optical maturity parameter (OMAT)[23] derived from Clementine multispectral data at Reiner Gamma. Compared with the scene average, swirl pixels have both lower overall CF and higher OMAT (lower maturity) values.

At Reiner Gamma, Airy, Firsov and Rima Sirsalis, the complex swirls have easily identifiable dark lanes between the high-albedo regions. At Airy, the background terrain albedo varies substantially, so it is difficult to determine if the dark lanes are, in fact, lower albedo than the surrounding terrain. However, at Firsov and Rima Sirsalis, the dark lanes (Fig. 4a,b) have demonstrably lower visible albedos than the surrounding terrain. In Fig. 4c,d, we plot the Clementine 950-nm/750-nm reflectance ratio against the 750-nm ratio for regions on the swirl, in the dark lanes, and on the local surrounding terrain. The dark lanes at these sites are the darkest sampled regions in the scene. At these sites, the averages and distributions of the CF values just off the swirls and within the dark lanes of each swirl are within 0.01 μm, and dark lane OMAT values are nearly identical to the surrounding terrain (Table 1). These data are consistent with the hypothesis that magnetic field lines over the dark lanes have a vertical orientation, allowing the solar wind to interact with the surface[24]. It has previously been suggested that the dark lanes and low-albedo off-swirl regions associated with some swirls result from an increased rate of space weathering as the charged solar wind particles deflected by local magnetic fields are funneled to these regions[14]. However, maturity enhancements in the dark lanes are not

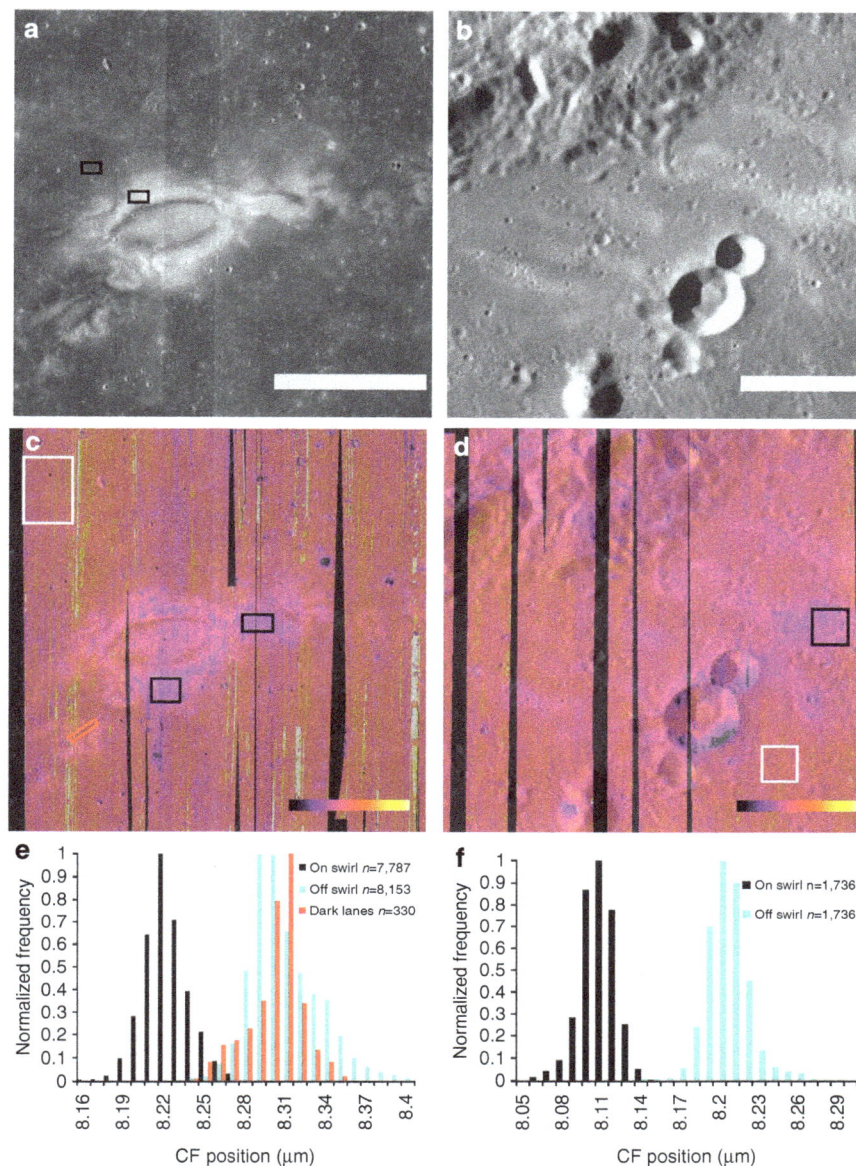

Figure 1 | Diviner CF position data for the Reiner Gamma and Van de Graaff crater swirls. (**a**) Wide Angle Camera (WAC) base map for Reiner Gamma. Black boxes indicate the regions used for surface roughness and thermal model analysis in Figs 5 and 8. Scale bar, 50 km. (**b**) WAC base map for Van de Graaff crater. Scale bar, 20 km. (**c**) CF position map of the Reiner Gamma swirl stretched from 8.15 to 8.45 μm overlayed on Lunar Reconnaissance Orbiter Camera (LROC) WAC mosaic. (**d**) CF position map of the Van de Graaff crater swirls stretched from 8.0 to 8.4 μm overlayed on LROC WAC mosaic. (**e**) Normalized histogram of CF values on the Reiner Gamma swirl (black boxes in **c**), off the swirl (white box in **1c**), and in the dark lanes (yellow boxes in **c**). (**f**) Normalized histogram of CF values on the Van de Graaff crater swirls (black box in **b**) and off the swirl (white box in **1b**).

detected in Diviner data. Rather, the nearly identical CF distributions within the dark lanes and the local terrains near the swirls (for example, Fig. 1) suggest that the main effects of space weathering at mid-infrared wavelengths (shift in CF position to longer wavelengths) reach a maximum early in the space-weathering process and are not enhanced on increased flux of solar wind particles.

The dust levitation model of lunar swirl formation[7] suggests that the spectral character of the swirls should be affected by a minor addition of feldspar, which is slightly enriched in the finest size fraction of the lunar soil[25]. Detailed analyses of the modal abundances of major minerals in sieved fractions of lunar soils show that feldspar is typically enriched by ~2–4 wt. % compared with coarser size fractions[26]. Therefore, the dust levitation model suggests an enrichment of feldspar in the optically active surface

layer of mare swirls by a comparable amount, assuming that all of the levitated dust is < 10 μm in diameter. Laboratory spectroscopic measurements of olivine-feldspar mixtures in a simulated lunar environment[27] (Fig. 5) show that ~16 wt.% feldspar would need to be added to account for a typical mare swirl CF shift from ~8.3 to ~8.2 μm. At the highlands sites, ~20 wt.% feldspar would need to be added to the swirl sites to account for a CF shift from ~8.2 to ~8.1 μm. While it is a simplification to model the swirl sites as olivine-feldspar mixtures, the laboratory data do suggest that the required feldspar contributions to the swirl CF shifts are ~5 to 10 times higher than the levels predicted by the dust levitation model. Swirls in both the highlands and mare consistently show CF shifts of ~0.04 to 0.09 μm to shorter wavelengths. Rather than an addition of a small amount of feldspar, this shift is comparable to CF shifts attributed to the optically immature surfaces. The Diviner data, therefore, suggest that a layer of fine particulate feldspar-enriched material is not responsible for the lunar swirl albedo anomalies and spectral properties. This interpretation supports visible and near-infrared measurements from the M^3 instrument[15].

Surface roughness at swirl sites. In addition to compositional analysis, we can use the daytime Diviner data to address the physical state of the lunar surface. We employ a roughness model[28] (Fig. 6) and Diviner daytime measurements in channels 3 and 6 (12.5–25 μm) to determine the spectral aniosthermality caused by surface roughness at Reiner Gamma. Surface roughness varies with length scale, and becomes greater with decreasing scale[29]. Previous lunar surface roughness estimates have included Lunar Orbiter Laser Altimeter data at metre to decametre scales[30], visible photometric studies typically at sub-mm scales[31,32] and additional sub-mm surface roughness estimates from in situ high-resolution imagery[29]. At sub-mm to cm scales, typical lunar surface roughness values, defined as a root mean square (RMS) slope are 16–25° (ref. 29). Previous thermal infrared observations of the Moon and Mars have shown substantial effects due to surface roughness at ~cm scales[33–37]. Diviner data offer the ability to characterize cm-scale surface roughness at the high spatial resolution (128–256 ppd (pixels per degree)) typical of the instrument. Figure 6 shows the modelled change in the channel 3 − channel 6 brightness temperature differences (ΔBT) on and off the Reiner Gamma swirl as a function of time of day for different RMS surface slope along with ΔBT data from 6 am to 6 pm local time. Within a standard

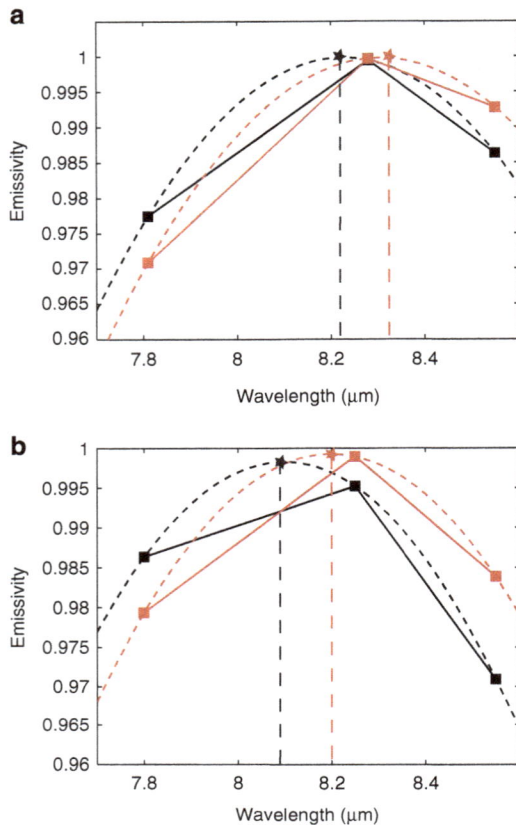

Figure 2 | Average Diviner spectra. Spectra are averaged using data from the white and black boxes outlined in Fig. 1. (**a**) Reiner Gamma swirl. (**b**) Van de Graaff crater swirl. Black spectra are average on-swirl spectra. Red spectra are average off-swirl spectra from the boxes outlined in Fig. 1. The parabolic traces represent the fitting function used to determine the CF position (stars) from the spectra.

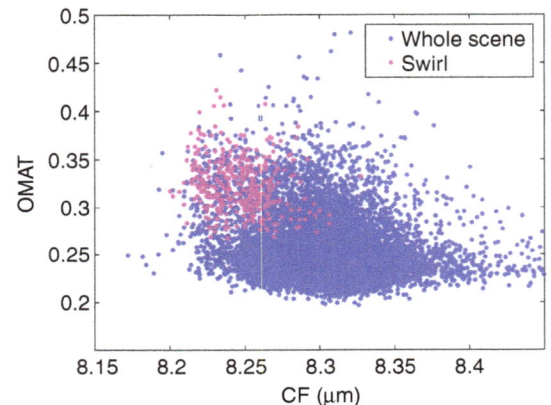

Figure 3 | Comparison of CF position and OMAT maturity index for Reiner Gamma. On-swirl pixels (taken from the regions outlined by white boxes in Fig. 1) clearly have lower CF and higher OMAT (lower maturity) values than the median scene.

Table 1 | CF and OMAT parameter values for each swirl examined in this study.

Swirl	Centre Location (lon., lat.)	On-swirl CF (μm)	Off-swirl CF (μm)	Dark lane CF (μm)	On-swirl OMAT	Off-swirl OMAT	Dark lane OMAT
Airy	3.5E, −18.5N	8.12 ± 0.02	8.16 ± 0.03	8.15 ± 0.03	0.30 ± 0.02	0.25 ± 0.03	0.26 ± 0.01
Descartes	15.9E, −10.6N	8.09 ± 0.01	8.18 ± 0.01	NA	0.31 ± 0.03	0.22 ± 0.01	NA
Firsov	113.5E, 0.5N	8.11 ± 0.02	8.19 ± 0.01	8.18 ± 0.02	0.31 ± 0.02	0.24 ± 0.01	0.24 ± 0.01
Gerasimovich	236.7E, −22.9N	8.09 ± 0.01	8.17 ± 0.01	NA	0.31 ± 0.01	0.22 ± 0.01	NA
Hopmann	159.4E, −50.8N	8.18 ± 0.01	8.23 ± 0.02	NA	0.30 ± 0.02	0.25 ± 0.01	NA
Ingenii mare	163.0E, −35.9N	8.25 ± 0.02	8.30 ± 0.03	NA	0.28 ± 0.02	0.22 ± 0.01	NA
Marginis highlands	96.0E, 20.4N	8.12 ± 0.02	8.20 ± 0.02	NA	0.26 ± 0.02	0.22 ± 0.01	NA
Marginis mare	84.9E, 13.4N	8.25 ± 0.01	8.31 ± 0.02	NA	0.22 ± 0.02	0.19 ± 0.01	NA
Moscoviense	148.9E, 25.4N	8.26 ± 0.02	8.30 ± 0.03	NA	0.21 ± 0.01	0.19 ± 0.01	NA
Reiner Gamma	301.0E, 7.4N	8.23 ± 0.02	8.31 ± 0.03	8.31 ± 0.03	0.35 ± 0.02	0.24 ± 0.01	0.26 ± 0.01
Rima Sirsalis	306.0E, −7.2N	8.29 ± 0.01	8.31 ± 0.01	8.31 ± 0.01	0.23 ± 0.01	0.22 ± 0.02	0.21 ± 0.01
Van de Graaff	171.1E, −27.4N	8.11 ± 0.01	8.20 ± 0.02	NA	0.30 ± 0.02	0.23 ± 0.01	NA

CF, Christiansen feature; E, East; lat, latitude; lon, longitude; N, North; NA, not available; OMAT, optical maturity.

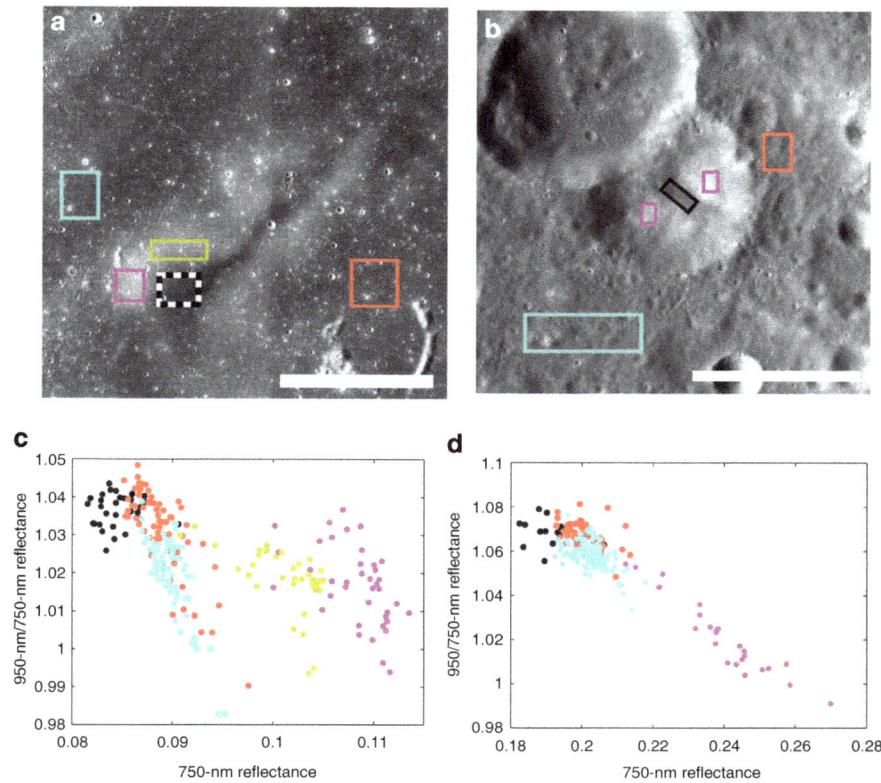

Figure 4 | Clementine albedo variations associated with swirl dark lanes. Dark lanes at Rima Sirsalis and Firsov crater have lower visible albedos than surrounding terrain. (**a**) Rima Sirsalis 750-nm reflectance. Scale bar, 25 km. (**b**) Firsov crater 750-nm reflectance. Scale bar, 25 km. (**c**) Rima Sirsalis 950/750-nm reflectance ratio versus 750-nm reflectance. Coloured points correspond to the coloured boxes in **a**. Black dots correspond to the white dotted box in **a**. (**d**) Firsov crater 950/750-nm reflectance ratio versus 750-nm reflectance. Coloured points correspond to the coloured boxes in **b**.

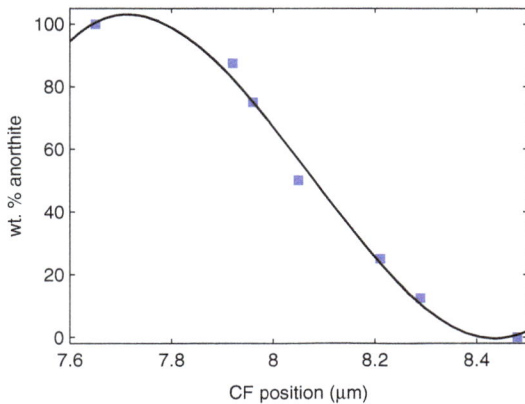

Figure 5 | CF positions of olivine-plagioclase mixture spectra acquired in a simulated lunar environment. Regression of the data using a degree 3 polynomial shows that to go from a CF of 8.3 to 8.2 μm, ~16 wt.% additional feldspar needs to be added to the mixture. To go from 8.2 to 8.1 μm, ~20 wt. % additional feldspar must be added.

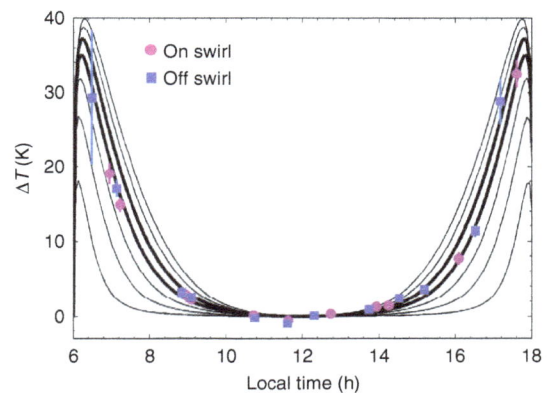

Figure 6 | Reiner Gamma roughness model. Modelled brightness temperature deviations between the channels 3 and 6 as function of the local time are plotted for RMS surface slopes varying in 5° increments from 10° (bottom line) to 40° (top line). Measured data from the Reiner Gamma swirl are displayed as magenta circles. Data from a nearby off-swirl surface are displayed as blue squares. Error bars are 1 s.d. Curves representing 25–30° roughness are bold. Diviner data from Reiner Gamma swirl indicate surface roughness in this range.

deviation of the temperature measurements, Diviner data indicate nearly identical surface roughnesses of ~25–30° both on and off the swirl (bold curves in Fig. 6). This surface roughness reflects the finely structured lunar regolith at cm scales and places tight constraints on geologic processes related to the formation of the swirls. That is, the viability of the dust levitation and comet swarm impact models hinges on the ability to demonstrate that the roughness characteristics of the swirls would not be altered by (1) piling of fine-grained dust, resulting in smoother surfaces at cm scales or (2) regolith disturbance and impact scouring resulting in a layer of fine particulates at cm scales.

Thermophysical characteristics of swirls. Diviner night-time data and thermal models can also help to differentiate between the competing formation mechanisms for the lunar swirls. Cooling of the lunar surface throughout the night provides information on the physical nature and degree of processing of the lunar regolith, including rock abundance and vertical structure and particle size of the regolith fines[21,22]. We constructed 128-ppd night-time temperature maps using Diviner channel 8 (50–100 μm) brightness temperatures acquired between local times of 19:30 and 05:30 (Fig. 7). Reported values are positive or negative deviations from the scene average normalized for local time variations. On-swirl and off-swirl temperature averages were determined using the same regions of interest shown in the boxes in Fig. 1. The average normalized temperature on the sampled portion of the Reiner Gamma swirl is -0.8 ± 1.5 K as opposed to the average off-swirl temperature of -0.3 ± 1.8 K (1-σ s.d.). The sampled portion of the swirl, on average, is only ~ 0.5 K colder than the off-swirl surface, but it is visible in the stretched night-time temperature image (white arrows in Fig. 7a). At the Van de Graaff crater site, the average on-swirl normalized night-time temperature is 0.4 K, while the off-swirl average is 0.0 K.

The measured temperature differences, though small, can be tied directly to the physical properties of the swirl surfaces. To do this, we use a thermal model to constrain the differences in regolith properties between the swirl and non-swirl surfaces. This standard lunar regolith model[22,38] uses Diviner night-time temperature data to constrain the upper regolith density profile. Using the on- and off-swirl surfaces delineated in Fig. 1a, and the Clementine 750-nm albedos of these regions, we show that the temperature difference between the on- and off-swirl surfaces can be accounted for completely by the albedo difference between the swirls (Fig. 8). In addition, we added a 2-mm low thermal inertia (~ 30 J m^{-2} K^{-1} s$^{-1/2}$, $\sim 50\%$ of the standard model) layer on top of the standard swirl model. Such a low thermal inertia layer would be expected from the admixture of additional fine particulates with typical lunar regolith[39]. It is possible, however, that this model does not capture the effects of the minor admixtures of fine particulates with the bulk lunar regolith. Nevertheless, our results show that even this 2-mm thick layer would produce night-time temperatures much lower than those that are observed. Put another way, the swirls are surficial features that, thermophysically, are nearly indistinguishable from the surrounding terrain. The surficial nature of the swirls as determined by Diviner is consistent with Mini-RF radar observations of these features[40].

The lunar regolith has been highly processed by impact gardening that produces a delicate vertical structure with unique thermophysical properties. Any deviation from this structure shows clear night-time temperature differences that are easily identified in Diviner data[21,22,28] and are absent from the swirl sites. For example, recent impacts by swarms of meteors or comets,[4–6] although they might not produce a 2-mm thick low thermal inertia layer as modelled in Fig. 8, would almost certainly disturb the regolith vertical structure and lead to clearly observable temperature and roughness phenomena in the Diviner data.

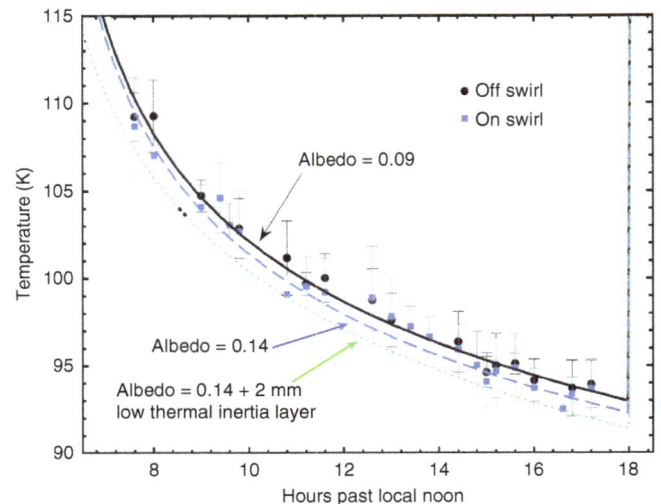

Figure 8 | Reiner Gamma thermal model. Diviner night-time temperature observations on (blue dots) and off (black dots) the Reiner Gamma swirl are plotted and compared with two thermal models. Error bars indicate standard 1 s.d. in temperature measurements within the spatially binned measurement regions (Fig. 1a). The black solid line and blue dashed line show the predicted temperatures using the standard regolith model[21,37]. Predicted night-time temperatures for a 2-mm thick low thermal inertia layer on top of the standard swirl thermal model (green dotted line) are too low to explain temperatures observed by Diviner.

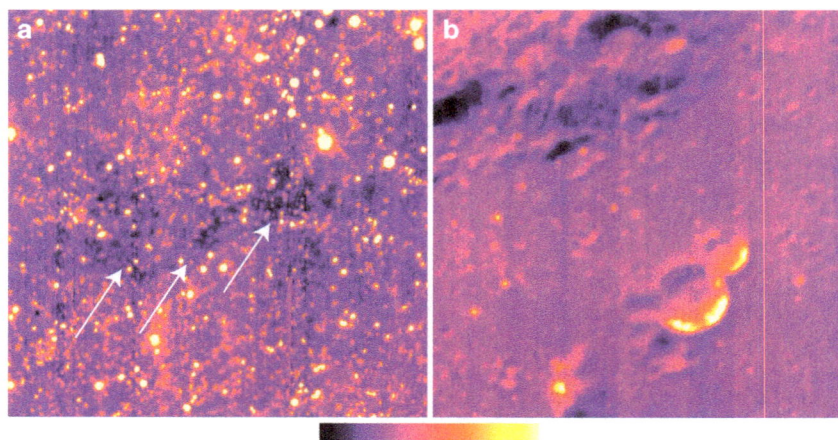

Figure 7 | Diviner night-time temperature data for the Reiner Gamma and Van de Graaff crater swirls. Area of each image is same as Wide Angle Camera base maps in Fig. 1a,b. (**a**) Average normalized night-time temperature for Reiner Gamma, stretched from -3.6 to 7.0 ΔK. Arrows show the colder temperatures associated with the swirl. (**b**) Average normalized night-time temperature for Van de Graaff crater, stretched from -7.0 to 15 ΔK.

Discussion

Previous spectroscopic observations of the swirls at visible and near-infrared wavelengths have been used to argue for the solar wind standoff model[3,14,24], the micrometeoroid/comet swarm model[4-6] or the feldspathic dust pile model. More recently, M^3 spectra have been used to propose a hybrid model for swirl formation that invokes both regolith disturbance and dust lofting[41]. The mid-infrared Diviner data presented here place tight constraints on the swirl formation process and must be accounted for by any proposed swirl formation mechanism. The proposed feldspathic dust lofting[7], meteoroid/comet swarm[4-6] and hybrid[41] swirl formation mechanisms would likely result in thermophysical and spectral properties that are inconsistent with the Diviner measurements. The proposed dust lofting and hybrid processes cannot account for the observed CF shift in Diviner data while the meteoroid/comet swarm and hybrid processes cause regolith disturbance that would result in a substantial thermal anomaly in the night-time Diviner data. On the other hand, the solar wind standoff mechanism for swirl formation would lead to a slight shift of the CF to shorter wavelengths compared with the surrounding regolith due to a reduction in space weathering, but show no difference in the unique lunar regolith thermophysical properties because impact gardening is otherwise unaffected. These properties are exactly what are observed, and the Diviner observations, in addition to M^3 and Mini-RF[14,15,40] observations and laboratory experiments[42], strongly support the solar wind standoff hypothesis for lunar swirl formation. Because they are clearly associated with magnetic field anomalies, lunar swirls present an opportunity to study and isolate the effects of solar wind sputtering/implantation and micrometeoroid bombardment processes on optical maturity. The immature nature of the swirls suggests that solar wind sputtering and implantation, which is prevented by local magnetic fields at the swirl sites, is the primary process for space weathering and optical maturation of airless body surfaces that do not have a global magnetic field, while micrometeoroid bombardment and resulting vapour deposition plays a subordinate role. The results of this study should motivate additional space-weathering experiments and work on the mature lunar regolith samples to test this hypothesis.

Methods

Diviner CF calculations and average spectra. The Diviner Lunar Radiometer Experiment on board the Lunar Reconnaissance Orbiter is an infrared radiometer with two solar reflectance channels and seven channels in the thermal infrared wavelength range, including three narrow band channels that span the wavelength ranges 7.55–8.05 μm, 8.10–8.40 μm and 8.38–8.68 μm. These '8 μm' channels can be used to calculate the position of the silicate CF[17] that is directly sensitive to the degree of silica tetrahedral polymerization and bulk SiO_2 content[18]. By viewing lunar surface targets multiple times under different viewing conditions, the precision of individual Diviner CF values, which are known to be systematically affected by illumination and viewing geometry, are estimated at < 0.02 μm (refs 17,43).

For this study, we used data collected between 8:30 and 16:30 local solar time, with emission angles < 5° from nadir. These restricted ranges minimize the effects of the observation conditions on Diviner CF estimates. All data used in the current study were reduced using the most recent available corrections[44].

CF values were modelled following the methods of Greenhagen et al.[17] In brief, the shapes of the CFs for the regolith surfaces of airless bodies can be approximated as parabolas. Each three-point Diviner spectrum is fit as a parabola by solving the following system of three equations to find the wavelength of the parabola maximum, λ_{Max}:

$$y_3 = A\lambda_3^2 + B\lambda_3 + C \quad (1)$$

$$y_4 = A\lambda_4^2 + B\lambda_4 + C \quad (2)$$

$$y_5 = A\lambda_5^2 + B\lambda_5 + C \quad (3)$$

$$\lambda_{Max} = -B/(2A) \quad (4)$$

In these equations, y_3, y_4 and y_5 are the emissivities for Diviner channels 3, 4 and 5, respectively, and A, B and C are constants. The calculated CF value is λ_{Max}.

To generate the representative three-point spectra, we averaged groups of 7,897 and 8,181 Diviner spectra to produce representative on-swirl and off-swirl spectra, respectively, at Reiner Gamma and model the CFs for these regions. At the Van de Graaff crater site, we averaged 1,736 spectra each for the on-swirl and off-swirl regions.

Thermal model. The one-dimensional thermal model employs standard forward difference approximations for spatial and temporal derivatives and incorporates important constraints and improvements based on fits to the Diviner data[22,38]. Most important of these are depth-dependent and temperature-dependent thermal conductivity, as well as spatially variable regolith porosity and rock abundance. Model layer thicknesses are typically < 1 mm, and the bottom boundary at 2 m is well below the ~10 cm skin depth of the diurnal thermal wave. We allow the model to fully equilibrate over at least 10 years before reporting the results.

Roughness model. The effects of roughness on thermal emission are simulated using a modified statistical shadowing model similar to several that have previously been described in the literature[29,36,37]. Using this model, we predicted the surface temperatures for slopes from 0 to 90° and azimuth orientations from 0 to 360°. We then convolved each calculated temperature with its expected fractional surface area for the given RMS slope, based on the assumed Gaussian height distribution[30]. Infrared emission at the given measurement wavelength is then calculated as a weighted sum of the contributions of all the surface elements (both illuminated and shadowed) within the field of view[28]. The model also incorporates a shadow approximation[31,33] that accounts for the statistical likelihood of a surface to be within a shadow cast by another surface.

Night-time data. Night-time (19:30 to 05:30 local time) temperature maps were constructed from the Diviner Level 2 Gridded Data Products available at the Planetary Data System. A total of 46 separate Diviner channel 8 (50–100 μm) brightness temperature maps were assembled from a single Lunar Reconnaissance Orbiter mapping cycle collected over the course of a lunar day. The average brightness temperature was subtracted from each map to normalize the temperature for the local time variations. The normalized maps were averaged to produce the final temperature deviation maps. This methodology ensures that any temperature differences between on swirl and off swirl are preserved.

References

1. El-Baz, F. The Alhazen to Abdul Wafa Swirl Belt: an extensive field of light-colored sinuous markings. *NASA Spec. Pub.* **315**, 29–93 (1972).
2. Blewett, D. T., Hawke, B. R. & Richmond, N. C. A magnetic anomaly associated with an albedo feature near Airy crater in the lunar nearside highlands. *Geophys. Res. Lett.* **34**, L24206 (2007).
3. Hood, L. L. & Schubert, G. Lunar magnetic anomalies and surface optical properties. *Science* **208**, 49–51 (1980).
4. Schultz, P. H. & Srnka, L. J. Cometary collisions on the Moon and Mercury. *Nature* **284**, 22–26 (1980).
5. Pinet, P. C., Shevchenko, V. V., Chevrel, S. D., Daydou, Y. & Rosemberg, C. Local and regional lunar regolith characteristics at Reiner Gamma formation: optical and spectroscopic properties from Clementine and Earth-based data. *J. Geophys. Res.* **105**, 9457–9476.
6. Starukhina, L. V. & Shkuratov, Y. G. Swirls on the Moon and Mercury: meteoroid swarm encounters as a formation mechanism. *Icarus* **167**, 136–147 (2004).
7. Garrick-Bethell, I., Head, III J. W. & Pieters, C. M. Spectral properties, magnetic fields, and dust transport at lunar swirls. *Icarus* **212**, 480–492 (2011).
8. Hood, L. L. & Williams, C. R. The lunar swirls—distribution and possible origins. *Proc. Lunar Planet Sci. Conf.* **19Th**, 99–113 (1989).
9. Richmond, N. C. et al. Correlations between magnetic anomalies and surface geology antipodal to lunar impact basins. *J. Geophys. Res.* **110**, E05011 (2005).
10. Blewett, D. T. et al. Lunar swirls: Examining crustal magnetic anomalies and space weathering trends. *J. Geophys. Res.* **116**, E02002 (2011).
11. Hapke, B. Space weathering from Mercury to the asteroid belt. *J. Geophys. Res.* **106**, 10,039–10,074 (2001).
12. Lin, R. P. et al. Lunar surface magnetic fields and their interaction with the solar wind: results from lunar prospector. *Science* **281**, 1480–1484 (1998).
13. Kurata, M. et al. Mini-magnetosphere over the Reiner Gamma magnetic anomaly region on the Moon. *Geophys. Res. Lett.* **32**, L24205 (2005).
14. Kramer, G. Y. et al. Characterization of lunar swirls at Mare Ingenii: a model for space weathering at magnetic anomalies. *J. Geophys. Res.* **116**, E04008 (2011).
15. Kramer, G. Y. et al. M3 spectral analysis of lunar swirls and the link between optical maturation and surface hydroxyl formation at magnetic anomalies. *J. Geophys. Res.* **116**, E00G18 (2011).
16. Denevi, B. W. et al. Characterization of space weathering from lunar reconnaissance orbiter camera ultraviolet observations of the Moon. *J. Geophys. Res.* **119**, 976–997 (2014).

17. Greenhagen, B. T. *et al.* Global silicate mineralogy of the Moon from the Diviner lunar radiometer. *Science* **329**, 1507–1509 (2010).

18. Logan, L. M., Hunt, G. R., Salisbury, J. W. & Balsamo, S. R. Compositional implications of Christiansen Frequency maximums for infrared remote sensing applications. *J. Geophys. Res.* **78**, 4983–5003 (1973).

19. Lucey, P. G., Paige, D. A., Greenhagen, B. T., Bandfield, J. L. & Glotch, T. D. Comparison of Diviner Christiansen feature position and visible albedo: composition and space weathering implications. *Lunar Planet Sci. XLI,* abstract 1600 (2010).

20. Glotch, T. D. *et al.* Highly silicic compositions on the Moon. *Science* **329**, 1510–1513 (2010).

21. Bandfield, J. L. *et al.* Lunar surface rock abundance and regolith fines temperatures derived from LRO Diviner Radiometer data. *J. Geophys. Res.* **116**, E00H02 (2011).

22. Vasavada, A. R. *et al.* Lunar equatorial surface temperatures and regolith properties from the Diviner lunar radiometer experiment. *J. Geophys. Res.* **117**, E00H18 (2012).

23. Lucey, P. G., Blewett, D. T. & Hawke, B. R. Imaging of lunar surface maturity. *J. Geophys. Res.* **105**, 20,377–20,386 (2000).

24. Hemingway, D. & Garrick-Bethell, I. Magnetic field direction and lunar swirl morphology: insights from Airy and Reiner Gamma. *J. Geophys. Res.* **117**, E10012 (2012).

25. Pieters, C. M., Fischer, E. M., Rode, O. & Basu, A. Optical effects of space weathering: the role of the finest fraction. *J. Geophys. Res.* **98**, 20817–20824 (1993).

26. Taylor, L. A., Pieters, C. M., Keller, L. P., Morris, R. V. & McKay, D. S. Lunar mare soils: space weathering and the major effects of surface-correlated nanophase Fe. *J. Geophys. Res.* **106**, 27,985–27,999 (2001).

27. Arnold, J. A., Glotch, T. D., Thomas, I. R. & Bowles, N. E. Plagioclase-olivine mixtures in a simulated lunar environment. *Lunar Planet. Sci. XLIV,* abstract 2972 (2013).

28. Hayne, P. *et al.* Thermophyscial properties of the lunar surface from Diviner observations. *EGU Gen. Assembly,* abstract EGU2013-10871-1 (2013).

29. Helfenstein, P. & Shepard, M. K. Submillimeter-scale topography of the lunar regolith. *Icarus* **72**, 342–357 (1999).

30. Rosenburg, M. A. *et al.* Global surface slopes and roughness of the Moon from the lunar orbiter laser altimeter. *J. Geophys. Res.* **116**, E02001 (2011).

31. Hapke, B. Bidirectional reflectance spectroscopy III—correction for macroscopic roughness. *Icarus* **59**, 41–59 (1984).

32. Shkuratov, Y. G. *et al.* Interpreting photometry of regolith-like surfaces with different topographies: shadowing and multiple scattering. *Icarus* **173**, 3–15 (2005).

33. Smith, B. G. Lunar surface roughness: shadowing and thermal emission. *J. Geophys. Res.* **72**, 4059–4067 (1967).

34. Johnson, P. E., Vogler, K. J. & Gardner, J. P. The effect of surface roughness on lunar thermal emission spectra. *J. Geophys. Res.* **98 (E11)**, 20825–20829 (1993).

35. Danilina, I. *et al.* Roughness effects on sub-pixel radiative temperature dispersion in a kinetically isothermal surface. In *Proceedings of Second Recent Advances in Quantitative Remote Sensing* (ed. Sobrino, J.) 13–18 (Publicacions de la Universitat de Valencia, 2006).

36. Bandfield, J. L. & Edwards, C. S. Derivation of martian surface slope characteristics from directional thermal infrared radiometry. *Icarus* **193**, 139–157 (2008).

37. Bandfield, J. L. Effects of surface roughness and graybody emissivity on martian thermal infrared spectra. *Icarus* **202**, 414–428 (2009).

38. Hayne, P. O. *et al.* Diviner lunar radiometer observations of the LCROSS impact. *Science* **330**, 477–479 (2010).

39. Presley, M. & Christensen, P. R. Thermal conductivity measurements of particulate materials. 2. Results. *J. Geophys. Res.* **102**, 6551–6566.

40. Neish, C. D. *et al.* The surficial nature of lunar swirls as revealed by the Mini-RF instrument. *Icarus* **215**, 186–196 (2011).

41. Pieters, C. M., Moriarty, D. P. & Garrick-Bethell, I. Atypical regolith processes hold the key to enigmatic lunar swirls. *Lunar Planet Sci. XLV,* abstract 1408 (2014).

42. Bamford, R. A. *et al.* Minimagnetospheres above the lunar surface and the formation of lunar swirls. *Phys. Rev. Lett.* **109**, 081101 (2012).

43. Allen, C. C., Greenhagen, B. T., Donaldson Hanna, K. L. & Paige, D. A. Analysis of lunar pyroclastic deposit FeO abundances by LRO Diviner. *J. Geophys. Res.* **117**, E00H28 (2012).

44. Greenhagen, B. T. *et al.* The Diviner lunar radiometer compositional data products: description and examples. *Lunar Planet Sci. XLII,* abstract 2679 (2011).

Acknowledgements

Funding for this work was provided by the Diviner Lunar Radiometer Experiment extended mission science investigation and the Remote, In Situ, and Synchrotron Studies (RIS⁴E) team of NASA's Solar System Research Virtual Institute (SSERVI). This is SSERVI publication number SSERVI-2014-158.

Author contributions

T.D.G. conceived of the study and performed the Diviner CF and temperature data analysis with the assistance of J.L.B., P.G.L. and B.T.G.. P.O.H. supplied the roughness and thermal models discussed in the paper and R.R.G. and D.A.P. assisted with their interpretation. J.A.A. acquired the simulated lunar environment spectra of plagioclase/olivine mixtures and provided interpretations. T.D.G. wrote the paper with the input of all the co-authors.

Additional information

15

A possible macronova in the late afterglow of the long–short burst GRB 060614

Bin Yang[1,2], Zhi-Ping Jin[1], Xiang Li[1,2], Stefano Covino[3], Xian-Zhong Zheng[1], Kenta Hotokezaka[4], Yi-Zhong Fan[1,5], Tsvi Piran[4] & Da-Ming Wei[1]

Long-duration (>2 s) γ-ray bursts that are believed to originate from the death of massive stars are expected to be accompanied by supernovae. GRB 060614, that lasted 102 s, lacks a supernova-like emission down to very stringent limits and its physical origin is still debated. Here we report the discovery of near-infrared bump that is significantly above the regular decaying afterglow. This red bump is inconsistent with even the weakest known supernova. However, it can arise from a Li-Paczyński macronova—the radioactive decay of debris following a compact binary merger. If this interpretation is correct, GRB 060614 arose from a compact binary merger rather than from the death of a massive star and it was a site of a significant production of heavy r-process elements. The significant ejected mass favours a black hole–neutron star merger but a double neutron star merger cannot be ruled out.

[1] Key Laboratory of Dark Matter and Space Astronomy, Purple Mountain Observatory, Chinese Academy of Sciences, Nanjing 210008, China. [2] University of Chinese Academy of Sciences, Yuquan Road 19, Beijing 100049, China. [3] INAF/Brera Astronomical Observatory, via Bianchi 46, I-23807 Merate, Italy. [4] Racah Institute of Physics, The Hebrew University, Jerusalem 91904, Israel. [5] Collaborative Innovation Center of Modern Astronomy and Space Exploration, Nanjing University, Nanjing 210046, China. Correspondence and requests for materials should be addressed to Y.-Z.F. (email: yzfan@pmo.ac.cn) or Z.-P.J. (email: jin@pmo.ac.cn) or T.P. (email: tsvi.piran@mail.huji.ac.il)

Long-duration (> 2 s) γ-ray bursts (GRBs) are believed to originate from Collapsars that involve death of massive stars and are expected to be accompanied by luminous super-novae (SNe). GRB 060614 was a nearby burst with a duration of 102 s at a redshift of 0.125(ref. 1). While it is classified as a long burst according to its duration, extensive searches did not find any SNe-like emission down to limits hundreds of times fainter[2-4] than SN 1998bw, the archetypal hypernova that accompanied long GRBs[5]. Moreover, the temporal lag and peak luminosity of GRB 060614 fell entirely within the short duration subclass and the properties of the host galaxy distinguish it from other long-duration GRB hosts. Thus, GRB 060614 did not fit into the standard picture in which long-duration GRBs arise from the collapse of massive stars while short ones arise from compact binary mergers. It was nicknamed the 'long–short burst' as its origin was unclear. Some speculated that it originated from compact binary merger and thus it is intrinsically a 'short' GRB[1,4,6-8]. Others proposed that it was formed in a new type of a Collapsar which produces an energetic γ-ray burst that is not accompanied by an SNe[2-4].

Two recent developments may shed a new light on the origin of this object. The first is the detection of a few very weak SNe (for example, SN 2008ha[9]) with peak bolometric luminosities as low as $L \sim 10^{41}$ erg s^{-1}. The second is the detection of an infrared bump, again with a $L \sim 10^{41}$ erg s^{-1}, in the late afterglow of the short burst GRB 130603B[10,11]. This was interpreted as a Li-Paczyński macronova (also called kilonova)[12-19]—a near-infrared/optical transient powered by the radioactive decay of heavy elements synthesized in the ejecta of a compact binary merger. Motivated by these discoveries, we re-examined the afterglow data of this peculiar burst searching for a signal characteristic to one of these events.

The X-ray and UV/optical afterglow data of GRB 060614, were extensively examined in the literature[20,21] and found to follow very well the fireball afterglow model up to $t \sim 20$ days[22]. The J-band has been disregarded because only upper limits ~ 19–20th mag with a sizeable scatter are available at $t > 2.7$ days, and these are too bright to significantly constrain even supernovae as luminous as SN 1998bw[23]. In this work we focus on the optical emission. We have re-analysed all the late time (that is, $t \geq 1.7$ days) very large telescope (VLT) V, R and I-band archival data and the Hubble space telescope (HST) F606W and F814W archival data, including those reported in the literature[3,4] and several unpublished data points. Details on data reduction are given in the Methods.

Results

The discovery of a significant F814W-band excess. Figure 1 depicts the most complete late-time optical light curves (see Supplementary Table 1; the late VLT upper limits are not shown in Fig. 1) of this burst. The VLT V, R and I-band fluxes decrease with time as $\propto t^{-2.30 \pm 0.03}$ (see Fig. 1, in which the VLT V/I band data have been calibrated to the F606W/F814W filters of HST with proper k-corrections), consistent with that found earlier[3,20,21]. However, the first HST F814W data point is significantly above the same extrapolated power-law decline. The significance of the deviation is $\sim 6\sigma$ (see the estimate in the Methods). No statistically significant excess is present in both the F606W and the R bands. The F814W-band excess is made most forcibly by considering the colour evolution of the transient, defined as the difference between the magnitudes in each filter, which evolves from V–$I \approx 0.65$ mag by the VLT (correspondingly for HST we have F606W–F814W ≈ 0.55 mag) at about $t \sim 1.7$ days to F606W–F814W ≈ 1.5 mag by HST at about 13.6 days after the trigger of the burst. With proper/minor extinction

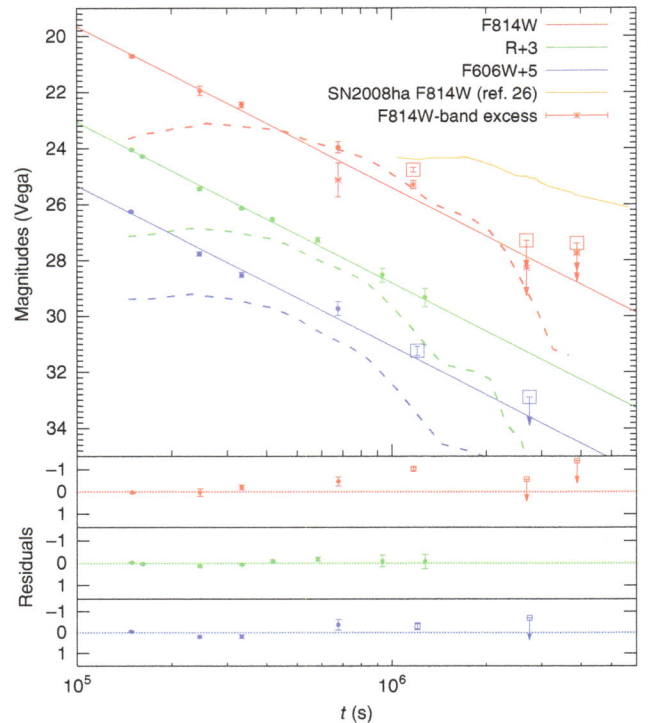

Figure 1 | The afterglow emission of GRB 060614. The VLT and HST observation vega magnitudes including their 1σ statistical errors of the photon noise and the sky variance and the 3σ upper limits (the downward arrows) are adopted from Supplementary Table 1. The small amounts of foreground and host extinction have not been corrected. Note that the VLT V/I band data have been calibrated to the HST F606W/F814W filters with proper k-corrections (see Methods). The VLT data (the circles) are canonical fireball afterglow emission while the HST F814W detection (marked in the square) at $t \sim 13.6$ days is significantly in excess of the same extrapolated power-law decline (see the residual), which is at odds with the afterglow model. The F814W-band light curve of SN 2008ha [27] expected at $z = 0.125$ is also presented for comparison. The dashed lines are macronova model light curves generated from numerical simulation [28] for the ejecta from a black hole–neutron star merger. Error bars represent s.e.

corrections, the optical to X-ray spectrum energy distribution for GRB 060614 at the epoch of ~ 1.9 days is nicely fitted by a single power law[3,20,21] $F_\nu \propto \nu^{-0.8}$. In the standard external forward shock afterglow model, the cooling frequency is expected to drop with time as[22] $\nu_c \propto t^{-1/2}$. Thus, it cannot change the optical spectrum in the time interval of 1.9–13.6 days. Hence, the remarkable colour change and the F814W-band excess of ~ 1 mag suggest a new component. Like in GRB 130603B this component was observed at one epoch only. After the subtraction of the power-law decay component, the flux of the excess component decreased with time faster than $t^{-3.2}$ for $t > 13.6$ days. Note that an unexpected optical re-brightening was also detected in GRB080503, another 'long–short' burst[24]. However, unlike the excess component identified here, that re-brightening was achromatic in optical to X-ray bands and therefore likely originated by a different process.

Discussion

Shortly after the discovery of GRB 060614 it was speculated that it is powered by an 'unusual' core collapse of a massive star[2,3]. We turn now to explore whether the F814W-band excess can be powered by a weak supernova. Figure 2 depicts the

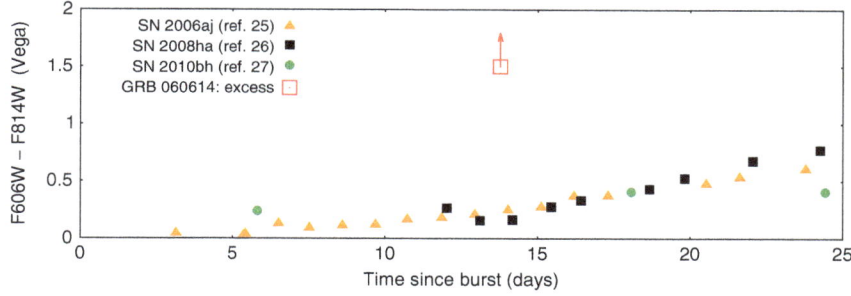

Figure 2 | The colour change of some supernovae in comparison with our excess component. The emission of SN 2006aj, SN 2008ha and SN 2010bh, accepted from the literature[25-27] has been shifted to $z = 0.125$, the redshift of GRB 060614, with corrections on the time, frequency and extinction. Note that the 'excess component' is much redder than them (the upward arrow represents a lower limit).

colour F606W–F814W of the excess component (we take F606W–F814W ≈ 1.5 mag as a conservative lower limit of the colour of the 'excess' component due to the lack of simultaneous excess in F606W-band) with that of SN 2006aj[25], SN 2008ha (i.e., the extremely dim event)[26] and SN 2010bh[27]. The excess component has a much redder spectrum than the three supernovae. If the 'excess component' was thermal it had a low effective temperature $T_{\rm eff} < 3,000$ K to yield the very soft spectrum. Such unusually low effective temperature is also needed to account for the very rapid decline of the excess component. The expansion velocity can be estimated as $v \sim 1.2 \times 10^4$ km s^{-1} $(L/10^{41}$ erg s$^{-1})^{1/2}(T_{\rm eff}/3,000$ K$)^{-2}(t/13.6$ days$)^{-1}$. The implied ^{56}Ni mass is $\sim 10^{-3} M_\odot$ if this was a supernova-like event that peaked at ~ 13.6 days[9]. We take a standard cosmology model with $H_0 = 71$ km s^{-1} Mpc^{-1}, $\Omega_{\rm M} = 0.32$ and $\Omega_\Lambda = 0.68$. Comparing with the extremely faint SN 2008ha after proper corrections to $z = 0.125$, the peak F814W-band emission of the 'excess component' is lower by ~ 1 mag and the decline is also much faster. Hence the 'excess component' is remarkably different from SN 2008ha.

The low luminosity as well as the low effective temperature of the transient emission are typical characteristics of a macronova, a transient arising from the radioactive β-decay of material ejected in a compact binary merger. The opacity of the macronova material is determined by the Lanthanides that are produced via r-process in the neutron-rich outflow. This opacity is very large ($\kappa \approx 10$ cm^2 g^{-1}) resulting in a weak, late and red emission. The emerging flux is greatly diminished by line blanketing, with the radiation peaking in the near-infrared and being produced over a timescale of ~ 1–2 weeks[17,18]. Simple analytic estimates, using a radioactive β-decay heating rate[16,28] of 10^{10} erg s^{-1} g$^{-1}[t/(1+z)1$ day$]^{-1.3}$, suggest that in order to explain the observed F814W-band excess, the required ejecta mass and expansion velocity are: $M_{\rm ej} \sim 0.13 M_\odot (L/10^{41}$ erg s$^{-1})(t/13.6$ day$)^{1.3}$ and $v \sim 0.1c$ $(L/10^{41}$ erg s$^{-1})^{1/2}(T_{\rm eff}/2,000$ K$)^{-2}(t/13.6$ days$)^{-1}$, respectively. Note that the macronova outflow is quite cold at such a late time[17,18]. The effective temperature is $T_{\rm eff} \approx 2,000$ K and the observer's F814W-band is above the peak of the black body spectrum. The emitting radius and the corresponding expansion velocity are much larger than in a supernova at this stage. Scaled up numerical simulations of lighter ejecta from black hole–neutron star mergers[28] suggest that $M_{\rm ej} \sim 0.1 M_\odot$ and a velocity $\sim 02c$ can account for the observed F814W-band excess. This numerical example is presented in Fig. 1 in dashed lines.

The implied ejecta mass is large compared with the mass ejection estimated numerically to take place in double neutron star mergers. However, it is within the possible range of dynamical ejecta of black hole–neutron star mergers with some extreme parameters (a large neutron star radius and a high black hole spin aligned with the orbital angular momentum)[14,29-32]. An accretion disk wind may contribute some additional mass as well[15,33,34]. However, the radioactive heating due to fission of the heavy r-process nuclei, which is quite uncertain and subdominant in current heating estimates[16], may play an important role in the energy deposition. It may increase the energy deposition rate at around 10 days by a significant factor[35]. This may reduce the required ejecta mass to $\sim 0.03 - 0.05 M_\odot$. This range of the ejecta masses is well within the range of the dynamical ejecta of black hole–neutron star mergers and it is even compatible with some estimates of double neutron star mergers.

We conclude that while a weak supernova cannot explain the observations, a high mass ejection macronova may. Like in GRB 130603B we must caution here that this interpretation is based on a single data point. However, if this interpretation is correct, it has far reaching implications. First, the presence of macronovae in both the canonical short burst GRB 130603B and in this 'long–short' one, GRB 060614, suggests that the phenomenon is common and the prospects of detecting these transients are promising. A more conclusive detection based on more than a single data point could be achieved in the future provided that denser HST observations are carried out. Moreover, as a black hole–neutron star merger is favoured in explaining the large ejected mass this implies that such binary systems may exist and their mergers are also responsible for GRBs. It also suggests that the 'long–short' burst was in fact 'short' in nature, namely, it arose from a merger and not from a Collapsar. The fact that a merger generates a 100 s long burst is interesting and puzzling by itself.

Clearly such events would contribute a significant fraction of the r-process material[36]. The actual contribution relative to the contribution of 130603B-like events is difficult to estimate as it is unclear which fraction of the macronovae/kilonovae behave as each type. Because of beaming most mergers will not be observed as GRBs. However, they emit omnidirectional gravitational radiation that can be detected by the upcoming Advanced LIGO/VIRGO/KAGRA detectors. These near-infrared/optical macronovae could serve as promising electromagnetic counterparts of gravitational wave triggers in the upcoming Advanced LIGO/VIRGO/KAGRA era.

Methods

Data reduction. We retrieved the public VLT imaging data of GRB 060614 from European Southern Observatory (ESO) Science Archive Facility (http://archive.e-so.org). The raw data were reduced following standard procedures, including bias subtraction, flat fielding, bad pixel removal and combination. Observations made with the same instrument and filter at different epochs are compared with that of the last epoch. The software package ISIS (http://www2.iap.fr/users/alard/pack-age.html) is used to subtract images and measure the GRB afterglow from the residual images. Photometric errors are estimated from the photon noise and the sky variance to 1σ confidence level. The 3σ of the background root mean square of the residual images is taken as the limiting magnitude. Finally, standard stars observed on 16 June 2006 were used for the absolute calibration. The results are presented in Supplementary Table 1, being well consistent with these given by other groups[3,21]. We assumed that the afterglow is characterized by the same power-law spectrum with index $\beta = 0.80$ during these observations[20], with which we get the k-corrections between the VLT V/I and HST F606W/F814W magnitudes, namely 0.12 mag and 0.02 mag, respectively. Such corrections have been taken into account in Fig. 1.

HST archive data of GRB 060614 are available from the Mikulski Archive for Space Telescopes (MAST; http://archive.stsci.edu), including one observation with the Wide Field and Planetary Camera 2 (WFPC2) and four observations with the Advanced Camera for Surveys (ACS) in F606W and F814W bands. The reduced data provided by MAST were used in our analysis. The last observation has been taken as the reference and the other images of the same filter are subtracted in order to directly measure fluxes of the afterglow from the residual images. Empirical point spread functions (PSFs) were built with bright stars in each image. Bright compact objects in the same field were used to align and relatively calibrate these images. WFPC2 image differs from ACS image in PSF. Before image subtraction, the WFPC2 and ACS images were matched to the same resolution by convolving each with the other's PSF. The PSF-matched WFPC2 and ACS images were aligned and subtracted. Aperture photometry was carried out for the afterglow in the residual image. The aperture correction derived from the empirical PSF was applied to yield the total flux. The host galaxy was used to relatively calibrate the afterglow between images, and the ACS zeropoints were used for absolute calibration. If the signal of the afterglow is too faint to be a secure detection, an upper limit of 3σ background root mean square is adopted. The results are reported in Supplementary Table 1, being well in agreement with these published in the literature[4]. The magnitudes of the host galaxy are measured in the last observation of all filters and can well be fitted by an Sc type galaxy template (Supplementary Fig. 1), demonstrating the self-consistence of our results.

VLT light curve decline rate and significance of the excess. As found in previous studies, the late-time optical/X-ray afterglow emission of GRB 060614 can be interpreted within the fireball forward shock model[20,21]. Motivated by such a fact, we assume that the I, R and V light curves follow the same power-law decline. In our fit there are four free parameters, three are related to the initial flux/magnitude in these three bands and the last is the decline rate needed in further analysis. We fitted all the VLT data (combined I, R and V band together) during the first 15 days (after which there are just upper limits) to determine these four parameters as well as their errors. The best-fit decline is found to be $\propto t^{-2.3 \pm 0.03}$, well consistent with that obtained in optical to X-ray bands in previous studies[3,20,21]. As a result of the propagation of uncertainties, the errors of the best-fit light curves are consequently inferred (the shadow regions in the residual plot of Fig. 2 represent the 1σ errors of the best-fit light curves). Please note that in Fig. 2 the VLT V/I band emission have been calibrated to HST F606W/F814W filters with proper k-corrections. The flux separation between the HST F814W-band data and the fitted curve at $t \sim 13.6$ days is $F_{excess} = 0.182\,\mu$Jy. The flux error of the F814W-band emission at $t \sim 13.6$ days is $\delta F_{obs} \approx 0.024\,\mu$Jy. The flux error of the best fitted F814W-band light curve at $t \sim 13.6$ days is $\delta F_{fit} \approx 0.012\,\mu$Jy. The significance of the excess component is estimated by $\mathcal{R} = F_{excess}/\sqrt{\delta F_{obs}^2 + \delta F_{fit}^2} \sim 6$. We therefore suggest that the excess component identified in this work is statistically significant at a confidence level of $\sim 6\sigma$.

References

1. Gehrels, N. *et al.* A new γ-ray burst classification scheme from GRB 060614. *Nature* **444**, 1044–1046 (2006).
2. Fynbo, J. P. U. *et al.* No supernovae associated with two long-duration gamma-ray bursts. *Nature* **444**, 1047–1049 (2006).
3. Della Valle, M. *et al.* An enigmatic long-lasting gamma-ray burst not accompanied by a bright supernova. *Nature* **444**, 1050–1052 (2006).
4. Gal-Yam, A. *et al.* A novel explosive process is required for the γ-ray burst GRB 060614. *Nature* **444**, 1053–1055 (2006).
5. Galama, T. J. *et al.* An unusual supernova in the error box of the gamma-ray burst of 25 April 1998. *Nature* **395**, 670–672 (1998).
6. Zhang, B. *et al.* Making a short gamma-ray burst from a long one: implications for the nature of GRB 060614. *Astrophys. J.* **655**, L25–L28 (2007).
7. Barkov, M. V. & Pozanenko, A. S. Model of the extended emission of short gamma-ray bursts. *Mon. Not. R. Astron. Soc.* **417**, 2161–2165 (2009).
8. Caito, L. *et al.* GRB060614: a 'fake' short GRB from a merging binary system. *Astron. Astrophys.* **498**, 501–507 (2011).
9. Valenti, S. *et al.* A low-energy core-collapse supernova without a hydrogen envelope. *Nature* **459**, 674–677 (2009).
10. Tanvir, N. R. *et al.* A 'kilonova' associated with the short-duration gamma-ray burst GRB 130603B. *Nature* **500**, 547–549 (2013).
11. Berger, E., Fong, W. & Chornock, R. An r-process kilonova associated with the short-hard GRB 130603B. *Astrophys. J.* **744**, L23 (2013).
12. Li, L. X. & Paczyński, B. Transient events from neutron star mergers. *Astrophys. J. Lett.* **507**, L59–L62 (1998).
13. Kulkarni, S. R. Modeling supernova-like explosions associated with gamma-ray bursts with short durations. Preprint at http://arxiv.org/abs/astro-ph/0510256 (2005).
14. Rosswog, S. Mergers of neutron star-black hole binaries with small mass ratios: nucleosynthesis, gamma-ray bursts, and electromagnetic transients. *Astrophys. J.* **634**, 1202–1213 (2005).
15. Metzger, B. D. *et al.* Electromagnetic counterparts of compact object mergers powered by the radioactive decay of r-process nuclei. *Mon. Not. R. Astron. Soc.* **406**, 2650–2662 (2010).
16. Korobkin, O., Rosswog, S., Arcones, A. & Winteler, C. On the astrophysical robustness of the neutron star merger r-process. *Mon. Not. R. Astron. Soc.* **426**, 1940–1949 (2012).
17. Barnes, J. & Kasen, D. Effect of a high opacity on the light curves of radioactively powered transients from compact object mergers. *Astrophys. J.* **773**, 18 (2013).
18. Tanaka, M. & Hotokezaka, K. Radiative transfer simulations of neutron star merger ejecta. *Astrophys. J.* **775**, 113 (2013).
19. Grossman, D., Korobkin, O., Rosswog, S. & Piran., T. The long-term evolution of neutron star merger remnants—II. Radioactively powered transients. *Mon. Not. R. Astron. Soc* **439**, 757–770 (2014).
20. Mangano, V. *et al.* Swift observations of GRB 060614: an anomalous burst with a well behaved afterglow. *Astron. Astrophys.* **470**, 105–118 (2007).
21. Xu, D. *et al.* In search of progenitors for supernovaless gamma-ray bursts 060505 and 060614: re-examination of their afterglows. *Astrophys. J.* **696**, 971–979 (2009).
22. Sari, R., Piran, T. & Narayan, R. Spectra and light curves of gamma-ray burst afterglows. *Astrophys. J. Lett.* **497**, L17–L20 (1998).
23. Cobb, B. E., Bailyn, C. D., van Dokkum, P. G. & Natarajan, P. Could GRB 060614 and its presumed host galaxy be a chance superposition? *Astrophys. J. Lett.* **651**, L85–L88 (2006).
24. Perley, D. A. *et al.* GRB 080503: implications of a naked short gamma-ray burst dominated by extended emission. *Astrophys. J.* **696**, 1871–1885 (2009).
25. Ferrero, P. *et al.* The GRB 060218/SN 2006aj event in the context of other gamma-ray burst supernovae. *Astron. Astrophys.* **457**, 857–864 (2006).
26. Foley, R. J. *et al.* Early- and late-time observations of SN 2008ha: additional constraints for the progenitor and explosion. *Astron. J.* **138**, 376–391 (2009).
27. Cano, Z. *et al.* XRF 100316D/SN 2010bh and the nature of gamma-ray burst supernovae. *Astrophys. J.* **740**, 41 (2011).
28. Tanaka, M. *et al.* Radioactively powered emission from black hole-neutron star mergers. *Astrophys. J.* **780**, 31 (2014).
29. Piran, T., Nakar, E. & Rosswog, S. The electromagnetic signals of compact binary mergers. *Mon. Not. R. Astron. Soc.* **430**, 2121–2136 (2013).
30. Lovelace, G. *et al.* Massive disc formation in the tidal disruption of a neutron star by a nearly extremal black hole. *Class. Quantum Grav.* **30**, 135004 (2013).
31. Foucart, F. *et al.* Neutron star-black hole mergers with a nuclear equation of state and neutrino cooling: dependence in the binary parameters. *Phys. Rev. D* **90**, 024026 (2014).
32. Kyutoku, K. *et al.* Dynamical mass ejection from black hole-neutron star binaries. Preprint at http://arxiv.org/abs/1502.05402 (2015).
33. Just, O., Bauswein, A., Pulpillo, R. A., Goriely, S. & Janka, H. T. Comprehensive nucleosynthesis analysis for ejecta of compact binary mergers. *Mon. Not. R. Astron. Soc.* **448**, 541–567 (2015).
34. Fernández, R., Quataert, E., Schwab, J., Kasen, D. & Rosswog, S. The interplay of disk wind and dynamical ejecta in the aftermath of neutron star—black hole mergers. *Mon. Not. R. Astron. Soc.* **449**, 390–402 (2015).
35. Wanajo, S. *et al.* Production of all the r-process nuclides in the dynamical ejecta of neutron star mergers. *Astrophys. J. Lett.* **789**, L39 (2014).
36. Lattimer, J. M. & Schramm, D. N. Black-hole-neutron-star collisions. *Astrophys. J. Lett.* **192**, L145–L147 (1974).

Acknowledgements

We thank Dr A. Gal-Yam for communication and the referees for helpful comments. The HST and VLT data presented in this work were obtained from the Mikulski Archive for Space Telescopes (MAST) and the ESO Science Archive Facility, respectively. This work was supported in part by the National Basic Research Programme of China (No. 2013CB837000 and No. 2014CB845800), NSFC under grants 11361140349, 11103084, 11273063, 11303098 and 11433009, the Foundation for Distinguished Young Scholars of Jiangsu Province, China (Grant No. BK2012047), the Chinese Academy of Sciences via the 100 talent programme and the Strategic Priority Research Program (Grant No. XDB09000000), the Israel ISF—China BSF grant and the I-Core center for excellence 'Origins' of the ISF. S.C. has been supported by ASI grant I/004/11/0.

Author contributions

Z.P.J., B.Y., X.L., X.Z.Z. (from PMO) and S.C. (from INAF/OAB) carried out the data analysis, following Y.Z.F.'s suggestion. K.H., T.P., Y.Z.F., D.M.W. and Z.P.J. interpreted the data. Y.Z.F. and T.P. prepared the paper and all authors joined the discussion.

Additional information

Competing financial interests: The authors declare no competing financial interest.

16

The solar magnetic activity band interaction and instabilities that shape quasi-periodic variability

Scott W. McIntosh[1], Robert J. Leamon[2], Larisza D. Krista[3], Alan M. Title[4], Hugh S. Hudson[5], Pete Riley[6], Jerald W. Harder[7], Greg Kopp[7], Martin Snow[7], Thomas N. Woods[7], Justin C. Kasper[8,9], Michael L. Stevens[8] & Roger K. Ulrich[10]

Solar magnetism displays a host of variational timescales of which the enigmatic 11-year sunspot cycle is most prominent. Recent work has demonstrated that the sunspot cycle can be explained in terms of the intra- and extra-hemispheric interaction between the overlapping activity bands of the 22-year magnetic polarity cycle. Those activity bands appear to be driven by the rotation of the Sun's deep interior. Here we deduce that activity band interaction can qualitatively explain the 'Gnevyshev Gap'—a well-established feature of flare and sunspot occurrence. Strong quasi-annual variability in the number of flares, coronal mass ejections, the radiative and particulate environment of the heliosphere is also observed. We infer that this secondary variability is driven by surges of magnetism from the activity bands. Understanding the formation, interaction and instability of these activity bands will considerably improve forecast capability in space weather and solar activity over a range of timescales.

[1] High Altitude Observatory, National Center for Atmospheric Research, PO Box 3000, Boulder, Colorado 80307, USA. [2] Department of Physics, Montana State University, Bozeman, Montana 59717, USA. [3] Cooperative Institute for Research in Environmental Sciences, University of Colorado, Boulder, Colorado 80205, USA. [4] Lockheed Martin Advanced Technology Center, 3251 Hanover Street, Building 252, Palo Alto, Colorado 94304, USA. [5] Space Sciences Laboratory, University of California, Berkeley, California 94720, USA. [6] Predictive Science Inc., 9990 Mesa Rim Road, Suite 170, San Diego, California 92121, USA. [7] Laboratory for Atmospheric and Space Physics, University of Colorado, 1234 Innovation Drive, Boulder, Colorado 80303, USA. [8] Harvard-Smithsonian Center for Astrophysics, 60 Garden Street, Cambridge, Massachusetts 02138, USA. [9] Department of Atmospheric, Oceanic and Space Sciences, University of Michigan, Ann Arbor, Michigan 48109, USA. [10] Division of Astronomy and Astrophysics, University of California, Los Angeles, Colorado 90095, USA. Correspondence and requests for materials should be addressed to S.M. (email: mscott@ucar.edu).

The obvious hemispheric asymmetry of the solar atmosphere over the past several years (2009–2014) has generated a significant amount of interest in the heliophysics community[1]. Indeed, the asymmetric magnetic evolution of the Sun's northern and southern hemispheres enabled the recent demonstration that the 22-year magnetic polarity cycle strongly influences the occurrence, and distribution of the sunspots which form the 11(-ish)-year solar activity cycle[2]—an observational result that challenges the current understanding of the Sun's magnetism factory, the solar dynamo[3].

McIntosh et al.[2] illustrated that the twisted toroidal bands of the 22-year magnetic polarity cycle are embedded in the Sun's convective interior and first appear at high latitudes ($\sim 55°$) before travelling equatorward. These bands interact with the oppositely polarized magnetic band from the previous cycle at lower latitudes in each hemisphere. The interaction of these activity bands is illustrated in Fig. 1 and modulates the occurrence of sunspots on the low-latitude bands (which have opposite magnetic polarity and sense of handedness) until they eventually cancel across the equator (as occurs in 1998). This equatorial cancellation signals the end of the sunspot cycle and leaves only the higher-latitude band in each hemisphere. Sunspots rapidly appear and grow on that band for several years until a new oppositely signed band appears at high latitude (for example, 2001 in the north, and 2003 in the south)—an occurrence that defines the maximum activity level of that new cycle and triggers a downturn in sunspot production. The perpetual interaction of these temporally offset 22-year activity bands drives the (quasi-) 11-year cycle of sunspots that form the decadal envelope of solar activity. The observational evidence presented by McIntosh et al.[2] points to the rotational energy at the bottom of our Star's convection zone as being the major driver of the Sun's long-term evolution.

Rotating atmospheres, like that of the Earth and the giant planets of our solar system, often exhibit shorter-timescale global-scale phenomena such as Kelvin and Rossby waves[4,5], which are important for the transport and regulation of energetics in those systems[6]. In the following analysis we argue, based on a host of observations displaying (quasi-)periodicities of significantly shorter—but commensurate amplitude—to the well-established decadal-scale 'solar cycle' variability, that the Sun is no different. It is possible that the convecting, magnetized, 'ocean' beneath the Sun's optical surface could exhibit similar global-scale wave behaviour to those readily observed in our atmosphere and other planetary atmospheres in the solar system[7,8]. Such phenomena could drive marked changes in the Sun's interior and the rate at which magnetic flux pierces our star's photosphere. Once forced into the outer solar atmosphere, that magnetic flux will strongly affect the radiative, particulate and eruptive output of the Sun.

Results

Activity band interaction and the Gnevyshev Gap. Variations of significantly shorter period than the canonical (11-year) envelope of solar variability are visible in the Sun's flaring activity (Fig. 2). The figure paints a canvas of the Sun's magnetism over the past 35 years—the last three-plus sunspot cycles. The dwindling number of sunspots and flares occurring on the Sun over that period is clearly indicating a net downturn in solar activity. We see that the peak flare rate occurs at a different time from the sunspot maximum—often a few years later—an observational phenomenon known as the 'Gnevyshev Gap'[9,10]. Superimposed on that decadal-scale envelope we see that flares and (the monthly number of) sunspots quasi-periodically surge in number. These well-documented[11,12] surges in solar activity, resulting from a 10–15% increase in sunspot numbers, result in a doubling or tripling of the flare rate over the course of several months. In addition, a latitude–time probability density function of flare activity from the National Oceanic and Atmospheric Administration (NOAA) record (Fig. 2c), indicates that the spatio-temporal clustering of strong flares shares a common origin with the 'herringbone' pattern seen in magnetic butterfly diagrams (Fig. 2d). That herringbone pattern appears to propagate from mid-to-high latitudes on a quasi-periodic basis from a common point of origin with the flare clusters. The relationship between the flare clusters and the root of the herringbone pattern is due to their association with sunspots. The correspondence of the two implies that the magnetic fields at the root of the system (in the magnetic activity bands) are being perturbed in a quasi-periodic manner by some physical process related to the evolution of the activity bands themselves. That process produces such a rapid and strong increase of magnetic flux emergence that is hard to reconcile with any known convective or shear phenomena occurring in the surface, or near-surface, layers.

Short-term variability in cycle 23. Studying the last solar cycle in more detail, Fig. 3 compares the daily coronal mass ejection (CME) rates inferred from the National Aeronautics and Space Administration (NASA) SOHO and STEREO spacecraft, the sunspot number and the flare rates determined from the GOES archive (Fig. 2) and the NASA RHESSI spacecraft. We see that two different CME detection algorithms[13,14] applied to the SOHO data set arrive at very similar whole-Sun statistics. Those also match the CME statistics derived from STEREO observations from late 2006 to the present[13]. An important detail to note here is that the STEREO spacecraft spent almost their entire mission time off of the Sun–Earth line, strengthening the perception that the phenomena driving the changes in CME rates are global in nature—being independent of the observer's specific (heliocentric) longitude. Due to uncertainties in identifying the absolute origin of CMEs on the solar disk, especially those from the far side (which nevertheless are detected by white-light coronagraphs), we do not attempt to identify the events from the

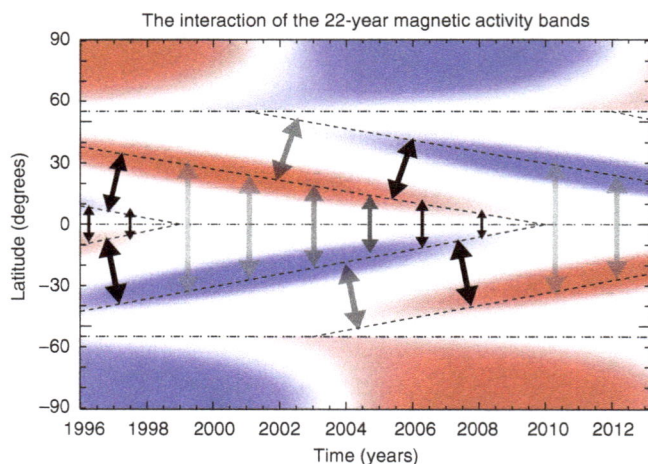

The interaction of the 22-year magnetic activity bands

Figure 1 | Activity band interaction. A data-derived schematic of the the the latitudinally interacting activity bands of the 22-year magnetic polarity cycle, as introduced by McIntosh et al.[2] The bands, visualized here in the radial component of the magnetic field, of opposite polarity start their migration towards the equator from high latitudes in each hemisphere and take ~ 19 years to reach their termination. The arrows illustrate some of the possible interactions between the bands within, and without, their hemisphere while the opacity of the arrows indicate the (potential) strength of the interaction between the two. This figure is adapted from Fig. 8 of ref. 2. Copyright 2014 by The American Astronomical Society.

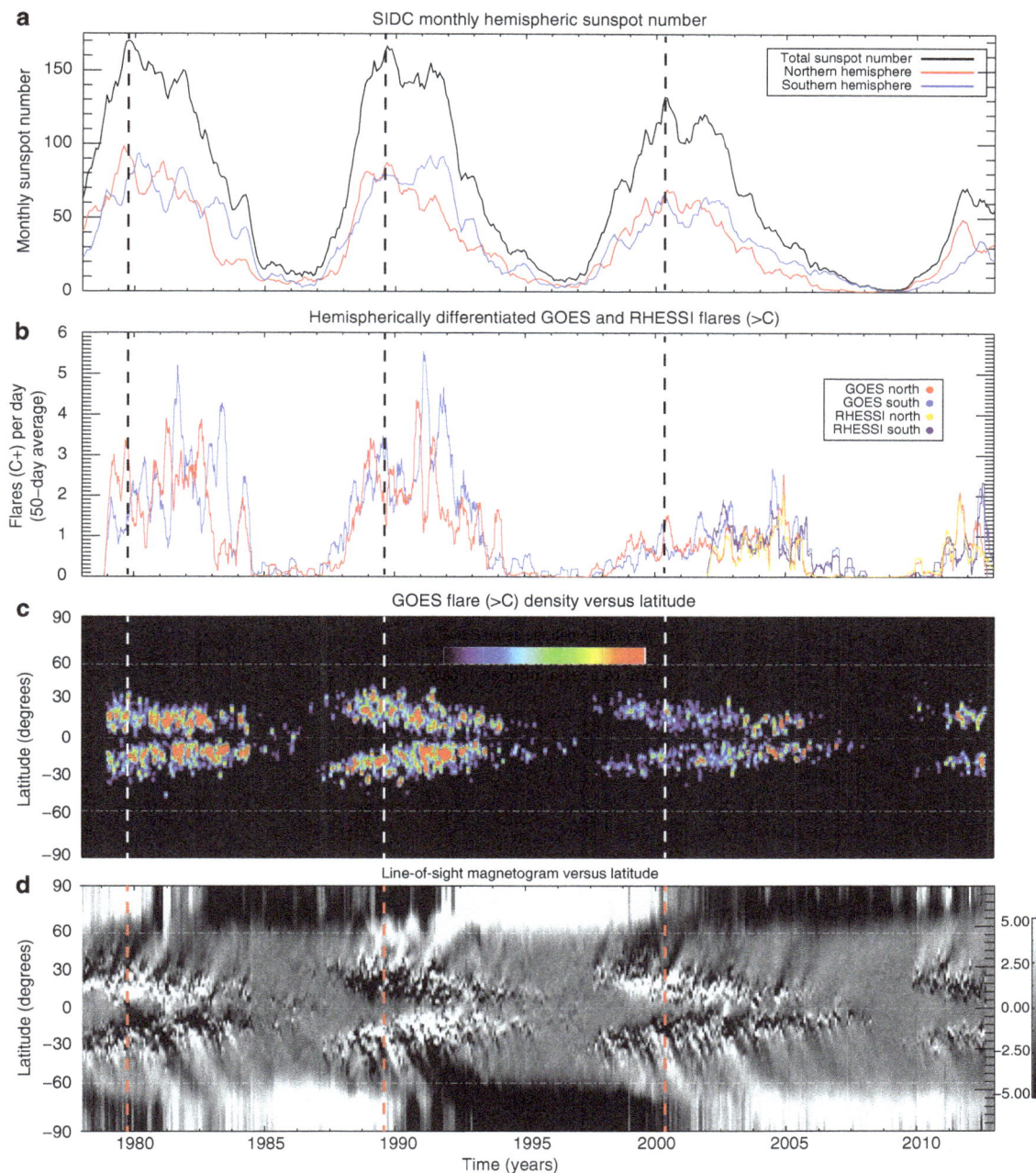

Figure 2 | Magnetic variability over the last three decades. Comparison of the variation in (monthly) sunspot number (SSN) and flare record with the 'butterfly' diagram of the photospheric magnetic field over the past three solar cycles. (**a**) Total (black) and hemispheric (red—north; blue—south) monthly sunspot numbers (hSSN) from the Solar influences data center (SIDC). (**b**) Variation of the hemispheric daily rate of flares larger than 'B' magnitude in the GOES (red—north; blue—south) and RHESSI (orange—north; purple—south) records. Note the strong modulation in the flare rate, the hemispheric differences in flare rates and that flare maximum does not occur at the same time as sunspot maximum—over the record shown, the flare activity maximum occurs several years post sunspot maximum. (**c**) Latitude–time distribution of the GOES flares of **b**. (**d**) Latitude–time variation of the photospheric magnetic field at the central meridian. Note the strong correspondence between the poleward pulses of photospheric magnetism and the surges in flare activity from **c** and **b**. All panels show a thick vertical dashed line indicating the time of sunspot maximum and the lower two panels show dot-dashed lines at 55° to delineate high- and low-latitude variation.

northern and southern hemisphere. As such, the CME statistics reflect the behaviour of the 'whole Sun'.

We see that the peaks in the total sunspot number have corresponding peaks in the CME rate. The surges in the daily sunspot number can be as large as 30% and they can lead to a 100% increase in the daily CME rate. The same strong correspondence is visible for the flare rate. The relationship between the disk-integrated CME rate and the hemispheric rates of sunspot and flare formation highlight a critical property in

disk-integrated quantities—they will typically exhibit shorter period variations than hemispherically resolved ones. The phase offset between the two hemispheres will determine the resulting 'hybrid' period observed. In this case, we see that marked increases in surface magnetism lead to a profound increase in the rate of eruptive phenomena.

It is not only eruptive phenomena that exhibit variations of similar magnitudes and timescales. Figure 4 shows the (disk-integrated) total solar irradiance (TSI) and components of the

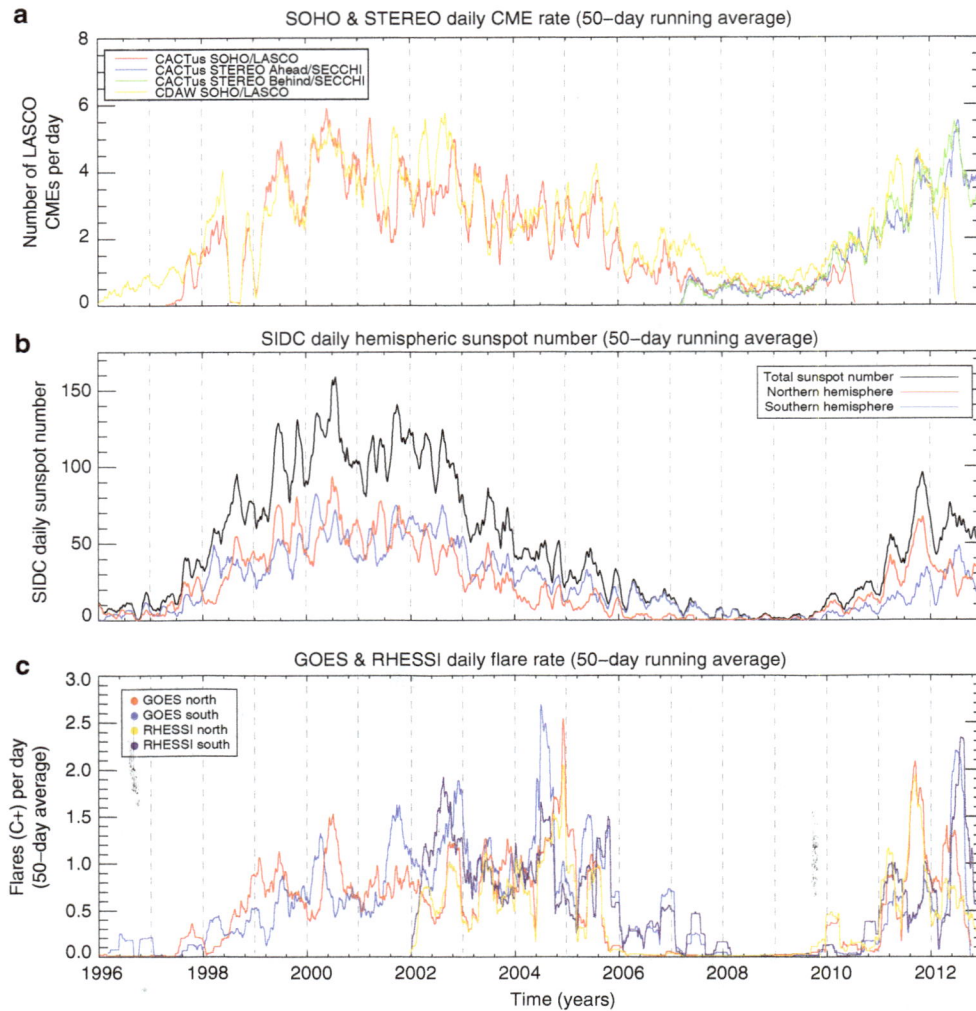

Figure 3 | Variability of the Sun's eruptive output over solar cycle 23. Comparison of the variation in the CME and flare rates over solar cycle 23 with the modulation in the (daily) sunspot number. (**a**) Variation in the (whole Sun) daily CME rates as detected by the CACTus[44] and CDAW[13] methods for the SOHO (red—CACTus; orange—CDAW) and the twin STEREO (blue—'ahead'; green—'behind') coronagraphic data sets. (**b**) SIDC- Solar influences data center. Total (black) and hemispheric (red—north; blue—south) daily sunspot numbers—compare with the monthly counterpart in Fig. 2. (**c**) Variation of the hemispheric daily rate of flares larger than 'B' magnitude in the GOES (red—north; blue—south) and RHESSI (orange—north; purple—south) records. As in Fig. 2, there is considerable lag between (total) sunspot maximum with the CME and flare series—occurring late in the descending phase. Almost every bump and wiggle in the sunspot number shows a corresponding surge in CME and flare activity—these surges can be as large amplitude as a doubling of the sunspot number or flare/CME rate over the course of only a few months before recovering. The panels of the figure show a set of dashed fine vertical lines that are 12 months apart and act as a timescale reference. Each timeseries shown in these panels is a 50-day running average over the original. The CME timeseries are not separated by hemisphere due to the uncertainty in determining the actual CME location from only plane-of-the-sky coronagraphic observations.

spectral solar irradiance measured from space. The variance in the TSI is visible over the entire record (Fig. 4a), but as the measurement has been refined and systematic errors in it have been reduced (especially with the addition of SOHO/VIRGO to the record)[15], we see that the amplitude of the short-term variability is $\sim 1\,\mathrm{Wm}^{-2}$—equivalent to the variation over the whole solar cycle (Fig. 4b). We see that the ultraviolet (Fig. 4c), extreme ultraviolet (Fig. 4d) and X-ray (Fig. 4e) components of the spectral solar irradiance (as measured by the SORCE spacecraft) show variability over the mean spectrum from a few to almost 100% during the activity surges.

Variability in the solar wind and fast wind source regions. Similarly, Fig. 5 shows another highly modulated facet of quiescent solar behaviour that illustrate these global surges in

magnetism—properties of the solar wind and its geomagnetic impact. The abundance of helium is a marker of magnetic activity in the solar atmosphere[16,17]. While the amount of helium in the fast and slow solar wind shows a strong decline over the past three decades[1]; (cf. Fig. 2b) we can also see the clear 20–50% swings of short-term variability. Short-term variability is also visible in the speed of the solar wind and the Ap geomagnetic activity index that it influences (Fig. 5b)—noting that solar wind characteristics are strongly impacted by the three-dimensional geometry of the heliosphere's magnetic field, and where the spacecraft sampling interplanetary space are situated. The coherent variation of the solar wind speed[18] indicates that the processes governing the shape of the magnetosphere[19] and heliosphere are being driven by the surges in magnetic variability.

These periodic changes in the morphology of the coronal (and heliospheric) magnetic field can be inferred from Fig. 6 where we

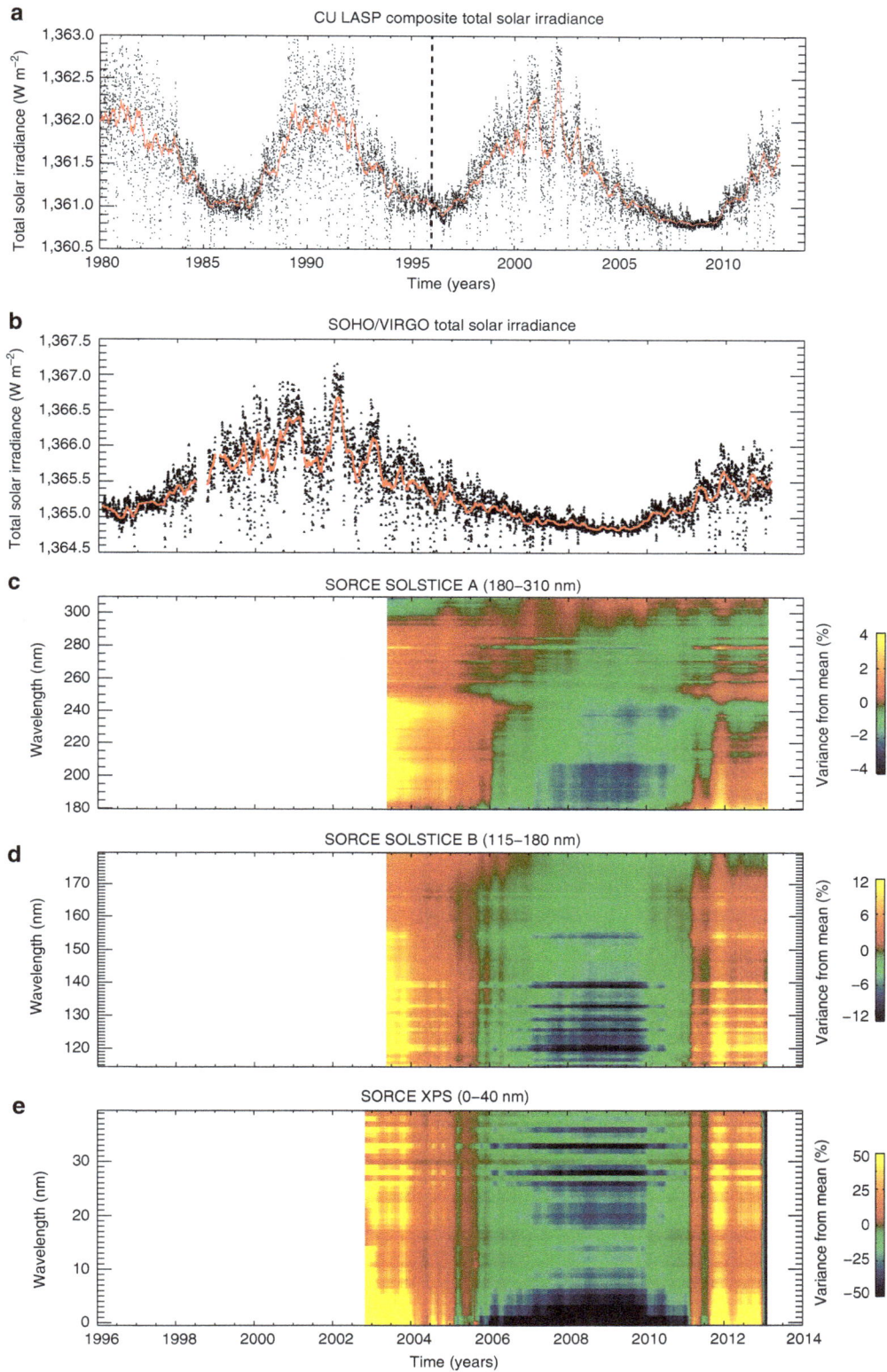

Figure 4 | Variability in the total solar irradiance over the past three decades in comparison with the variance in components of the solar spectral irradiance over solar cycle 23. (**a**) The University of Colorado TSI composite[10] in comparison with (**b**) the SOHO/VIRGO TSI over solar cycle 23—the thick vertical dashed line marks the start of the SOHO/VIRGO record used. In both cases, the thick red lines are the 50-day running average over the measurements. While the mean solar minimum to solar maximum change in TSI is ~1Wm^{-2}, there is a shorter-period modulation variation visible in the ¯SI over the entire time frame. That variation, of the same magnitude as the decadal variation, is better defined in solar cycle 23 due to refinement in instrument design and calibration[10]. (**c-e**) Percentage variation in different bands (relative to the mean spectrum) of the solar spectral irradiance from the SORCE spacecraft from the far-ultraviolet, ultraviolet SOLSTICE measurements. As we move to shorter wavelengths, the degree of variation in one of the surges in solar radiation increases from a few to 50%. XPS, X-ray photoelectron spectroscopy.

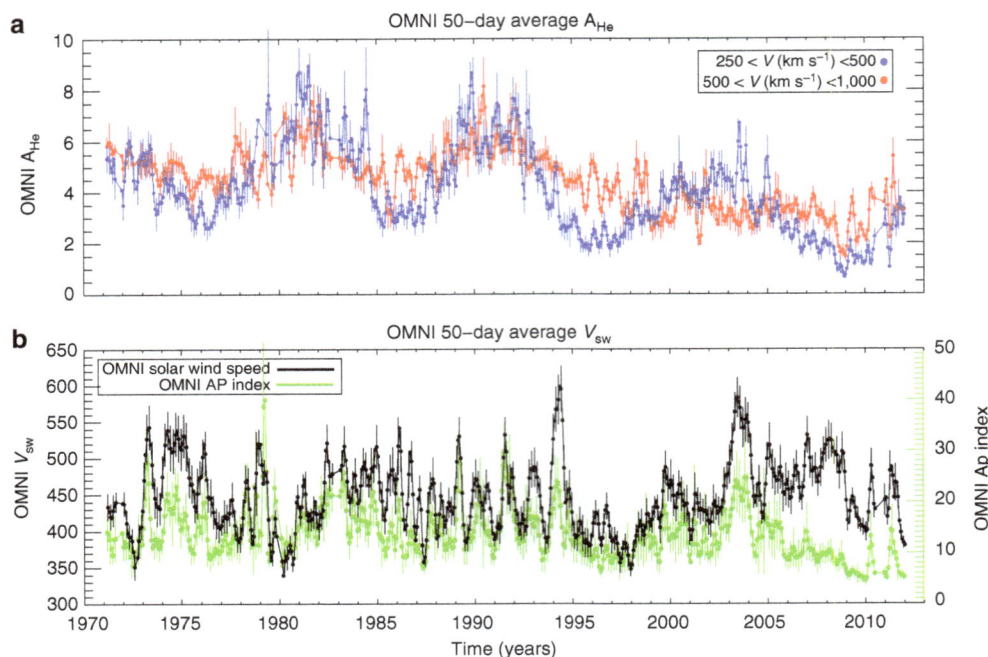

Figure 5 | Variation of solar wind and geomagnetic properties over the past four decades. The data presented are from the NASA/GSFC Space Physics Data Facility OMNI database (http://omniweb.gsfc.nasa.gov/). (**a**) Variation in 50-day running averages of the fast (red) and slow (blue) solar wind helium abundance (A_{He}; ref. 17)—a proxy of plasma heating at the base of the solar wind[16]. (**b**) Variation in the 50-day running averages of the solar wind speed (V_{sw}; black) and geomagnetic storm Ap index (green). Note the steady drop in A_{He} over the time frame and the strongly correlated quasi-periodicities in all four quantities where the surges in V_{sw}, A_{He} and Ap are of the order 100 km s^{-1}, 15 and 50%, respectively. The error bars in the plot reflect the variance of the signal over the 50-day running window.

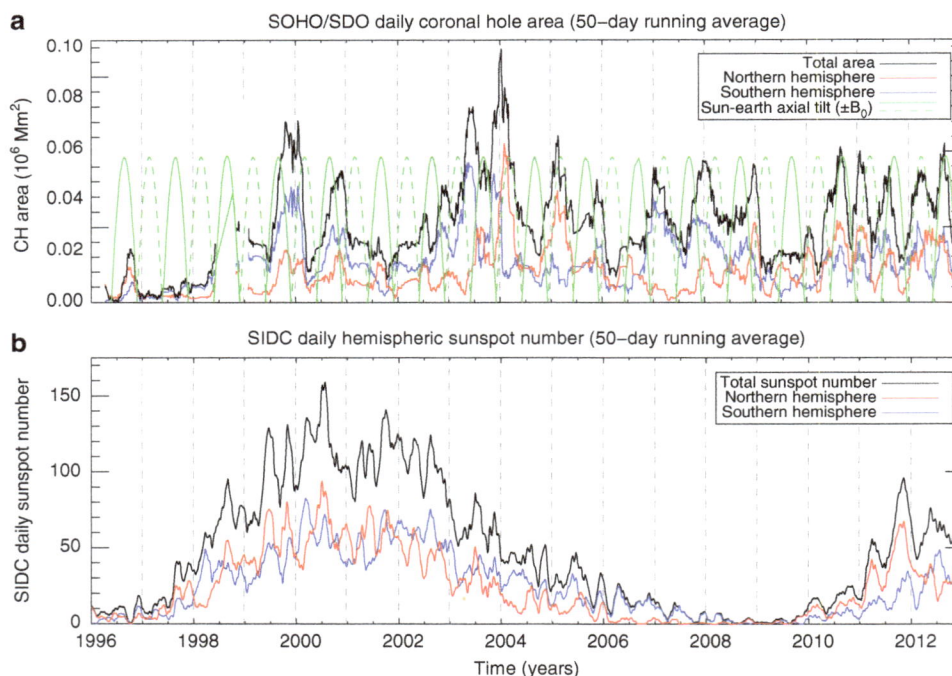

Figure 6 | Variation in the low-latitude coronal hole area and the sunspot number over solar cycle 23. (**a**) Using the CHARM[21] automated SOHO/EIT and SDO/AIA coronal hole-detection algorithm, we show the variation of the 50-day running average in the total (black), northern (red) and southern (blue) hemispheric coronal hole areas below 55° latitude. For reference, the seasonal variation in the Sun's axial tilt relative to the Sun-Earth line is shown (green—the dashed green line shows amplitude of the negative tilt). (**b**) Solar influences data center (SIDC). Total (black) and hemispheric (red—north; blue—south) daily sunspot numbers as in Fig. 3. Like the flare and CME timeseries shown above, the coronal hole areas peak after solar sunspot maximum (~2004). A comparison of the hemispheric sunspot and coronal hole areas shows a systematic lag in the peaks of the latter (~6 months). The 2004 peak in coronal hole area corresponds to the peaks in the Ap index and solar wind speed of Fig. 5. The panels of the figure show a set of dashed fine vertical lines that are 12 months apart and act as a timescale reference.

contrast the variation of total and hemispheric areas of low-latitude ($<55°$) coronal hole areas with hemispheric sunspot numbers (cf Fig. 3) and the solar B_0-angle (the heliographic latitude of the central point of the solar disk). The seasonal variation of the latter modulates the visible area of the solar disk[20], but cannot exclusively explain the strong periodicities in the hemispheric coronal hole areas or their time-varying phase relationship. Further, the hemispheric coronal hole areas appear (on-average) to lag the sunspot numbers by a few months—the former are typically a higher-latitude phenomena than the sunspot band[2]. This strengthens the premise that at least some of the magnetic flux which forms coronal holes is the result of active region flux diffusion[21]. The increase in coronal hole area during the declining phase of the sunspot cycle (with noticeable peaks in 2003–2004) is another well-observed phenomenon[22] but is more related to the gross interaction of the 22-year activity bands that we have discussed earlier.

Quasi-periodic variability of solar magnetism. Magnetic fields on smaller spatial scales than coronal holes and sunspots display similar periodicities to their larger brethren throughout the solar cycle. Figure 7a shows the evolution in number density of the magnetic elements associated with the vertices of the giant convective scale[2,23]. This convective scale is driven by the rotation of the deep radiative interior, and these 'g-nodes'[23] are believed to be anchored close to the bottom of the convection zone's boundary with the radiative interior. The number of g-nodes in each hemisphere waxes and wanes over the course of solar cycle 23, in addition to being strongly variable over shorter timescales. g-Node densities also display a varying phase offset between the two solar hemispheres. The Fourier power spectra (Fig. 7b) of the hemispheric g-node density and (daily) sunspot timeseries have very similar characteristic timescales as indicated by the grey-shaded regions in the figure. The short-period (higher frequency) envelope peak of 11–16 days is approximately one half of the rotational period (24–35 days). This indicates that magnetic patterns do not diffuse immediately on the Sun's surface. The slight offset between peaks in the low- (28 days) and high-latitude (30 days) period is consistent with observed solar differential rotation[24]. The broad peak centred around 330 days is common to the timeseries, although the southern hemisphere seems to be shifted further and is consistent with the analysis of Getko[25,26]. This appears to be the primary (quasi-)periodicity of the magnetic surges that shape the heliosphere and drive the host of energetic phenomena observed as described above. Wavelet analyses of these timeseries (see the Methods section; Supplementary Figs 1–3) demonstrate that the aforementioned peaks occur with a 99% confidence level.

Discussion
The physical origin of these strong quasi-periodic surges in the Sun's magnetism is not known. However, their effect on the outer solar atmosphere and on the geospace environment is profound. Their existence has been documented extensively since the start of the space-age. For example, strong quasi-periodicities that are longer than the Sun's rotation rate have been amply documented in the literature for sunspot areas[27], flares[11], CMEs[28] and major geomagnetic storms[29,30], but it is likely that any property of the outer solar atmosphere that is dependent on magnetism will show a response of varying degree[1] and that extends to the interplanetary magnetic field[31].

As we have noted above, it is unlikely that a strong modulation in the number of sunspots can be easily explained by processes in the near-surface layers of the Sun. However, considering a spatio-temporal decomposition of solar surface magnetism[32] can

provide some interpretative guidance. Figure 8 shows Ulrich's decomposition of the photospheric butterfly diagram (Fig. 2d) into a long-term smoothed radial field and a residual. The latter reveals poleward-propagating features in each hemisphere. The primary signal in the (smoothed) butterfly diagram is divided into high- and low-latitude evolution at $\sim 55°$ latitude[2], both alternate in sign and are long lived—the lower-latitude pattern propagates equatorially. This pattern is associated with the interacting activity bands of the 22-year magnetic polarity cycle described by McIntosh et al.[2] The secondary pattern, visible in the residual between the primary pattern and original data set, is poleward propagating, is not symmetric across the equator and has a much shorter timescale than the former. Ulrich[32] notes that the latter pattern is not compatible with simple (single meridional cell) surface advection of magnetic flux.

We infer that the interaction of the oppositely signed, long-lived activity bands in each hemisphere as discussed by McIntosh et al.[2] can help explain why the flare, CME and coronal hole timeseries peak so long after (total) sunspot maximum. The latitudinal interaction—via flux emergence—of the activity bands in each hemisphere must peak at some point after the time it starts propagating equatorward, the time that defines solar maximum[2]. Such an interaction of the activity bands, combined with the phase difference of hemispheric evolution[1], can explain Gnevyshev's observational findings[10] where hemispheric asymmetry alone cannot[33]. Substantial numerical simulations of the interaction between deep-rooted magnetic flux and convection[34] are positive initial steps in exploring the range of variability in decadal-scale solar output by placing magnetic flux systems in a rotating convective envelope.

In addition to the decadal envelope of solar activity, there is a clear, strong, variability of the magnetic flux in each solar hemisphere of approximately 1 (terrestrial) year. We propose that the process at the root of the short-term propagating pattern shown in Fig. 8 is responsible for the surges in solar activity and the latitudinal variation in the proxies that we have noted above.

Figure 7 permits a phenomenological explanation of the quasi-periodicities (of order 150 days) that have been observed in a large number of heliospheric quantities by Rieger and others[9–11,25–29,35]. Those are 'hybrid' periodicities—a consequence of the phase relationship between the short-term variability in each solar hemisphere. In short, the longer-period hemispheric timeseries from each hemisphere will combine to produce a shorter period (higher frequency) whole-sun timeseries—consider our earlier example for flares and CMEs. Indeed, the same principle can possibly explain the quasi-periodicities seen in helioseismic measurements of the deep convection zone[36]—if our assertion that the phenomena at the root of this problem occur on the activity bands, near the base of the convection zone, is correct. In this case, noting that (standard) global helioseismology analyses impose hemispheric symmetry, the phase of the timeseries in each hemisphere is critical. Only in the earlier part of cycle 23 (1998–2002) would the two hemispheres constructively create a signal that can be detected using this method, as the hemispheres were then approximately in phase.

So, what are the poleward-propagating excursions seen in Fig. 8 and how are they driven? The simplest possible explanation is one where the surges in solar magnetism periodically load more flux into the Sun's surface layers. Once those magnetic regions begin to decay and diffuse over time[37], the surface meridional circulation[21] is loaded with magnetic flux that is then carried poleward. While this appears straightforward, it does not answer the second and most important part of the question—what drives the surges of magnetism?

One possible explanation follows from the deliberations of Howe et al.[36] Howe et al. indicate that there are global-scale

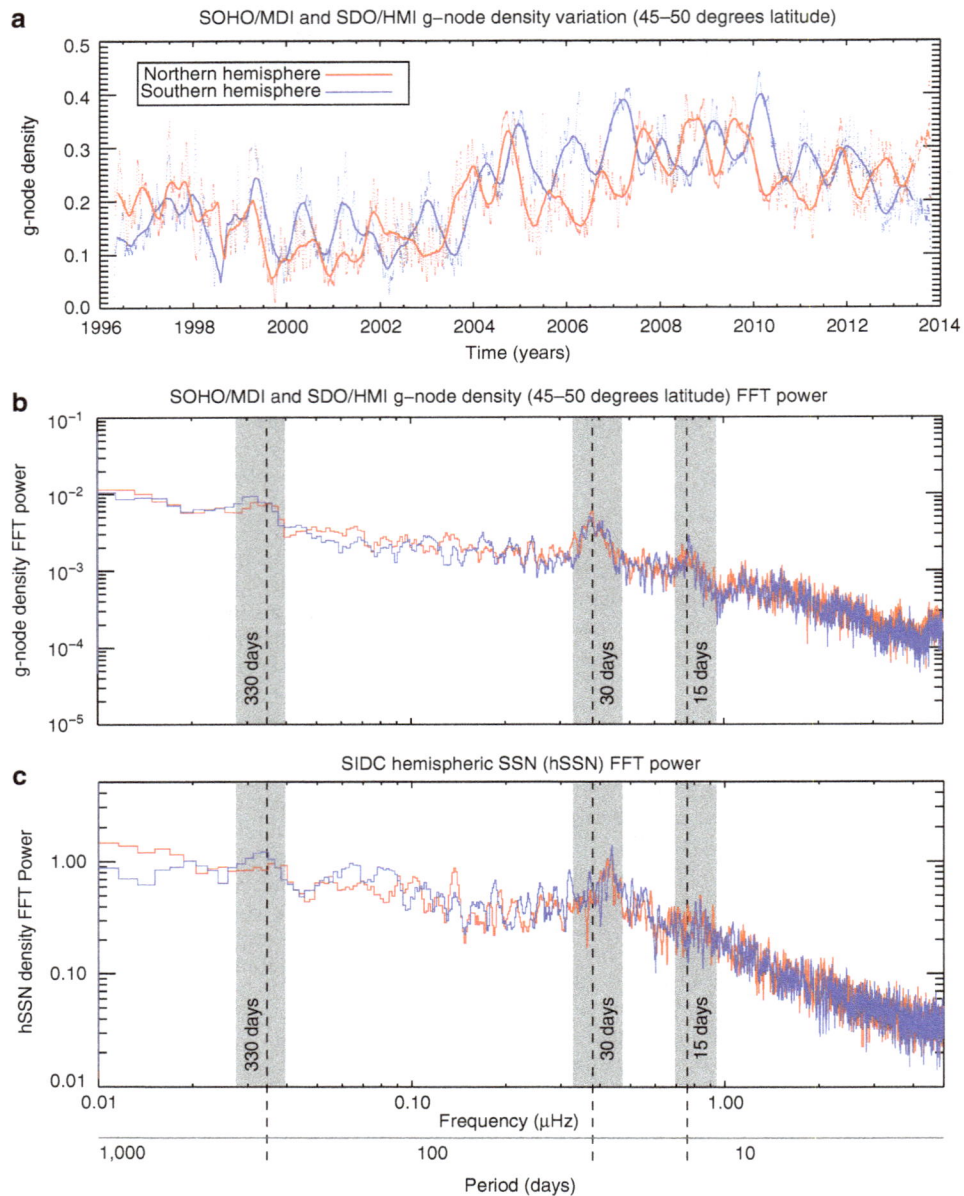

Figure 7 | Variation in the deep-rooted small-scale solar magnetic features over solar cycle 23. (**a**) Variation in the density of giant convective cell vertices (g-nodes) averaged over 45–50° latitude in the northern (red) and southern (blue) hemispheres from the SOHO Michelson Doppler Imager and SDO Helioseismic Magnetic Imager—markers of deep-rooted solar magnetism that belong to the toroidal magnetic flux systems of the 22-year magnetic activity cycle[2]. The small dots are individual daily averages, while the thick lines are the corresponding 50-day running average. As in Fig. 6, the variable phase of the timeseries in each hemisphere is strongly indicative of a solar origin for these phenomena and not some orbital or Sun–spacecraft distance variability. The periods where the hemispheres vary in phase correspond to the times of strongest modulation in the energetic parameters shown in the figures above. (**b,c**) Fast Fourier transform (FFT) power spectra of the northern and southern hemispheric g-node timeseries, when compared with counterparts for the daily hemispheric sunspot number, respectively, (Fig. 3) show broad peaks of significant power occurring throughout the timeseries, especially those centred on 330 days, 30 days and 15 days in the shaded regions.

waves and instabilities that propagate in the shear layer known as the tachocline[3] at the bottom of the convection zone—where the activity bands appear to be rooted. We then observe the impact of those waves and instabilities on the surface magnetism of our star through their modulation of the global magnetic flux emergence process[38,39].

What could these perturbations to the magneto-convective system be? The Earth's mantle, ocean and thermosphere/stratosphere exhibit global-scale waves that are driven by the rotation of the planet at shear interfaces, or Rossby waves[40,41], like the tachocline. The action of such energetic interface waves in

the solar interior could dynamically modify the buoyancy characteristics of the flux tubes present in the region above[39]. Theoretical efforts indicate that magnetized Rossby waves with periods of order several hundred days are highly likely[42,43] in a non-zero thickness tachocline[38]. Whether or not the surges of magnetism are caused by large-scale Rossby-like waves in the Sun's convective interior, we have seen that they force large upswings in solar activity of quiescent and explosive nature. The period of the surges in each solar hemisphere is close to 1 terrestrial year, and the hemispheric phase relationship influences the period of the disturbances felt in the heliosphere. Significant

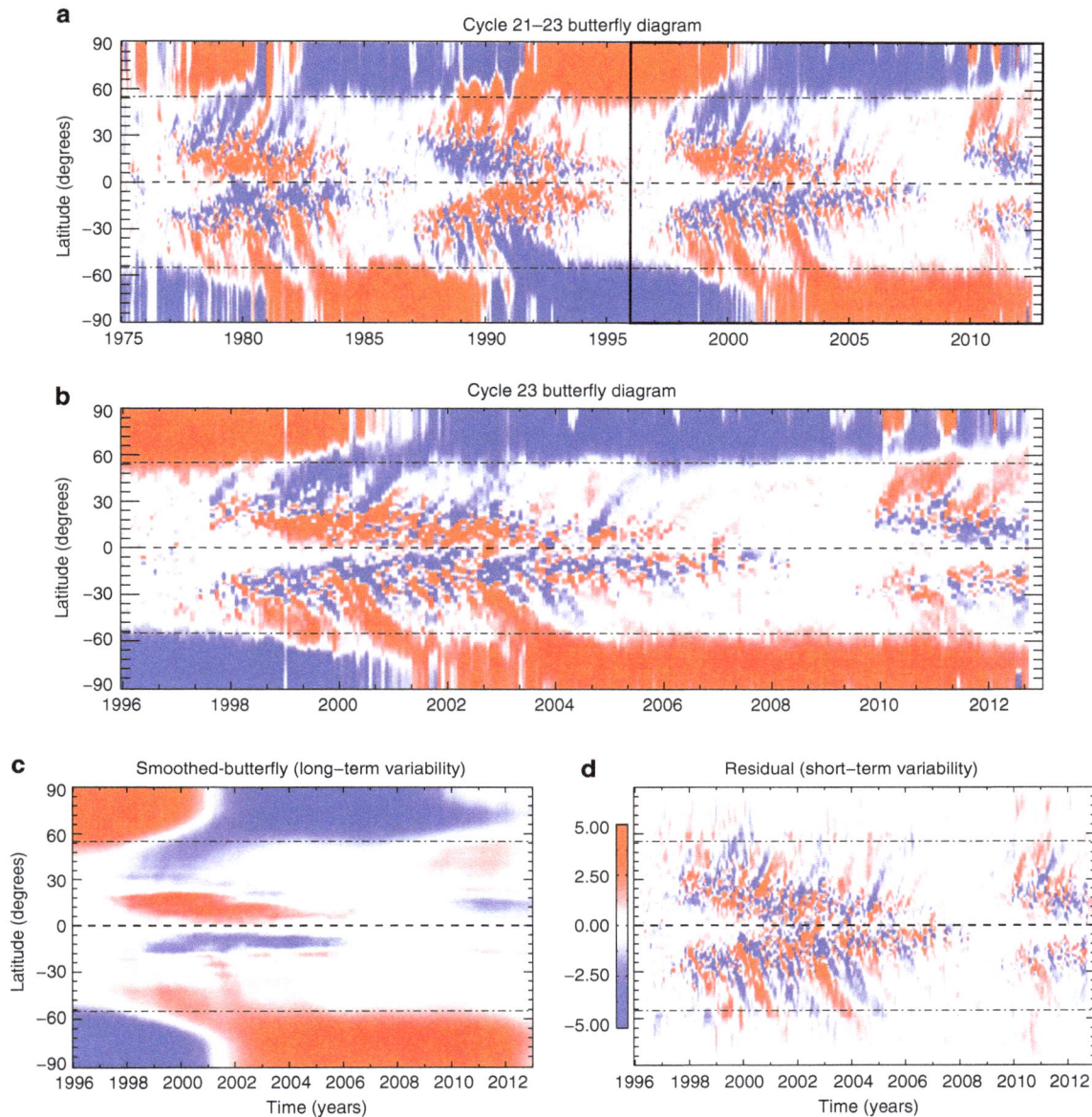

Figure 8 | Gross decomposition of the surface magnetism of the past four decades into short-term and long-term variability components. The decomposition follows the method of Ulrich[30]. (**a**) The pattern of photospheric magnetic field in a latitude–time plot constructed using Carrington rotation (28-day) sampling of the central meridian field. The inset region shown as a black rectangle outlines the latitude–time plot shown in **b**. (**c**) Hundred-day average field from **b** and the residual between that 100-day average and the original latitudinal variation (**d**). The average and residual correspondingly decompose the surface magnetism into the space climate and space weather modulations that bathe the earth in radiation, particles and disruptive events. The poleward surges of magnetism shown in **d** are directly related to the strong modulation shown in the figures above. For illustration, the equator and 55° lines are shown as black dashed and dot-dashed lines, respectively.

research remains to be done to determine whether the apparent periodicity is a fundamental characteristic of our star's deep interior, and to understand the processes responsible for producing it.

To summarize, we have inferred that the interaction of the activity bands belonging to the Sun's relentless 22-year magnetic polarity cycle shape the decadal-scale variability of solar activity[1,2]. In addition, there is a quasi-annual modulation of solar activity—with a magnitude commensurate to that of the decadal variability—which appears to be driven by surges of magnetic flux originating in those activity bands.

The growing dependence of our civilization on technology susceptible to space weather should motivate investigations into the rotational forcing of the Sun's deep convection zone by the

radiative zone. Specifically, challenging simulations of activity band formation, intra- and extra-hemispheric activity band interaction and the zoo of rotational-gravity-buoyancy waves that interact with those activity bands are required. These factors appear to be key drivers of solar variability on decadal and annual timescales. A better understanding of the processes responsible for modulating the decadal variability and the (quasi-)annual 'seasons' of solar activity will yield a significantly increased forecast skill for solar activity in parallel with continued observational monitoring.

Methods

Periodicities. Figure 7 presents a Fourier analysis of the hemispheric g-node and sunspot variability. The figure indicates the prevalence of a ∼330-day quasi-periodicity in those timeseries. In the following discussion we employ wavelet transform techniques, as presented by Torrence and Compo[43], as a means to demonstrate the significance of the timescales in the hemispheric g-node (Supplementary Fig. 1), daily hemispheric sunspot number (Supplementary Fig. 2) and the SOHO/VIRGO TSI measurements (Supplementary Fig. 3) that appear in Fig. 4b.

Using the Morlet wavelet as a model representation of the oscillatory signals observed, we construct the wavelet power spectra of each timeseries. In the most general sense a wavelet power spectrum can be thought of as an image that indicates the strength and the duration that an oscillation of a particular period is present in the timeseries. It is a particularly powerful tool for the analysis of timeseries that exhibit mixed periods and quasi-periodicities as is often the case in many solar phenomena.

For the wavelet power spectra shown in Supplementary Figs 1–3, we can clearly see that periodicities of order 330 days are present at an 99% confidence level. The 99% confidence level in the wavelet power spectra shown is indicated by a solid, thick, closed contour. Statistical confidence in this case is computed with respect to a red noise model of the spectral background—a spectral background that has increasing power with decreasing frequency. This model is common for most solar and geophysical data sets[43] and is adequate for the present application. It appears that the ~30- and ~15-day quasi-periodicities also have strong wavelet power in the timeseries studied, although the wavelet power does not always meet the 95% confidence criteria. This can be most easily seen in Supplementary Fig. 2 for the hemispheric daily sunspot number. However, the signal and wavelet power spectra are likely impacted at shorter periods by our earlier choice to study a 50-day running average of the sunspot timeseries.

The cross-hatched areas in the wavelet power spectrum define the 'cone of influence.' The interpretation of the cone of influence is relatively straightforward—the signal in the cross-hatched area may not be entirely reliable because of the influence of edge effects of the timeseries. Because these are finite-length timeseries, errors will occur at the beginning and end of the wavelet power spectrum, as the Fourier transform used in the Wavelet method assumes that the data are cyclic[43]. The solution used in these particular wavelet methods pads the end of the timeseries with zeroes before performing the wavelet transform and removes them afterwards. In the examples shown, the timeseries are padded with sufficient zeroes to bring the total length N up to the next-higher power of two—this limits any edge effects and speeds up the Fourier transform at the core of the computation.

Data sources. The data used in this paper are openly available from the NGDC, SOHO, SDO and the Virtual Solar Observatory (http://virtualsolar.org) data archives.

References

1. McIntosh, S. W. et al. Hemispheric asymmetries of solar photospheric magnetism: radiative, particulate, and heliospheric impacts. Astrophys. J. 765, 146 (2013).
2. McIntosh, S. W. et al. Deciphering solar magnetic activity I: on the relationship between the sunspot cycle and evolution of small magnetic features. Astrophys. J. 792, 12 (2014).
3. Charbonneau, P. Dynamo models of the solar cycle. Living Rev. Solar Phys. 7, 3 (2010).
4. Thomson, W. On gravitational oscillations of rotating water. Proc. Roy. Soc. Edinburgh 10, 92–100 (1879).
5. Rossby, C. G. et al. Relation between variations in the intensity of the zonal circulation of the atmosphere and the displacements of the semi-permanent centers of action. J. Mar. Res. 2, 38–55 (1939).
6. Dickinson, R. E. Rossby waves—long-period oscillations of oceans and atmospheres. Ann. Rev. Fluid Mech. 10, 159–195 (1978).
7. Gilman, P. A. A Rossby-wave dynamo for the Sun I. Sol. Phys. 8, 316–330 (1969).
8. Gilman, P. A. A Rossby-wave dynamo for the Sun II. Sol. Phys. 9, 3–18 (1969).
9. Gnevyshev, M. N. The corona and the 11-year cycle of solar activity. Sov. Astron. 7, 311–318 (1963).
10. Gnevyshev, M. N. Essential features of the 11-year solar cycle. Sol. Phys. 51, 175–183 (1977).
11. Rieger, E. et al. A 154-day periodicity in the occurrence of hard solar flares? Nature 312, 623–625 (1984).
12. Bai, T. Distributions of flares on the Sun: superactive regions and active zones of 1980–1985. Astrophys. J. 314, 795–807 (1987).
13. Robbrecht, E. & Berghmans, D. Automated recognition of coronal mass ejections (CMEs) in near-real-time data. Astron. Astrophys. 425, 1097–1106 (2004).
14. Gopalswamy, N. et al. The SOHO/LASCO CME Catalog. Earth Moon Planets 104, 295–313 (2009).
15. Kopp, G. & Lean, J. L. A new, lower value of total solar irradiance: evidence and climate significance. Geo. Res. Lett. 38, L01706 (2011).
16. McIntosh, S. W., Kiefer, K. K., Leamon, R. J., Kasper, J. C. & Stevens, M. L. Solar cycle variations in the elemental abundance of helium and fractionation of iron in the fast solar wind: indicators of an evolving energetic release of mass from the lower solar atmosphere. Astrophys. J. 740, 23 (2011).
17. Kasper, J. C. et al. Evolution of the relationships between helium abundance, minor ion charge state, and solar wind speed over the solar cycle. Astrophys. J. 745, 162 (2012).
18. Richardson, J. D., Paulerina, K. I., Belcher, J. W. & Lazarus, A. J. Solar wind oscillations with a 1.3 year period. J. Geophys. Res. Lett. 21, 1559–1560 (1994).
19. Paulerina, K. I., Szabo, A. & Richardson, J. D. Coincident 1.3-year periodicities in the ap geomagnetic index and the solar wind. J. Geophys. Res. Lett. 22, 3001–3004 (1995).
20. Krista, L. D. & Gallagher, P. T. Automated coronal hole detection using local intensity thresholding techniques. Sol. Phys. 256, 87–100 (2009).
21. Sheeley, N. R. Surface evolution of the Sun's magnetic field: a historical review of the flux-transport mechanism. Liv. Rev. Solar Phys. 2, 5 (2005).
22. Hundhausen, A. J., Hansen, R. T. & Hansen, S. F. Coronal evolution during the sunspot cycle: coronal holes observed with the Mauna Loa K-coronameters. J. Geophys. Res. 86, 2079–2094 (1981).
23. McIntosh, S. W., Wang, X., Leamon, R. J. & Scherrer, P. H. On the signature of giant convective scales in photospheric magnetograms and counterpart coronal emission. Astrophys. J. Lett. 784, 32 (2014).
24. Snodgrass, H. B. & Ulrich, R. K. Rotation of Doppler features in the solar photosphere. Astrophys. J. 351, 309–316 (1990).
25. Getko, R. The intermediate-term quasi-cycles of wolf number and group sunspot number fluctuations. Sol. Phys. 238, 187–206 (2006).
26. Getko, R. The ten-rotation quasi-periodicity in sunspot areas. Sol. Phys. 289, 2269–2281 (2014).
27. Chowdhury, P., Khan, M. & Ray, P. C. Intermediate-term periodicities in sunspot areas during solar cycles 22 and 23. Mon. Not. Astron. Soc. 392, 1159–1180 (2009).
28. Riley, P., Schatzman, C., Cane, H. V., Richardson, I. G. & Gopalswamy, N. On the rates of coronal mass ejections: remote solar and in situ observations. Astrophys. J. 647, 648–653 (2006).
29. Zhang, J. et al. Solar and interplanetary sources of major geomagnetic storms (Dst ≤ -100 nT) during 1996–2005. J. Geophys. Res. Phys. 112, A10102 (2007).
30. Le, G.-M., Cai, Z.-Y., Wang, H.-N., Yin, Z.-Q. & Li, P. Solar cycle distribution of major geomagnetic storms. Res. Astron. Astrophys. 13, 739–748 (2013).
31. Owens, M. J. & Forsyth, R. J. The heliospheric magnetic field. Liv. Rev. Sol. Phys. 10, 5 (2013).
32. Ulrich, R. K. & Tran, T. The global solar magnetic field—identification of traveling, long-lived ripples. Astrophys. J. 768, 189 (2013).
33. Norton, A. A. & Gallagher, J. C. Solar-cycle characteristics examined in separate hemispheres: phase, Gnevyshev gap, and length of minimum. Sol. Phys. 261, 193–207 (2010).
34. Weber, M. A., Fan, Y. & Miesch, M. S. Comparing simulations of rising flux tubes through the solar convection zone with observations of solar active regions: constraining the dynamo field strength. Sol. Phys. 287, 239–263 (2013).
35. Lou, Y.-Q. Rossby-type wave-induced periodicities in the flare activities and sunspot areas or groups during solar maxima. Astrophys. J. 540, 1102–1108 (2000).
36. Howe, R. et al. Dynamic variations at the base of the solar convection zone. Science 287, 2456–2460 (2000).
37. Castenmiller, M. J. M., Zwaan, C. & van der Zalm, E. B. J. Sunspot nests—manifestations of sequences in magnetic activity. Sol. Phys. 105, 237–255 (1986).
38. Miesch, M. Large-scale dynamics of the convection zone and tachocline. Liv. Rev. Sol. Phys. 2, 1 (2005).
39. Cagliari, P. et al. Emerging flux tubes in the solar convection zone. 1: Asymmetry, tilt, and emergence latitude. Astrophys. J. 441, 886–902 (1995).
40. Hide, R. On the Earth's core–mantle interface. Quart. J. R. Met. Soc. 96, 579–590 (1970).
41. Hoskins, B. J. & Karoly, D. J. The steady linear response of a spherical atmosphere to thermal and orthographic forcing. J. Atmos. Sci. 38, 1179–1196 (1981).
42. Spruit, H. C. in The Sun, the Solar Wind, and the Heliosphere. (eds Miralles, M. P. & Sánchez Almeida., J.) Proceedings of the conference held 23–30 August 2009 in Sopron, Hungary. IAGA Special Sopron Book Series vol. 4, 39–54 (Springer, 2011) (2010).
43. Zaqarashvili, T., Carbonell, M., Oliver, R. & Ballester, J. L. Quasi-biennial oscillations in the solar tachocline by magnetic Rossby wave instabilities. Astrophys. J. Lett. 724, 95–98 (2010).
44. Torrence, C. & Compo, G. P. A practical guide to wavelet analysis. Bull. Am. Met. Soc. 79, 61–78 (1998).

Acknowledgements

S.W.M. was partly funded by NASA grants (NNX08AU30G, NNX08AL23G, NNM07AA01C—Hinode and NNG09FA40C—IRIS). J.C.K. and M.L.S. acknowledge the support of NASA grant NNX13AQ26G. SOHO is a project of an international collaboration between ESA and NASA. NCAR is sponsored by the National Science Foundation.

Author contributions

S.W.M. performed data analysis, figure construction and wrote the manuscript. R.J.L. assisted S.W.M. with writing, data analysis and discussion. L.D.K. developed the coronal hole-detection algorithm and related statistics. A.M.T., H.S.H. and P.R. made significant contributions to the structure and discussion presented in the manuscript. J.W.H., G.K., M.S. and T.N.W. developed and assisted in the construction of the TSI/SSI timeseries presented. J.C.K. and M.L.S. performed the original analysis shown in Fig. 5, and R.K.U. assisted in the construction of Fig. 8.

Additional information

Competing financial interests: The authors declare no competing financial interests.

17

Ultraviolet luminosity density of the universe during the epoch of reionization

Ketron Mitchell-Wynne[1], Asantha Cooray[1], Yan Gong[2,1], Matthew Ashby[3], Timothy Dolch[4], Henry Ferguson[5], Steven Finkelstein[6], Norman Grogin[5], Dale Kocevski[7], Anton Koekemoer[5], Joel Primack[8] & Joseph Smidt[9]

The spatial fluctuations of the extragalactic background light trace the total emission from all stars and galaxies in the Universe. A multiwavelength study can be used to measure the integrated emission from first galaxies during reionization when the Universe was about 500 million years old. Here we report arcmin-scale spatial fluctuations in one of the deepest sky surveys with the Hubble Space Telescope in five wavebands between 0.6 and 1.6 µm. We model-fit the angular power spectra of intensity fluctuation measurements to find the ultraviolet luminosity density of galaxies at redshifts greater than 8 to be $\log \rho_{UV} = 27.4^{+0.2}_{-1.2}$ ergs^{-1}Hz^{-1}Mpc^{-3} (1σ). This level of integrated light emission allows for a significant surface density of fainter primeval galaxies that are below the point-source detection level in current surveys.

[1] Department of Physics & Astronomy, University of California, Irvine, California 92697, USA. [2] National Astronomical Observatories, Chinese Academy of Sciences, 20A Datun Road, Chaoyang District, Beijing 100012, China. [3] Harvard-Smithsonian Center for Astrophysics, 60 Garden St., Cambridge, Massachusetts 02138, USA. [4] Department of Astronomy, Cornell University, Ithaca, New York 14853, USA. [5] Space Telescope Science Institute, 3700 San Martin Dr., Baltimore, Maryland 21218, USA. [6] Department of Astronomy, The University of Texas at Austin, Austin, Texas 78712, USA. [7] Department of Physics and Astronomy, University of Kentucky, Lexington, Kentucky 40506, USA. [8] Physics Department, University of California Santa Cruz, Santa Cruz, California 95064, USA. [9] Theoretical Division, Los Alamos National Laboratory, Los Alamos, New Mexico 87545, USA. Correspondence and requests for materials should be addressed to A.C. (email: acooray@uci.edu).

The formation and early evolution of the first galaxies in the universe occurred some time after the dark ages, when the coalescence of gravitationally bound masses formed in complex structures, with a spatial distribution that can be traced back to primordial overdensities[1,2]. The ultraviolet (UV) photons from these first sources initiated the reionization of the surrounding neutral medium, thus ending the dark ages and beginning the era of a transparent cosmos, which we are increasingly familiar with today. The luminosity per unit volume of these ultraviolet photons at a rest wavelength around 1,500 Å (ρ_{UV}) during this reionization period is an important quantity to measure, as it traces the star formation and evolution of these ionizing sources. The traditional method to measure the ultraviolet luminosity density of the universe, ρ_{UV}, during the epoch of reionization, involves searching for candidate galaxies at $z > 6$ through their Lyman-dropout signature[3–7] and then constructing the luminosity function of those detected galaxies based on the observed number counts. This luminosity function is then extrapolated to a fainter absolute magnitude and integrated in luminosity to calculate ρ_{UV}.

There is a second way to quantify ρ_{UV}. This involves a measurement of the extragalactic background light (EBL) and, in particular, the angular power spectrum of the EBL intensity fluctuations. Because these intensity fluctuations are the result of emissions throughout the cosmic time, the signal we measure today is the sum of many different emission components, from nearby in our Galaxy to distant sources. If the integrated intensity from reionization can be reliably separated from that of foreground signals, we may be able to make an accounting of the total luminosity density of UV photons from reionization. Just as Lyman-dropout galaxies are detected in deep sky surveys, there is a way to achieve such a separation. Due to redshifting of the photons arising from sources present during reionization, their emission, as seen today, is expected to peak between 0.9 and 1.1 µm. This assumes that the reionization occurred around $z \sim 7$ to 9, consistent with optical depth to electron scattering as measured by Planck[8]. Due to absorption of ionizing ultraviolet photons, there is no contribution shortward of the redshifted Lyman break around 0.8 µm (refs 9,10). Spatial fluctuations of the EBL centred around 1 µm thus provide the best mechanism to discriminate the signal generated by galaxies present during reionization[11,12] from those at lower redshifts, based on the strength of the drop-out signature in the fluctuations measured in different bands.

There are existing measurements of the EBL fluctuations though their origin remain uncertain. This is mostly due to the fact that the previous measurements of EBL fluctuations have until now been limited to wavelengths > 1.1 µm, with the best measurements performed at 3.6 µm (refs 13–17). These studies have been interpreted with models involving populations of sources present during reionization at redshift $z > 8$, direct collapse and other primordial blackholes at $z > 12$ (refs 18,19), and with stellar emission from tidally stripped intergalactic stars residing in dark matter halos, or the 'intrahalo light' (IHL)[15] at $z < 3$. The IHL is diffuse stars in dark matter halos due to galaxy mergers and tidal interactions. While the relative strengths of these various contributions are still unknown, we expect the signal from high-redshift galaxies to be separable from low-redshift contributions, including those from faint nearby dwarf galaxies[20], through a multiwavelength fluctuation study spanning the 1-µm range, including in the optical ($\lambda \lesssim 1\mu m$) and near-IR ($\lambda \geq 1\mu m$) wavelengths.

Here we present results from a multiwavelength fluctuation study using data from the Hubble Space Telescope (HST) that span across the interesting wavelength range centred at 1 µm. Through models for multiple sources of intensity fluctuations,

from diffuse Galactic Light to primordial faint galaxies, we are able to describe the five-band fluctuation measurements in optical and near-infrared wavelengths to obtain a constraint on the ultraviolet luminosity density of galaxies present during reionization at a redshift above 8. We compare our measurement with existing constraints on the quantity and we also discuss the implication of our measurement.

Results

Fluctuation power spectra. We make use of imaging data from the Cosmic Assembly Near-Infrared Deep Extragalactic Legacy Survey[21,22] (CANDELS), a legacy program of the HST (see Fig. 1). Due to extensive data in the Hubble archive, we selected the southern area of the Great Observatories Origins Deep Survey (GOODS)[23,24] for the measurements (see the Methods section for details on our field selection process). This field contains HST observations with two instruments (Wide Field Camera 3 and Advanced Camera for Surveys) that have the deepest and most continuous coverage. Our total data set is comprised of observations that were taken between 2002 and 2012, with exposure times ranging between 180 and 1,469 s per frame. We avoided the background gradients evident in the publicly available WFC3 and ACS mosaics by creating our own self-calibrated mosaics using custom software[25], to produce 120-square arcmin mosaics combined from 234 to 428 (depending on the passband) individual flux-calibrated, flat-fielded (FLT) frames (Fig. 2). These mosaics are publicly available (see Supplementary Note 1). We mask stars and galaxies using an internally developed masking algorithm, facilitated by a public multiwavelength catalogue spanning from the ultraviolet to the mid-infrared[26]. The auto- (Fig. 3) and cross-power spectra (Fig. 4) are computed using standard Fourier Transform techniques on the masked images, which retain 53% of their pixels after masking. A number of corrections are performed on the power spectra. Details regarding these corrections can be found in the Methods section.

Our measurements continue to show the significant excess in the fluctuation amplitude at 30 arcsec and larger angular scales, when compared with the clustering of faint, low-redshift galaxies[20]. The large-scale fluctuations correlate between filters (Fig. 4). The excess in the amplitude of fluctuations relative to faint low-redshift galaxies is consistent with previous measurements at 3.6 µm (refs 13,15). Our HST-based power spectra probe deeper into the fluctuations and have shapes departing from the fluctuations measured with the CIBER sounding rocket experiment[16]. Due to the shallow depth of the CIBER imaging data, the measured fluctuations there are dominated by the shot noise of the residual galaxies at the arcmin angular scales that we probe here with Hubble data (Fig. 5).

At angular scales of tens of arcmin and above, CIBER detected an up-turn in the fluctuations with an amplitude well above the level expected from instrumental systematics and residual flat-field errors[16]. However, as shown in Fig. 5, the combination of CIBER and Hubble fluctuations is consistent with a power-law clustering signal out to the largest angular scales probed by CIBER. If the power-law signal is of the form $C_\ell \propto \ell^\alpha$, the best-fit slope to combined CIBER and Hubble measurements at 1.6 µm is $\alpha = -3.05 \pm 0.07$. This slope is consistent with Galactic dust, which in emission at 100 µm has a power-law of -2.89 ± 0.22 (ref. 27). At the largest angular scales we could be detecting interstellar light scattered off of Galactic dust or diffuse Galactic light (DGL). The overall amplitude of fluctuations we measure at 1.6 µm is consistent with 10% DGL fluctuations at tens of arcmin angular scales, given the 100 µm surface brightness of GOODS-S

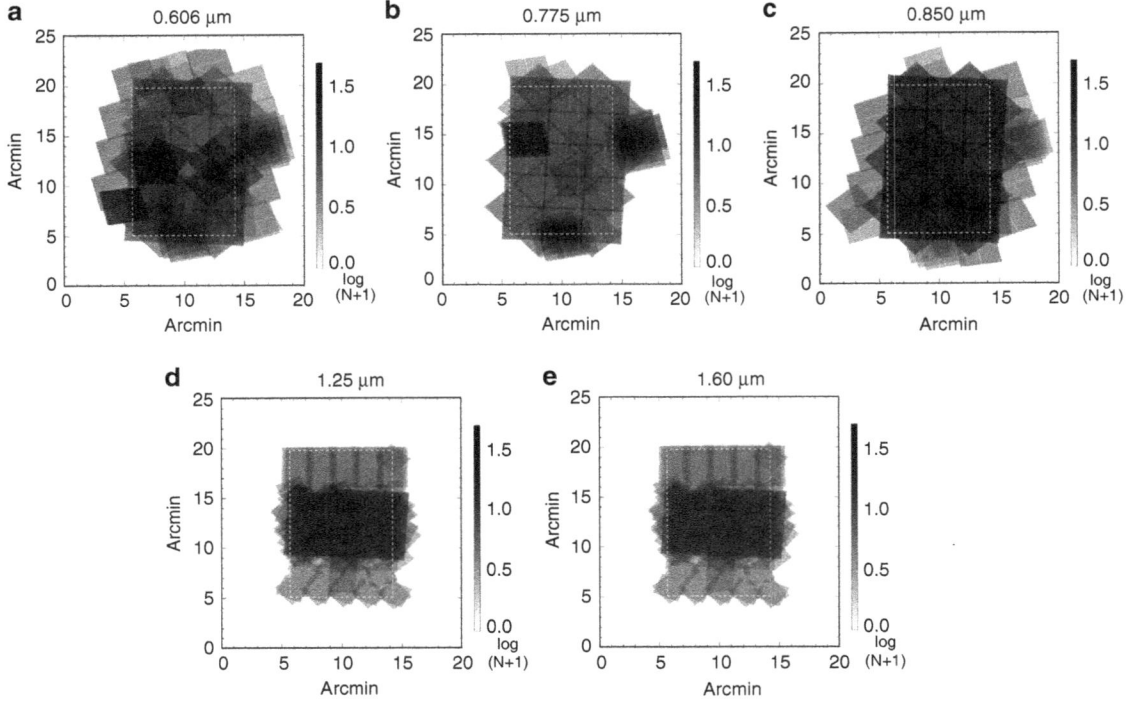

Figure 1 | Summary of tile patterns and their data archive identifications. (a) Proposal ID's for each filter in the GOODS-S field. The ACS and WFC3 rows show the proposals which are common between all the bands in each instrument. For each proposal we did not necessarily use all the frames, specifically those from deep surveys. Also show are the tiling patterns for all the bands: 0.606 μm (F606W); **(b)** 0.775 μm (F775W); **(c)** 0.850 μm (F850LP); **(d)** 1.25 μm (F125W); **(e)** and 1.60 μm (F160W); **(f)** The units of the tile pattern figures are $\log_{10}(N+1)$, where N is the number of frames overlapping. The dashed white line indicates the cropped region where the fluctuation analysis was performed.

of ~0.5 MJy/sr, and existing DGL intensity measurements (Fig. 6). Future measurements in the optical wavelengths over a wider area are necessary to confirm the Galactic nature of fluctuations at angular scales >1°.

Multicomponent model. For our theoretical interpretation, we invoke a model which involves four main components: (a) intrahalo light (IHL) following ref. 15, (b) diffuse galactic light (DGL) due to interstellar dust-scattered light in our Galaxy, (c) low-redshift residual faint galaxies[20]; and (d) the high-redshift signal. We assume the flat ΛCDM model with $\Omega_M = 0.27$, $\Omega_b = 0.046$, $\sigma_8 = 0.81$, $n_s = 0.96$ and $h = 0.71$ in our theoretical modelling[28]. We summarize the basic ingredients of our model now while providing references for further details.

For IHL, we follow the model developed in ref. 15. The mean luminosity of the IHL at rest-frame wavelength λ for a halo with mass M at z is described as

$$\bar{L}_\lambda^{\text{IHL}}(M,z) = f_{\text{IHL}}(M)L_{\lambda_p}(M)(1+z)^\alpha F_\lambda(\lambda_0/(1+z)), \quad (1)$$

where $\lambda_0 = \lambda(1+z)$ is the observed wavelength, α is the power-law index which takes account of the redshift-evolution effect and $f_{\text{IHL}}(M)$ is the IHL luminosity fraction of the total halo, which takes the form

$$f_{\text{IHL}}(M) = A_{\text{IHL}}\left(\frac{M}{M_0}\right)^\beta. \quad (2)$$

Here A_{IHL} is the amplitude factor, $M_0 = 10^{12}M_\odot$, and β is the mass power index. In equation (1), $L_{\lambda p}(M) = L_0(M)/\lambda_p$ is the total halo luminosity at 2.2 μm and at $z=0$, where $\lambda_p = 2.2$ μm and L_0

is given by[29]

$$L_0(M) = 5.64\times10^{12}h_{70}^{-2}\left(\frac{M}{2.7\times10^{14}h_{70}^{-1}M_\odot}\right)^{0.72}L_\odot. \quad (3)$$

Here $H_0 = 70h_{70}$ km s^{-1} Mpc^{-1} is the present Hubble constant. The F_λ term is the IHL spectral energy distribution (SED), which can transfer L_{λ_p} to the other wavelengths and is normalized to be 1 at 2.2 μm (see the discussion in ref. 15 for details). We assume the IHL SED to be the same as the SED of old elliptical galaxies, which are comprised of old red stars[30].

The angular power spectrum of IHL fluctuations, C_ℓ^{IHL}, is calculated via a halo model approach[31] and involves both a one-halo term associated with spatial distribution inside a halo and a two-halo term involving clustering between halos. The details are provided in ref. 15. In the clustering calculation, we assume the IHL density profile follows the Navarro–Frenk–White (NFW) profile[15,32]. We set the maximum IHL redshift at $z_{max} = 6$. The M_{min} and M_{max} are fixed to be 10^9 and 10^{13} M$_\odot$ h^{-1}, and the power-law index β is fixed to be 0.1 in this work[15]. The IHL model is then described with two parameters: A_{IHL} in equation (2) and the power-law index α in equation (1).

A simple test of IHL is to grow the source mask and study how the fluctuation power spectrum varies as a function of the mask size. However, we note that our model for IHL involves clustering at large angular scales. That is, in our description IHL is not restricted to regions near galactic disks only. The dependence of the power spectrum with mask size is studied in refs 33,34. These studies find that the fluctuations do not vary strongly with the mask radius, though such studies ignored the mode-coupling effects associated with the mask as the mask radius is varied. As discussed in ref. 15, to test IHL one has to grow the masking area to a factor of 10 larger than the typical mask radius used in the

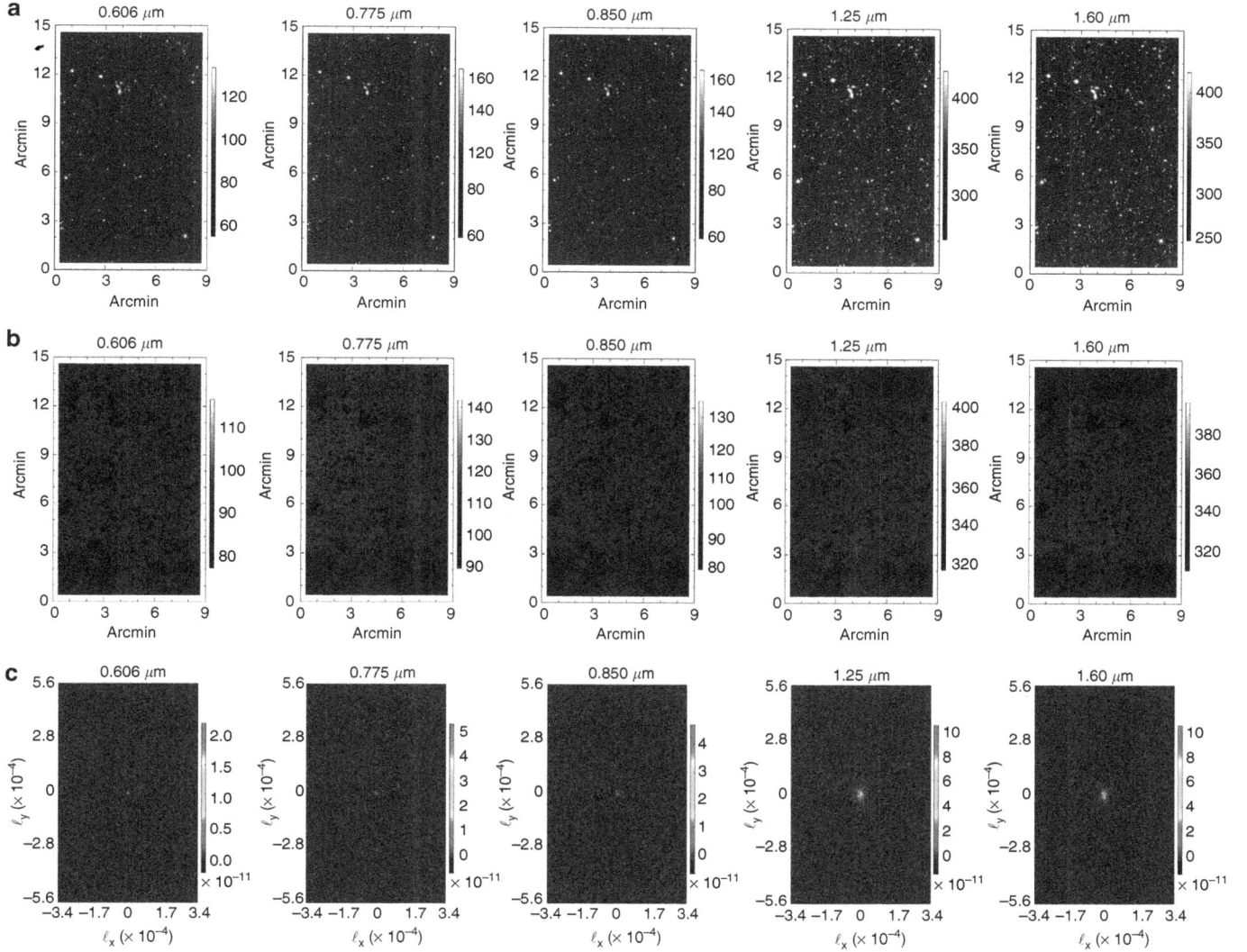

Figure 2 | Self-calibrated mosaics. (a) GOODS-S SelfCal mosaics for each band. We astrometrically align each map and crop the outer regions so as to include only sections that have been observed in all five bands. The units of the maps are nW m^{-2} sr^{-1}. **(b)** The same as **a** except with the source mask applied. **(c)** The fast fourier transform (FFT) of each of the maps in **b** which is what is used to measure the angular power spectrum. This is plotted in Fourier space as a function of modes ℓ_x and ℓ_y. The FFT of each map is structureless and contains only Gaussian noise, which is indicative of high-quality mosaics. By definition, the units of the FFT are the same as the units of the map. Each column is filter specific, plotted as 0.606, 0.775, 0.850, 1.25 and 1.60 µm.

current analyses of fluctuations. In the case of Spitzer data, where we expect fluctuations to be dominated by IHL, the relatively large 2 arcsec point spread function makes it close to impossible to test IHL directly with a varying mask. However, additional tests of IHL exist in the literature[33]. These include correlations between artificial halos and masked sources, and correlations between masked sources and foreground galaxies. Such tests have not ruled out the IHL component. Furthermore, without such a component we are not able to explain the fluctuations measured at wavelengths below 0.8 µm, as residual faint galaxy clustering[20] is not adequate to explain the measurements.

The DGL component involves dust-scattered light and it is likely that the same dust is observed by IRAS at far-infrared wavelengths through thermal emission. The DGL component was considered in ref. 16 and an upper limit on the expected amplitude was included based on the cross-correlation with 100 µm IRAS map of the same fields. The CIBER final model results focussing on IHL to explain the fluctuations did not allow the DGL fluctuations amplitude to vary as a free parameter. With

Hubble data, we find stronger evidence for DGL or a DGL-like signal once combined with CIBER, and with an (root mean squared) r.m.s. amplitude for fluctuations that is at least a factor of 3 larger than the upper limit used in ref. 16 based on the cross-correlation with CIBER. Moreover, we find that the angular power spectrum is proportional to $\sim \ell^{-3}$ over the degree scales measured from CIBER to tens of arcmin scales of CANDELS measurements. So we model it with an amplitude factor A_{DGL} as

$$C_\ell^{DGL} = A_{DGL}\ell^{-3}. \qquad (4)$$

To validate this ℓ^{-3} DGL slope dependence, we perform a linear fit in log space at low multipoles of the HST 1.6-µm data simultaneously with the CIBER 1.6-µm data. We measure a slope of -3.05 ± 0.07, so the functional form of our DGL model with ℓ^{-3} is appropriate (Fig. 5). The power-law behaviour of the DGL signal is consistent with Galactic dust emission power spectra in far-infrared and sub-mm surveys[27], and dust polarization measurements in all-sky experiments like Planck[35]. We summarize a comparison of our DGL intensity measurements

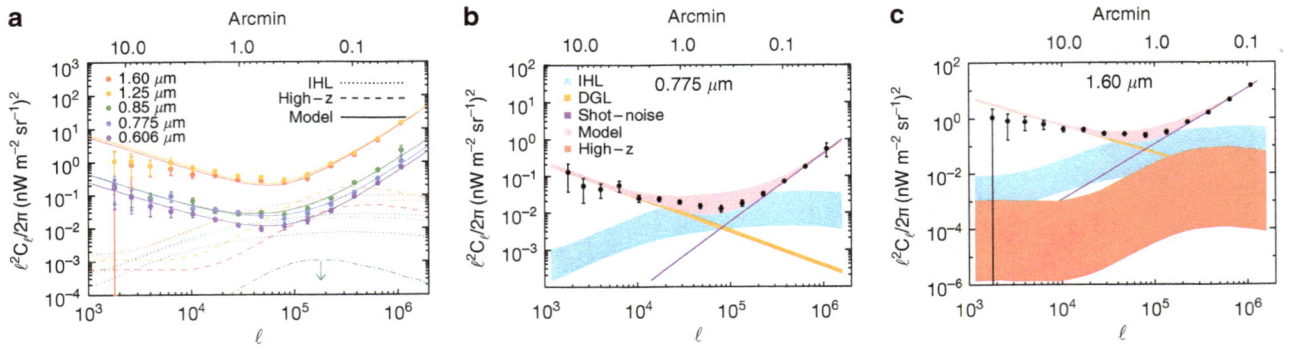

Figure 3 | Angular power spectra of optical and near-infrared background intensity fluctuations. (**a**) Multiwavelength auto power spectra of optical to near-infrared intensity fluctuations in the GOODS-South field using Hubble Space Telescope data (see Supplementary Methods for data selection details). The error bars are calculated by adding in quadrature the errors from the beam transfer function, map-making transfer function and calibration errors, to the s.d. at each multipole, δC_ℓ, described in equation 11. Thus the 1σ uncertainties account for all sources of noise and error, including map-making, calibration, detector noise and cosmic variance associated with finite size of the survey. We show the best-fit model which makes use of four components: (**a**) $z > 8$ high-redshift galaxies; (**b**) intrahalo light (IHL)[15]; (**c**) faint low-redshift galaxies[20]; and (**d**) diffuse Galactic light. At 1.25 and 1.6 μm, the best-fit high-redshift galaxy signal is shown as dashed lines. The signal is zero in the optical bands. We show the upper limit (denoted by a downward facing arrow) of fluctuations generated by $6 < z < 8$ galaxies as a dot-dashed line. Fluctuation power spectra and the best-fit models with 1σ error bounds for the model components are shown at 0.775 μm in **b** and 1.60 μm in **c** The dominant model contributors to the total power spectrum are DGL at low multipoles, or angular scales greater than a few arcmin, IHL at intermediate multipoles corresponding to angular scales of about an arcmin, and shot noise associated with faint low-redshift dwarf galaxies dominating the high multipoles or sub-arcmin angular scales. The clustering signal of low-z galaxies is more than an order of magnitude below the lower limit plotted here, thus we did not include a low-z component in our modelling.

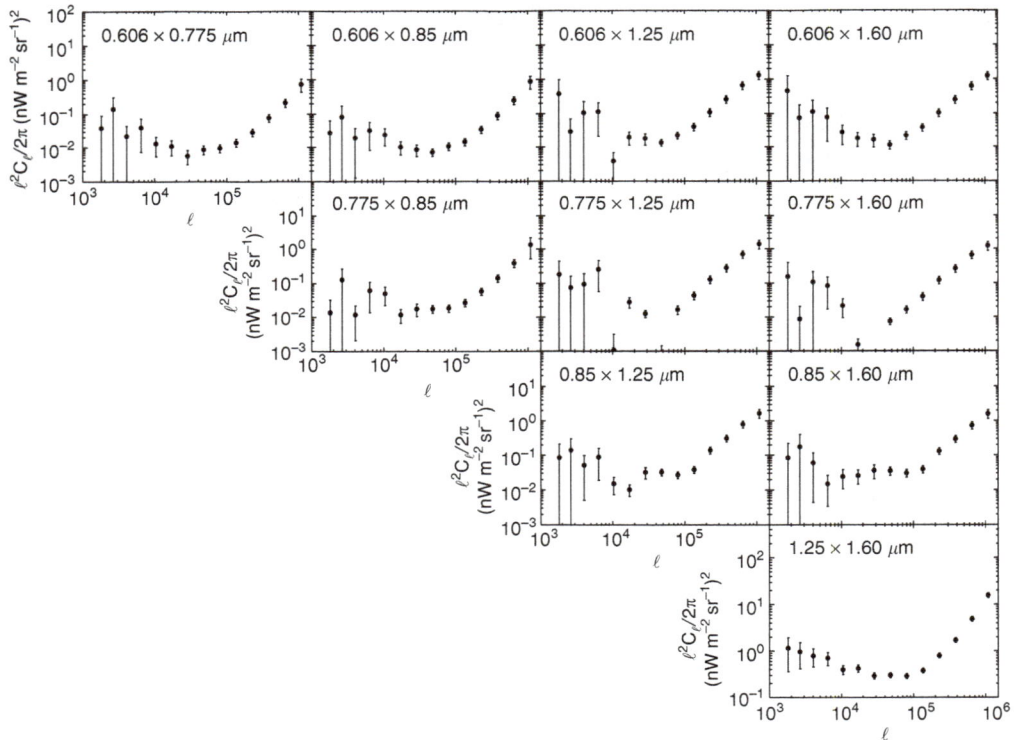

Figure 4 | Angular cross- power spectra of optical and near-infrared background. Ten cross-correlations between the HST bands. Excess signal is detected in the cross-correlations. The error bars are 1σ uncertainties which are calculated in a similar way as in Fig. 3, which accounts for all sources of noise and error, including map-making, calibration, detector noise, and cosmic variance. However, the noise power spectra for the cross-correlations are calculated slightly differently. For each filter we have two maps, so for each cross-correlation between bands we have four maps (label them A and B for the first filter, and C and D for the second). This enables us to generate a noise power spectrum by computing $(A-B) \times (C-D)$, as opposed to taking the auto-spectrum of $(A-B)$ for the autocorrelations. The first row corresponds to all correlations with the 0.606-μm band, the second for all correlations with the 0.775-μm band not found in the first row, the third row corresponds to all correlations with the 0.850-μm band not found in any of the preceding rows, and the last row corresponds to correlations at 1.25 μm. The columns similarly increase in wavelength as you move across the page.

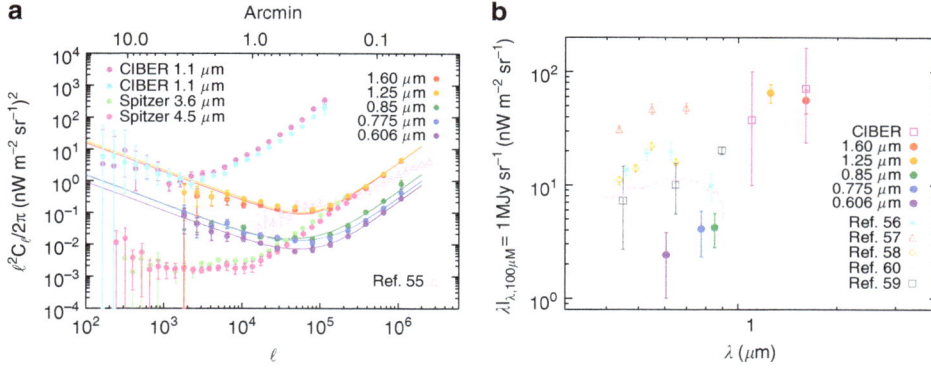

Figure 5 | Various autospectra and the spectral energy distribution of diffuse galactic light. (**a**) CANDELS corrected power spectra plotted against the C BER[16], Spitzer[15] and NICMOS[54] measurements (see also ref. 34 for more recent NICMOS measurements). The power spectrum resulting from the NICMOS analysis was measured from a MultiDrizzle map and has not been corrected for the transfer function and mode-coupling matrix resulting from source masking as discussed in our Methods section. Therefore we show it as a comparison but do not use it in our modelling. The error bars are 1σ uncertainties that account for all sources of noise and error, including map-making, calibration, detector noise, and cosmic variance associated with finite size of the survey. (**b**) Optical and infrared diffuse galactic light (DGL) spectrum. The CANDELS points are taken from the DGL model components at $10^4 \leq \ell \leq 3 \times 10^4$, and the CIBER points are taken directly from Fig. 2. of ref. 16 where they subtract off the shot noise component from their data. The galactic latitude for the optical points are $|b| \simeq 39°, 32°, 41°, 40°$ for the points labelled Witt[55], Paley[56], Ienaka[57] and Guhathakurta[58]. The Brandt[59] points are modeled over the full sky. GOODS-S is at a galactic latitude of $|b| = 54°$.

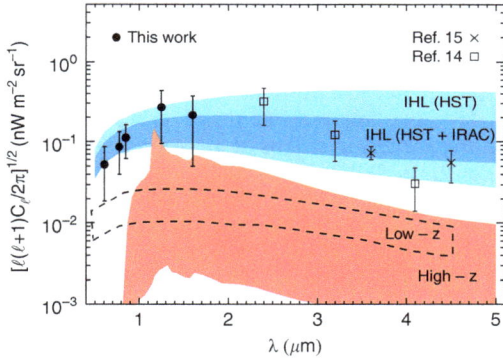

Figure 6 | Spectral energy distribution of optical and infrared fluctuations at arcmin angular scale. The Hubble/CANDELS points are averaged over $10^4 \leq \ell \leq 3 \times 10^4$, with the best-fit shot noise and DGL components subtracted. Our model fits for the high-redshift and IHL components, with their 1σ bounds, are shown as the filled regions. The errors here are propagated from the errors on the auto spectrum at the same ℓ range. The light blue region shows the 1σ confidence bound for the IHL component when we use only the HST data in our model fitting; the dark blue region shows the 1σ confidence bound for the IHL component when use both the HST and Spitzer IRAC data in our model fitting. The light red coloured region signifies the 1σ error bound for the high-redshift model component. The dashed line corresponds to the 1σ bound for the low-redshift component. The Spitzer[15] and AKARI[14] data are taken from previous measurements at $\ell = 3,000$. Note the spectral dependence difference between the high-redshift signal and IHL. Below 0.8 μm we do not expect any signal from $z > 8$ galaxies. The presence of fluctuations at optical wavelengths requires a low-redshift signal in addition to high-redshift sources to explain combined optical and infrared background intensity fluctuations.

with those of existing measurements as a function of wavelength in Fig. 5.

The clustering of low-redshift faint galaxies at $z < 5$, where reliable luminosity functions exist in the literature, is based on the detailed models in ref. 20. We follow the calculations presented

there to establish the expected level of low-redshift clustering, and the uncertainty of that expectation at the depth to which we have masked foreground galaxies. Because the low-redshift luminosity functions are not steep, unless there is a break or steepening in the luminosity function—which is not supported by the halo model—these low-redshift populations do not dominate the clustering we have measured. Given that our fluctuation measurements reach the deepest depths provided by both Hubble and Spitzer/IRAC, it is also unlikely that populations such as extreme red galaxies at $z < 2$ are responsible for the measured fluctuations. If there are populations at low redshift responsible for the SED of the fluctuations we measure, then they would need to have individual SEDs that are consistent with a sharp break redshifted between 0.8 and 1.25 μm. While fluctuation measurements in just two bands cannot separate galaxies that have redshifted 4,000-Å break, or galaxies that have red-shifted Lyman-α break between those two bands, with five bands we have adequate knowledge on the SED of fluctuations, and the shape of the clustering over two decades in angular scales to separate high-z galaxies from low-z faint interlopers. The low-z interlopers are also likely captured by our IHL model as we cannot distinguish between diffuse stars and faint, dwarf galaxies that happen to be a satellite of a large dark matter halo in our modelling description.

Galaxies during reionization. The final and critical component in our model is the signal from $z > 6$. We break this signal into two redshift intervals given the placement of the five ACS and WFC3 bands, based on the Lyman-dropout signal that moves across these bands. In particular, we consider $8 < z < 13$ and $6 < z < 8$ as the two windows. As we discuss later, given the availability of SFR density measurements in the $6 < z < 8$ interval, we mostly allow the signal in that redshift interval to be constrained by the existing data, and model-fit independently the SFR density in the higher redshift interval. We do not have a strong independent constraint on the $6 < z < 8$ signal since it is only a Lyman-dropout in our shortest-wavelength band at 0.6 μm. This allows a better separation of the $8 < z < 13$ from the rest of the signals discussed above. To measure $6 < z < 8$ independently, we would need at least one more band below 0.6 μm. The signal from $8 < z < 15$ disappears from the three optical bands (0.6–0.85 μm) and is present in the two IR bands at 1.25 and 1.6 μm.

To model the high-redshift signals, we adopt an analytic model[19,36] based on the work of ref. 37. It involves a combination of two separate classifications of stars—moderate-metallicity, second-generation or later stars (PopII) and the first generation of stars ever formed in the Universe, hence zero metallicity (PopIII). These are modelled with Salpeter[38] and Larson[39] initial mass functions (IMFs) for PopII and PopIII stars, respectively. The calculation related to direct stellar emission and the associated nebular lines, including especially Lyman-α emission, follows the work of Fernandez and Komatsu[37]. The total integrated intensity from $z_{min} < z < z_{max}$ is

$$\nu I_\nu = \int_{z_{min}}^{z_{max}} dz \, \frac{c}{H(z)} \frac{\nu(z)\bar{j}_\nu(z)}{(1+z)^2}, \tag{5}$$

where $\nu(z) = (1+z)\nu$. The comoving specific emissivity, as a function of the frequency is composed of both PopII and PopIII stars with an assumed z-dependent fraction as discussed in ref. 36 with the form given by

$$f_p(z) = \frac{1}{2}\left[1 + \mathrm{erf}\left(\frac{z-10}{\sigma_p}\right)\right], \tag{6}$$

with $\sigma_p = 0.5$. The model thus assumes most of the halos have PopIII stars at $z > 10$, while PopII stars dominate at redshifts lower than that.

There are a number of theoretical parameters related to this model, especially the escape fraction of the Lyman-α photons f_{esc}, the star-formation efficiency denoting the fraction of the baryons converted to stars in high-redshift dark matter halos, or f_*, and the minimum halo mass to host galaxies, or M_{min}. The overall quality of the data is such that we are not able to independently constrain all of the parameters related to the high-redshift intensity fluctuation signal. Moreover these parameters are degenerate with each other (that is, changing f_* can be compensated by a change in M_{min}, for example). Thus given that we do not have the ability to constrain multiple parameters, we simply model-fit a single parameter A_{high-z} that scales the overall amplitude from the default model, interpret that scaling through a variation in f_*, and subsequently convert that to a constraint on the SFRD. We fix our default model to a basic set of parameters, and assume $f_{esc} = 0.2$, $M_{min} = 5 \times 10^7 \, M_\odot$, and $f_* = 0.03$. The resulting optical depth to reionization of this default model is 0.07, consistent with the optical depth measured by Planck[8]. Among all these parameters, the most significant change (over the angular scales on which we measure the fluctuations) comes effectively from f_*, or the overall normalization of $\bar{j}_\nu(z)$, given that it is directly proportional to f_*. This can in turn be translated to a direct constraint on the SFRD, $\psi(z)$, since with f_* we are measuring the integral of the halo mass function such that

$$\psi(z) = f_* \frac{\Omega_b}{\Omega_m} \frac{d}{dt} \int_{M_{min}}^{\infty} dM M \frac{dn}{dM}(M,z) \tag{7}$$

where dn/dM is the halo mass function[40].

Finally, to calculate the angular power spectrum of fluctuations, we also need to assign galaxies and satellites to dark matter halos. For that we make use of the halo model[31]. We make use of the same occupation number distribution as in ref. 36 where the central and satellite galaxies are defined following ref. 41. However, departing from the low-redshift galaxy models, we take a steep slope for the satellite counts in galaxies with $\alpha_s = 1.5$. The low-redshift galaxy clustering and luminosity functions are consistent with $\alpha_s \sim 1$ (ref. 41), but such a value does not reproduce the steep faint-end slopes of the Lyman-break galaxy

(LBG) luminosity functions[3-7]. Such a high slope for the satellites also boost the non-linear clustering or the 1-halo term of the fluctuations. We do not have the ability to independently constrain the slope of satellites from our fluctuation measurements. In the future a joint analysis of fluctuations and LBG luminosity functions may provide additional information on the parameters of the galaxy distribution that is responsible for fluctuations. It may also be that the models can be improved with additional external information, such as the optical depth to reionization. We also note that other sources at high redshift include direct collapse black holes (DCBHs[19]), but we do not explicitly account for them here as the existing DCBH model is finely tuned to match Spitzer fluctuations, and the low signal-to-noise ratio of the Chandra–Spitzer cross-correlation results in them residing primarily at $z > 12$. DCBHs at such high redshifts will not contribute to Hubble fluctuations.

Finally, at smaller angular scales, the shot noise dominates the optical and infrared background intensity fluctuation. Since it is scale independent, we set it as a free variable parameterized as

$$C_\ell^{shot} = A_{shot}. \tag{8}$$

This noise term in the fluctuation power spectrum arises because of the Poisson behaviour of the galaxies at small angular scales, a product of the finite number of galaxies. Our measured shot noise comes from a combination of the unmasked, faint low-redshift dwarf galaxies and the high-redshift population. We do not use the information related to the shot noise in our models, but instead treat it as a free independent parameter, since we cannot separate the high-redshift shot noise from the shot noise produced by faint, low-redshift dwarf galaxies. Here we focus mainly on the clustering at tens of arcseconds and larger angular scales to constrain SFRD during reionization. In the future, with either a precise model for the low-redshift galaxies or a model for high-redshift galaxies that determines their expected number counts as a function of the free parameters such as M_{min} and f_*, it may be possible to separate the overall shot noise associated with reionization sources from that of the low-redshift faint galaxies. If this is the case, then it might also be possible to improve the overall constraints on the high-redshift population. It may also be that under an improved model, shot noise may end up providing complementary information to galaxy clustering to break certain degeneracies in model parameters.

Our overall model for the optical and infrared background fluctuations is

$$C_\ell = \begin{cases} C_\ell^{IHL} + C_\ell^{DGL} + C_\ell^{low-z} + C_\ell^{shot} + C_\ell^{6<z<8} + C_\ell^{8<z<13} & \text{F125W and above} \\ C_\ell^{IHL} + C_\ell^{DGL} + C_\ell^{low-z} + C_\ell^{shot} + C_\ell^{6<z<8} & \text{F775W and F850LP} \\ C_\ell^{IHL} + C_\ell^{DGL} + C_\ell^{low-z} + C_\ell^{shot} & \text{F606W} \end{cases} \tag{9}$$

Given that we are not able to constrain the amplitude of $C_\ell^{6<z<8}$ given the degeneracies with the parameters involving the IHL model, and the fact that we only have a single band below it, we set $C_\ell^{6<z<8}$ based on the default prediction of our model, but allow the overall amplitude $A_{6<z<8}$ to vary such that it uniformly samples the SFRD between $[0.003, 0.2] \, M_\odot \, yr^{-1} \, Mpc^{-3}$. The range is fully consistent with the existing measurements on the SFRD between $z = 6$ and 8 (refs 3–6). Our constraint on $A_{8<z<13}$ is mostly independent of this parameter since we can safely constrain the Lyman-dropout signal between 0.8 and 1.25 μm with our existing data.

We also included the CIBER[16] data at 1.1 and 1.6 μm and Spitzer[15] data at 3.6 μm in our fitting process. When compared with the Hubble data at 1.25 and 1.6 μm, we find the CIBER data are likely dominated by the emission from a DGL-like signal at large angular scales, and low-z faint galaxies at $z < 5$ at small angular scales (Fig. 5). For the fluctuations from faint, low-z galaxies, we adopt a model of residual galaxies that is derived from the observations of the luminosity function for different near-infrared bands[20]. This model already includes the shot-noise term, and we add a scale factor $f_{\text{low-}z}$ to vary the low-z angular power spectrum, $C_\ell^{\text{low-}z}$, in 1σ uncertainty. For the DGL component, we use the C_ℓ^{DGL} of Hubble data at 1.25 and 1.6 µm to fit the CIBER data at 1.1 and 1.6 µm.

We perform joint fits for Hubble, CIBER and Spitzer data with the Markov Chain Monte Carlo (MCMC) method. The Metropolis–Hastings algorithm is used to find the probability of acceptance of a new MCMC chain point[42,43]. We estimate the likelihood function as $\mathscr{L} \propto \exp(-\chi^2/2)$, where χ^2 is given by

$$\chi^2 = \sum_{i=1}^{N_d} \frac{\left(C_\ell^{\text{obs}} - C_\ell^{\text{th}}\right)^2}{\sigma_\ell^2}. \qquad (10)$$

Here C_ℓ^{obs} and C_ℓ^{th} are the observed and theoretical angular power spectra for HST, Spitzer or CIBER data, respectively. σ_ℓ is the error for each data point at ℓ, and N_d is the number of data points. The total χ^2 of HST, Spitzer and CIBER is $\chi_{\text{tot}}^2 = \chi_{\text{HST}}^2 + \chi_{\text{CIBER}}^2 + \chi_{\text{Spitzer}}^2$.

We assume a flat prior probability distribution for the free parameters; see Table 1 for prior information. Both A_{DGL}, C_ℓ^{shot} vary as independent parameters for each band. Both the A_{DGL} and C_ℓ^{shot} parameters are sixfold, with one for each HST band and for Spitzer/IRAC 3.6 µm. (we combined the two CIBER bands with two of the HST bands). We have two parameters for IHL and one parameter for the normalization of the reionization galaxies with $A_{8<z<13}$. We have two more parameters that we vary, $A_{6<z<8}$ and $f_{\text{low-}z}$. We set a uniform prior on $A_{6<z<8}$ in the SFRD following the existing measurements to be between

0.003 and 0.2 $M_\odot$ yr^{-1} Mpc^{-3}. We also set a uniform prior on $f_{\text{low-}z}$ over a reasonable range of models to account for the overall uncertainty in the models of ref. 20 to describe the $z < 5$ faint galaxy clustering at the same masking depth as our measurements. We marginalize over both $A_{6<z<8}$ and $f_{\text{low-}z}$ as well as all other parameters when quoting results for an individual parameter. We have a total of 14 free parameters in our MCMC fitting procedure that we extract from the data. Among these parameters, 12 of them simply describe the small and large angular scale fluctuations in each of the bands we have performed the measurements. These parameters are summarized in Table 1. We generate twenty MCMC chains, where each chain contains about 100,000 points after convergence. After thinning the chains, we merge all chains and collect about 10,000 points for illustrating the probability distributions of the parameters. Contour maps for each of the fitted model parameters are shown in Fig. 7. Our best-fit model with 14 free parameters have a minimum χ^2 value of 278 for a total degrees of freedom of $N_{\text{dof}} = 104$.

Discussion

Our results are summarized in Fig. 3, where we show the best-fit model curves. While the dominant contribution to the excess fluctuations comes from DGL at $\ell < 10^4$, at intermediate scales we find the IHL and reionization contributions to be roughly comparable. In Fig. 6 we show the r.m.s. fluctuation amplitude at ~ 5 arcmin angular scales over the interval $10,000 < \ell < 30.000$. We find a spectral energy distribution that is consistent with Rayleigh–Jeans from 4.5 to 2.4 µm, but diverges between 2.4 and 1.6 µm, and even more rapidly between 1.25 and 0.85 µm. The fluctuations can be explained with a combination of IHL and

Table 1 | Proposal ID's for each filter.

Parameter	Best fit	Best fit (no high-z)	Prior min, max
$\log_{10}(A_{8 \leq z \leq 13})$	$1.19^{+0.27}_{-2.62}$	—	$-5, 7$
$\log_{10}(A_{IHL})$	$-3.23^{+0.14}_{-0.12}$	$-3.32^{+0.25}_{-0.09}$	$-6, 10$
α	$1.00^{+0.14}_{-0.99}$	$1.35^{+0.39}_{-0.73}$	$-5, 5$
$f_{\text{low-}z}$	0.47 ± 0.03	0.47 ± 0.03	$0.1, 10$
$A_{\text{DGL}}^{1.6}$	$(3.74^{+0.30}_{-0.45}) \times 10^4$	$(3.72^{+0.35}_{-0.38}) \times 10^4$	$10^3, 10^5$
$A_{\text{DGL}}^{1.1,\ 1.25}$	$(4.35^{+0.54}_{-0.79}) \times 10^4$	$(4.72^{+0.42}_{-0.48}) \times 10^4$	$10^3, 10^5$
$A_{\text{DGL}}^{0.850}$	$(2.83^{+0.40}_{-0.42}) \times 10^3$	$(2.77^{+0.32}_{-0.34}) \times 10^3$	$10^2, 10^4$
$A_{\text{DGL}}^{0.775}$	$(2.74^{-0.36}_{-0.38}) \times 10^3$	$(2.65^{+0.38}_{-0.48}) \times 10^3$	$10^2, 10^4$
$A_{\text{DGL}}^{0.606}$	$(1.61^{+0.20}_{-0.40}) \times 10^3$	$(1.43^{+0.23}_{-0.22}) \times 10^3$	$10^2, 10^4$
$C_{\ell,\text{shot}}^{1.6}$	$(7.54 \pm 0.13) \times 10^{-11}$	$(7.54 \pm 0.13) \times 10^{-11}$	$10^{-11}, 10^{-10}$
$C_{\ell,\text{shot}}^{1.25}$	$(7.77^{+0.21}_{-0.28}) \times 10^{-11}$	$(7.77 \pm 0.14) \times 10^{-11}$	$10^{-11}, 10^{-10}$
$C_{\ell,\text{shot}}^{0.850}$	$(7.73^{+0.75}_{-0.45}) \times 10^{-12}$	$(8.10 \pm 0.45) \times 10^{-12}$	$10^{-12}, 10^{-11}$
$C_{\ell,\text{shot}}^{0.775}$	$(4.60^{+0.50}_{-0.30}) \times 10^{-12}$	$(4.65 \pm 0.30) \times 10^{-12}$	$10^{-12}, 10^{-11}$
$C_{\ell,\text{shot}}^{0.606}$	$(3.27^{+0.24}_{-0.21}) \times 10^{-12}$	$(3.39 \pm 0.15) \times 10^{-12}$	$10^{-13}, 10^{-11}$

Filter	Proposal	IDs							
F506W	9500	9978	11563	12007	12060	12062	12099	12461	12534
WFC3	11359	12060	12061	12062					
ACS	9425	9803	10189	10258					
F350LP	9500	9978	10086						
F775W	9575								

Summary of free model parameters. The best-fit values are quoted with 1σ errors. $\log_{10}(A_{8 \leq z \leq 13})$ is the high-redshift component used to constrain the SFRD during the reionization epoch, which is fit to the 1.25 and 1.60-µm bands. $\log_{10}(A_{\text{IHL}})$ and α are the two parameters necessary to describe the IHL component, C_ℓ^{IHL}, in equation 9. $f_{\text{low-}z}$ is the low-redshift scaling factor which varies the low redshift power spectrum within a 1σ uncertainty. A_{DGL}^i and $C_{\ell,\text{shot}}^i$ are, respectively, the DGL amplitude scaling factor and shot noise at wavelength i. All parameter values have units of (nW m^{-2} sr^{-1})^2. The best-fit model where $\log_{10}(A_{8 \leq z \leq 13})$ is non-zero with 14 free parameters (second column) have a minimum χ^2-value of 278 for a total degrees of freedom of $N_{\text{dof}} = 104$. The case where $\log_{10}(A_{8 \leq z \leq 13}) = 0$ with 13 free parameters (third column) has a minimum χ^2-value of 283 for a total degrees of freedom of $N_{\text{dof}} = 105$. The ACS and WFC3 rows show the proposals that are common between all the bands in each instrument. For each proposal we did not necessarily use all the frames, specifically those from deep surveys.

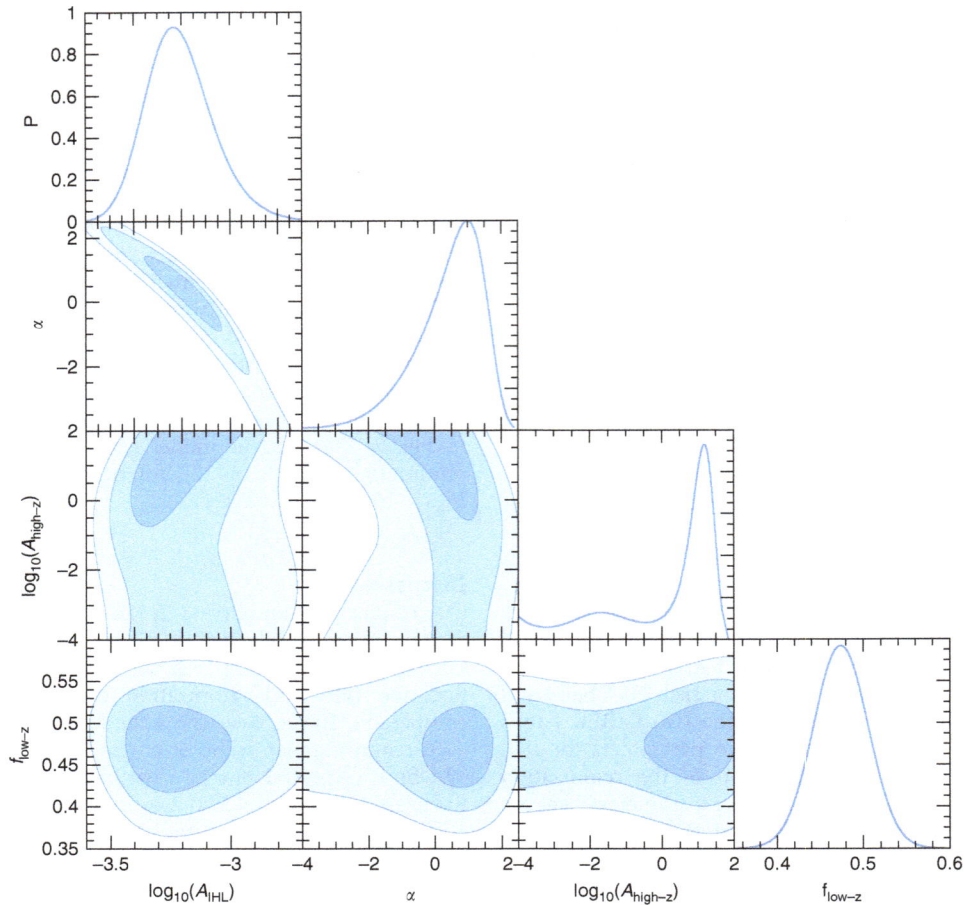

Figure 7 | Probability distributions of fitted model parameters. Here we show the probablity density distributions for our fitted model parameters $\log_{10}(A_{IHL})$, α, f_{low-z}, and $\log_{10}(A_{high-z})$ corresponding to the distribution from $8 \leq z \leq 13$. The single curves on the outermost column of each row, labelled with a 'P', show the marginalized probability distribution for each parameter labelled on the bottom of the figure. Contour regions to the left of these probability distributions show how the parameters scale with one another. Each of the shaded regions in the contours correspond to the 1, 2 and 3σ uncertainty ranges.

high-redshift galaxies. The residual low-z galaxy signal is small but non-negligible. We find that it is mostly degenerate with IHL, especially if we allow its amplitude to vary more freely than the range allowed by the existing models based on $z < 5$ galaxy luminosity functions[20]. Thus, modelling uncertainities related to the low-z galaxy confusion do not contaminate our statements about reionization. Assuming the existing low-z galaxy model[20], the best-fit model is such that the IHL intensity peaks at lower redshifts with decreasing wavelength (Fig. 8). At 3.6 μm, the IHL signal is associated with galaxies at $z \sim 1$, while at 0.6 μm over 80% of the signal is associated with galaxies at $z < 0.5$. The total intensities are $0.13^{+0.08}_{-0.05}$, $0.23^{+0.17}_{-0.11}$, $0.27^{+0.21}_{-0.13}$, $0.45^{+0.43}_{-0.24}$ and $0.54^{+0.58}_{-0.31}$ nW m^{-2} sr^{-1} at 0.60, 0.77, 0.85, 1.25 and 1.6 μm, respectively. We find that the implied IHL intensities at 1.25 and 1.6 μm are a factor of 10 lower than the implied IHL intensities for a model of CIBER fluctuations with IHL alone. The difference is due to the CIBER model that only included IHL and ignored the presence of DGL.

The drop in the fluctuation amplitude from 1.25/1.6 μm to 0.85 μm allows for a signal from reionization, but the presence of fluctuations at shorter wavelengths, such as 0.6 μm, rules out a scenario in which reionization sources are the sole explanation for the fluctuations at wavelengths at 1 μm and above. The 3.6 μm and X-ray cross-correlation[18] was explained with primordial direct collapse blackholes at $z > 12$ (ref. 19). In our multicomponent model we are able to account for the presence

of fluctuations at short wavelengths with IHL, DGL and faint low-redshift galaxies, while a combination of those components and high-redshift galaxies are preferred to account for fluctuations at 1.25 and 1.6 μm. The high-z signal is modelled following the calculations in ref. 36. The signal has an overall amplitude scaling that is related to the star-formation rate during reionization. The bright end of the counts are normalized to existing luminosity function measurements, and the faint-end of the luminosity functions to have a steeper slope than measured with counts extending down to arbitrarily low luminosities. To test whether a component at high redshift is required to explain the measurements, we also re-ran the MCMC model fits but with A_{high-z} fixed at 0. In this case our best-fit model with 13 free parameters has a minimum χ^2 value of 283 for a total degrees of freedom of $N_{dof} = 105$. The difference in the best-fit χ^2 values with and without a model for high-redshift galaxies suggests a P value of 0.025. This is consistent with the 2σ to 3σ detection of $8 < z < 13$ signal in the fluctuations (Fig. 6).

With multiwavelength measurements extending down to the optical, we are now able to constrain the amplitude of that signal with a model that also accounts for low-redshift sources in a consistent manner. This improves over previous qualitative arguments that have been made, or models involving high-redshift sources alone that have been presented, for the presence of a signal from reionization in the infrared background fluctuations[13,14,19]. In our models, the total intensity arising

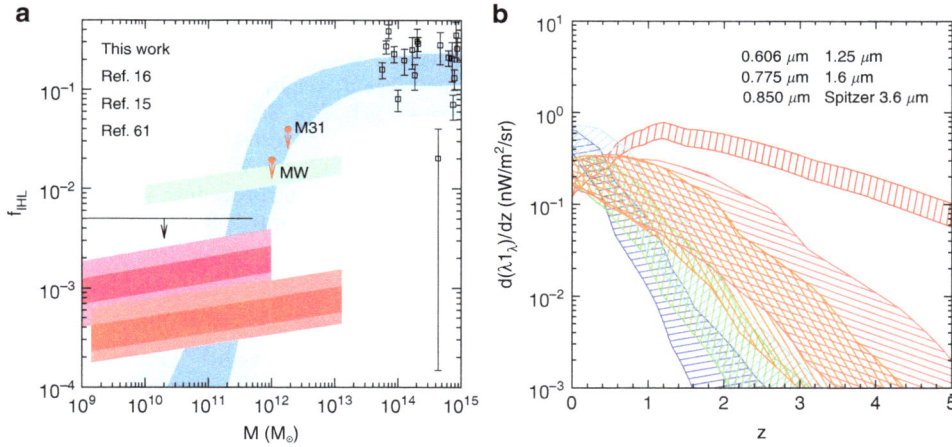

Figure 8 | Intrahalo light fraction and model intensities. (a) f_{IHL}, the intrahalo light fraction, as a function of halo mass. The dark and light shaded regions show the 95 and 68% ranges of f_{IHL} from anisotropy measurements, and from an analytical prediction[60] (blue). Intracluster measurements are shown as boxes[61], with 1σ errors. The red downward arrows denote the 95% confidence upper limit on f_{IHL} estimated for Andromeda (M31) and our Milky Way (MW), following Fig. 2 of ref. 15. (b) $d(\lambda I_\lambda)/dz$ from the model, as a function of redshift. We show the 68% confidence uncertainties derived from MCMC fitting of the data at 0.606, 0.775, 0.850, 1.25 and 1.6 μm. The total IHL intensity is $0.13^{+0.08}_{-0.05}, 0.24^{+0.17}_{-0.11}, 0.28^{+0.21}_{-0.13}, 0.45^{+0.43}_{-0.24}$, and $0.54^{+0.58}_{-0.31}$ nW m^{-2} sr^{-1} for 0.606, 0.775, 0.850, 1.25 and 1.6 μm, respectively.

from all galaxies at $z > 6$ is $\log \nu I_\nu = -0.32 \pm 0.12$ in units of nW m^{-2} sr^{-1} at 1.6 μm. At 1.6 μm the intensity from high-redshift sources is dominated by $z > 8$ galaxies, while at 0.85 μm we find an intensity $\log \nu I_\nu = -0.75 \pm 0.05$ in units of nW m^{-2} sr^{-1} for $6 < z < 8$ galaxies. The total intensity from $z > 8$ galaxies in the 1.6-μm band is comparable to the IHL intensity at the same wavelength (Fig. 8). However, at 3.6 μm, the IHL signal is a factor of about five times brighter than the $z > 8$ galaxies. At 1.6 μm the total of the IHL, high-z, and integrated galaxy light[44] of $10.0^{+2.7}_{-1.8}$ nW m^{-2} sr^{-1} is comparable to the EBL intensity inferred by gamma-ray absorption data[45] of 15 ± 2(stat) ± 3(sys).

Using the best-fit model and uncertainties as determined by MCMC model fits, we also convert the $A_{8<z<13}$ constraint to a measure of the luminosity density of the universe at $z > 8$ (Fig. 9). The resulting constraint is $\log \rho_{UV} = 27.4^{+0.2}_{-1.2}$ in units of erg s^{-1} Hz^{-1} Mpc^{-3} at (1σ). As shown in Fig. 9, the 68% confidence level constraint on ρ_{UV} is higher than the existing results from Lyman drop-out galaxy surveys during reionization at $z > 8$ (refs 46,47), and especially at $z \sim 10$ (ref. 4). At the 95% confidence level, our measurement is fully consistent with the existing results at $z \sim 10$ (ref. 48). Our constraint allows for the possibility that a substantial fraction of the ultraviolet photons from the reionization era is coming from fainter sources at depths well below the detection threshold of existing Lyman-dropout surveys, as is indeed anticipated from the steep measured slopes of the ultraviolet luminosity functions from detected galaxies. Despite their lack of detections in the deepest surveys with HST, the majority of the faint sources responsible for both fluctuations and reionization should be detectable in deep surveys with JWST centred at 1 μm.

Methods

Field selection. To obtain angular power spectra over large angular scales, individual exposures must be combined into one or more mosaiced images. We generate our own mosaics using the self-calibration technique[25] (SelfCal), instead of using the publicly available mosaiced images produced by astrodrizzle (available at http://candels.ucolick.org/data_access/ Latest_Release.html). Foreground emissions, predominantly that

of Zodiacal light, are particularly pernicious at infrared wavelengths[49], so care must be taken when producing mosaics which combine observations taken at different times, especially at the WFC3/IR channels. Offsets between frames will lead to a fictitious anisotropy signal if care is not taken to properly model and remove those offsets, which is what SelfCal was designed to do.

Although the CANDELS observations cover multiple fields, only the two deep fields — the Great Observatories Origins Deep Survey-South and -North (GOODS-S and GOODS-N)[23]—have sufficient overlap between frames to perform a self-calibration. The GOODS-N data set has a larger number of frames with clear overall offsets resulting from scattered light than does GOODS-S. Therefore we have restricted our analysis to GOODS-S, which has an area of approximately 120 square arcmins. The wider CANDELS fields are composed of much poorer tile patterns, as can be seen in Fig. 18 of ref. 21, which are a significant drawback for fluctuation studies, since one cannot calibrate the full mosaic to a consistent background level without introducing artificial gradients to the background intensity. Such dithering patterns were pursued by CANDELS to maximize the total area covered with WFC3. The increase in area is of benefit to studies that aim to detect rare galaxies, such as LBGs at $z > 5$.

Initial data reduction and map making. In addition to the data collected by CANDELS, the GOODS-S field has a wealth of HST archive data, publicly available on the Barbara A. Mikulski Archive for Space Telescopes (MAST; located at https://archive.stsci.edu/hst/search.php). We assembled our own collection of calibrated, FLT frames from the MAST archive, comprised of some or all of the data from ten different HST proposals[21–23,50]. These data are also supplemented by the Early Release Science observations[24]. The tile patterns of these observations can be found in Fig. 1.

In addition to selecting frames with a favourable tile pattern appropriate for self-calibration, we also had to take two additional potential issues into account. After the replacement of the ACS CCD Electronics Box during the fourth Hubble servicing mission (SM4), ACS imaging data are plagued with horizontal striping dominated by $1/f$ noise. Furthermore, ACS

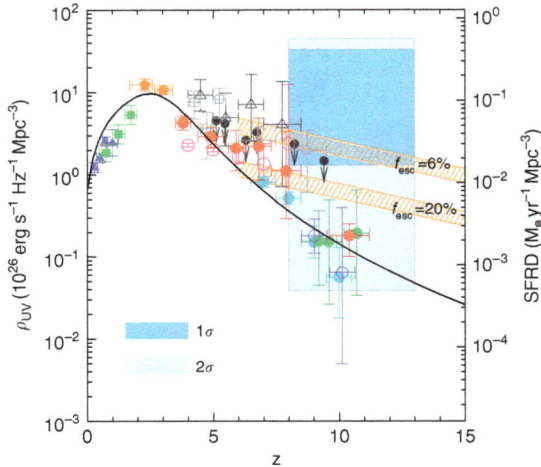

Figure 9 | The ultraviolet luminosity density and star-formation rate density as measured with intensity fluctuations. Plotted here is the specific ultraviolet luminosity density (left axis), with the equivalent star formation rate density (SFRD, right axis), as a function of the redshift z. We show the 1σ and 2σ error bounds in our redshift bin as the light and dark blue regions. Results from low-redshift surveys are shown as blue triangles[62], bright green squares[63], and orange pentagons[64]. At $z \sim 4$ to 10 the star formation rate density is shown to decrease with increasing redshift as measured by previous works, plotted as filled cyan circles[65], filled red circles[6], open red circles[7] filled green circles[4,5] and open blue circles[3]. Gamma ray burst (GRB) studies are shown as grey triangles[66], squares[67] and dark grey circles[68]. Except for ref. 65, other estimates are luminosity function extrapolations and integrations down to $M_{UV} = -13$. Our measured star formation rate densities are consistent with previous works between $z \sim 8$ to 10, however only extremely bright galaxies are directly detected in the aforementioned works with extrapolations down to $M_{UV} = -13$ involves the measured faint-end slope of the luminosity function. For reference we plot the theoretically expected relation[69] between ultraviolet luminosity density and redshift to reionize the universe and/or to maintain reionization using an optical depth to reionization of $\tau = 0.066 \pm 0.012$ (ref. 8). We take a gas clumping factor of $C = 3$ and show two cases where the escape fraction of galaxies f_{esc} is 6 and 20% (see also ref. 47).

frames have a tendency to introduce a Moiré pattern (correlated noise) when the pixel scale is modified in a low signal-to-noise area. Both of these characteristics can potentially contaminate an angular power spectrum to such an extent that the systematics dominate the measurement. With simulations, we found that with an increased number of ACS frames taken with varying position angles effectively removes the Moiré pattern upon repixelization, and the bias-striping issue is ameliorated by simply omitting a large percentage of post-SM4 frames. Any given collection of ACS frames we used contained <27% post-SM4 frames.

Our MAST archive data were initially reduced with PyRAF version 2.1.1. MAST queries are reprocessed 'on-the-fly', which entails using the most recent calibration files. Thus the FLT frames we retrieved from the archive had standard calibrations of bias and dark frame subtraction, along with flat-field correction, already performed. We identified cosmic rays in the the FLT frames with the CRCLEAN PyRAF module; sub-arcsecond astrometric alignment against the publicly available CANDELS mosaics(http://candels.ucolick.org/data_access/GOODS-S.html) was achieved with tweakreg. All ACS data were charge transfer efficiency corrected, and post-SM4 ACS frames were destriped prior to charge transfer efficiency correction.

We generate mosaics from the reduced FLT frames using the same SelfCal model as in ref. 51, an example where HST data has already been self-calibrated; details of the model can be found there. We deweight bad pixels and cosmic rays, and iterate three times to find a SelfCal solution. Our input FLT frames are geometrically distorted with a pixel size of $0.''0498 \times 0.''0502$ for ACS, and $0.''1354 \times 0.''1210$ for WFC3. We remove the distortion in the map making procedure and produce mosaics with a slightly larger pixel size of $0.''140$ (geometrically square).

Note that the algorithm we use is same as the one used to generate self-calibrated maps of IRAC in the Spitzer fluctuation studies[13,15,52]. The same method was implemented by the Herschel SPIRE Instrument Science team to generate wide area mosaics of the Herschel-SPIRE data, resulting in far-infrared fluctuations[27]. The algorithm originates from the time of FIRAS[25] and has wide applications. In the future we expect it will be used to combine frames and produce stable wide-area mosaics from JWST, Euclid, and WFIRST, among others.

We generate two maps per band so we can use the differences and sums to study systematics and noise biases, as was done in previous studies[15]. The data are sorted by observation date and every other FLT frame was used for each half map. HST data generally have two or more exposures per pointing, so this results in two maps per band of the same or similar dither pattern and exposure time per pointing. One map from each band can be found in Fig. 2. Multiple maps of the same band enable us to do cross-correlations, which ensures a removal of uncorrelated noise in the auto-spectra. This jack-knife process is similar to all other analyses related to large-scale structure and CMB angular power spectra from maps.

Generation of resolved source mask. We utilized existing multi-wavelength catalogues of detected sources from the ultraviolet to mid-infrared (CITO/MOSAIC, VLT/VIMOS, VLT/ISAAC, VLT/HAWK-I, and *Spitzer*/IRAC)[26]. In addition, all the sources from the CANDELS, HUDF and Early Release Science surveys (F435W, F606W, F775W, F814W, F850LP, F098M, F105W, F125W, and F160W) are also present in the catalogue. The 50% completeness limit for F160W in the catalogue is $m_{AB} = 25.9, 26.6$ and 28.1 for the CANDELS wide, deep and HUDF regions respectively; the 5σ limiting magnitudes are 27.4, 28.2 and 29.7. For each source detected by any of the aforementioned instruments, we apply an elliptical mask with parameters corresponding to the SEXTRACTOR Kron elliptical aperture. This catalogue simplifies our masking procedure and ensures we are masking sources detected at other wavelengths, which may otherwise be undetected in our five bands.

In addition to the source mask generated from sources detected by other instruments, we also generate our own internal masks. We run SEXTRACTOR on a coadded map in each of our five filters and apply the same elliptical masking procedure to incorporate the shapes of sources. Next, we take the union of all five internal source masks, plus the source mask we made from the pre-existing catalogue. After applying this union mask to each band, we clip 5σ outliers and visually inspect each masked map. Any residual sources are masked by hand. We verified that all sources detected above 5σ in any of the bands, including deep IRAC data at 3.6 μm, are masked. This process yields 53% of the pixels unmasked for the Fast Fourier Transform (FFT) computation. Note that tests can be performed which expand and shrink the source mask to further test the IHL model (ref. 33).

Absolute flux calibration. SelfCal achieves relative calibration between frames, so in general, the absolute flux calibration, or gain, of SelfCal output maps needs to be determined from a standard flux reference. The multi-wavelength catalogue used in the masking procedure is in principle a good enough reference, however, the photometry in the public CANDELS source catalogues were obtained with a private version of SExtractor, and has aperture corrections applied to the flux densities which were extracted from PSF-matched maps. We instead generate internal catalogues from public CANDELS MultiDrizzled mosaics using the same procedure as we used on our mosaics.

In each band, we repixelize the MultiDrizzle maps to our SelfCal pixel scale, and perform source extraction with SEx-TRACTOR on both our mosaics and the MultiDrizzle mosaics. We use the same parameter files for all the source extractions. We then astrometrically match the resultant catalogues in each band and keep those sources that are common within a radius of $0''.1$ (our FLT frames were aligned in tweakreg with MultiDrizzle mosaics as the astrometric reference, so our astrometry is similar to the MultiDrizzle maps at sub-arcsecond scales). Counts in electron per second are converted to μJy using the current HST magnitude zero points[22]. The calibration introduces a 4–5% error which is propagated into our final error bars.

Power spectrum evaluation. All the statistical information contained in any one of our maps is summarized by its angular power spectrum, C_ℓ, which is just the variance of the a_{lm}'s. We used standard FFT techniques to estimate the C_ℓ's of our masked maps[13,15,16,27]. The mosaics can be seen in Fig. 2, both in real space and Fourier space. For each band we have two half maps, A and B. To measure the inherent noise in the data, we compute the noise power spectrum as the auto power spectrum of $(A - B)/2$. To measure the raw auto-spectrum, we compute the cross spectrum of the two half maps, $A \times B$, which eliminates any uncorrelated noise in our power spectrum estimate. In general, the standard deviation at each multipole is

$$\delta C_\ell = \sqrt{\frac{2}{f_{\text{sky}}(2\ell+1)\Delta\ell}}(C_\ell^{\text{auto}} + C_\ell^{\text{noise}}), \qquad (11)$$

where $\Delta\ell$ is the bin width for the given C_ℓ, and f_{sky} is the fractional sky coverage from all the pixels used in the FFT (excluding zeros). In our case, we have some error associated with the absolute calibration of the maps, so we take the total error budget of our raw power spectra as the quadratic sum of the calibration errors with the variance in equation (1). The final measured auto-spectra and associated errors can be found in Supplementary Table 1.

To account for the effects that the source mask, tile pattern and finite beam size introduce into the power spectrum, we employ the correction techniques of the MASTER algorithm[53], and closely follow the implementation procedures explained in Section 4, 5 and 6 of the Supplementary Information of ref. 15 (including Supplementary Information Fig. 1 of ref. 15). Among the procedures listed there, we have only slightly modified the way we generate our transfer function, T(ℓ). In addition to adding instrumental noise (step 2 in Section 6 of the Supplementary Information of ref. 15), we also add an offset to each tile equal to the median of the given FLT frame. This additional step should in principle be a good indicator of how well SelfCal is performing in offset removal, unique to each observation. Transfer functions for each of our five bands can be found in Supplementary Figure 2; measurements of the beam transfer function can also be found in Supplementary Figure 2.

References

1. Loeb, A. & Barkana, R. The reionization of the universe by the first stars and quasars. *Ann. Rev. Astron. Astrophys.* **39**, 19–66 (2001).

2. Fan, X., Carilli, C. L. & Keating, B. Observational constraints on cosmic reionization. *Ann. Rev. Astron. Astrophys.* **44**, 415–462 (2006).

3. Oesch, P. A. *et al.* The most luminous z ~ 9-10 galaxy candidates yet found: the luminosity function, cosmic star-formation rate, and the first mass density estimate at 500 Myr. *Astrophys. J.* **786**, 108 (2014).

4. Zheng, W. *et al.* A magnified young galaxy from about 500 million years after the Big Bang. *Nature* **489**, 406–408 (2012).

5. Coe, D. *et al.* CLASH: Three Strongly Lensed Images of a Candidate z 11 Galaxy. *Astrophys. J.* **762**, 32 (2013).

6. Bouwens, R. J. *et al.* A census of star-forming galaxies in the z ~ 9-10 universe based on HST + spitzer observations over 19 clash clusters: three candidate z ~ 9-10 galaxies and improved constraints on the star formation rate density at Z ~ 9.2. *Astrophys. J.* **795**, 126 (2014).

7. Finkelstein, S. L. *et al.* The evolution of the galaxy rest-frame ultraviolet luminosity function over the first two billion years. Preprint at *arXiv*:1410.5439 (2014).

8. Planck Collaboration *et al.* Planck 2015 results. XIII. Cosmological parameters. Preprint at *arXiv*:1502.01589 (2015).

9. Santos, M. R., Bromm, V. & Kamionkowski, M. The contribution of the first stars to the cosmic infrared background. *Mon. Not. R. Astron. Soc.* **336**, 1082–1092 (2002).

10. Salvaterra, R. & Ferrara, A. The imprint of the cosmic dark ages on the near-infrared background. *Mon. Not. R. Astron. Soc.* **339**, 973–982 (2003).

11. Cooray, A., Bock, J. J., Keatin, B., Lange, A. E. & Matsumoto, T. First star signature in infrared background anisotropies. *Astrophys. J.* **606**, 611–624 (2004).

12. Fernandez, E. R., Komatsu, E., Iliev, I. T. & Shapiro, P. R. The cosmic near-infrared background. II. fluctuations. *Astrophys. J.* **710**, 1089–1110 (2010).

13. Kashlinsky, A. *et al.* New measurements of the cosmic infrared background fluctuations in deep spitzer/IRAC survey data and their cosmological implications. *Astrophys. J.* **753**, 63 (2012).

14. Matsumoto, T. *et al.* AKARI Observation of the Fluctuation of the Near-infrared Background. *Astrophys. J.* **742**, 124 (2011).

15. Cooray, A. *et al.* Near-infrared background anisotropies from diffuse intrahalo light of galaxies. *Nature* **490**, 514–516 (2012).

16. Zemcov, M. *et al.* On the origin of near-infrared extragalactic background light anisotropy. *Science* **346**, 732–735 (2014).

17. Seo, H. J. *et al.* AKARI Observation of the Sub-degree Scale Fluctuation of the Near-infrared Background. *Astrophys. J.* **807**, 140 (2015).

18. Cappelluti, N. *et al.* Cross-correlating Cosmic Infrared and X-Ray Background Fluctuations: Evidence of Significant Black Hole Populations among the CIB Sources. *Astrophys. J.* **769**, 68 (2013).

19. Yue, B., Ferrara, A., Salvaterra, R. & Chen, X. The contribution of high-redshift galaxies to the near-infrared background. *Mon. Not. R. Astron. Soc.* **431**, 383–393 (2013).

20. Helgason, K., Ricotti, M. & Kashlinsky, A. Reconstructing the near-infrared background fluctuations from known galaxy populations using multiband measurements ofluminosity functions. *Astrophys. J.* **752**, 113 (2012).

21. Grogin, N. A. *et al.* CANDELS: the cosmic assembly near-infrared deep extragalactic legacy survey. *Astrophys. J. Suppl. S.* **197**, 35 (2011).

22. Koekemoer, A. M. *et al.* CANDELS: the cosmic assembly near-infrared deep extragalactic legacy survey-the hubble space telescope observations. *Astrophys. J. Suppl.* **197**, 36 (2011).

23. Giavalisco, M. *et al.* The great observatories origins deep survey: initial results from optical and near-infrared imaging. *Astrophys. J.* **600**, L93–L98 (2004).

24. Windhorst, R. A. *et al.* The hubble space telescope wide field camera 3 early release science data: panchromatic faint object counts for 0.2-2 fim Wavelength. *Astrophys. J. Suppl.* **193**, 27 (2011).

25. Fixsen, D. J., Moseley, S. H. & Arendt, R. G. Calibrating Array Detectors. *Astrophys. J. Suppl.* **128**, 651–658 (2000).

26. Guo, Y. *et al.* CANDELS multi-wavelength catalogs: source detection and photometry in the GOODS-south field. *Astrophys. J. Suppl.* **207**, 24 (2013).

27. Amblard, A. *et al.* Submillimetre galaxies reside in dark matter haloes with masses greater than3x10[11] solar masses. *Nature* **470**, 510–512 (2011).

28. Komatsu, E. *et al.* Seven-year wilkinson microwave anisotropy probe (WMAP) observations: cosmological interpretation. *Astrophys. J. Suppl.* **192**, 18 (2011).

29. Lin, Y.-T., Mohr, J. J. & Stanford, S. A. K-Band Properties of Galaxy Clusters and Groups. Luminosity function, radial distribution, and halo occupation number. *Astrophys. J.* **610**, 745–761 (2004).

30. Krick, J. E. & Bernstein, R. A. Diffuse Optical Light in Galaxy Clusters. II. Correlations with Cluster Properties. *Astron. J.* **134**, 466–493 (2007).

31. Cooray, A. & Sheth, R. Halo models of large scale structure. *Phys. Phys. Rep.* **372**, 1–129 (2002).

32. Navarro, J. F., Frenk, C. S. & White, S. D. M. A universal density profile from hierarchical clustering. *Astrophys. J.* **490**, 493–508 (1997).

33. Arendt, R. G., Kashlinsky, A., Moseley, S. H. & Mather, J. Cosmic Infrared Background Fluctuations in Deep Spitzer Infrared Array Camera Images: Data Processing and Analysis. *Astrophys. J. Suppl.* **186**, 10–47 (2010).

34. Donnerstein, R. L. The contribution of faint galaxy wings to source-subtracted near-infrared background fluctuations. *Mon. Not. R. Astron. Soc.* **449**, 1291–1297 (2015).

35. Planck Collaboration *et al.* Planck intermediate results. XXX. The angular power spectrum of polarized dust emission at intermediate and high Galactic latitudes. Preprint at *arXiv*:1409.5738 (2014).

36. Cooray, A., Gong, Y., Smidt, J. & Santos, M. G. The near-infrared background intensity and anisotropies during the epoch of reionization. *Astrophys. J.* **756**, 92 (2012).

37. Fernandez, E. R., Iliev, I. T., Komatsu, E. & Shapiro, P. R. The Cosmic near Infrared Background. III. Fluctuations. *Astrophys. J.* **750**, 20 (2012).

38. Salpeter, E. E. The luminosity function and stellar evolution. *Astrophys. J.* **121**, 161 (1955).

39. Larson, R. B. in *The Stellar Initial Mass Function*. (ed. Nakamoto, T.) 336–340 (Star Formation, 1999).

40. Tinker, J. *et al.* Toward a halo mass function for precision cosmology: the limits of universality. *Astrophys. J.* **688**, 709–728 (2008).

41. Zheng, Z. *et al.* Theoretical models of the halo occupation distribution: separating central and satellite galaxies. *Astrophys. J.* **633**, 791–809 (2005).

42. Metropolis, N., Rosenbluth, A. W., Rosenbluth, M. N., Teller, A. H. & Teller, E. Equation of state calculations by fast computing machines. *J. Chem. Phys.* **21**, 1087 (1953).

43. Hastings, W. K. Monte Carlo sampling methods using Markov chains and their applications. *Biometrika* **57**, 97 (1970).

44. Franceschini, A., Rodighiero, G. & Vaccari, M. Extragalactic optical-infrared background radiation, its time evolution and the cosmic photon-photon opacity. *Astron. Astraphys.* **487**, 837–852 (2008).

45. H.E.S.S. Collaboration *et al.* Measurement of the extragalactic background light imprint on the spectra of the brightest blazars observed with H.E.S.S. *Astron. Astraphys.* **550**, A4 (2013).

46. Atek, H. *et al.* New constraints on the faint end of the UV luminosity function at z ∼ 7-8 using the gravitational lensing of the hubble frontier fields cluster A2744. *Astrophys. J.* **800**, 18 (2015).

47. Robertson, B. E. *et al.* New constraints on cosmic reionization from the 2012 hubble ultra deep field campaign. *Astrophys. J.* **768**, 71 (2013).

48. Bouwens, R. J. *et al.* UV luminosity functions at redshifts z ∼ 4 to z ∼ 10: 10,000 galaxies from HST legacy fields. *Astrophys. J.* **803**, 34 (2015).

49. Kelsall, T. *et al.* The COBE diffuse infrared background experiment search for the cosmic infrared background. ii. model of the interplanetary dust cloud. *Astrophys. J.* **508**, 44–73 (1998).

50. Beckwith, S. V. W. *et al.* The hubble ultra deep field. *Astron. J.* **132**, 1729–1755 (2006).

51. Arendt, R. G., Fixsen, D. J. & Moseley, S. H. in *A practical demonstration of self-calibration of NICMOS HDF North and South Data.* vol. 281 (eds Bohlender, D. A., Durand, D. & Handley, T. H.) 217 (Astronomical Society of the Pacific Conference Series, 2002).

52. Kashlinsky, A., Arendt, R. G., Mather, J. & Moseley, S. H. Tracing the first stars with fluctuations of the cosmic infrared background. *Nature* **438**, 45–50 (2005).

53. Hivon, E. *et al.* MASTER of the cosmic microwave background anisotropy power spectrum: a fast method for statistical analysis of large and complex cosmic microwave background data sets. *Astrophys. J.* **567**, 2–17 (2002).

54. Thompson, R. I., Eisenstein, D., Fan, X., Rieke, M. & Kennicutt, R. C. Constraints on the cosmic near-infrared background excess from NICMOS deep field observations. *Astrophys. J.* **657**, 669–680 (2007).

55. Witt, A. N., Mandel, S., Sell, P. H., Dixon, T. & Vijh, U. P. Extended red emission in high galactic latitude interstellar clouds. *Astrophys. J.* **679**, 497–511 (2008).

56. Paley, E. S., Low, F. J., McGraw, J. T., Cutri, R. M. & Rix, H.-W. An infrared/optical investigation of 100 micron 'cirrus'. *Astrophys. J.* **376**, 335–341 (1991).

57. Ienaka, N. *et al.* Diffuse galactic light in the field of the translucent high galactic latitude cloud MBM32. *Astrophys. J.* **767**, 80 (2013).

58. Guhathakurta, P. & Tyson, J. A. Optical characteristics of Galactic 100 micron cirrus. *Astrophys. J.* **346**, 773–793 (1989).

59. Brandt, T. D. & Draine, B. T. The Spectrum of the diffuse galactic light: the milky way in scattered light. *Astrophys. J.* **744**, 129 (2012).

60. Purcell, C. W., Bullock, J. S. & Zentner, A. R. Shredded Galaxies as the Source of Diffuse Intrahalo Light on Varying Scales. *Astrophys. J.* **666**, 20–33 (2007).

61. Gonzalez, A. H., Zabludoff, A. I. & Zaritsky, D. Intracluster light in nearby galaxy clusters: relationship to the halos of brightest cluster galaxies. *Astrophys. J.* **618**, 195–213 (2005).

62. Schiminovich, D. *et al.* The GALEX-VVDS measurement of the evolution of the far-ultraviolet luminosity density and the cosmic star formation rate. *Astrophys. J.* **619**, L47–L50 (2005).

63. Oesch, P. A. *et al.* The evolution of the ultraviolet luminosity function from z ∼ 0.75 to z ∼ 2.5 using HST ERS WFC3/UVIS observations. *Astrophys. J.* **725**, L150–L155 (2010).

64. Reddy, N. A. & Steidel, C. C. A steep faint-end slope of the uv luminosity function at z ∼ 2-3: implications for the global stellar mass density and star formation in low-mass halos. *Astrophys. J.* **692**, 778–803 (2009).

65. McLure, R. J. *et al.* A new multifield determination of the galaxy luminosity function at z = 7-9 incorporating the 2012 Hubble ultra-deep field imaging. *Mon. Not. R. Astron. Soc.* **432**, 2696–2716 (2013).

66. Kistler, M. D., Yuksel, H., Beacom, J. F., Hopkins, A. M. & Wyithe, J. S. B. The star formation rate in the reionization era as indicated by gamma-ray bursts. *Astrophys. J.* **705**, L104–L108 (2009).

67. Robertson, B. E. & Ellis, R. S. Connecting the gamma ray burst rate and the cosmic star formation history: implications for reionization and galaxy evolution. *Astrophys. J.* **744**, 95 (2012).

68. Tanvir, N. R. *et al.* Star formation in the early universe: beyond the tip of the iceberg. *Astrophys. J.* **754**, 46 (2012).

69. Madau, P., Haardt, F. & Rees, M. J. Radiative transfer in a clumpy universe. iii. the nature of cosmological ionizing sources. *Astrophys. J.* **514**, 648–659 (1999).

Acknowledgements

This work is based on observations taken by the CANDELS Multi-Cycle Treasury Program with the NASA/ESA HST, which is operated by the Association of Universities for Research in Astronomy, Inc., under NASA contract NAS5-26555. A.C. acknowledges support from NSF CAREER AST-06455427, AST-1310310, and STScI Archival Research program. We thank R. Arendt for useful discussions pertaining to the CANDELS map-making process.

Author contributions

K.M.W. collected the data from the archive, developed the reduction and analysis pipeline, and performed the power spectrum analysis. A.C. developed the model, supervised the research of K.M.W. and Y.G., and wrote much of the text in the Article. Y.G. interpreted the power spectrum measurements with models. M.A., A.C., T.D., H.F., N.G., D.K., A.K. and J.P. are members of the CANDELS project (led by H.F. as a Co-PI, co-principal investigator) and obtained the necessary key data used in the study. J.S. provided suggestions for the power spectrum analysis. All coauthors provided feedback and comments on the paper.

Additional information

Competing financial interests: The authors declare no competing financial interest.

18

Observing the release of twist by magnetic reconnection in a solar filament eruption

Zhike Xue[1,2], Xiaoli Yan[1], Xin Cheng[3], Liheng Yang[1,2], Yingna Su[4], Bernhard Kliem[1,5], Jun Zhang[2], Zhong Liu[1], Yi Bi[1], Yongyuan Xiang[1], Kai Yang[3] & Li Zhao[1]

Magnetic reconnection is a fundamental process of topology change and energy release, taking place in plasmas on the Sun, in space, in astrophysical objects and in the laboratory. However, observational evidence has been relatively rare and typically only partial. Here we present evidence of fast reconnection in a solar filament eruption using high-resolution H-alpha images from the New Vacuum Solar Telescope, supplemented by extreme ultraviolet observations. The reconnection is seen to occur between a set of ambient chromospheric fibrils and the filament itself. This allows for the relaxation of magnetic tension in the filament by an untwisting motion, demonstrating a flux rope structure. The topology change and untwisting are also found through nonlinear force-free field modelling of the active region in combination with magnetohydrodynamic simulation. These results demonstrate a new role for reconnection in solar eruptions: the release of magnetic twist.

[1] Yunnan Observatories, Chinese Academy of Sciences, Kunming, Yunnan 650216, China. [2] Key Laboratory of Solar Activity, National Astronomical Observatories, Chinese Academy of Sciences, Beijing 100012, China. [3] School of Astronomy and Space Science, Nanjing University, Nanjing, Jiangsu 210093, China. [4] Key Laboratory for Dark Matter and Space Science, Purple Mountain Observatory, Chinese Academy of Sciences, Nanjing, Jiangsu 210008, China. [5] Institute of Physics and Astronomy, University of Potsdam, Potsdam 14476, Germany. Correspondence and requests for materials should be addressed to X.Y. (email: yanxl@ynao.ac.cn).

It is widely accepted that magnetic reconnection plays an important role in plasmas, particularly in solar eruptive events, such as flares[1-5], filament eruptions[6], coronal mass ejections[7,8] and jets[9]. In the models of magnetic reconnection, magnetic field lines with an oppositely directed component approach each other in a current sheet or at a magnetic null point, break up and reconnect to form new magnetic lines. Magnetic energy is thereby released into thermal and kinetic energy of the plasma, potentially leading to large-scale phenomena. Evidence of reconnection on the Sun has been mostly limited to single aspects and has been indirect, showing changes of magnetic connections[10], reconnection inflows[11-13] and outflows[14-17], hot cusp-shaped structures at their interface[11,13,14], supra-arcade downflows[15,18], loop shrinkage[19,20], sudden brightenings[21,22], current sheets[15,20,23], plasmoid ejections[24,25], loop-top hard X-ray sources[20,26], pulsating radio emissions[27], and coronal heating in the interface between emerging and ambient magnetic flux[28].

In recent years, direct and more comprehensive evidence of reconnection has been discovered in a couple of energetic events on the Sun. Two very clear cases of reconnection, including inflowing cool loops, outflowing hot loops and plasma heated to > 10 MK, were observed in solar flares through extreme ultraviolet (EUV) and X-ray data[12,13]. The reconnection inflow and outflow velocities could be inferred in these flares to lie in the ranges of 10–90 and 90–460 km s^{-1}, respectively, constraining the reconnection rate to the range of 0.05–0.5. Inflows, outflows and the formation of new loops could also be studied in a well-observed case of reconnection between two sets of small-scale chromospheric loops, imaged at high resolution in H-alpha[22]. This study showed a transition from slow to fast reconnection. The three-dimensional (3D) topology of reconnecting loops and their heating has been inferred using the combined perspectives and multiple EUV channels of two spacecraft[29].

Here we present comprehensive observational evidence of reconnection between a set of chromospheric fibrils and the threads of an erupting filament, which gave rise to a small flare. We observe the in- and outflows and hot cusp-shaped structures at the ends of a small-scale reconnecting current sheet, as well as newly formed loops that demonstrate the change of magnetic connectivity. We estimate that the reconnection is fast. The intriguing rotational motion of the erupted filament is found to show the untwisting of a flux rope, enabled by the reconnection with the chromospheric fibrils, which extend to essentially current-free magnetic flux in the corona. Our detailed study is made possible using H-alpha images from the New Vacuum Solar Telescope (NVST)[30] and EUV images from the Solar Dynamics Observatory (SDO)[31] (see the Methods section for details). Hot coronal plasma is also imaged using the X-Ray Telescope (XRT) on board Hinode[32] and the Soft X-ray Imager (SXI) on board the Geostationary Operational Environmental Satellite (GOES)[33]. Photospheric vector magnetograms taken by the Helioseismic and Magnetic Imager (HMI)[34] on board SDO allow us to obtain the 3D field structure of the reconnection region, independently demonstrating the change of topology. The dynamics of this source region model are studied in a data-constrained magnetohydrodynamic (MHD) simulation, which reproduces the observed features of the eruption and confirms that the untwisting of the erupted flux is enabled by reconnection with ambient current-free flux.

Results

Confined partial eruption of the filament. On 3 October 2014, a filament is observed in active region (AR) 12178 (as designated by the National Oceanic and Atmospheric Administration [NOAA])

near the disk centre. The filament is composed of a western and an eastern section, which join smoothly at a bend point near solar $(x,y) = (-70, -185)$. Many thin threads extending along the filament spine make up its fine structure (see the red arrows in Fig. 1a). The threads in the western section show indications of twist of approximately one turn, consistent with the widely but not universally adopted assumption that the magnetic structure of filaments is that of a weakly twisted magnetic flux rope[35-37]. Figure 1b shows the position of the filament in the magnetogram. Positive photospheric polarity is given at the western footpoint and southern side of the filament, and negative polarity is given at the eastern footpoint and northern side, so the filament has sinistral chirality[38]. This corresponds to positive (right handed) magnetic helicity, because the axial current of the filament must also point eastward for a force-free equilibrium to exist in the given ambient flux distribution.

Figure 1c–h and the related Supplementary Movie 1 show the eruption of the filament in H-alpha images taken by the NVST. The eruption commences near the bend point at $\sim 07{:}26$ UT, as a motion of the dark threads with a southwestward projected direction, which later turns more southward. Associated brightenings, indicating heating due to reconnection, appear soon but only after the onset of the motion. The eruption comprises the whole western filament section and part of the eastern section. Some of the threads in the western section follow the main eruption with a small delay. The threads in the eastern section's part adjacent to the bend point ($-100 \lesssim x \lesssim -70$) connect smoothly to the threads in the western section and move jointly with them (see the blue-dotted lines in Fig. 1c,d). Subsequently, other threads become visible in their place; these, as well as all other threads in the eastern filament section, remain stable (see the red arrows in Fig. 1f–h).

We conclude that part of the magnetic flux in the filament experiences an instability, which causes the motion and subsequently triggers reconnection. The unstable flux comprises the whole western section and part of the eastern section, where unstable flux lies on top of the stable flux. The unstable flux extends into the eastern section at least up to approximately $x = -100$ (the apparent end region of the erupting threads), but possibly up to approximately $x = -120$ (the extension of the triggered brightenings); the stable flux extends from the bend point to the edge of the active region. Because the indicated twist of approximately one turn lies below the threshold of the helical kink instability[39,40] and the erupted filament does not build up a clear helical shape as a whole (see the blue-dotted line in Fig. 1c–h), the torus instability[41] is the primary candidate mechanism for the onset of eruption.

All southward motions end by 07:50 UT, and all material is subsequently seen to slide down towards the western end point of the filament in this confined eruption, which does not produce a coronal mass ejection. Obviously, the southward end region of the erupted filament threads is the highest part of the visible erupted structure. The data do not definitely reveal whether the erupted flux is still magnetically rooted in the eastern filament section or if it now connects to other negative flux in the photosphere. However, the erupted threads point approximately towards the neighbouring AR 12179 in the southeast, away from the direction of the eastern filament section, but similar to pre-existing interconnecting loops between the two ARs. This is suggestive of reconnection between the erupting filament and these loops. After $\sim 09{:}30$ UT, a new filament forms along the original western filament section, initially separate from the surviving filament in the eastern section, but 2 h later beginning to join it.

The eruption gives rise to two signatures of magnetic reconnection. The two prominent brightenings that develop at the sides of the eastern filament section (Fig. 1c–h) represent an

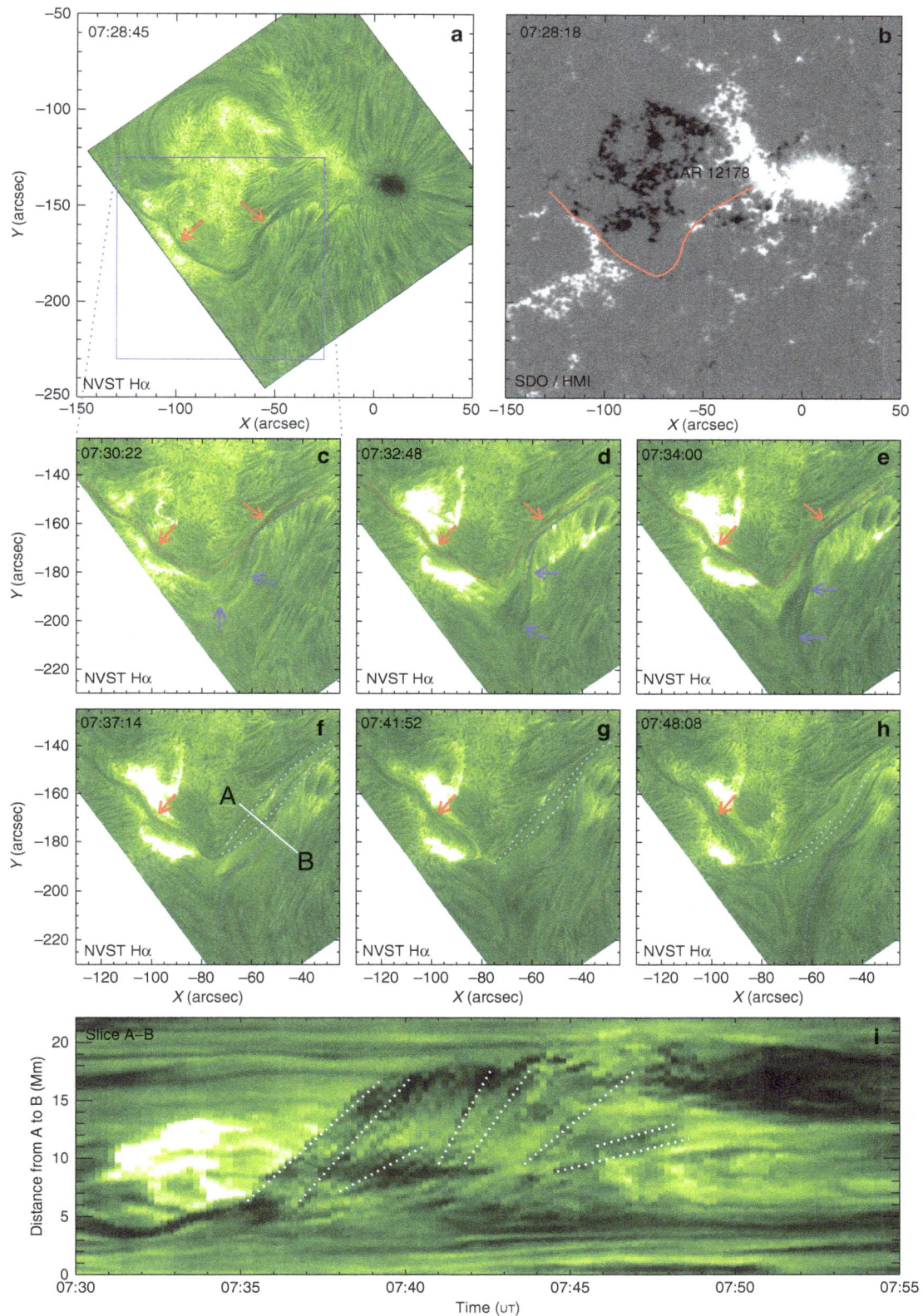

Figure 1 | Structure and eruption of the filament. (**a**) NVST H-alpha image showing the general appearance of the filament (indicated by the red arrows) at the onset of the eruption. (**b**) HMI line-of-sight magnetogram of AR 12178, with positive (negative) photospheric flux shown in white (black) and the red line representing the position of the filament. (**c**–**h**) The process of filament eruption in H-alpha images. The original position is marked by the red-dotted line. The erupting part is indicated by the blue-dotted lines and blue arrows. Cyan-dotted lines in **f**–**h** indicate some of the filament threads that rise with a delay. The stable eastern filament section is marked by the red arrows in **f**–**h**. (**i**) Time-slice plot acquired at the position A–B, with the white-dotted lines marking the filament threads. The motion of these threads indicates the untwisting of the filament during its eruption.

indirect signature, which agrees perfectly with the reconnection assumed to release the energy in the so-called standard flare model[1-5]. Ambient flux passing over the filament is lifted; subsequently, its legs approach each other and reconnect in a vertical current sheet that is known to form under the erupting flux, but not imaged in the present event. The released energy is channelled downward along the field lines, producing the chromospheric brightenings where the reconnecting ambient flux is rooted. This can likewise be seen under the western filament section (Fig. 1d,e), but it is much weaker there, in agreement with the much smaller amount of ambient flux in this area (Fig. 1b).

After ∼ 07:38 UT, a second, direct signature of reconnection is revealed. Chromospheric fibrils south of the bend point reconnect with the trailing threads of the erupting western filament section on the north side of the bend point. The observations of the reconnecting current sheet are analysed in detail in the following section, where we find that it also forms under the erupting flux. However, here it is the erupting flux that reconnects with ambient flux, a process not envisioned in the standard flare model.

In addition, the erupted filament displays an intriguing motion that is highly suggestive of a rotation about its main direction (Supplementary Movie 1). The motion is also seen in the Atmospheric Imaging Assembly (AIA)[42] data (Supplementary Movie 2). At least one full turn is indicated. The motion of the better visible threads on the upper side is also shown in a time-slice plot (white-dotted lines in Fig. 1i). When looking along the filament toward the western footpoint, the rotation is clockwise. Erupted flux is generally assumed to have the structure of a flux rope[5-8]. The positive helicity inferred above implies that the twist of the rope is right handed. The clockwise rotation thus represents an untwisting that is equivalent to the relaxation of magnetic tension and supports the conjecture of a flux rope structure for the erupted filament.

Erupting flux ropes show an apparent untwisting simply as the result of their expansion. While the total number of field line turns is preserved (in the absence of reconnection), the twist per unit length decreases. The stretching can appear as a propagation of twist if only a part of the rope displays a twist pattern. Although the threads of the erupting filament do not display a pronounced twist pattern in images (Fig. 1c-h), the animation in Supplementary Movie 1 is consistent with this effect in the southward expanding end region of the threads. The stretching can also mimic a rotation in a time-slice plot, as in Figure 1i. However, it cannot explain two effects visible in the present event: many threads shift nearly completely to the other side of the rope (cyan lines in Fig. 1g,h) and their southern end points appear to rotate about the rope axis.

True untwisting results from the conversion of twist into writhe of the flux rope axis and from reconnection with less twisted flux. The first effect is negligible here because a clear helical shape does not develop (Fig. 1c-h). Following reconnection with less twisted flux, the twist tends to equilibrate along the new structure. Again, the total number of turns present after the reconnection is preserved, but the equilibration occurs as a result of a true propagation of twist from the more twisted to the less twisted part (as a torsional Alfvén wave packet) and involves a true rotation about the axis of the new structure. Hence, the observed rotational motion indicates that the erupted filament flux reconnects with ambient flux, which usually has no twist. The H-alpha observations reveal only a small part of that flux (the chromospheric fibrils south of the bend point of the filament channel), but a consistent whole picture is provided below by our models for the coronal field of the AR and by the evolution of one of these models in an MHD simulation.

Imaging small-scale magnetic reconnection. The reconnection process and formation of the associated current sheet are shown in Fig. 2. Their full evolution can be better seen in Supplementary Movie 1. Figure 2a shows the positions of the original structures in an H-alpha image overlaid by a line-of-sight magnetogram. The reconnection occurs between two sets of magnetic loops with opposite directions (see the arrows in Fig. 2a). One set of magnetic loops, that is, the filament threads whose eruption is delayed, is indicated by the red arrow before the reconnection. The other set, that is, the chromospheric fibrils, is indicated by the white arrow. The reconnection occurs in a small region (marked by the black rectangle in Fig. 2b). In Fig. 2c, two typical loops (red- and white-dotted lines with red and white arrows, respectively) are indicated just before their reconnection. They further approach each other, and their interaction results in a reconnection at ∼ 07:38 UT. During the reconnection process, two cusp-shaped structures are formed (blue-dotted lines in Fig. 2d). Simultaneously, new loops appear on the other side of the cusp structure northeast of the reconnection region (the black arrows in Fig. 2d,e). Subsequently, these loops move away from the cusp. With the two sets of outer loops continuing to move towards each other, the reconnection continues from ∼07:38 to ∼07:50 UT. After the reconnection ceases, the newly formed loops can be seen more obviously (marked by black arrows in Fig. 2d-h,). They have accumulated in the northeast reconnection outflow region. The new loops, which form early in the southwest of the reconnection region (at the right cusp), follow the filament eruption (Supplementary Movie 1), while those that form there later do not rise further and can be clearly seen near the end of the event (indicated by the yellow arrow in Fig. 2h).

During the eruptive process the reconnection region is stretched, apparently to form a current sheet, which is visible in H-alpha as a bright linear structure extending between the tips of the two cusps (Fig. 2f). The average length and width of the current sheet are estimated to be 4.3×10^3 and 1.06×10^3 km, respectively. At the same time, we find that the plasma at the tip of the northeast cusp structure is significantly heated, causing the brightening (Fig. 2g). The cusp and current sheet can be seen in multiple channels (Fig. 3a-d; Supplementary Movie 2). The northeastern bright cusp appears very clearly in all EUV channels of AIA ($T \approx 0.02 - 10$ MK). The southwestern cusp structure is weaker, as is to be expected from lower plasma densities at the upper end of a current sheet that is formed and stretched out upward by the eruption. This also shows up only intermittently owing to absorption by the moving threads of the erupted filament, which demonstrates that the current sheet forms under the erupting flux. The current sheet appears as a bright linear structure in all AIA channels, most clearly at the highest temperatures (131 Å; $T = 11$ MK). The hot cusp structure can also be clearly seen in the XRT and SXI images during the reconnection process (Fig. 3e,f). The unusual simultaneous visibility of the current sheet and cusp in H-alpha and EUV emissions (Figs 2f and 3a-d) is probably due to the fact that cool and dense threads and fibrils are embedded in the reconnection inflow. The major brightening of the current sheet and cusp in H-alpha immediately follows the inflow and fading of a major filament thread (prominent in the current sheet in Fig. 2c-e), suggesting heating of the thread by the reconnection.

The emission measure (EM) maps at the different temperature bins (Fig. 3g-i) show that the northeastern, newly formed loops are dominated by EM at temperatures of 4–8 MK, several times higher than the unperturbed corona. On the other hand, the tips

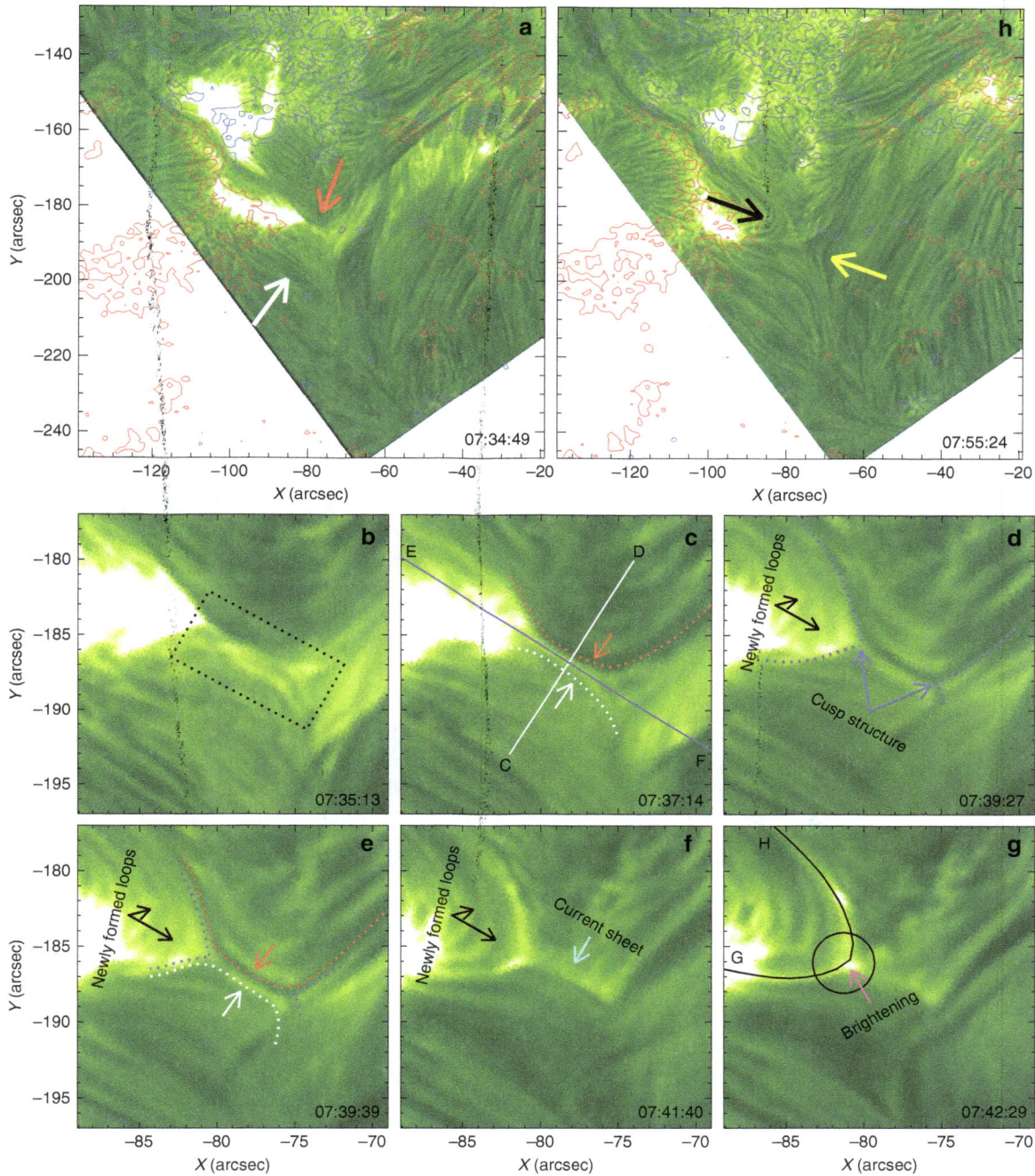

Figure 2 | Evolution of reconnection. (**a,h**) NVST H-alpha images in the early and late phases of the reconnection event, overlaid by red (blue) contours representing positive (negative) photospheric polarity from near-simultaneous HMI data. (**b–g**) NVST H-alpha images of the reconnection process in which the red- and white-dotted lines indicate the magnetic loops before the reconnection (with red arrows marking the filament threads and white arrows marking the chromospheric fibrils), the blue-dotted lines mark the cusp-shaped structures, and the black (yellow), cyan and pink arrows point to newly formed loops, current sheet and brightening, respectively. The three time-slice plots shown in Fig. 4 are made along the white solid line C–D, the blue solid line E–F and the black solid line G–H, respectively.

of the two cusp structures (circles A and B) and the current sheet between them are most prominent at even higher temperatures, that is, $T > 10$ MK (Supplementary Movie 3). This is similar to earlier observations of a pair of hard X-ray sources located at the two ends of the conjectured current sheet in flares[20]. The EUV images at high temperatures (Fig. 3d) and the EM maps (Fig. 3i) also show a diffuse area of enhanced temperature in the northwest inflow region after ~07:35 UT, which begins to cool down ~10 min later and then transiently brightens in H-alpha

(Fig. 1h). The magnetic field inhibits heat conduction from the current sheet into the inflow region. Guided by the MHD simulation, which shows the formation of additional weak current layers north of the erupting filament, we interpret these enhancements as a signature of additional energy release in the complex ambient coronal field when it is perturbed by the eruption, followed by downward heat conduction.

To quantitatively investigate the motions of the plasma in the reconnection region, we construct the three time-slice plots in

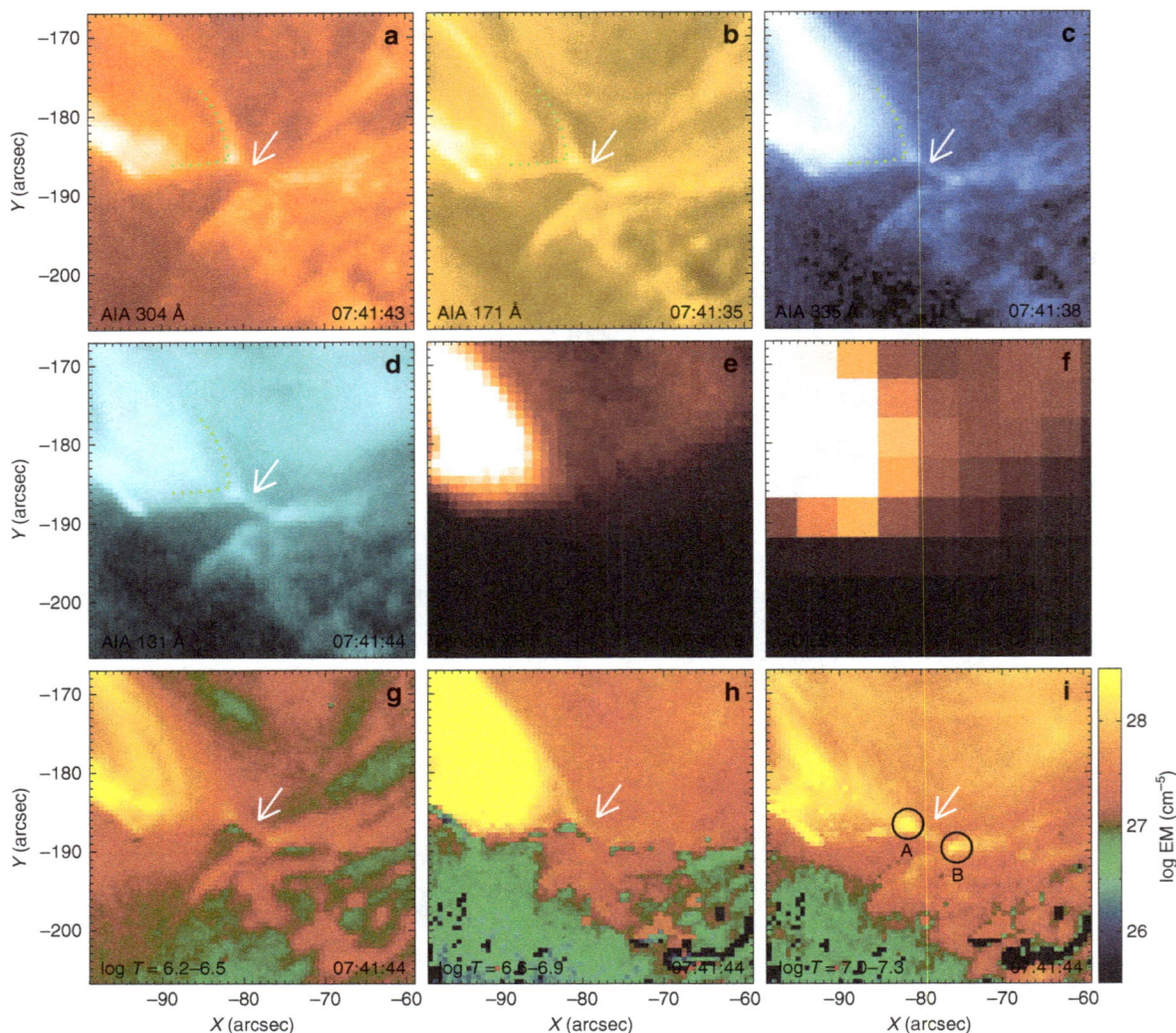

Figure 3 | EUV/X-ray images and EM maps. (a–d) AIA EUV images at 304, 171, 335 and 131 Å channels, in which the cusp-shaped structures are marked by the dotted lines and the current sheet is indicated by the white arrows. (e,f) Cusp structure seen in X-ray images observed by Hinode/XRT and GOES/SXI, respectively. (g–i) EM maps in three temperature ranges showing the dominant temperatures of the inflow region, current sheet, cusps (encircled and marked 'A' and 'B' in i), and reconnected loops.

Fig. 4a–c, which yield projected velocities along the lines marked in Fig. 2. Figure 4a shows the inward motions of loops, that is, the reconnection inflow. The apparent velocities of these loops are in the range of 3.7–25.0 km s^{-1} for the various filament threads on the northwest side and 6.3–16.8 km s^{-1} for the various fibrils on the southeast side. Figure 4b displays outward motions of bright (hot) plasma, that is, the reconnection outflow, seen immediately downstream of the northeastern cusp. The outflow velocities lie in the range of 41.7–43.7 km s^{-1}, with an average value of 42.7 km s^{-1}. At the same time, two newly formed loops containing cooled plasma, dark in H-alpha (the white-dotted lines), can be seen in Fig. 4b. These shrink slowly, with an average velocity of 2.6 km s^{-1}. In addition, a mixture of bright and dark plasma moves down the legs of the cusp structure, that is, in the interface between the reconnection inflow and outflow (Fig. 4c). This apparent mix of inflowing and outflowing plasma, as well as the intermittently irregular structure of the cusp legs (Supplementary Movie 1), whose observation is made possible by the high resolution of the NVST, provide support for the recent numerical finding of turbulence in reconnection developing at the separatrices[43]. The downflows at the cusps show velocities in the range of 17.6–80.2 km s^{-1} with a decreasing trend.

The brightenings at the end of the current sheet were observed in multiple channels. To investigate their evolution, Fig. 4d displays the intensities in the H-alpha, 304, 171 and 335 Å channels, integrated over the area of the circle in Fig. 2g. The early brightenings are caused by the eruption of the filament. The onset of reconnection, as marked by the rise of the EUV emissions at 07:38 UT, coincides with the arrival of the first inflow trace at the forming current sheet (Fig. 4a) and with the onset of outflows from the current sheet (Fig. 4b,c). The bright structures and the outflows show some degree of intermittency, which is a characteristic of plasmoid-mediated reconnection in very long current sheets, but here is probably caused mainly by the inhomogeneity of the inflowing plasma. All tracers of the reconnection process decay after ∼07:45 UT, by which time the length of the current sheet has shortened considerably because of the accumulation of new loops in the downward (northeast) outflow region and the approximate stationarity of the upper tip of the sheet under the erupted but no longer rising filament.

3D NLFFF configuration of the reconnection region. To investigate the 3D magnetic field structure of the reconnection

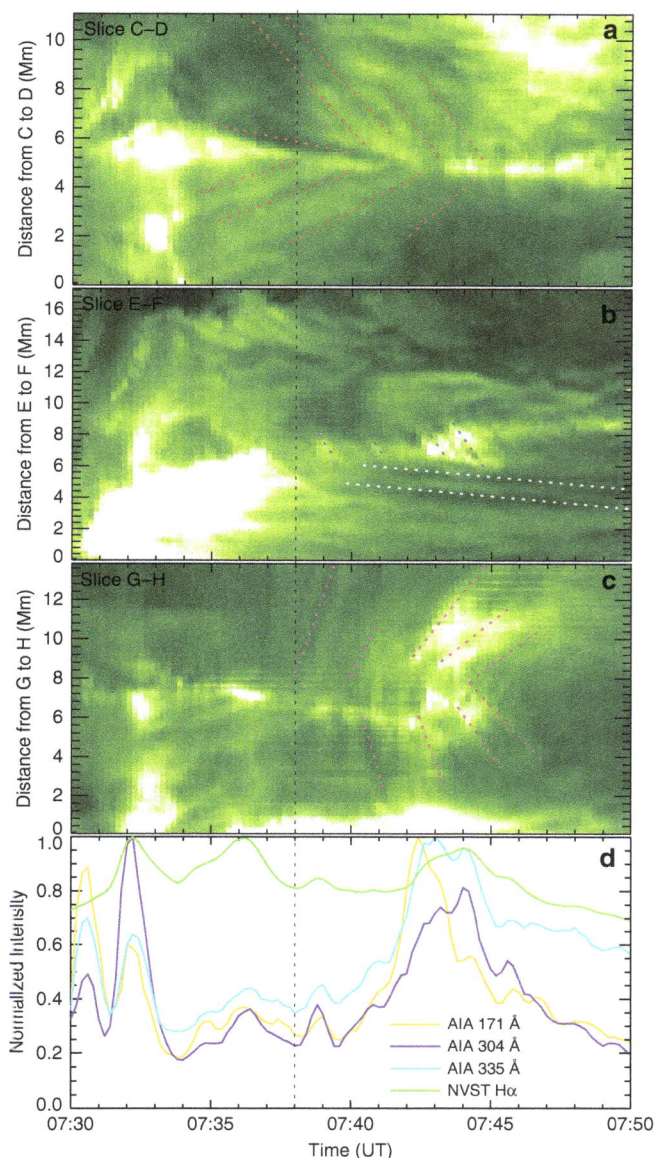

Figure 4 | Time-slice plots and EUV brightening. (a–c) H-alpha time-slice plots made by stacking the slits along the slices C–D, E–F and G–H in Fig. 2, respectively. The red-, blue-, white- and pink-dotted lines represent reconnection inflows, outflows, shrinking loops, and downward motions, respectively. (d) Temporal evolution of the brightenings in the region marked by the black circle in Fig. 2g, using H-alpha, 304, 171 and 335 Å images, where each light curve is normalized by its maximum. The vertical black-dotted line in each panel indicates the onset of the observed fast reconnection.

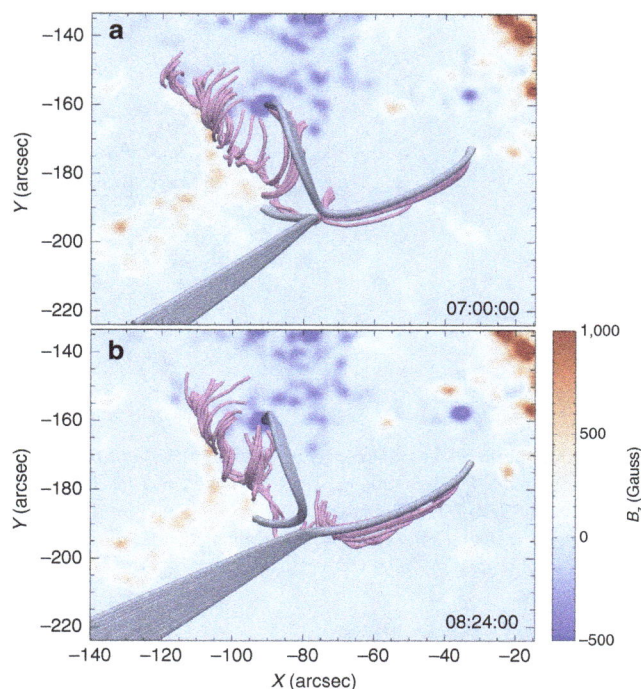

Figure 5 | 3D configuration of the reconnection region from NLFFF extrapolation. (a,b) 3D NLFFF configuration before (07:00 UT on 3 October 2014) and after (08:24 UT on 3 October 2014) the reconnection. The background images indicate the radial component of the vector magnetic field in the photosphere. The grey lines represent the magnetic loops involved in the reconnection. The pink lines show the magnetic structures of the filament.

region, we carry out extrapolations of the HMI vector magnetograms at 07:00 and 08:24 UT based on the nonlinear force-free field (NLFFF) assumption for the coronal field (Fig. 5). In the eastern section of the filament, an incoherent flux rope structure is obtained, similar to several recent extrapolation results for active regions before an eruption[44,45]. In the western section, the strong axial flux expected in a filament is found, but the poloidal (twist) field component is largely missing. We expect that the latter results from the low signal-to-noise ratio in the extended area of the weak field around the western filament section (Fig. 1b). The overall shape of the filament is nevertheless well matched. The extrapolation also correctly indicates that the field

in the area of the western filament section has a similar structure sometime after the eruption, providing the prerequisite for the observed formation of a new filament in the western section. These findings yield confidence in the large-scale structure of the extrapolated NLFFF.

The extrapolations demonstrate the change of magnetic connections in agreement with the observed change. They further show that the ambient field rooted in the southeast reconnection inflow region (grey field lines in Fig. 5a) first follows the low-lying reconnecting fibrils seen in H-alpha in the westward direction, but then bends strongly upward, forming high-reaching connections to the neighbouring AR 12179 (see Fig. 6). These field lines exchange footpoints, that is, reconnect, with a set of lower field lines running in the northwest inflow region and the western section of the filament channel (also coloured grey). The resulting low-lying loop (Fig. 5b) corresponds very well to the newly formed H-alpha loops downstream of the observed cusp (Fig. 2d–g). The other resulting set of field lines indicates that flux in the erupting western part of the filament reconnects with the high-reaching field lines that extend to the neighbouring active region. This corresponds very well to the southward bending of the erupting flux (blue-dotted lines in Fig. 1c–h) and supports our conjecture above that the strong untwisting motion is realized by a torsional Alfvén wave packet propagating to ambient, essentially untwisted flux.

MHD simulation. To further substantiate the occurrence of reconnection triggered by the erupting filament, we model the event in an MHD simulation whose initial and boundary conditions are constrained by the HMI data. Because the

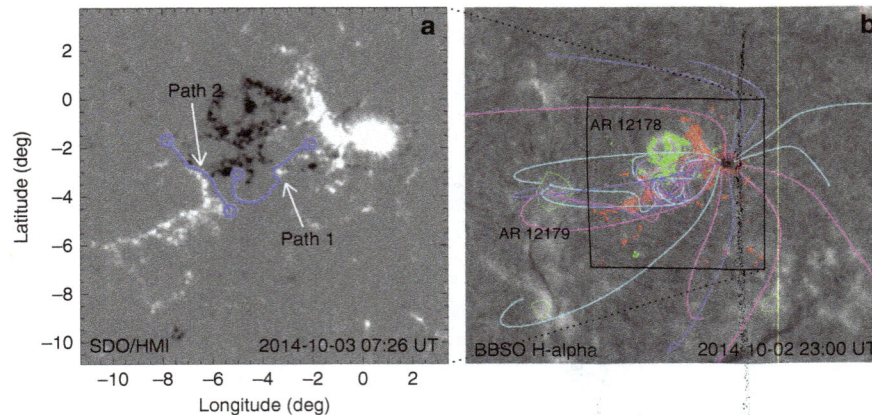

Figure 6 | Magnetic field model constructed using the Flux Rope Insertion method. (a) Line-of-sight magnetogram observed by SDO/HMI at 07:26 UT on 3 October 2014. The two blue curves with a circle at each end show the paths of the inserted flux ropes. **(b)** H-alpha image taken by Big Bear Solar Observatory at 23:00 UT on 2 October 2014, overlaid with red (positive) and green (negative) contours representing the magnetic field shown in **a**. The colour lines show selected field lines from the magnetic field model, and the horizontal size of the MHD computation box (height of $0.75R_\odot$ and fully within the high-resolution region) is indicated by the black square.

extrapolation largely fails to reproduce the twist in the western part of the filament, we construct a new NLFFF model of the AR from the pre-eruption magnetogram using the Flux Rope Insertion method, as detailed in the Methods section. On the basis of the observation that the western filament section erupts fully and the eastern section only partly, two flux ropes are inserted along the path of the filament. A low-lying rope is inserted along the eastern section. A higher-lying rope models the western section and extends into the eastern section as far as the moving threads at the onset of the event (Fig. 6a). Numerical relaxation partly merges them into a smooth configuration, which is essentially a flux rope split into two branches in the eastern part. From models for a range of values of the inserted fluxes, we select the one that yields an unstable western section, with approximately one field line turn, and a stable eastern section to serve as the initial condition of the MHD simulation (Figs 6b and 7a–d).

The model also includes the high-reaching field lines anchored in positive flux southeast of the filament (grey in Fig. 5a and left set of open field lines in Fig. 7c), the bottom part of which represents the reconnecting chromospheric fibrils (white arrows in Fig. 2a,c,e). Figure 6b shows that these field lines connect to the neighbouring AR 12179. Similarly, the low-lying field lines north of the filament bend point (grey in Fig. 5a) are here also seen to be part of the flux in the western filament section (right set of weakly twisted field lines in Fig. 7c). This flux models the trailing threads of the erupting western filament section (red arrows in Fig. 2a,c,e), which reconnect with the chromospheric fibrils.

A wide range of agreement with the observations is obtained. The simulation shows a confined eruption in the southern direction, comprising the full western and part of the eastern filament section (Fig. 7; Supplementary Movie 4). Magnetic reconnection is triggered very similarly to that shown by the NVST and SDO data. A current sheet is dragged upward into the corona between the rising and remaining flux along the filament section eastward of the original bend point (Fig. 7, fourth column). The ambient flux on the southeast side of the filament moves towards the sheet to reconnect at the observed position with approaching flux from the western section of the flux rope (Fig. 7, third column). As shown above, while being in agreement with the NVST observations, this differs partly from the standard view of reconnection in solar eruptions, which supposes that only ambient flux reconnects under a rising flux rope[5–8,46]. The reconnection forms small loops that accumulate under the

current sheet at the position of the newly formed H-alpha loops (Figs 2d–g, 5b and 7o). The other parts of the reconnected field lines yield new, high-reaching magnetic connections from the western footpoint of the erupted flux to AR 12179 (Fig. 7, left two columns; compare with the evolution of the blue-dotted line in Fig. 1c–h). The untwisting of the erupted flux via propagation of twist along these new connections is very clear. These results are similar to the indications from the dynamics of the erupted filament in the H-alpha data (the draining of filament material in the westward direction along the whole length of the erupted filament and the indication of a rotational motion). They are also in agreement with the new connections in the extrapolated field (Fig. 5b). The accumulation of reconnected flux causes a gradual rise of the cusp that forms at the bottom tip of the current sheet (Fig. 7h,l,p). As the rise of the erupted flux is stalled, but the bottom cusp continues to rise as a result of the reconnection, the current sheet weakens and eventually decays.

The simulation shows further reconnection in a current sheet forming in the western filament section in the interface between rising and overlying flux. This rebuilds connections from the positive footpoint area of the flux rope to the strong negative photospheric flux patches adjacent to the eastern filament section (Fig. 7i,m). The current sheet is not imaged by NVST and SDO because it is not aligned with the line of sight; however, the resulting connections match the filament that reforms in the western section after 09:30 UT.

Discussion

In this paper, we study in detail a small-scale magnetic reconnection event triggered by a filament eruption and obtain a comprehensive set of solid observational evidence for the occurrence of reconnection which is unprecedented. This includes the reconnection inflows at both sides of the current sheet at projected velocities of 3.7–25.0 km s^{-1}; the reconnection outflow at one side of the current sheet, moving downward at projected velocities of 41.7–43.7 km s^{-1}; the formation of a long, thin structure suggestive of a current sheet and of two cusp-shaped structures at its end points, both occurring simultaneously with the onset of reconnection; newly formed loops in both outflow regions, demonstrating the change of magnetic connectivity, those on the downward side shrinking with an average projected velocity of 2.6 km s^{-1}; and fast downward motion at

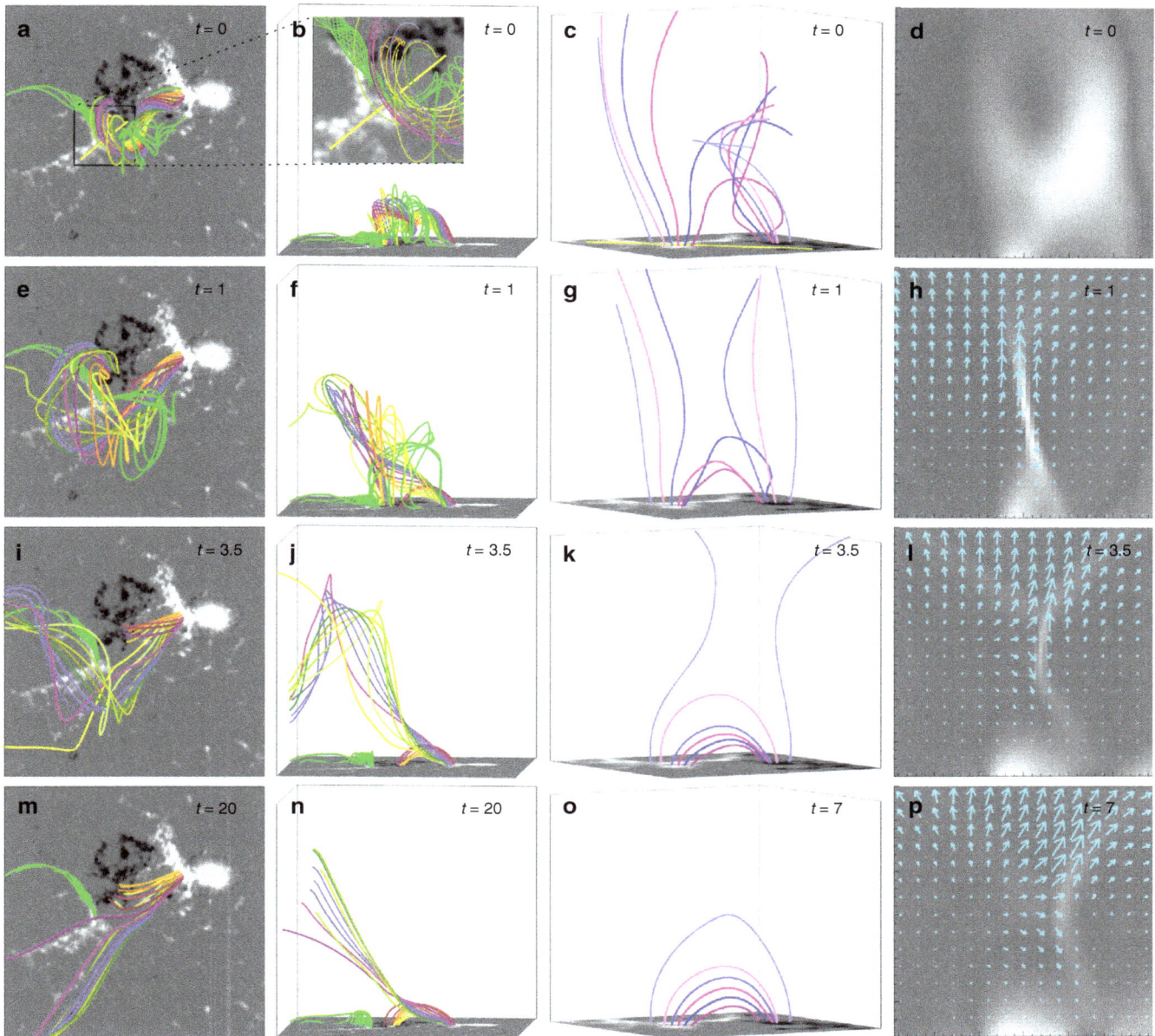

Figure 7 | MHD simulation of the confined filament eruption and magnetic reconnection. The left two columns show field lines of the fully erupting western flux rope section (rainbow colours) and the largely stable eastern flux rope section (green) and the line-of-sight magnetogram in a cube of $0.25R_\odot$ per side. The inset at $t = 0$ indicates the area shown in the third column, and the yellow line shows the position of the vertical cut plane in the fourth column. The third column shows selected field lines in the area of the observed reconnection in a cube of $0.06R_\odot$ per side. Field lines rooted in positive photospheric flux (white) initially mostly extend to AR 12179 eastward of AR 12178 (Fig. 6b), while field lines rooted in negative flux (black) initially mostly join the unstable western flux rope section. The fourth column displays the current density and in-plane velocity vectors in a vertical cut through the reconnecting current sheet ($0.06R_\odot$ per side). At $t = 0$, the inserted, unstable flux rope is seen, with velocities set to zero. Times are given in Alfvén time, which corresponds to ~3 min when scaled to the observations.

projected velocities of $17.6–80.2$ km s^{-1} of cool and hot plasma at the cusp-shaped boundary between inflow and outflow. More-over, the newly formed loops, current sheet and cusps show enhanced emission measure in a broad temperature range, with the loops being dominated by plasma at 4–8 MK and the current sheet and cusps by plasma at $T > 10$ MK. The length, L, and width, d, of the current sheet are estimated to be 4.3×10^3 and 1.06×10^3 km, respectively.

The reconnection rate is an important physical parameter in the theory of reconnection and can be estimated from our observations. It can be expressed as the Alfvénic Mach number,

M_A, of the inflow velocity because the reconnection outflow generally approaches the Alfvén velocity, V_A (ref. 47). Assuming that the observed outflow velocity is of the order of V_A and neglecting projection effects, $M_A \approx V_{in}/V_{out}$, which lies in the range of $\sim 0.08 – 0.60$ for our measurements. In a steady-state reconnection, the reconnection rate also equals the inverse aspect ratio of the current sheet[47], which is obtained as $d/L \approx 0.25$, consistent with the estimated range for M_A. Both indicate that the reconnection is fast. Our estimate tends to be relatively high, but is consistent with the range of previous estimates for solar events, for example, 0.001–0.03 (ref. 11), 0.01–0.23 (ref. 48), 0.055–0.20

(ref. 12), 0.16 (ref. 17) and 0.05–0.5 (ref. 13). Because the measured reconnection outflow in our event hits the accumulating newly formed loops after only a short distance (Fig. 2), it may not have reached the Alfvén velocity. Furthermore, the current sheet width obtained from the images, although resolved by the NVST, is considered an upper limit for the actual width of the field reversal, which tends to occur at microscopic scales[47]. Therefore, the estimated reconnection rate is also considered an upper limit.

Finally, the confinement of the eruption may be due to a strong restraining force of the overlying flux or due to the cancelation of the upward Lorentz force in the filament by its reconnection with ambient flux. The former effect is equivalent to saturation of the torus instability in an ambient field that decreases too slowly with height, which can be quantified by the decay index, $n(z) = -d\log B_{p,hor}(z)/d\log z$, where $B_{p,hor}(z)$ is the horizontal component of the potential field at the x–y position of interest[41]. The decay index profile at the observed position of the current sheet indicates instability ($n > 3/2$) in a small interval around $z = 0.03R_\odot$, where the upper flux rope is located in our NLFFF model, and at $z > 0.13R_\odot$, with the intermediate height range being stable. In this stable range, the flux rope reconnects in our MHD simulation. Thus, both potential reasons for confinement may be relevant in the event.

Methods

Differential emission measure calculation. We calculate the differential emission measure (DEM) using the almost simultaneous observations of six AIA EUV lines (131, 94, 335, 211, 193 and 171 Å) formed at coronal temperatures. The DEM is determined by

$$I_i = \int Ri(T) \times \mathrm{DEM}(T)\mathrm{d}T, \tag{1}$$

where I_i is the observed intensity of the waveband i, $R_i(T)$ represents the temperature response function of waveband i, and $\mathrm{DEM}(T)$ is the DEM of coronal plasma, which is computed using the routine xrt_dem_iterative2.pro in the Solar Software package. This code was first written for the Hinode/X-ray telescope data[49,50], and then modified for the SDO/AIA data[51]. In this work, $\log T$ is set in the range of 5.5–7.5, where the DEM is generally well constrained[52,53]. To obtain the emission measure in the temperature range (T_{min}, T_{max}), we evaluate

$$\mathrm{EM} = \int_{T_{min}}^{T_{max}} \mathrm{DEM}(T)\mathrm{d}T. \tag{2}$$

NLFFF extrapolation. The 3D NLFFF structures of the reconnection region and associated filament (Fig. 5) are reconstructed using an optimization algorithm[54,55]. A preprocessing procedure is first used to deal with the bottom vector data to remove most of the net force and torque that usually results in an inconsistency between the photospheric magnetic field and the force-free assumption in the NLFFF models[56]. The visualization of 3D magnetic field is realized by the software Paraview. The field lines shown are traced from nearly identical points in the photosphere.

Flux rope insertion method and MHD simulation. The magnetic field model used as the initial condition of the MHD simulation is constructed through Flux Rope Insertion. This method involves computing the potential field in the volume of interest by extrapolation, inserting a magnetic flux rope along the observed path of the filament and numerically relaxing the configuration to an NLFFF[36,57,58]. The active region of interest (Fig. 6a) is modelled with high spatial resolution ($0.002R_\odot$), and the more distant regions are modelled with a lower-resolution (1°) global potential field. The high-resolution region is derived from the line-of-sight photospheric magnetogram taken by SDO/HMI on 3 October 2014 at 07:26 UT, and the global field is constructed based on the HMI synoptic map. The high-resolution computational domain extends about 25° in longitude, 15° in latitude and up to 1.75 $R_\odot$ from Sun centre. Two flux ropes are inserted along the path of the filament (blue lines) and anchored in the photosphere (at the blue circles), with path 1 chosen to run higher than path 2. The axial and poloidal fluxes along paths 1 and 2 are 4×10^{20} Mx and 10^{11} Mx cm^{-1}, and 10^{20} Mx and 10^{10} Mx cm^{-1}, respectively, giving the western section higher free magnetic energy and twist. The numerical relaxation partly merges the flux ropes into a smooth configuration and changes the end positions slightly (Fig. 6b). All flux connecting to photospheric sources outside AR 12178 corresponds well to the large-scale coronal loops imaged

in the AIA 171-Å channel (which are not shown here, but can be accessed at the public source http://helioviewer.org/).

The MHD simulation code[59] is used with numerical settings analogous to an earlier data-constrained simulation of a solar eruption[60]. In particular, the velocity in the bottom plane is set to zero, to model the inertia of the photosphere. The vertical component of the HMI magnetogram thus remains invariant.

Software availability. The Solar Software package is available at http://www.lmsal.com/solarsoft/.

Data availability. The NVST is a ground-based telescope with a 986-mm clear aperture in the Fuxian Solar Observatory of the Yunnan Observatories, Chinese Academy of Sciences. It is designed to observe the fine structures on the Sun and their activities in multiple layers (photosphere and chromosphere) with high spatial resolution (0.165 arcsec per pixel) and high temporal resolution (~ 12 s). We use data from the H-alpha line-centre channel at 6,562.8 Å, corresponding to chromospheric temperatures. These data can be accessed at http://fso.ynao.ac.cn/dataarchive_ql.aspx. EUV and far-ultraviolet images obtained by the AIA on board SDO, including the 304, 171, 193, 335, 211, 94, 131 and 1,600 Å channels, display magnetic reconnection and the associated current sheet at higher temperature, that is, primarily in the corona. Photospheric magnetograms are provided by the HMI Instrument on SDO. The SDO data are publicly available at http://jsoc.stanford.edu/ajax/lookdata.html and at http://helioviewer.org/. Soft X-ray images of the Sun by Hinode/XRT and GOES/SXI are available at http://darts.isas.jaxa.jp/solar/hinode/query.php?A01=Go%20to%20Search/ and http://sxi.ngdc.noaa.gov/sxi/servlet/sxisearch/, respectively.

References

1. Carmichael, H. A process for flares. *NASA Spec. Publ.* **50**, 451–456 (1964).
2. Sturrock, P. A. Model of the high-energy phase of solar flares. *Nature* **211**, 695–697 (1966).
3. Hirayama, T. Theoretical model of flares and prominences. I: evaporating flare model. *Sol. Phys.* **34**, 323–338 (1974).
4. Kopp, R. A. & Pneuman, G. W. Magnetic reconnection in the corona and the loop prominence phenomenon. *Sol. Phys.* **50**, 85–98 (1976).
5. Shibata, K. Evidence of magnetic reconnection in solar flares and a unified model of flares. *Astrophys. Space Sci.* **264**, 129–144 (1999).
6. van Tend, W. & Kuperus, M. The development of coronal electric current systems in active regions and their relation to filaments and flares. *Sol. Phys.* **59**, 115–127 (1978).
7. Lin, J. & Forbes, T. G. Effects of reconnection on the coronal mass ejection process. *J. Geophys. Res.* **105**, 2375–2392 (2000).
8. Antiochos, S. K., DeVore, C. R. & Klimchuk, J. A. A model for solar coronal mass ejections. *Astrophys. J.* **510**, 485–493 (1999).
9. Shibata, K. et al. Observations of X-ray jets with the YOHKOH Soft X-ray Telescope. *Publ. Astron. Soc. Japan* **44**, L173–L179 (1992).
10. van Driel-Gesztelyi, L. et al. Coronal magnetic reconnection driven by CME expansion-the 2011 June 7 event. *Astrophys. J.* **788**, 85 (2014).
11. Yokoyama, T., Akita, K., Morimoto, T., Inoue, K. & Newmark, J. Clear evidence of reconnection inflow of a solar flare. *Astrophys. J. Lett.* **546**, L69–L72 (2001).
12. Takasao, S., Asai, A., Isobe, H. & Shibata, K. Simultaneous observation of reconnection inflow and outflow associated with the 2010 August 18 solar flare. *Astrophys. J. Lett.* **745**, L6 (2012).
13. Su, Y. et al. Imaging coronal magnetic-field reconnection in a solar flare. *Nat. Phys.* **9**, 489–493 (2013).
14. Tsuneta, S. et al. Observation of a solar flare at the limb with the YOHKOH Soft X-ray Telescope. *Publ. Astron. Soc. Japan* **44**, L63–L69 (1992).
15. McKenzie, D. E. & Hudson, H. S. X-ray observations of motions and structure above a solar flare arcade. *Astrophys. J. Lett.* **519**, L93–L96 (1999).
16. Wang, T., Sui, L. & Qiu, J. Direct observation of high-speed plasma outflows produced by magnetic reconnection in solar impulsive events. *Astrophys. J. Lett.* **661**, L207–L210 (2007).
17. Tian, H. et al. Imaging and spectroscopic observations of magnetic reconnection and chromospheric evaporation in a solar flare. *Astrophys. J. Lett.* **797**, L14 (2014).
18. Asai, A., Yokoyama, T., Shimojo, M. & Shibata, K. Downflow motions associated with impulsive nonthermal emissions observed in the 2002 July 23 solar flare. *Astrophys. J. Lett.* **605**, L77–L80 (2004).
19. Forbes, T. G. & Acton, L. W. Reconnection and field line shrinkage in solar flares. *Astrophys. J.* **459**, 330–341 (1996).
20. Sui, L. & Holman, G. D. Evidence for the formation of a large-scale current sheet in a solar flare. *Astrophys. J. Lett.* **596**, L251–L254 (2003).
21. Li, L. P. & Zhang, J. Observations of the magnetic reconnection signature of an M2 flare on 2000 March 23. *Astrophys. J.* **703**, 877–882 (2009).
22. Yang, S. H., Zhang, J. & Xiang, Y. Y. Magnetic reconnection between small-scale loops observed with the New Vacuum Solar Telescope. *Astrophys. J. Lett.* **798**, L11 (2015).
23. Liu, R. et al. A reconnecting current sheet imaged in a solar flare. *Astrophys. J. Lett.* **723**, L28–L33 (2010).

24. Shibata, K. et al. Hot-plasma ejections associated with compact-loop solar flares. Astrophys. J. Lett. **451,** L83–L85 (1995).

25. Ohyama, M. & Shibata, K. X-ray plasma ejection associated with an impulsive flare on 1992 October 5: physical conditions of X-ray plasma ejection. Astrophys. J. **499,** 934–944 (1998).

26. Masuda, S., Kosugi, T., Hara, H., Tsuneta, S. & Ogawara, Y. A loop-top hard X-ray source in a compact solar flare as evidence for magnetic reconnection. Nature **371,** 495–497 (1994).

27. Kliem, B., Karlický, M. & Benz, A. O. Solar flare radio pulsations as a signature of dynamic magnetic reconnection. Astron. Astrophys. **360,** 715–728 (2000).

28. Zhang, J. et al. Coronal heating by the interaction between emerging active regions and the quiet Sun observed by the Solar Dynamics Observatory. Astrophys. J. Lett. **799,** L27 (2015).

29. Sun, J. Q. et al. Extreme ultraviolet imaging of three-dimensional magnetic reconnection in a solar eruption. Nat. Commun. **6,** 7598 (2015).

30. Liu, Z. et al. New Vacuum Solar Telescope and observations with high resolution. Res. Astron. Astrophys. **14,** 705–718 (2014).

31. Pesnell, W. D., Thompson, B. J. & Chamberlin, P. C. The Solar Dynamics Observatory (SDO). Sol. Phys. **275,** 3–15 (2012).

32. Golub, L. et al. The X-Ray Telescope (XRT) for the Hinode mission. Sol. Phys. **243,** 63–86 (2007).

33. Hill, S. M. et al. The NOAA GOES-12 Solar X-ray Imager (SXI) 1. Instrument, operations, and data. Sol. Phys. **226,** 255–281 (2005).

34. Scherrer, P. H. et al. The Helioseismic and Magnetic Imager (HMI) investigation for the Solar Dynamics Observatory (SDO). Sol. Phys. **275,** 207–227 (2012).

35. Mackay, D. H., Karpen, J. T., Ballester, J. L., Schmieder, B. & Aulanier, G. Physics of solar prominences: II-magnetic structure and dynamics. Space Sci. Rev. **151,** 333–399 (2010).

36. Su, Y., Surges, V., van Ballegooijen, A., DeLuca, E. & Golub, L. Observations and magnetic field modeling of the flare/coronal mass ejection event on 2010 April 8. Astrophys. J. **734,** 53 (2011).

37. Yan, X. L. et al. The formation and magnetic structures of active-region filaments observed by NVST, SDO, and Hinode. Astrophys. J. Suppl. Ser. **219,** 17–35 (2015).

38. Martin, S. F. Conditions for the formation and maintenance of filaments. Sol. Phys. **182,** 107–137 (1998).

39. Hood, A. W. & Priest, E. R. Critical conditions for magnetic instabilities in force-free coronal loops. Geophys. Astrophys. Fluid Dyn. **17,** 297–318 (1981).

40. Török, T., Kliem, B. & Titov, V. S. Ideal kink instability of a magnetic loop equilibrium. Astron. Astrophys. **413,** L27–L30 (2004).

41. Kliem, B. & Török, T. Torus instability. Phys. Rev. Lett. **96,** 255002 (2006).

42. Lemen, J. R. et al. The Atmospheric Imaging Assembly (AIA) on the Solar Dynamics Observatory (SDO). Sol. Phys. **275,** 17–40 (2012).

43. Daughton, W. et al. Role of electron physics in the development of turbulent magnetic reconnection in collisionless plasmas. Nat. Phys. **7,** 539–542 (2011).

44. Guo, Y. et al. Coexisting flux rope and dipped arcade sections along one solar filament. Astrophys. J. **714,** 343–354 (2010).

45. Guo, Y., Ding, M. D., Cheng, X., Zhao, J. & Pariat, E. Twist accumulation and topology structure of a solar magnetic flux rope. Astrophys. J. **779,** 157 (2013).

46. Janvier, M. et al. Electric currents in flare ribbons: observations and three-dimensional standard model. Astrophys. J. **788,** 60–70 (2014).

47. Priest, E. & Forbes, T. Magnetic Reconnection (Cambridge Univ. Press, 2000).

48. Lin, J. et al. Direct observations of the magnetic reconnection site of an eruption on 2003 November 18. Astrophys. J. **622,** 1251–1264 (2005).

49. Golub, L., Deluca, E. E., Sette, A. & Weber, M. Differential emission measure reconstruction with the Solar-B X-ray telescope. In ASP Conf. Ser. 325, The Solar-B Mission and the Forefront of Solar Physics. (eds Sakurai, T. & Sekii, T.) 217–225 (San Francisco, 2004).

50. Weber, M. A., DeLuca, E. E., Golub, L. & Sette, A. L. Temperature diagnostics with multichannel imaging telescopes. In IAU Symposium, Vol. 223, Multi-Wavelength Investigations of Solar Activity. (eds Stepanov, A. V., Benevolenskaya, E. E. & Kosovichev, A. G.) 321–328 (Cambridge Univ. Press, 2004).

51. Cheng, X., Zhang, J., Saar, S. H. & Ding, M. D. Differential emission measure analysis of multiple structural components of coronal mass ejections in the inner corona. Astrophys. J. **761,** 62 (2012).

52. Aschwanden, M. J. & Boerner, P. Solar corona loop studies with the Atmospheric Imaging Assembly. I. Cross-sectional temperature structure. Astrophys. J. **732,** 81 (2011).

53. Hannah, I. G. & Kontar, E. P. Differential emission measures from the regularized inversion of Hinode and SDO data. Astron. Astrophys. **539,** A146 (2012).

54. Wheatland, M. S., Sturrock, P. A. & Roumeliotis, G. An optimization approach to reconstructing force-free fields. Astrophys. J. **540,** 1150–1155 (2000).

55. Wiegelmann, T. Optimization code with weighting function for the reconstruction of coronal magnetic fields. Sol. Phys. **219,** 87–108 (2004).

56. Wiegelmann, T., Inhester, B. & Sakurai, T. Preprocessing of vector magnetograph data for a nonlinear force-free magnetic field reconstruction. Sol. Phys. **233,** 215–232 (2006).

57. van Ballegooijen, A. A. Observations and modeling of a filament on the Sun. Astrophys. J. **612,** 519–529 (2004).

58. Su, Y. et al. Observations and nonlinear force-free field modeling of active region 10953. Astrophys. J. **691,** 105–114 (2009).

59. Török, T. & Kliem, B. The evolution of twisting coronal magnetic flux tubes. Astron. Astrophys. **406,** 1043–1059 (2003).

60. Kliem, B., Su, Y., van Ballegooijen, A. A. & DeLuca, E. E. Magnetohydrodynamic modeling of the solar eruption on 2010 April 8. Astrophys. J. **779,** 129 (2013).

Acknowledgements

We thank the NVST, SDO, GOES/SXI, Hinode and BBSO teams for providing the data. SDO is a mission of NASA's Living With a Star Program. Hinode is a Japanese mission developed and launched by ISAS/JAXA, with NAOJ as a domestic partner, and NASA and STFC (UK) as international partners. It is operated by these agencies in co-operation with ESA and NSC (Norway). This work is sponsored by the National Science Foundation of China under the grant numbers 11503080, 11373066, 11573012, 11373065, 11303016, 11533008 and 11473071, Key Laboratory of Solar Activity of Chinese Academy of Sciences (CAS) under numbers KLSA201412, KLSA201407, Yunnan Science Foundation of China under number 2013FB086, CAS 'Light of West China' Program, Youth Innovation Promotion Association CAS (no. 2011056), and the national basic research program of China (973 program, 2011CB811400). B.K. is supported by the CAS under grant no. 2012T1J0017 and by the DFG. Y.N.S is supported by the One Hundred Talent Program of Chinese Academy of Sciences.

Author contributions

Z.K.X. developed the ideas, performed analysis of the data, discussed the results and wrote the first manuscript. X.L.Y. developed the ideas and lead the discussion of the manuscript. X.C. and K.Y. carried out the NLFFF extrapolation, DEM analysis and discussion. L.H.Y. carried out the analysis of the XRT and GOES data, and contributed to the discussion. Y.N.S. performed NLFFF modelling of the AR to provide the initial condition for the MHD simulations carried out by B.K.; Y.N.S. and B.K. also contributed to the discussion. B.K. and X.L.Y made major revisions of the manuscript. J.Z. developed ideas and contributed to the discussion. Z.L., Y.B., Y.Y.X and L.Z. contributed to the original observational data. All authors reviewed the manuscript.

Additional information

19

Laboratory measurements of resistivity in warm dense plasmas relevant to the microphysics of brown dwarfs

N. Booth[1], A.P.L. Robinson[1], P. Hakel[2,†], R.J. Clarke[1], R.J. Dance[3], D. Doria[4], L.A. Gizzi[5], G. Gregori[6], P. Koester[5], L. Labate[5], T. Levato[5,†], B. Li[1], M. Makita[4], R.C. Mancini[2], J. Pasley[1,3], P.P. Rajeev[1], D. Riley[4], E. Wagenaars[3], J.N. Waugh[3] & N.C. Woolsey[3]

Since the observation of the first brown dwarf in 1995, numerous studies have led to a better understanding of the structures of these objects. Here we present a method for studying material resistivity in warm dense plasmas in the laboratory, which we relate to the microphysics of brown dwarfs through viscosity and electron collisions. Here we use X-ray polarimetry to determine the resistivity of a sulphur-doped plastic target heated to Brown Dwarf conditions by an ultra-intense laser. The resistivity is determined by matching the plasma physics model to the atomic physics calculations of the measured large, positive, polarization. The inferred resistivity is larger than predicted using standard resistivity models, suggesting that these commonly used models will not adequately describe the resistivity of warm dense plasma related to the viscosity of brown dwarfs.

[1] Central Laser Facility, STFC Rutherford Appleton Laboratory, Didcot OX11 0QX, UK. [2] Department of Physics, College of Science, University of Nevada, Reno, Nevada 89557-0208, USA. [3] Department of Physics, York Plasma Institute, University of York, Heslington York YO10 5DD, UK. [4] School of Mathematics and Physics, Queen's University Belfast, Belfast BT1 4NN, UK. [5] Intense Laser Irradiation Laboratory, Istituto Nazionale di Ottica, Area della Ricerca del CNR, 56124 Pisa, Italy. [6] Department of Physics, University of Oxford, Oxford OX4 3PU, UK. † Present addresses: Division of Computational Physics, Los Alamos National Laboratory, Los Alamos, New Mexico 87545, USA (P.H.); ELI Beamlines, Fyzikalni Ustav AV CR Vvi, 182 21 Prague, Czech Republic (T.L.). Correspondence and requests for materials should be addressed to N.B. (email: nicola.booth@stfc.ac.uk) or to N.C.W. (email: nigel.woolsey@york.ac.uk).

Intense laser interactions with matter provide us with the ability to probe the conditions present at the core of dense stars in the laboratory[1]. Brown dwarfs are an ideal candidate for study with laboratory plasmas as their cores are at temperatures and mass densities of ~ 2–3×10^6 K (~ 200 eV) and 10^2–10^3 g cm^{-3}, respectively[2], which are comparable to conditions in laser-produced plasmas. The viscosity and equation of state of these systems are uncertain and requires detailed modelling[3]. The heat transport and electron propagation models at high density and pressure are complex[4–7] and need benchmarking. Laser-produced plasmas are a method of creating matter at conditions that approach those believed to occur in brown dwarfs[8,9] enabling laboratory studies of the plasma in these systems[10].

Plasma polarization spectroscopy is an *in-situ* method for probing the resistivity inside warm dense plasma environments[11–13]. The technique is sensitive to the complex electron microphysics of collisions in a solid, allowing us to observe the non-equilibrium states in the electron distributions. Other measurements have provided an indirect method of examining the sensitivity of electron transport to low-temperature resistivity in thick, cold, targets (for example, refs 14–19) and electron conductivity models of aluminium have been examined through exploding wire experiments to determine the expansion rate through X-ray backlighter radiographs[20]. As the resistivity of a laboratory plasma and the viscosity of the plasma of a brown dwarf are both dependent upon electron collisions[21] and the Coulomb logarithm, ln Λ, examining the resistivity models in laser-produced plasmas allows us to provide tests of the viscosity of matter found in brown dwarfs, which can be considered to be in local thermodynamic equilibrium. The laboratory measurements of return current distributions are characterized by a Maxwellian distribution of isotropic equilibrium (f_M), which gives these measurements relevance to local thermodynamic equilibrium conditions, and a beam component, which is a small perturbation from the isotropic equilibrium distribution (f_b), given by $f = f_M + f_b\cos\theta$.

In the case of brown dwarfs, the diffusion coefficient is inversely proportional to the viscosity, and as such any changes in the understanding of the viscosity model will therefore influence the temperatures of the brown dwarf. The temperature has an effect on the timescale and mixing in the radiative part of the atmosphere[22–25], affecting the brown dwarf evolutionary models. As both systems are dependent on the Coulomb logarithm, a laboratory measurement of the resistivity would therefore constrain the viscosity in a brown dwarf.

In ultra-intense laser-plasma interactions, beams of relativistic 'hot' electrons are driven from the interaction region[6] into materials of much higher density. Anisotropies in the speed distributions of both hot and return currents generate polarized X-ray line emission. As such the experimental measurements of the degree of polarization give us a unique ability to model both the non-equilibrium anisotropy in the return current electron distribution function[12,13,26] and in turn allows us to evaluate the resistivity generated in the warm dense matter regime applicable to the study of the viscosity of brown dwarfs. It is the non-equilibrium, or anisotropic, parts of the return current distribution that yield information about material resistivity. The return current density J_{rc} is related to the resistivity, η, by the resistive electric field $E = -\eta J_{rc}$, which converts the energy of the hot electrons into Ohmic heating[16,27]. We can infer the resistivity from electron beam anisotropy, as the fast electrons are generated with an anisotropic distribution function, and this is enhanced by their propagation into the target[16].

Here we demonstrate experimental measurements of the polarization of X-rays produced in intense laser-plasma

interactions as a method of determining resistivity models in warm dense matter. Through doping a low-atomic number (plastic) foil target with sulphur and observing the polarization of emitted X-rays, we observe anisotropic current distributions that drive the heat transport. Our experimental measurements show a high, positive, X-ray polarization arising from anisotropic electron transport. By matching the measured polarization to detailed modelling we show that the commonly used resistivity models do not adequately describe these amorphous materials.

Results

Experimental set-up. Our experiment was performed using the Vulcan Petawatt laser at the Central Laser Facility; with on target intensities of $\sim 5 \times 10^{20}$ W cm^{-2} with sulphur-doped plastic targets (polysulphone, $C_{27}H_{26}O_6S$; for the full description of the experimental set-up, including laser parameters and diagnostics, see Fig. 1 and Methods section). The measurements were performed with an orthogonal pair of highly oriented pyrolytic graphite (HOPG) crystals at a Bragg angle, $\vartheta_B = 45°$ (where the intensity of the p-polarized reflection is given by $I_P \approx R(\theta)\cos^2(2\vartheta_B)$), which allows us to observe time-integrated measurements of both polarizations of emitted X-rays in a single shot. The degree of polarization of X-rays emitted is calculated by $P = \frac{I_\pi - I_\sigma}{I_\pi + I_\sigma}$ (ref. 11), and the sign of the polarization indicates whether it arises due to beam-directional electrons with energies close to or far exceeding the excitation potential. We define a quantization axis as the direction of the free electron current. When a free electron with energy close to the excitation potential excites a ground state atomic electron, the atomic electron oscillates mostly parallel to the quantization axis. As a result, the atomic electron selects certain excited states emitting π-polarized dipole radiation (that is, $P > 0$) as it de-excites. In contrast, an energetic free electron will exert a pulse of electric field that is perpendicular to the quantization axis causing the atomic electron to oscillate and emit mostly σ-polarized radiation (that is, $P < 0$).

Experimental polarization measurements. The degree of polarization in the Ly-α transitions of sulphur, obtained from the data shown in Fig. 2, was measured to be $P = +0.16 \pm 0.04$. This value for the polarization indicates a beam of free electrons in the

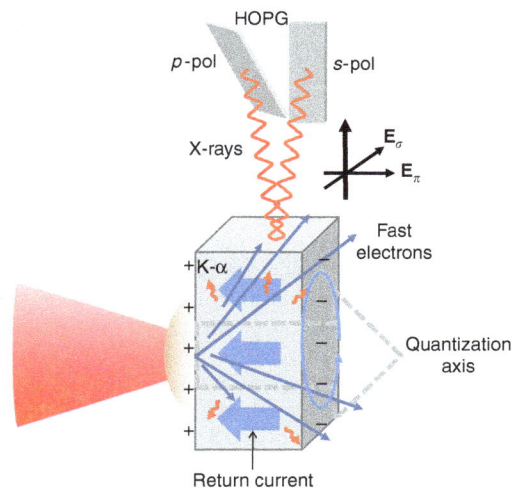

Figure 1 | The generation of polarized X-rays in a laser-plasma interaction. The target geometry of a high-intensity pulse, incident on a foil target produces polarized X-rays. These polarized X-rays are recorded by the orthogonal pair of HOPG crystals positioned above the target (not to scale) to measure each polarization independently in a single shot.

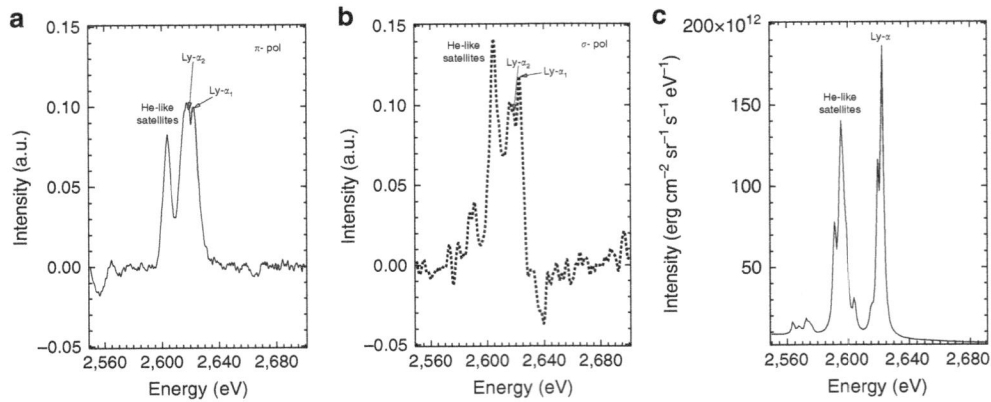

Figure 2 | X-ray spectrum from a 25-μm-thickness sulphur-doped plastic (polysulphone) target. The two plots show the (**a**) π-polarized X-ray (—) and (**b**) σ-polarized X-ray (......) spectrum from the same shot. The degree of polarization observed in the He-α lines differ from those of the Ly-α, as these lines are a result of interplay between direct collisional excitation within the He-like ion as well as electron capture from the ground state[29] and provide a future avenue for study. The third plot (**c**) is the simulated emission spectra of polysulphone modelled using the collisional-radiative spectral analysis code PrismSPECT[37] with a temperature of $T_b = 150$ eV and $\alpha = 0.01$.

return current close to the excitation potential (≥ 2.8 keV). This has led us to use the non-equilibrium, anisotropic, velocity distribution of the return current to explore the resistivity in warm dense plasmas. The experimental measurement of the polarization is time-integrated, and whilst laboratory plasmas do change on very fast timescales, it does not prevent a comparison with theoretical models. Uncertainties in the measurement are discussed in the Methods section.

Simulations of the electron transport and polarization. These measurements are modelled using the ZEPHYROS[28] and POLAR[29] codes to calculate plasma hybrid electron kinetics and atomic magnetic sub-level population kinetics respectively. POLAR requires three electron populations and temperatures[30] to calculate the polarization, and so by varying the input to ZEPHYROS to give us values for background temperature T_b, return current temperature T_{rc} and hot electron fraction α, we can iterate between the two codes to find the conditions that most closely match the experimental observations. Using this process we find that there is a balance to be achieved between T_b, T_{rc} and α that maximizes the degree of polarization. If T_b is too high, the depolarizing Maxwellian electrons will equilibrate P towards zero and equally, if T_{rc} is too low, the beam is ineffective in driving anisotropy.

On taking some typical values from ZEPHYROS: $T_b = 200$ eV, $T_{rc} = 600$ eV and hot temperature $T_h = 7$ MeV (from Wilks' scaling[31]) with $\alpha = 0.0032$, we obtain a polarization value from the POLAR model of $P = +0.14$. The bulk temperature T_b is in the typical temperature region in the core of brown dwarfs. The ZEPHYROS simulations using the above electron distribution parameters are shown in Fig. 3. In the region shown, heating is dominated by resistive heating of the target by the anisotropic component of the return current.

To compare this model combination to the experimental results we carried out temporal and spatial averaging of the post-processed ZEPHYROS output. Two resistivity models were compared: the Lee-More[5] model, and the Spitzer[4] model (with an initial temperature of 50 eV). These two resistivity models primarily differ at low temperatures, where the Spitzer resistivity curve is significantly higher. The simulation results for the different resistivity models are shown in Fig. 4. The figures show the plasma material with the region with the strongest Ly-α emission at 2.6 keV highlighted, which arises from a region in the

bulk plasma heated to approximately $T_b = 200$ eV. These plots of background electron temperature show that the target is heated to higher average temperature in the case where the Spitzer resistivity curve was employed compared with that where the Lee-More resistivity curve was used. When these results are processed with POLAR we obtained an average polarization of $P = +0.275 \pm 0.1$ for the Lee-More case and $P = +0.163 \pm 0.03$ for the Spitzer case (T init $= 50$ eV). This indicates that a model with somewhat higher resistivity at low temperatures than predicted by the Lee-More model is necessary for there to be good quantitative agreement with the experimental results. The simulated polarization is sensitive to T_h for both the Lee-More and the Spitzer (T init $= 50$ eV) models. For our simulations we use the Wilks scaling which gives $T_h = 7$ MeV, rather than the scaling laws of Beg[32], Haines[33] or Sherlock[34] which predict lower temperatures. The calculations show Lee-More over predicts the polarization, whilst Spitzer is more consistent with our experimental result. How the polarization calculation varies with T_h for these two models are shown in Supplementary Table 1 and discussed in Supplementary Note 1. These results clearly show that conclusions of this work are not dependent on the T_h model or scaling law. Indeed a factor of > 2 reduction in T_h is needed for the Lee-More model to match observation. This reduction cannot be justified using accepted scaling laws.

Discussion

Resistivity models with a low-temperature (< 100 eV) resistivity equal to or lower than the standard Lee-More model with a minimum electron scattering mean free path of $2 \times r_s$ do not agree well with our experimental results. To be clear: our analysis is not suggesting that the Spitzer model is a 'good' model of these plasmas. Rather it suggests that accurate computation of the resistivity at low-temperature (< 100 eV) requires careful examination.

Measuring the polarization state of X-rays to give *in-situ* information about return current electron transport in warm dense matter shows a large degree of anisotropy in the return current distributions. The use of our combination of experiment and computational models allows us to observe a beam-like return current, and use this to interrogate resistivity models of warm dense matter, in this case plastic. We find a model which is highly resistive at low temperatures is necessary to obtain the level of polarization that we observe. This indicates that the

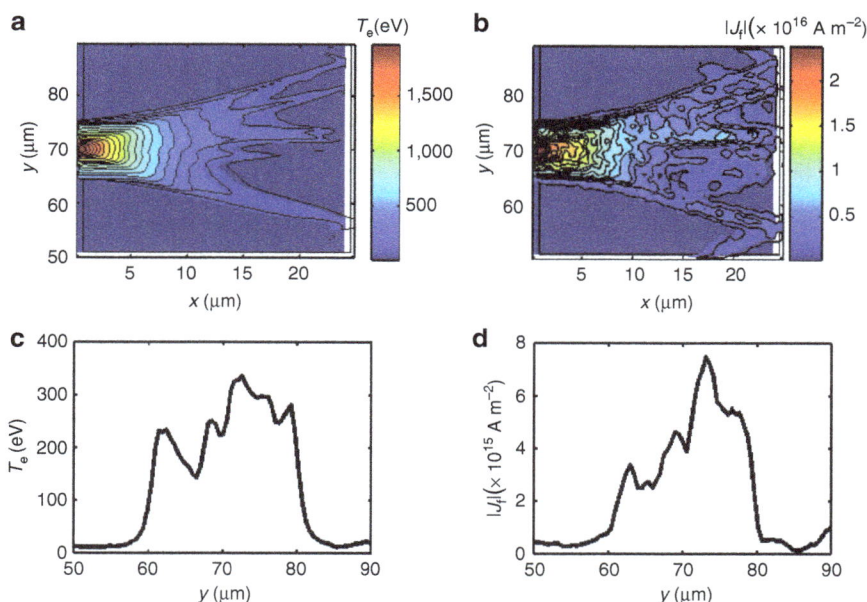

Figure 3 | ZEPHYROS simulations of the sulphur-doped plastic target conditions. The simulations use a combination of plasma hybrid electron kinetics and atomic magnetic sub-level population kinetics to obtain the plasma parameters $T_b = 200$ eV, $T_{rc} = 600$ eV and $\alpha = 0.0032$. The simulation results show the simulated electron temperature (**a**) and current density (**b**) and are taken in the mid-plane of the interaction, with x and y the horizontal and vertical axes through the target respectively, and lineouts of each distribution are shown below the simulation taken through middle of the target at $x = 12.5$ μm (**c** and **d** respectively). The parameters selected for the hot electron beam match the experiment and initially use the resistivity model of Davies[36].

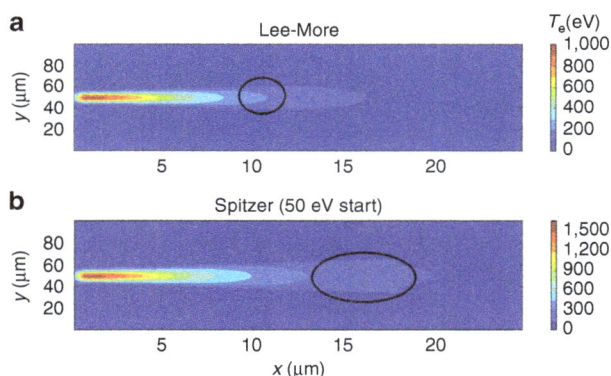

Figure 4 | Simulations comparing the electron temperature profiles. The simulations show the plasma conditions for plasma parameters $T_b = 200$ eV, $T_{rc} = 600$ eV and $\alpha = 0.0032$. The simulation results are again taken in the mid-plane of the interaction, with x and y the horizontal and vertical axes through the target respectively. (**a**) shows the expected temperature profile with the commonly used Lee-More model starting with a minimum mean free path of $2 \times r_s$. This resistivity condition produces a small volumetric region heated to 200 eV and leads to a calculated polarization of $P = +0.275 \pm 0.1$ (s.d.). Whilst it is possible that the observed polarizations are produced in this small region, the simulation from the Spitzer model with a 50 eV starting temperature in the case shown in **b** has a much larger region heated to 200 eV, which dominates over any other single temperature region, leading to a calculated polarization of $P = +0.163 \pm 0.03$ (s.d.). In both figures the region heated to 200 eV is highlighted, demonstrating that the region heated to the necessary temperature to generate the observed polarization is larger in the case of the higher resistivity Spitzer model.

resistivity in warm dense plasmas is higher than one would determine from commonly used models, and needs much more careful consideration. Since our resistivity models have to be

significantly altered to understand the energy transport through plasmas, it is likely that the viscosity of the brown dwarf would also need to be modified, potentially leading to a change in temperature which would alter the evolutionary models and luminosity of these objects. With further study, through examining the resistivity of plasmas in the laboratory, it is possible to quantify the coulomb logarithm for brown dwarfs. The modelling of warm dense matter resistivity, and viscosity in the case of brown dwarfs, is so complex, it is clearly necessary to consider in more detail the resistivity curves used in such modelling.

Methods

Laser-target interaction. The experiment was performed with the Vulcan Petawatt system at the Central Laser Facility, Rutherford Appleton Laboratory. The laser pulses contained ~ 250 J in 600 fs on target, at 1,053 nm. The beam is focussed with an $f/3$ parabola and focuses to a best spot of $\approx 6 \times 6$ μm^2 with $\sim 30\%$ of the total energy in the full width at half maximum, with peak intensities of 5×10^{20} W cm^{-2}. The nanosecond amplified spontaneous emission contrast ratio was measured on each shot with a fast photodiode at 10^{-6}.

Targets. The targets were 25-μm-thick polysulphone foils ($C_{27}H_{26}O_6S$) cut to 100×100 μm^2 squares mounted on copper stalks.

X-ray polarization spectrometer. Two flat, mosaic, $25 \times 50 \times 2$ mm^3, grade ZYA (mosaic spread $0.4° \pm 0.1°$) HOPG (002) Bragg reflecting crystals were placed above the target aligned either parallel or perpendicular to the target surface, perpendicular to the general direction of hot electrons (the quantization axis). Diffracted X-rays (with $\vartheta_B \sim 45°$) were recorded on image plates. The use of two high-reflectivity HOPG crystals was essential to enable single-shot spectroscopy measurements in an environment of high-level electromagnetic and particle noise, with relatively weak X-ray signal. (The reflectivity and rocking curves of the HOPG crystals in this context are discussed further in ref. 35).

Experimental degree of polarization. The degree of polarization is extracted by comparing Ly-α_1 integrated line intensities from the orthogonal HOPGs. To unfold the relative contributions of the Ly-α components and assess the level of uncertainty, a multi-line fitting procedure is applied. This procedure enables a cross calibration of the HOPG spectra by normalizing the integrated intensity of the Ly-α_2 and then extracting the relative integrated intensities of the Ly-α_1 emission. Sources of error and uncertainty in instrumentation and analysis are scrutinized. The experimental environment is noisy and given the crystals are positioned next to each other above the target and orientated to diffract in different directions the image plates are exposed to different levels of background. This is accounted for in the data analysis as are the uncertainty that arises from the differences in positioning.

Numerical modelling. The plasma electron kinetics simulations have been carried out using the three-dimensional particle hybrid code ZEPHYROS[28]. The Lyman-α sub-level population calculations were performed with the POLAR[29] model, which is a sub-level population, zero-dimensional atomic kinetics code. Hybrid simulations treat the hot electrons kinetically, but the background plasma is treated as a fluid. The parameters selected for the hot electron beam match the experiment and use the resistivity model of Davies' heuristic curve for CH (ref. 36), and modified to give the Spitzer resistivity for sulphur or nickel in the high-temperature limit. From this we extract spatial profiles of the hot electron current density and the 'average' background temperature and density (T_b and n_b), which are used as input to POLAR to calculate the polarization. POLAR includes background electrons as an isotropic Maxwellian population and a return current as a beamed electron population with a longitudinal temperature (T_{rc}) and density αn_b, where α is the return current fraction.

Errors in the calculated polarizations were calculated by performing a study varying the simulation inputs; intensity, pulse duration and absorption. These parameters were varied between upper and lower limits of possible experimental values, to find the likelihood of errors between the experimental data and our simulated polarizations.

References

1. Chabrier, G. & Baraffe, I. Theory of low-mass stars and substellar objects. *Ann. Rev. Astron. Astrophys.* **38**, 337–377 (2000).
2. Ichimaru, S. & Kitamura, H. Pycnonuclear reactions in dense astrophysical and fusion plasmas. *Phys. Plasmas* **6**, 2649–2671 (1999).
3. Baraffe, I., Chabrier, G. & Barman, T. The physical properties of extra-solar planets. *Rep. Prog. Phys.* **73**, 016901 (2010).
4. Spitzer, L. & Harm, R. Transport phenomena in a completely ionized gas. *Phys. Rev.* **89**, 977–981 (1953).
5. Lee, Y. T. & More, R. M. An electron conductivity model for dense plasmas. *Phys. Fluids* **27**, 1273–1286 (1984).
6. Bell, A. R., Davies, J. R., Guerin, S. & Ruhl, H. Fast electron transport in high-intensity, short-pulse laser-solid experiments. *Plasma Phys. Control. Fusion* **39**, 653–659 (1997).
7. Evans, R. G. Modelling electron transport for fast ignition. *Plasma Phys. Control. Fusion* **49**, B87–B93 (2007).
8. Burrows, A. & Liebert, J. The science of brown dwarfs. *Rev. Mod. Phys.* **65**, 301–336 (1993).
9. Saumon, D., Chabrier, G. & Van Horn, H. M. An equation of state for low-mass stars and giant planets. *Astrophys. J. Suppl.* **99**, 713–741 (1995).
10. Remington, B. A., Arnett, D., Drake, R. P. & Takabe, H. Modelling astrophysical phenomena in the laboratory with intense lasers. *Science* **284**, 1488–1493 (1999).
11. Kieffer, J. C. *et al.* Electron distribution anisotropy in laser-produced plasmas from x-ray line polarization measurements. *Phys. Rev. Lett.* **68**, 480–483 (1992).
12. Inubushi, Y. *et al.* X-ray line polarization spectroscopy to study hot electron transport in ultra-short laser produced plasma. *J. Quant. Spectrosc. Radiat. Transf.* **99**, 305–313 (2006).
13. Kawamura, T. *et al.* Polarization of He-α radiation due to anisotropy in fast electron transport in ultra-intense laser-produced plasmas. *Phys. Rev. Lett.* **99**, 115003 (2007).
14. Milchberg, H. M., Freeman, R. R. & Davey, S. C. Resistivity of a simple metal from room temperature to 10^6 K. *Phys. Rev. Lett.* **61**, 2364–2367 (1988).
15. Davies, J. R., Bell, A. R. & Tatarakis, M. Magnetic focussing and trapping of high-intensity laser-generated fast electrons at the rear of solid targets. *Phys. Rev. E* **59**, 6032–6036 (1999).
16. Bell, A. R. & Kingham, R. J. Resistive collimation of electron beams in laser produced plasmas. *Phys. Rev. Lett.* **91**, 035003 (2003).
17. MacLellan, D. A. *et al.* Annular fast electron transport in silicon arising from low-temperature resistivity. *Phys. Rev. Lett.* **111**, 095001 (2013).
18. DeSilva, A. W. & Katsouros, J. D. Electrical conductivity of dense copper and aluminum plasmas. *Phys. Rev. E* **57**, 5945–5951 (1998).
19. Brown, C. R. D. *et al.* Measurements of electron transport in foils irradiated with a picosecond timescale laser pulse. *Phys. Rev. Lett.* **106**, 185003 (2011).
20. Sinars, D. B. *et al.* Exploding aluminium wire expansion rate with 1-4.5kA per wire. *Phys. Plasmas* **7**, 1555–1563 (2000).
21. Mohanty, S. *et al.* Activity in very cool stars: Magnetic dissipation in late M and L dwarf atmospheres. *Astrophys. J.* **571**, 469–486 (2002).
22. Saumon, D. *et al.* Physical parameters of very cool T dwarfs. *Astrophys. J.* **656**, 1136–1149 (2007).
23. Leggett, S. K. *et al.* The physical properties of four $\sim$600K T dwarfs. *Astrophys. J.* **695**, 1517–1526 (2009).
24. Stephens, D. C. *et al.* The 0.8 – 14.5μm spectra of mid-L to mid-T dwarfs: diagnostics of effective temperature, grain sedimentation, gas transport and surface gravity. *Astrophys. J.* **702**, 154–170 (2009).
25. Leggett, S. K. *et al.* Properties of the T8.5 dwarf wolf 940B. *Astrophys. J.* **720**, 252–258 (2010).
26. Inubushi, Y. *et al.* Analysis of x-ray polarization to determine three-dimensionally anisotropic velocity distributions of hot electrons in plasma produced by ultrahigh intensity lasers. *Phys. Rev. E* **75**, 026401 (2007).
27. Huang, X. & Cumming, A. Ohmic dissipation in the interiors of hot Jupiters. *Astrophys. J.* **757**, 47 (2012).
28. Kar, S. *et al.* Guiding of relativisitic electron beams in solid targets by resistively controlled magnetic fields. *Phys. Rev. Lett.* **102**, 055001 (2009).
29. Hakel, P. & Mancini, R. C. X-ray line polarization of He-like Si satellite spectra in plasmas driven by high-intensity ultrashort pulsed lasers. *Phys. Rev. E* **69**, 056405 (2004).
30. Hakel, P. Polarization properties of the Ly-α line from sulphur plasmas driven by high-intensity, ultrashort-duration laser pulses. *Can. J. Phys.* **89**, 509–511 (2011).
31. Wilks, S. C. & Kruer, W. L. Absorption of ultrashort, ultra-intense laser light by solids and overdense plasmas. *IEEE J. Quantum Electron.* **33**, 1954–1968 (1997).
32. Beg, F. N. *et al.* A study of picosecond laser–solid interactions up to 1×10^{19} W cm^{-2}. *Phys. Plasmas* **4**, 447–457 (1997).
33. Haines, M. G. *et al.* Hot-electron temperature and laser-light absorption in fast ignition. *Phys. Rev. Lett.* **102**, 045008 (2009).
34. Sherlock, M. Universal scaling of the electron distribution function in one-dimensional simulations of relativistic laser-plasma interactions. *Phys. Plasmas* **16**, 103101 (2009).
35. Woolsey, N. C. *et al.* Precision x-ray spectroscopy of intense laser-plasma interactions. *High Energy Density Phys.* **7**, 105–109 (2011).
36. Davies, J. R. How wrong is collisional Monte Carlo modelling of fast electron transport in high-intensity, laser-solid interactions? *Phys. Rev. E* **65**, 026407 (2002).
37. McFarlane, J. J. *et al.* Simulation of the ionization dynamics of aluminum irradiated by intense short-pulse lasers. *Proc. Inertial Fusion and Sciences Applications 2003. Amer. Nucl. Soc.* 457–469 (2004).

Acknowledgements

We would like to acknowledge the support and expertise of the staff of the Central Laser Facility. This work was partially supported by the Extreme Light Infrastructure Project, the HiPER project, the United Kingdom Engineering and Physical Sciences Research Council (grant number EP/H012605/1) and the Science and Technology Facilities Council. The work of GG has received partial funding from the European Research Council under the European Community's Seventh Framework Programme (FP7/2007–2013)/ERC grant agreement no. 256973.

Author contributions

The experiment was conceived by N.C.W., G.G. and P.P.R. The experiment, planning and analysis were carried out by N.B. and N.C.W. with contributions from R.J.C., D.D., L.A.G., G.G., P.K., L.L., T.L., B.L., M.M., J.P., P.P.R., D.R., E.W., R.J.D. and J.N.W. Simulations were carried out by A.P.L.R., P.H. and R.C.M., N.B., P.H., A.P.L.R., P.P.R. and N.C.W. prepared the manuscript.

Additional information

20

Demonstration of a near-IR line-referenced electro-optical laser frequency comb for precision radial velocity measurements in astronomy

X. Yi[1], K. Vahala[1], J. Li[1], S. Diddams[2,3], G. Ycas[2,3], P. Plavchan[4], S. Leifer[5], J. Sandhu[5], G. Vasisht[5], P. Chen[5], P. Gao[6], J. Gagne[7], E. Furlan[8], M. Bottom[9], E.C. Martin[10], M.P. Fitzgerald[10], G. Doppmann[11] & C. Beichman[8]

An important technique for discovering and characterizing planets beyond our solar system relies upon measurement of weak Doppler shifts in the spectra of host stars induced by the influence of orbiting planets. A recent advance has been the introduction of optical frequency combs as frequency references. Frequency combs produce a series of equally spaced reference frequencies and they offer extreme accuracy and spectral grasp that can potentially revolutionize exoplanet detection. Here we demonstrate a laser frequency comb using an alternate comb generation method based on electro-optical modulation, with the comb centre wavelength stabilized to a molecular or atomic reference. In contrast to mode-locked combs, the line spacing is readily resolvable using typical astronomical grating spectrographs. Built using commercial off-the-shelf components, the instrument is relatively simple and reliable. Proof of concept experiments operated at near-infrared wavelengths were carried out at the NASA Infrared Telescope Facility and the Keck-II telescope.

[1] Department of Applied Physics and Materials Science, Pasadena, California 91125, USA. [2] National Institute of Standards and Technology, 325 Broadway, Boulder, Colorado 80305, USA. [3] Department of Physics, University of Colorado, 2000 Colorado Avenue, Boulder, Colorado 80309, USA. [4] Department of Physics, Missouri State University, 901 S National Avenue, Springfield, Missouri 65897, USA. [5] Jet Propulsion Laboratory, California Institute of Technology, 4300 Oak Grove Drive, Pasadena, California 91109, USA. [6] Division of Geological and Planetary Sciences, California Institute of Technology, Pasadena, California 91125, USA. [7] Department of Terrestrial Magnetism, Carnegie Institution of Washington, 5241 Broad Branch Road, Washington, District of Columbia 20015, USA. [8] NASA Exoplanet Science Institute, California Institute of Technology, Pasadena, California 91125, USA. [9] Department of Astronomy, California Institute of Technology, Pasadena, California 91125, USA. [10] Department of Physics and Astronomy, University of California Los Angeles, Los Angeles, California 90095, USA. [11] W.M. Keck Observatory, Kamuela, Hawaii 96743, USA. Correspondence and requests for materials should be addressed to K.V. (email: vahala@caltech.edu) or to C.B. (email: chas@ipac.caltech.edu).

The earliest technique for the discovery and characterization of planets orbiting other stars (exoplanets) is the Doppler or radial velocity (RV) method whereby small periodic changes in the motion of a star orbited by a planet are detected via careful spectroscopic measurements[1]. The RV technique has identified hundreds of planets ranging in mass from a few times the mass of Jupiter to less than an Earth mass, and in orbital periods from less than a day to over 10 years (ref. 2). However, the detection of Earth-analogues at orbital separations suitable for the presence of liquid water at the planet's surface, that is, in the 'habitable zone'[3], remains challenging for stars like the Sun with RV signatures $<0.1 \, \mathrm{m \, s^{-1}}$ ($\Delta V/c < 3 \times 10^{-10}$) and periods of a year ($\sim 10^8$ sec to measure three complete periods). For cooler, lower luminosity stars (spectral class M), however, the habitable zone moves closer to the star which, by application of Kepler's laws, implies that a planet's RV signature increases, $\sim 0.5 \, \mathrm{m \, s^{-1}}$ ($\Delta V/c < 1.5 \times 10^{-9}$), and its orbital period decreases, ~ 30 days ($\sim 10^7$ s to measure three periods). Both of these effects make the detection easier. But for M stars, the bulk of the radiation shifts from the visible wavelengths, where most RV measurements have been made to date, into the near-infrared. Thus, there is considerable interest among astronomers in developing precise RV capabilities at longer wavelengths.

Critical to precision RV measurements is a highly stable wavelength reference[4]. Recently a number of groups have undertaken to provide a broadband calibration standard that consists of a 'comb' of evenly spaced laser lines accurately anchored to a stable frequency standard and injected directly into the spectrometer along with the stellar spectrum[5-9]. While this effort has mostly been focused on visible wavelengths, there have been successful efforts at near-IR wavelengths as well[10-12]. In all of these earlier studies, the comb has been based on a femtosecond mode-locked laser that is self-referenced[13-15], such that the spectral line spacing and common offset frequency of all lines are both locked to a radio frequency standard. Thus, laser combs potentially represent an ideal tool for spectroscopic and RV measurements.

However, in the case of mode-locked laser combs, the line spacing is typically in the range of 0.1–1 GHz, which is too small to be resolved by most astronomical spectrographs. As a result, the output spectrum of the comb must be spectrally filtered to create a calibration grid spaced by >10 GHz, which is more commensurate with the resolving power of a high-resolution astronomical spectrograph[8]. While this approach has led to spectrograph characterization at the $\mathrm{cm \, s^{-1}}$ level[16], it nonetheless increases the complexity and cost of the system.

In light of this, there is interest in developing photonic tools that possess many of the benefits of mode-locked laser combs, but that might be simpler, less expensive and more amenable to 'hands-off' operation at remote telescope sites. Indeed, in many RV measurements, other system-induced errors and uncertainties can limit the achievable precision, such that a frequency comb of lesser precision could still be equally valuable. For example, one alternative technique recently reported is to use a series of spectroscopic peaks induced in a broad continuum spectrum using a compact Fabry–Perot interferometer[17-19]. While the technique must account for temperature-induced tuning of the interferometer, it has the advantage of simplicity and low cost. Another interesting alternative is the so-called Kerr comb or microcomb, which has the distinct advantage of directly providing a comb with spacing in the range of 10–100 GHz, without the need for filtering[20]. While this new type of laser comb is still under development, there have been promising demonstrations of full microcomb frequency control[21,22] and in the future it could be possible to fully integrate such a microcomb on only a few square centimetres of silicon, making a very robust

and inexpensive calibrator. Another approach that has been proposed is to create a comb through electro-optical modulation of a frequency-stabilized laser[23,24].

In the following, we describe a successful effort to implement this approach. We produce a line-referenced, electro-optical modulation frequency comb (LR-EOFC) ~ 1559.9 nm in the astronomical H band (1,500–1,800 nm). We discuss the experimental set-up, laboratory results and proof of concept demonstrations at the NASA Infrared Telescope Facility (IRTF) and the W. M. Keck observatory (Keck) 10 m telescope.

Results

Comb generation. A LR-EOFC is a spectrum of lines generated by electro-optical modulation of a continuous-wave laser source[25-29] which has been stabilized to a molecular or atomic reference (for example, $f_0 = f_{atom}$). The position of the comb teeth ($f_N = f_0 \pm N f_m$, N is an integer) has uncertainty determined by the stabilization of f_0 and the microwave source that provides the modulation frequency f_m. However, the typical uncertainty of a microwave source can be sub-Hertz when synchronized with a compact Rb clock and moreover can be global positioning system (GPS)-disciplined to provide long-term stability[12]. Thus, the dominant uncertainty in comb tooth frequency in the LR-EOFC is that of f_0.

The schematic layout for LR-EOFC generation is illustrated in Fig. 1 and a detailed layout is shown in Fig. 2. All components are commercially available off-the-shelf telecommunications components. Pictures of the key components are shown in the left column of Fig. 1. The frequency-stabilized laser is first pre-amplified to 200 mW with an Erbium-Doped Fibre Amplifier (EDFA, model: Amonics, AEDFA-PM-23-B-FA) and coupled into two tandem lithium niobate ($LiNbO_3$) phase modulators ($V_\pi = 3.9$ V at 12 GHz, RF input limit: 33 dBm). The phase modulators are driven by an amplified 12 GHz frequency signal at 32.5 and 30.7 dBm, and synchronized by using microwave phase shifters. This initial phase modulation process produces a comb having ~ 40 comb lines ($\approx 2\pi \times V_{drive}/V_\pi$), or equivalently 4 nm bandwidth. This comb is then coupled into a $LiNbO_3$ amplitude modulator with 18–20 dB distinction ratio, driven at the same microwave frequency by the microwave power recycled from the phase modulator external termination port. The modulation index of $\pi/2$ is set by an attenuator and the phase offset of the two amplitude modulator arms is set and locked to $\pi/2$. Microwave phase shifters are used to align the drive phase so that the amplitude modulator gates-out only those portions of the phase modulation that are approximately linearly chirped with one sign (that is, parabolic phase variation in time). A nearly transform-limited pulse is then formed when this parabolic phase variation is nullified by a dispersion compensation unit using a chirped fibre Bragg grating with 8 ps nm^{-1} dispersion. A 2 ps full-width at half-maximum pulse is measured after the fibre grating using an autocorrelator. Owing to this pulse formation, the duty cycle of the pulse train reaches below 2.5%, boosting the peak intensity of the pulses. These pulses are then amplified in a second EDFA (IPG Photonics, EAR-5 K-C-LP). For an average power of 1 W, peak power (pulse energy) is 40 W (83 pJ). The amplified pulses are then coupled into a 20 m length of highly nonlinear fibre with 0.25 ± 0.15 ps nm^{-1} km^{-1} dispersion and dispersion slope of 0.006 ± 0.004 ps nm^{-2} km^{-1}. Propagation in the highly nonlinear fibre causes self-phase modulation and strong spectral broadening of the comb[30]. Comb spectra span and envelope can be controlled by the pump power launched into the highly nonlinear fibre. A typical comb spectrum with >600 mW pump power from the 1,559.9 nm laser is shown in Fig. 3a, with >100 nm spectral span. Moreover, by using various nonlinear fibre and spectral flattening methods, broad combs with level power are possible[31].

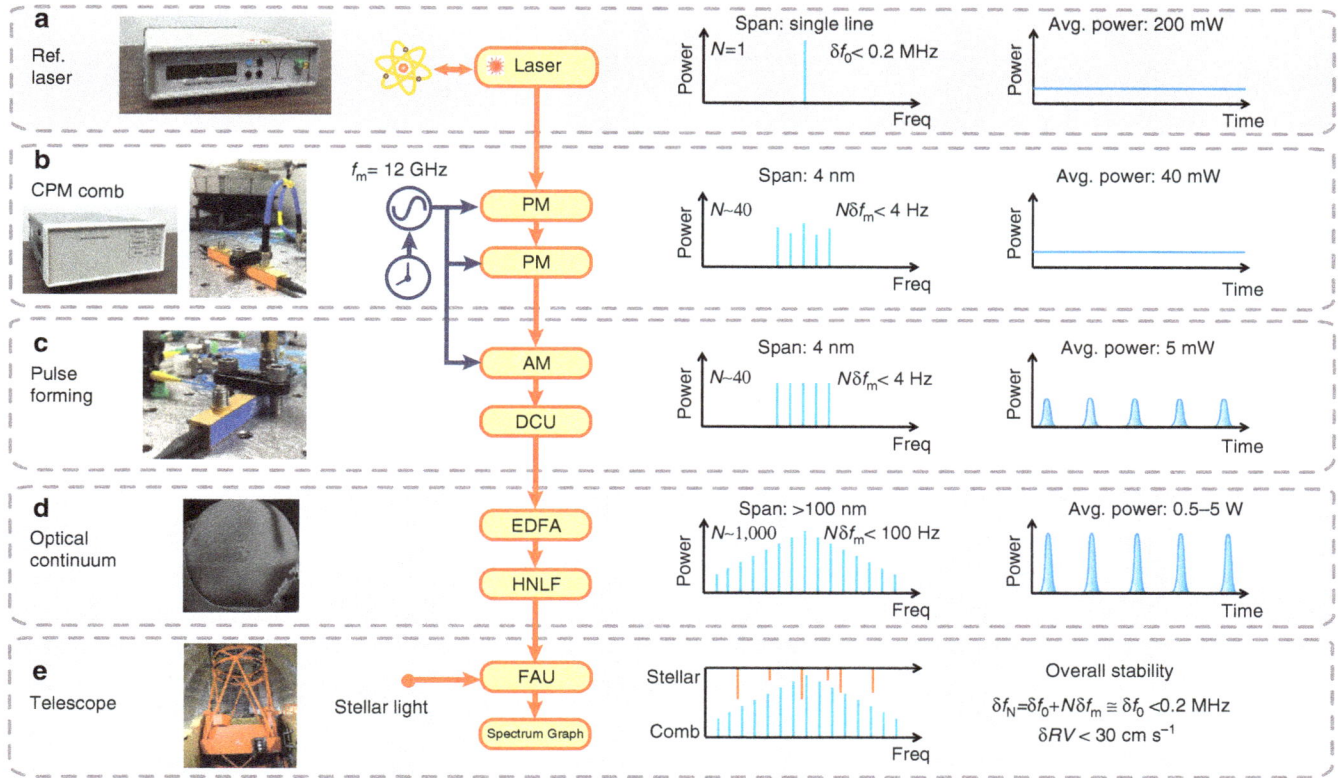

Figure 1 | Conceptual schematics of the line-referenced electro-optical frequency comb for astronomy. Vertically, the first column contains images of key instruments. (**a–e**) The images are reference laser, Rb clock (left) and phase modulator (right), amplitude modulator, highly nonlinear fibre and telescope. A simplified schematic set-up is in the second column. Third and fourth columns present the comb state in the frequency and temporal domains. The frequency of N-th comb tooth is expressed as $f_N = f_0 + N \times f_m$, where f_0 and f_m are the reference laser frequency and modulation frequency, respectively. N is the number of comb lines relative to the reference laser (taken as comb line $N = 0$), RV is radial velocity and δf_N, δf_0 and δf_m are the variance of f_N, f_0 and f_m. (**a**) The reference laser is locked to a molecular transition, acquiring stability of 0.2 MHz, corresponding to 30 cm s^{-1} RV. (**b**) Cascaded phase modulation (CPM) comb: the phase of the reference laser is modulated by two phase modulators (PM), creating several tens of sidebands with spacing equal to the modulation frequency. The RF frequency generator is referenced to a Rb clock, providing stability at the sub-Hz level ($\delta f_m < 0.03$ Hz at 100 s). (**c**) Pulse forming is then performed by an amplitude modulator (AM) and dispersion compensation unit (DCU), which could be a long single mode fibre (SMF) or chirped fibre Bragg grating (FBG). (**d**) After amplification by an erbium-doped fibre amplifier (EDFA), the pulse undergoes optical continuum broadening in a highly nonlinear fibre (HNLF), extending its bandwidth >100 nm. (**e**) Finally the comb light is combined with stellar light using a fibre acquisition unit (FAU) and is sent into the telescope spectrograph. The overall comb stability is primarily determined by the pump laser.

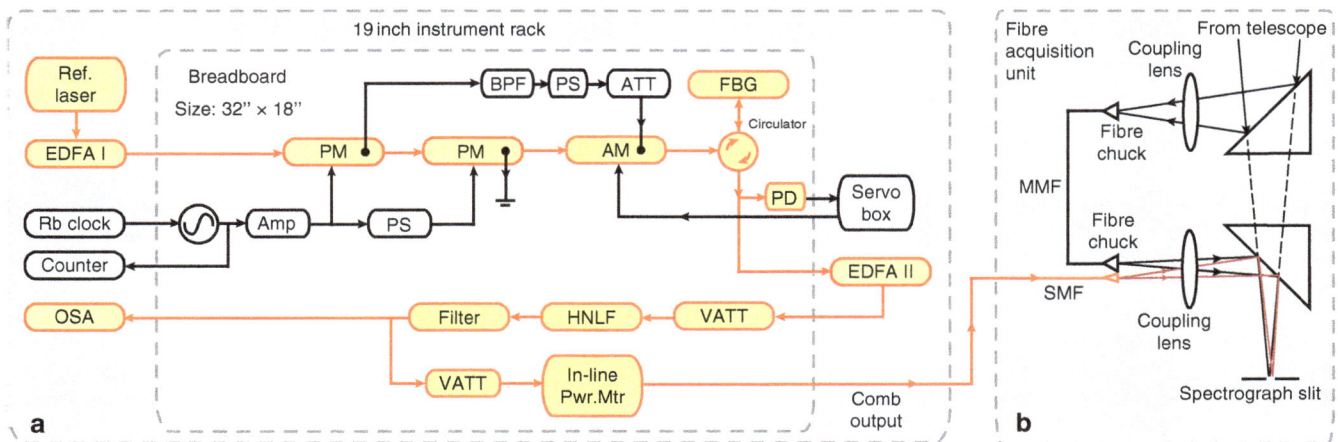

Figure 2 | Detailed set-up of line-referenced electro-optical frequency comb. (**a**) The entire LR-EOFC system sits in a 19 inch instrument rack. Optics and microwave components in the rack are denoted in orange and black, respectively. Small components were assembled onto a breadboard. These included the phase modulators (PM), amplitude modulator (AM), fibre Bragg grating (FBG), photodetector (PD), variable attenuator (VATT), attenuator (ATT), highly nonlinear fibre (HNLF), microwave source, microwave amplifier (Amp), phase shifter (PS) and band-pass filter (BPS). The reference laser, erbium-doped fiber amplifier (EDFA), rubidium (Rb) clock, counter, optical spectrum analyser (OSA) and servo lock box are separately located in the instrument rack. (**b**) A simplified schematic of the fibre acquisition unit (FAU) is also shown. Stellar light is focused and coupled into a multimode fibre (MMF). The comb light from a single mode fibre (SMF), together with the stellar light in the MMF, are focused on the spectrograph slit and sent into the spectrograph.

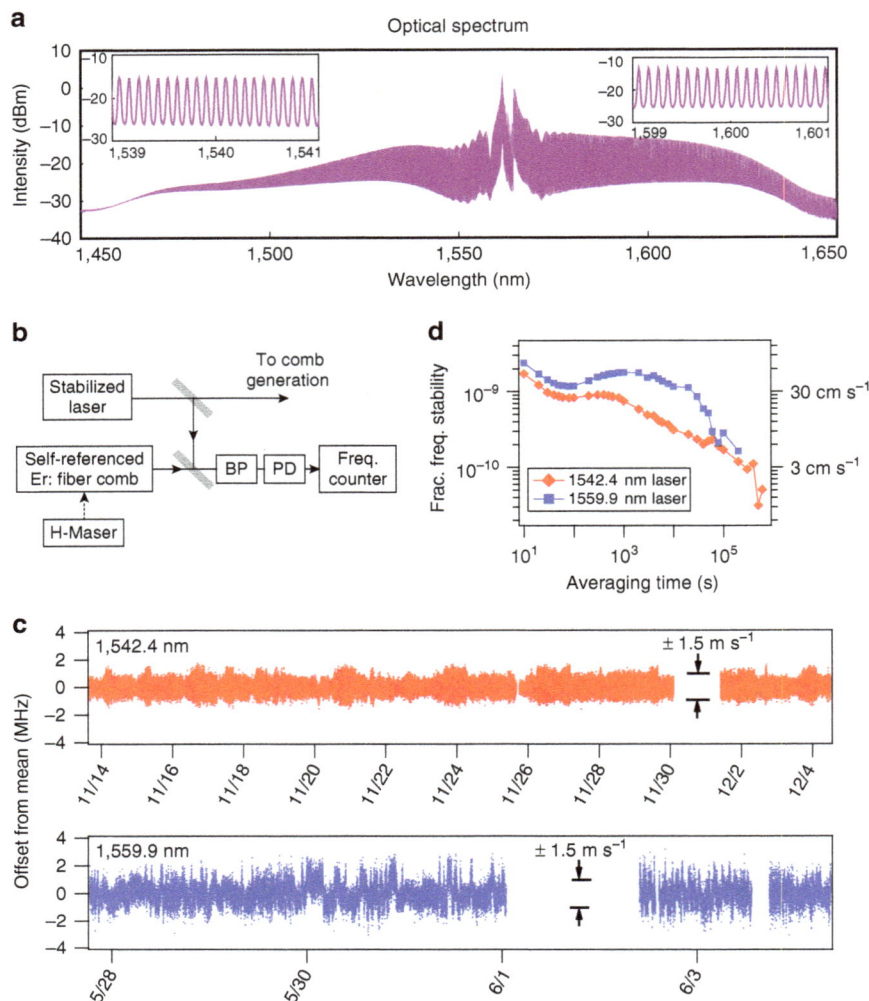

Figure 3 | Comb spectra and stability of the C$_2$H$_2$ and HCN reference lasers. (**a**) A typical comb spectrum from the 1,559.9 nm laser with >100 nm span generated with 600 mW pump power. The insets show the resolved line spacing of 12 GHz or ~0.1 nm. (**b**) Experimental set-up: BP, optical band-pass filter; PD, photodiode. All beam paths and beam combiners are in single mode fibre. (**c**) Time series of measured beat frequencies for the two frequency-stabilized lasers with 10 s averaging per measurement. The x axes are the dates in November of 2013 and May/June of 2014, respectively. (**d**) Allan deviation, which is a measure of the fractional frequency stability, computed from the time series data of **c**. Right-side scale gives the radial velocity precision.

The LR-EOFC system is mounted on an aluminum breadboard (18" × 32", or equivalently 45.7 × 81.3 cm) in a standard 19-inch instrument rack (see Fig. 2) for transport and implementation with the spectrograph at the NASA IRTF and at Keck II on Mauna Kea in Hawaii. The system is designed to provide operational robustness matching the requirements of astronomical observation. All optical components before the highly nonlinear fibre are polarization maintaining fibre-based, so as to eliminate the effect of polarization drift on spectral broadening in the highly nonlinear fibre. Moreover, no temperature control is required at the two telescope facilities. As a result, the comb is able to maintain its frequency, bandwidth and intensity without the need to adjust any parameters. During a 5 day run at IRTF, the comb had zero failures and the intensity of individual comb teeth was measured to deviate less than 2 dB, including multiple power-off and on cycling of the optical continuum generation system (see Fig. 4b).

Comb stability. As noted above, the frequency stability of the LR-EOFC is dominated by the stability of the reference laser frequency f_0. We explored the use of two different commercially available lasers (Wavelength References) that were stabilized, respectively, to Doppler- and pressure-broadened transitions in acetylene (C$_2$H$_2$) at 1,542.4 nm, and in hydrogen cyanide (H^{13}C^{15}N) at 1,559.9 nm. We note that the spectroscopy related to the locking of the reference laser to the molecular resonances is done internally to the laser system, so that our experiments only assess the stability of these commercial off-the-shelf lasers. To assess the stability, the stabilized laser frequencies were measured relative to an Er:fibre-based self-referenced optical frequency comb[11,32]. Fibre-coupled light from a reference laser was combined into a common optical fibre with light from the Er:fibre comb. Then the heterodyne beat between a single-comb line and the line-stabilized reference was filtered, amplified and counted with a 10 s gate time using a frequency counter that was referenced to a hydrogen maser (see Fig. 3b). The Er:fibre comb was stabilized relative to the same hydrogen maser, such that the fractional frequency stability of the measurement was <2 × 10^{-13} at all averaging times. The drift of the hydrogen maser frequency is <1 × 10^{-15} per day, thereby providing a stable reference at levels corresponding to a RV uncertainty ≪ 1 cm s^{-1}. Thus, the frequency of the counted heterodyne beat accurately represents the fluctuations in the reference laser.

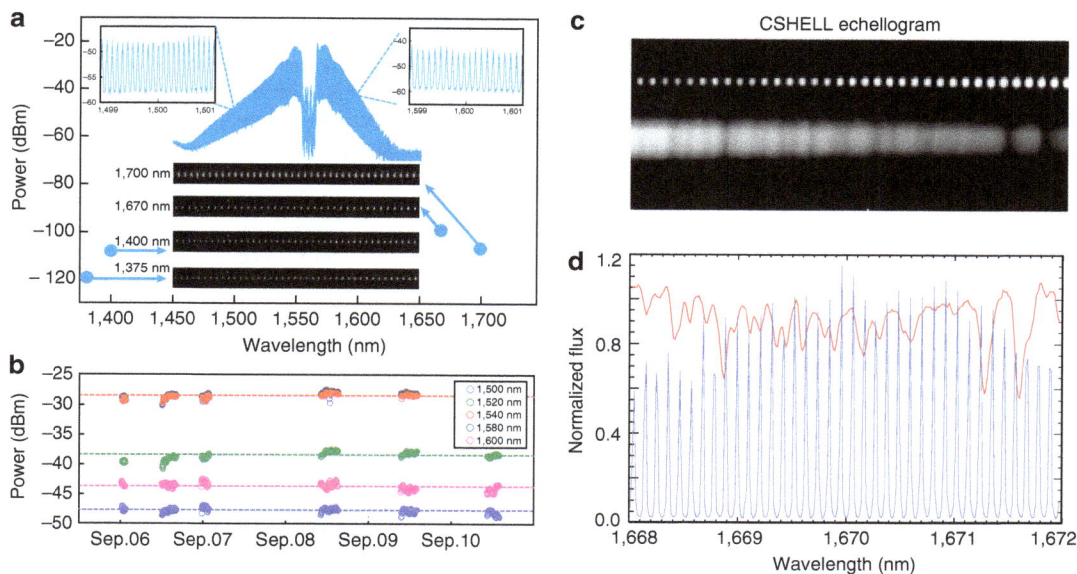

Figure 4 | **Experimental results at IRTF.** (**a**) Comb spectrum produced using 1,559.9 nm reference laser. The insets on top left and right show the resolved comb lines on the optical spectrum analyser. Comb spectra taken by the CSHELL spectrograph at 1,375, 1,400, 1,670 and 1,700 nm are presented as insets in the lower half of the figure. The blue circles mark the estimated comb line power and centre wavelength for these spectra. Comb lines are detectable on CSHELL at fW power levels. (**b**) Comb spectral line power versus time is shown at five different wavelengths. During the 5 day test at IRTF, no parameter adjustment was made, and comb intensity was very stable even with multiple power-on and -off cycling of the optical continuum generation system. (**c**) An image of the echelle spectrum from CSHELL on IRTF showing a 4 nm portion of spectrum ∼1,670 nm. The top row of dots are the laser comb lines, while the broad spectrum at the bottom is from the bright M2 II–III giant star β Peg seen through dense cloud cover. (**d**) Spectra extracted from **c**. The solid red curve denotes the average of 11 individual spectra of β Peg (without the gas cell) obtained with CSHELL on the IRTF. The regular sine-wave like blue lines show the spectrum from the laser comb obtained simultaneously with the stellar spectrum. The vertical axis is normalized flux units.

The series of 10 s measurements of the heterodyne beat was recorded over 20 days in 2013 for the case of the 1,542.4 nm laser and more than 7 days in 2014 for the case of the 1,559.9 nm laser, as shown in Fig. 3c. Gaps in the measurements near 11/31 and 6/4 are due to unlocking of the Er:fibre comb from the hydrogen maser reference. From these time series, we calculate the Allan deviation, which is a measure of the fractional frequency fluctuations (instability) of the reference laser as a function of averaging time. As seen in Fig. 3d, the instability of the 1,542.2 nm laser is $<10^{-9}$ (30 cm s^{-1} RV, or corresponding to 200 kHz in frequency) at all averaging times greater than ∼30 s. The 1,559.9 nm laser is less stable, but provides a corresponding RV precision of <60 cm s^{-1} for averaging times greater than 20 s. This different instability was to be expected because of the difference in relative absorption line strength between the acetylene and HCN-stabilized lasers. In both cases, the stability improves with averaging time, although at a rate slower than predicted for white frequency noise. As an aside, we note that despite the lower stability of the 1,559.9 nm laser, this wavelength ultimately produced wider and flatter comb spectra owing to the better gain performance of the fibre amplifier used in this work. We did not explore the noise mechanisms that lead to the observed Allan deviation, as they arise from details of the spectroscopy internal to the commercial off-the-shelf laser, to which we did not have access.

Additional analysis included an estimate of the drift of the frequencies of the two reference lasers obtained by fitting a line to the full multi-day counter time series. From these linear fits, an upper limit of the drift over the given measurement period was determined to be $<9 \times 10^{-12}$ per day for the acetylene-referenced laser and $<4 \times 10^{-11}$ for the hydrogen cyanide-referenced laser. This corresponds to equivalent RV drifts of <0.27 and <1.2 cm s^{-1} per day for the two references. Finally, we attempted to place a bound on the repeatability of the

1,542.4 nm reference laser during re-locking and power cycling. Although only evaluated for a limited number of power cycles and re-locks, in all cases, we found that the laser frequency returned to its predetermined value within <100 kHz, or equivalently, with a RV precision of <15 cm s^{-1}.

While these calibrations are sufficient for the few-day observations reported below, confidence in the longer term stability of the molecularly referenced continuous-wave lasers would be required for observations that could extend over many years. Likewise, frequency uncertainty of the molecular references should be examined. Properly addressing the potential frequency drifts on such a multi-year time scale would require a more thorough investigation of systematic frequency effects due to a variety of physical and operational parameters (for example, laser power, pressure, temperature and electronic offsets). Alternatively, narrower absorption features, as available in nonlinear Doppler-free saturation spectroscopy, could provide improved performance. For example, laboratory experiments have shown fractional frequency instability at the level of 10^{-12} and reproducibility of 1.5×10^{-11} for lasers locked to a Doppler-free transition in acetylene[33]. Most promising of all, self-referencing of an EOFC comb has been demonstrated recently[34], enabling full stabilization of the frequency comb to a GPS-disciplined standard. This would eliminate the need for the reference laser to define f_0, and thereby provide comb stability at the level of the GPS reference (for example, $<10^{-11}$ or equivalently <0.3 cm s^{-1}) on both long and short timescales.

IRTF telescope demonstration. To demonstrate that the laser comb is portable, robust and easy-to-use as a wavelength calibration standard, we shipped the laser comb to the NASA IRTF. IRTF is a 3 m diameter infrared-optimized telescope located at the summit of Mauna Kea, Hawaii. The telescope is equipped with a cryogenic echelle spectrograph (CSHELL) operating from 1–5.4 µm. CSHELL is a cryogenic, near-infrared traditional slit-fed spectrograph, with a resolution[35,36] of $R \sim \lambda/\Delta\lambda = 46,000$ and it images an adjustable single $\sim$5-nm-wide order spectrum on a 256×256 InSb detector. We have modified the CSHELL spectrograph to permit the addition of a fibre acquisition unit for the injection of starlight and laser frequency comb light into a fibre array and focusing on the spectrograph entrance slit. A simple schematic of the fibre acquisition unit is shown in Fig. 2 and the details are described elsewhere[37,38]. Before the starlight reaches the CSHELL entrance slit, it can be switched to pass through an isotopic methane absorption gas cell to introduce a common optical path wavelength reference[38]. A pickoff mirror is next inserted into the beam to re-direct the near-infrared starlight to a fibre via a fibre-coupling lens. A dichroic window re-directs the visible light to a guide camera to maintain the position of the star on the entrance of the fibre tip. For the starlight, we made use of a specialized non-circular core multi-mode fibre, with a 50×100 µm rectangular core. These fibres 'scramble' the near-field spatial modes of the fibre, so that the spectrograph is evenly illuminated by the output from the fibre, regardless of the alignment, focus or weather conditions of the starlight impinging upon the input to the fibre. We additionally made use of a dual-frequency agitator motor to vibrate the 10 m length of the fibre to provide additional mode mixing, distributing the starlight evenly between all modes. Finally, a lens and a second pickoff mirror are used to relay the output of the starlight from the fibre output back to the spectrograph entrance slit. A single-mode fibre carrying the laser comb is added next to the non-circular core fibre carrying the starlight. This was accomplished by replacing the output single-fibre SMA-fibre chuck with a custom three-dimensional printed V-groove array ferrule. This allowed us to send the light from both the star and frequency comb to the entrance slit of the CSHELL spectrograph when rotated in the same orientation as the slit.

Finally, the laser comb and associated electronics rack were set-up in the room temperature ($\sim \pm 5\,^{\circ}$C) control room. A 50 m length of single mode fibre was run from the control room to the telescope dome floor, and along the telescope mount to the CSHELL spectrograph to connect to the V-groove array and the fibre acquisition unit. The unpacking, set-up and integration of the comb fibre with CSHELL were straightforward, and required only 2 days working at an oxygen-deprived elevation of 14,000 feet in preparation for the observing run. Because the CSHELL spectrometer has a spectral window $<$5 nm, there was no effort made to generate spectrally flat combs. Comb lines are well resolved on CSHELL from 1,375 to 1,700 nm (Fig. 4a), with power adjusted by tunable optical attenuators to match the power of starlight and 6.7 pixels per comb line spacing at 1,670 nm wavelength. Also, comb line power was monitored (Fig. 4b) periodically during the observing run and was stable.

Three partial nights of CSHELL telescope time in September 2014 were used for this first on-sky demonstration of the laser comb. Unfortunately, the observing run was plagued by poor weather conditions, with 5–10 magnitudes of extinction because of clouds. Consequently, we observed the bright M2 II–III star β Peg (H $=-2.1$ mag), which is a pulsating variable star ($P = 43.3$ days). Typical exposure times were 150 s, and multiple exposures were obtained in sequence.

The star was primarily observed at 1,670 nm, with and without the isotopic methane gas cell to provide a wavelength calibration comparison for the laser comb. Other wavelengths were also observed to demonstrate that the spectral grasp of the comb is much larger than the spectral grasp of the spectrograph itself. Given the low SNR (signal-to-noise ratio) on β Peg from the high extinction because of clouds and CSHELL's limited spectral grasp, the SNR of these data is inadequate to demonstrate that the comb is more stable than the gas cell, as shown above.

One critical aspect of demonstrating the usability of the comb for astrophysical spectrographs is the comb line spacing. As seen in Fig. 4a,c,d, the spectra clearly demonstrate that the individual comb lines are resolved with the CSHELL spectrograph without the need for additional line filtering[39]. Thus this comb operates at a frequency that is natively well-suited for astronomical applications with significantly less hardware complexity compared with 'traditional' laser frequency combs.

Keck telescope demonstration. We were able to use daytime access to the near-infrared cryogenic echelle spectrograph

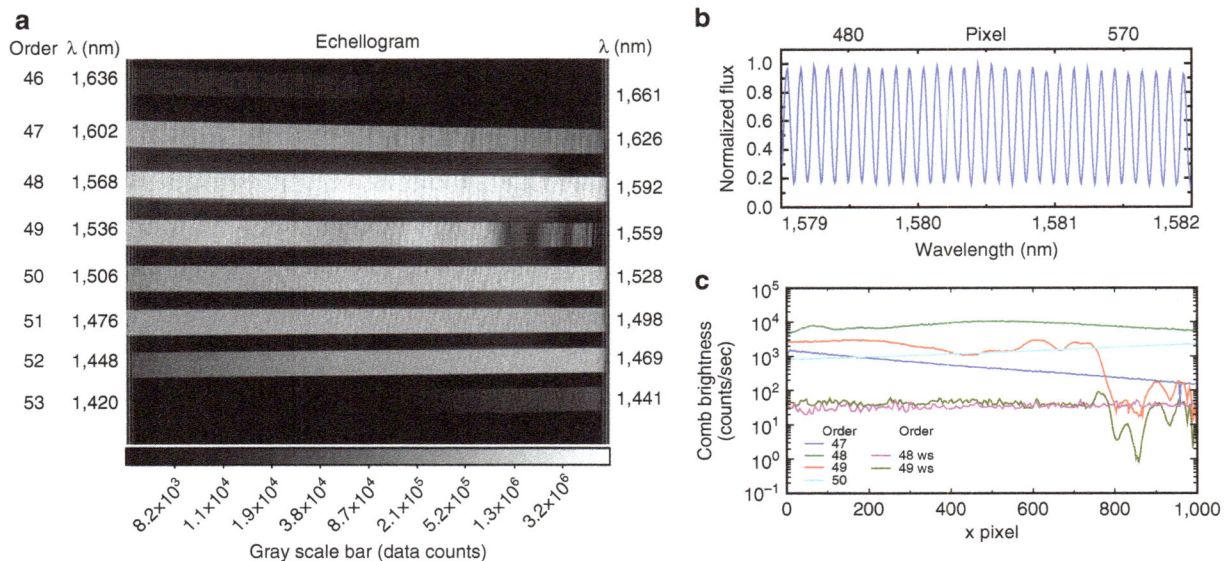

Figure 5 | Data from testing at Keck II. (a) Reduced NIRSPEC image from echelle order 46–53, displaying the stabilized laser comb using the 1,559.9 nm reference laser. Line brightness represents data counts. **(b)** A portion of the extracted comb spectrum from order 48 is plotted versus wavelength. **(c)** Comb brightness envelope of orders 47–50 and orders 48 and 49 when flattened by a waveshaper (ws).

(NIRSPEC) on the Keck-II telescope[40] to demonstrate our laser comb. NIRSPEC is a cross-dispersed echelle capable of covering a large fraction of the entire H-band in a single setting with a spectral resolution of R~25,000. Observations were taken on 18 and 19 May 2015, with the comb set-up in the Keck-II control room in the same configuration as at the IRTF. The apparatus was reassembled after almost 8 months of storage from the time of the IRTF experiment and was fully operational within a few hours. The fibre output from the comb was routed through a cable wrap up to the Nasmyth platform where NIRSPEC is located. We injected the comb signal using a fibre feed into the integrating sphere at the input to the NIRSPEC calibration subsystem. While this arrangement did not allow for simultaneous stellar and comb observations, we were able to measure the comb lines simultaneously with the arc lamps normally used for wavelength calibration and to make hour-long tests of the stability of the NIRSPEC instrument at the sub-pixel level.

Figure 5a shows the laser comb illuminating more than six orders of the high-resolution echellogram. The echelle data were reduced in standard fashion, correcting for dark current and flat-field variations. Under this comb setting, a spectral grasp of ~200 nm is covered, from 1,430 to 1,640 nm. A zoomed-in spectral extraction (Fig. 5b) shows that individual comb lines are well resolved at NIRSPEC's resolution and spaced approximately 4 pixels apart (0.1 nm), consistent with the higher resolution IRTF observations described above. The spectral intensity of the comb lines can be made more uniform with a flattening filter to allow constant illumination over the entire span. In this demonstration, we were also able to implement a programmable optical filter (Waveshaper 1000s) from 1,530 to 1,600 nm, greatly reducing comb intensity variation (Plots 48ws and 49ws in Fig. 5c). If desired, a customized filter could increase the bandwidth of the flattened regime to cover the entire comb span.

We used a series of 600 spectra taken over a ~2 h time period to test the instrumental stability of NIRSPEC. Order 48, which had the highest SNR comb lines, was reduced following a standard procedure to correct for dark current and flat-field variations. Due to the quasi-Littrow configuration of the instrument, the slits appear tilted on the detector and the spectra have some curvature. We performed a spatial rectification using a flat-field image taken with a pinhole slit to mimic a bright

compact object on the spectrum in order to account for this curvature. Wavelength calibration and spectral rectification to account for slit tilting were applied using the Ne, Kr, Ar and Xe arc lamps and the rectification procedure in the REDSPEC software written for NIRSPEC.

Instrumental stability was tested by performing a cross-correlation between the first comb spectrum in the 600 image series and each successive comb spectrum. The peak of the cross-correlation function corresponded to the drift, measured in pixels, between the images. Figure 6a demonstrates the power of the laser comb to provide a wavelength standard for the spectrometer. Over a period of roughly an hour the centroid of each comb line in Order 48 moved by about 0.05 pixel, equivalent to 0.0114 Å. By examining various internal NIRSPEC temperatures it is possible to show that this drift correlates to changes inside the instrument. Figure 6b shows changes in the temperatures measured at five different points within the instrument: the grating mechanism motor, an optical mounting plate, the top of the grating rotator mechanism, the base of the (unused) LN_2 container and the three mirror anastigmat assembly[40]. At these locations the temperatures range from 50 to 75 K and have been standardized to fit onto a single plot: $\Theta_i(t) = (T_i(t) - <T>)/\sigma(T)$. Average values of each temperature are given in Table 1 and show drifts of order 15–35 mK over this 1 h period. In its present configuration NIRSPEC is cooled using a closed cycle refrigerator without active temperature control—only the detector temperature is maintained under closed cycle control to ~1 mK.

Examination of the wavelength and temperature drifts in the two figures reveals an obvious correlation. A simple linear fit of the wavelength drift to the five standardized temperatures reduces the temperature-induced wavelength drifts from 0.05 pixel per hour to a near-constant value with a s.d. of $\sigma = 0.0017$ pixel for a single-comb line (bottom curve in Fig. 6a). While other

Table 1 | Internal NIRSPEC temperatures (K).

Rotator motor	54.944 ± 0.015	Optics plate	52.887 ± 0.023
Top of rotator	74.778 ± 0.035	LN_2 Can	53.663 ± 0.021
TMA	53.866 ± 0.022		

NIRSPEC, near-infrared cryogenic echelle spectrograph; TMA, three mirror anastigmat.

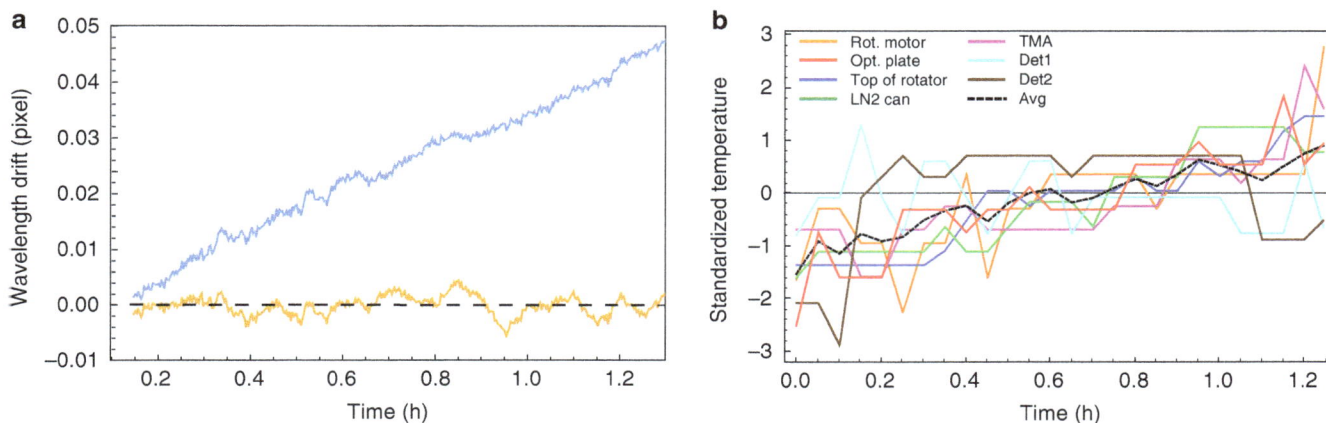

Figure 6 | Measurement of wavelength and temperature drift on the Keck II NIRSPEC spectrometer. (a) The blue curve shows the drift in the pixel location of individual comb lines in order 48 as measured with the cross-correlation techniques described in the text. The yellow curve shows the residual shifts after de-correlating the effects of the internal NIRSPEC temperatures. **(b)** Five internal NIRSPEC temperatures are shown as a function of time. For ease of plotting, the individual temperatures have been standardized with respect to the means and s.d. of each sensor (Table 1). The black dashed curve shows the average of these standardized temperatures. The effect of the quantization of the temperature data at the 10 mK level (as recorded in the available telemetry) is evident in the individual temperature curves.

mechanical effects may manifest themselves in other or longer time series, this small data set indicates the power of the laser comb to stabilize the wavelength scale of the spectrometer. At the present spectral resolution of NIRSPEC, $R \sim 25{,}000$, and with over 240 comb lines in just this one order, we can set a limit on the velocity drift due to drifts within NIRSPEC of $\sim c/R \times \sigma / \sqrt{\# \, lines} < 1.5 \, \mathrm{m \, s^{-1}}$ where c is the speed of light.

Thus, operation with a laser comb covering over 200 nm with more than 2,000 lines in the H-band would allow much higher RV precision than is presently possible using, for example, atmospheric OH lines, as a wavelength standard. NIRSPEC's ultimate RV precision will depend on many factors, including the brightness of the star, NIRSPEC's spectral resolution (presently 25,000 but increasing to 37,500 after a planned upgrade) and the ability to stabilize the input stellar light against pointing drifts and line profile variations. We anticipate that in an exposure of 900 s NIRSPEC should be able to achieve an RV precision $\sim 1 \, \mathrm{m \, s^{-1}}$ for stars brighter than $H = 7 \, \mathrm{mag}$ and $< 3 \, \mathrm{m \, s^{-1}}$ for a stars brighter than $H < 9 \, \mathrm{mag}$. A detailed discussion of the NIRSPEC error budget is beyond the scope of this paper, but a stable wavelength reference, observed simultaneously with the stellar spectrum, is critical to achieving this precision.

Discussion

Many challenges remain to achieving the high precision RV capability needed for the study of exoplanets orbiting late M dwarfs, jitter-prone hotter G and K spectral types, or young stars exhibiting high levels of RV noise in the visible. Achieving adequate signal-to-noise on relatively faint stars requires a large spectral grasp on a high-resolution spectrometer on a large aperture telescope. Injecting both the laser comb and starlight into the spectrograph with a highly stable line spread function demands carefully designed interfaces between the comb light and starlight at the entrance to the spectrograph. Extracting the data from the spectrometer requires careful attention to flat-fielding and other detector features. Finally, reducing the extracted spectra to produce RV measurements at the required level of precision requires sophisticated modelling of complex stellar atmospheres and telluric atmospheric absorption. The research described here addresses only one of these steps, namely the generation of a highly stable wavelength standard in the near IR suitable for sub m s^{-1} RV measurements.

References

1. Perryman, M. *The Exoplanet Handbook* (Cambridge University Press, 2011).
2. Marcy, G. & Howard, A. The astrophysics of planetary systems: formation, structure, and dynamical evolution. in *Proceedings IAU Symposium*. vol. 276, 3–12 (2011).
3. Kasting, J. F., Whitmire, D. P. & Reynolds, R. T. Habitable zones around main sequence stars. *Icarus* **101**, 108–128 (1993).
4. Pepe, F., Ehrenreich, D. & Meyer, M. R. Instrumentation for the detection and characterization of exoplanets. *Nature* **513**, 358–366 (2014).
5. Murphy, M. *et al.* High-precision wavelength calibration of astronomical spectrographs with laser frequency combs. *Mon. Not. R. Astron. Soc.* **380**, 839–847 (2007).
6. Osterman, S. *et al. Optical Engineering + Applications* 66931G–66931G (International Society for Optics and Photonics, 2007).
7. Li, C.-H. *et al.* A laser frequency comb that enables radial velocity measurements with a precision of 1 cm/s. *Nature* **452**, 610–612 (2008).
8. Braje, D., Kirchner, M., Osterman, S., Fortier, T. & Diddams, S. Astronomical spectrograph calibration with broad-spectrum frequency combs. *Eur. Phys. J. D* **48**, 57–66 (2008).
9. Glenday, A. G. *et al.* Operation of a broadband visible-wavelength astro-comb with a high-resolution astrophysi-cal spectrograph. *Optica* **2**, 250–254 (2015).
10. Steinmetz, T. *et al.* Laser frequency combs for astronomical observations. *Science* **321**, 1335–1337 (2008).
11. Yeas, G. G. *et al.* Demonstration of on-sky calibration of astronomical spectra using a 25 Ghz near-IR laser frequency comb. *Opt. Express* **20**, 6631–6643 (2012).
12. Quinlan, F., Yeas, G., Osterman, S. & Diddams, S. A 12.5 Ghz-spaced optical frequency comb spanning > 400 nm for near-infrared astronomical spectrograph calibration. *Rev. Sci. Instrum.* **81**, 063105 (2010).
13. Jones, D. J. *et al.* Carrier-envelope phase control of femtosecond mode-locked lasers and direct optical frequency synthesis. *Science* **288**, 635–639 (2000).
14. Cundiff, S. T. & Ye, J. Colloquium: femtosecond optical frequency combs. *Rev. Mod. Phys.* **75**, 325 (2003).
15. Diddams, S. A. The evolving optical frequency comb [invited]. *JOSA B* **27**, B51–B62 (2010).
16. Wilken, T. *et al.* A spectrograph for exoplanet observations calibrated at the centimetre-per-second level. *Nature* **485**, 611–614 (2012).
17. Wildi, F., Pepe, F., Chazelas, B., Curto, G. L. & Lovis, C. *SPIE Astronomical Telescopes I Instrumentation* 77354X–77354X (International Society for Optics and Photonics, 2010).
18. Halverson, S. *et al.* Development of fiber fabry-perot interferometers as stable near-infrared calibration sources for high resolution spectrographs. *Publ. Astron. Soc. Pac.* **126**, 445–458 (2014).
19. Bauer, F. F., Zechmeister, M. & Reiners, A. Calibrating echelle spectrographs with fabry-perot etalons, Astronomy & Astrophysics. **581**, A117 (2015).
20. Kippenberg, T. J., Holzwarth, R. & Diddams, S. Microresonator-based optical frequency combs. *Science* **332**, 555–559 (2011).
21. Del'Haye, P., Arcizet, O., Schliesser, A., Holzwarth, R. & Kippenberg, T. J. Full stabilization of a microresonator-based optical frequency comb. *Phys. Rev. Lett.* **101**, 053903 (2008).
22. Papp, S. B. *et al.* Microresonator frequency comb optical clock. *Optica* **1**, 10–14 (2014).
23. Suzuki, S. *et al. Nonlinear Optics* NM3A–NM33 (Optical Society of America, 2013).
24. Kotani, T. *et al. SPIE Astronomical Telescopes + Instrumentation* 914714–914714 (International Society for Optics and Photonics, 2014).
25. Imai, K., Kourogi, M. & Ohtsu, M. 30-thz span optical frequency comb generation by self-phase modulation in an optical fiber. *IEEE J. Quantum Electron.* **34**, 54–60 (1998).
26. Fujiwara, M., Kani, J., Suzuki, I. I., Araya, K. & Teshima, M. Flattened optical multicarrier generation of 12.5 Ghz spaced 256 channels based on sinusoidal amplitude and phase hybrid modulation. *Electron. Lett.* **37**, 967–968 (2001).
27. Huang, C.-B., Park, S.-G., Leaird, D. E. & Weiner, A. M. Nonlinearly broadened phase-modulated continuous-wave laser frequency combs characterized using dpsk decoding. *Opt. Express* **16**, 2520–2527 (2008).
28. Morohashi, I. *et al.* Widely repetition-tunable 200 fs pulse source using a mach-zehnder-modulator-based fiat comb generator and dispersion-flattened dispersion-decreasing fiber. *Opt. Lett.* **33**, 1192–1194 (2008).
29. Ishizawa, A. *et al.* Phase-noise characteristics of a 25-ghz-spaced optical frequency comb based on a phase-and intensity-modulated laser. *Opt. Express* **21**, 29186–29194 (2013).
30. Dudley, J. M., Genty, G. & Coen, S. Supercontinuum generation in photonic crystal fiber. *Rev. Mod. Phys.* **78**, 1135 (2006).
31. Mori, K. Supercontinuum lightwave source employing fabry-perot filter for generating optical carriers with high signal-to-noise ratio. *Electron. Lett.* **41**, 975–976 (2005).
32. Ycas, G., Osterman, S. & Diddams, S. Generation of a 660–2100 nm laser frequency comb based on an erbium fiber laser. *Opt. Lett.* **37**, 2199–2201 (2012).
33. Edwards, C. S. *et al.* Absolute frequency measurement of a 1.5-μm acetylene standard by use of a combined frequency chain and femtosecond comb. *Opt. Lett.* **29**, 566–568 (2004).
34. Beha, K. *et al.* Self-referencing a continuous-wave laser with electro-optic modulation. Preprint at arXiv:1507.06344 (2015).
35. Greene, T. P., Tokunaga, A. T., Toomey, D. W. & Carr, J. B. in *Optical Engineering and Photonics in Aerospace Sensing* 313–324 (International Society for Optics and Photonics, 1993).
36. Tokunaga, A. T., Toomey, D. W., Carr, J. B., Hall, D. N. & Epps, H. W. *Astronomy'90, Tucson AZ, 11-16 Feb 90*, 131–143 (International Society for Optics and Photonics, 1990).
37. Plavchan, P. P. *et al. SPIE Optical Engineering + Applications* 88641J–88641J (International Society for Optics and Photonics, 2013).
38. Plavchan, P. P. *et al. SPIE Optical Engineering + Applications* 88640G–88640G (International Society for Optics and Photonics, 2013).
39. Osterman, S. *et al. EPJ Web of Conferences* vol. 16, 02002 (EDP Sciences, 2011).
40. McLean, I. S. *et al. Astronomical Telescopes and Instrumentation* 566–578 (International Society for Optics and Photonics, 1998).

Acknowledgements

Three IRTF nights were donated in September 2014 to integrate and test the laser comb with CSHELL. One of these nights came from IRTF engineering time and the other two came from Peter Plavchan's CSHELL program to observe nearby M dwarfs with the absorption gas cell to obtain precise radial velocities. We are grateful to the leadership of the IRTF, Director Alan Tokunaga and Deputy Director John Rayner, as well as to the daytime and night time staff at the summit for their support. We further thank Jeremy Colson at Wavelength References for his assistance with the molecular-stabilized lasers. On-sky observations were obtained at the Infrared Telescope Facility, which is operated by the University of Hawaii under Cooperative Agreement no. NNX-08AE38A with the National Aeronautics and Space Administration, Science Mission Directorate, Planetary Astronomy Program. Daytime operations at the Keck-II telescope were carried out with the assistance of Sean Adkins and Steve Milner. We greatfully acknowledge the support of the entire Keck summit team in making these tests possible. We recognize and acknowledge the very significant cultural role and reverence that the summit of Mauna Kea has always had within the indigenous Hawaiian community. We are most fortunate to have the opportunity to conduct observations from this mountain. The data presented herein were obtained at the W.M. Keck Observatory, which is operated as a scientific partnership among the California Institute of Technology, the University of California and the National Aeronautics and Space Administration. The Observatory was made possible by the generous financial support of the W.M. Keck Foundation. We also acknowledge support from NIST and the NSF grant AST-1310875. This research was carried out at the Jet Propulsion Laboratory and the California Institute of Technology under a contract with the National Aeronautics and Space Administration and funded through the President's and Director's Fund Program. Copyright 2014 California Institute of Technology. All rights reserved.

Author contributions

X.Y., K.V., J.L., S.D., P.P., S.L., G.V., P.C. and C.B. conceived the experiments. All co-authors designed and performed experiments. X.Y. and K.V. prepared the manuscript with input from all co-authors.

Additional information

Competing financial interests: The authors declare no competing financial interests.

21

Data-driven magnetohydrodynamic modelling of a flux-emerging active region leading to solar eruption

Chaowei Jiang[1,2], S.T. Wu[2], Xuesheng Feng[1] & Qiang Hu[2]

Solar eruptions are well-recognized as major drivers of space weather but what causes them remains an open question. Here we show how an eruption is initiated in a non-potential magnetic flux-emerging region using magnetohydrodynamic modelling driven directly by solar magnetograms. Our model simulates the coronal magnetic field following a long-duration quasi-static evolution to its fast eruption. The field morphology resembles a set of extreme ultraviolet images for the whole process. Study of the magnetic field suggests that in this event, the key transition from the pre-eruptive to eruptive state is due to the establishment of a positive feedback between the upward expansion of internal stressed magnetic arcades of new emergence and an external magnetic reconnection which triggers the eruption. Such a nearly realistic simulation of a solar eruption from origin to onset can provide important insight into its cause, and also has the potential for improving space weather modelling.

[1] SIGMA Weather Group, State Key Laboratory for Space Weather, National Space Science Center, Chinese Academy of Sciences, No.1 Nan-Er-Tiao, Zhong-Guan-Cun, Hai-Dian District, Beijing 100190, China. [2] Center for Space Plasma & Aeronomic Research, The University of Alabama in Huntsville, Huntsville, Alabama 35899, USA. Correspondence and requests for materials should be addressed to C.J. (email: cwjiang@spaceweather.ac.cn) or to X.F. (email: fengx@spaceweather.ac.cn).

Although manifested diversely as flares, eruptive prominences and coronal mass ejections (CMEs), solar eruptions are essentially explosive release of excess magnetic energy of the Sun's corona. Observations show that solar eruptions can occur abruptly after a quasi-static evolution phase of a few hours to even days during which the magnetic free energy is accumulated[1-3]. There has been an intense debate for decades about what causes such catastrophic disruption of the coronal magnetic field. It is not only a fundamental question in astrophysics, but also has unique importance for space weather, in which solar eruptions play a significant role. Over the past 40 years, a variety of models have been proposed to explain the initiation mechanism of solar eruptions[4-8]. Some researchers[9,10] emphasize the importance of ideal magnetohydrodynamic (MHD) instabilities[11], in particular, the unstable nature of the pre-existing magnetic flux rope[12-14], which is a volumetric channel of electric current emerging from the convection zone[15-17] or formed *in situ* in the corona[18]. Others[19-21] stress the primary role of magnetic reconnection[22], and believe that without reconnection the eruptions can never happen even if the magnetic energy is excessively supplied. The theoretical models complemented with numerical MHD simulations[9,23-27] have greatly improved our understanding of those most violent space weather drivers. All of these models are, however, idealized or hypothetical simplification of the realistic case that is much more complex and elusive in observation.

Existing models that attempt to characterize the realistic magnetic environment for studying solar eruptions are mostly restricted to static reconstruction of the near force-free coronal magnetic field[28]. In this category, the mechanism of eruption can only be investigated tentatively because no dynamics is included. Even a time-sequence of magnetic fields reconstructed following the coronal evolution does not reflect its intrinsic dynamics because these magnetic fields are treated as being independent of each other. There are models[29-32] using the reconstructed coronal field immediately preceding eruption (thus the unstable nature of the field has already well-developed) as the initial condition for MHD simulation, which prove to be able to reproduce the fast dynamic phase of the erupting field[30]. However, these kinds of simulations do not self-consistently show how the pre-eruptive field is formed and how the eruption is triggered. Thus such models may not identify the true trigger mechanism.

Here we present a self-consistent MHD simulation of the whole process from the formation to initiation of a coronal eruptive field in a complex multi-polar active region (AR). The event is characterized by a fast magnetic flux emergence of over 2 days leading to an M-class eruptive flare on the 3rd day. Distinct from the aforementioned works, we use a unified MHD model and start it from a very stable state when the coronal field is still near potential (that is, current-free). A 3-day sequential data of surface vector magnetograms are used to drive the coronal magnetic field evolution all the way from its initial potential state to eruption. It is found that the modelled magnetic field evolves stably in the non-eruptive duration of 2 days and becomes unstable at a time instant in good agreement with that of the observed flare eruption. Moreover, the continuously evolving coronal field presents good morphological similarities with the extreme ultraviolet (EUV) emissions. From the simulated magnetic field, we further identify the important role played by magnetic topology changes and magnetic reconnection in leading to the eruption. Detailed analyses are to be presented in the following sections.

Results

Overview of the event. NOAA AR 11283 is one of the very flare-productive ARs in solar cycle 24. From 6 to 8 September 2011, four Geostationary-Operational-Environmental-Satellite (GOES) M- and X-class flares occurred successively in this AR, roughly separated by 20 h between one another[33]. Here we follow the evolution of a flux-emerging region (FER, see Fig. 1) in this AR early from 4 September 2011 (day 1) to the onset of its first flare and CME on 6 September (day 3). In this time period, the AR was passing the central meridian of the solar disk as viewed by the Solar Dynamics Observatory (SDO) spacecraft, thus providing an uninterrupted window for measuring the changes of the photospheric magnetic field by the Helioseismic and Magnetic Imager (HMI)[34] instrument onboard SDO. The basic magnetic configuration of the AR, as shown in Fig. 1, consists of two main polarities, a positive one in the east (P) and a negative one in the west (N). Part of the negative flux also connects to a positive polarity remotely in the northwest (P1). In addition, a global coronal-field extrapolation using the potential-field-source-surface model[35] indicates the probable presence of open flux (field lines extending beyond the corona to interplanetary space) from N. Starting from day 1, evolution of the photospheric

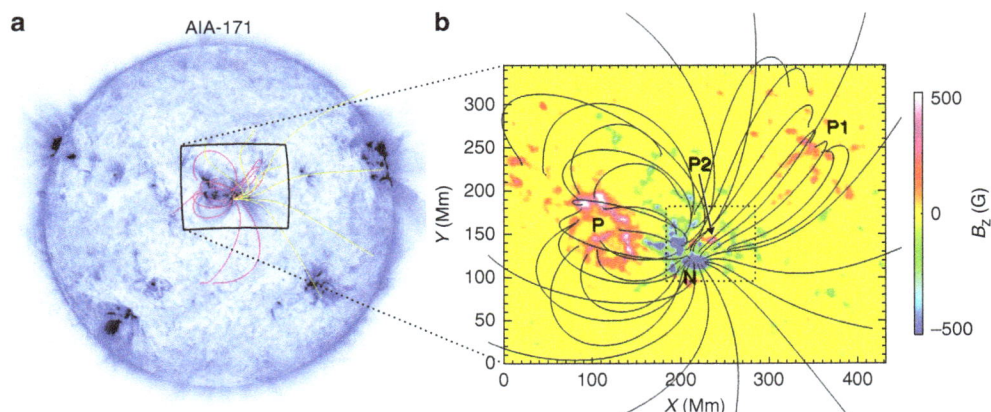

Figure 1 | Location of AR 11283 and its basic magnetic topology. (**a**) A full-disk SDO AIA 171 Å image of the Sun observed near the end of 5 September 2011 (day 2). Overlaid on the image are selected magnetic field lines of global potential-field-source-surface model with the pink (yellow) colour denoting closed (open) flux. (**b**) Magnetic environment associated with the flux-emerging region (FER) which is denoted by the dashed box. The magnetic field lines as shown are calculated by the potential field model in a local Cartesian coordinate system. The background image is the map of photospheric magnetic flux with the main polarities labelled as N, P, P1 and P2, where P2 is the newly emerging one and surrounded by negative flux of N. Temporal evolution of photospheric magnetic field in the FER is shown in Fig. 2. The full simulation volume has a slightly larger size of $460(x) \times 460(y)$ Mm^2 and a vertical extent of $368(z)$ Mm.

magnetic field is dominated by a parasitic positive polarity (P2) emerging into N (Fig. 2 and Supplementary Movie 1). New flux is injected mainly in day 1, then followed by a fast shearing motion of P2 with respect to N. At the beginning of day 1, magnetic configuration of the FER is close to a potential-field state as the electric current crossing the photospheric surface is very small. Also a non-linear force-free reconstruction shows that its free magnetic energy accounts for a tiny fraction of its total magnetic energy[36].

During the first 2 days there is no eruption from the AR as observed by the Atmospheric Imaging Assembly (AIA) telescope onboard SDO. Early on day 3 a major flare occurs (Fig. 3 and Supplementary Movie 2), which starts at 01:35 UT and peaks at 01:50 UT, reaching a magnitude M5.3 as recorded by GOES. Interestingly, the flare emission consists of a quasi-circular ribbon[37] enclosing the newly emerged polarity P2 and two small remote brightening patches outside of the circular ribbon, one at polarity P and the other at P1. A slow CME is initiated immediately after the flare peak time from the AR as observed by the Solar Terrestrial Relations Observatory (STEREO) spacecraft in side views of the Sun. Also there are two filament ejections: the first one starts at the flare peak time from the

southern corner of the circular ribbon, and the second one starts at 02:30 UT from the north around the circular ribbon. The apparent path of these filament ejections is nearly linear without twist or rotation and it co-aligns well with that of the open flux from N, suggesting that the open flux might be involved with the eruption and provides a channel for the escape of the ejecta. There appears to be secondary EUV brightenings in the declining phase of the main flare corresponding to the small bumps of the GOES X-ray flux (for example, at 02:10 UT), while our study will focus on the main flare event.

Data-driven simulation. Our simulation starts from the beginning ($t=0$) of day 1, when the FER is almost current-free. The MHD model is initialized with a potential field extrapolation[38] from the vertical component of the photospheric field (Fig. 1) and a highly tenuous plasma in hydrostatic, isothermal state (with solar gravity) to approximate the coronal low-β plasma condition[39]. Then we drive the model continuously by supplying the bottom boundary with data stream of photospheric vector magnetograms from day 1 to 3. The HMI provides routinely high-quality vector magnetograph data at the photosphere with spatial resolution of 1 arcsec and cadence of 12 m, which is adequate for tracking the relatively long-term (hours to days) evolution of AR magnetic structures from formation to eruption. To ensure the input of boundary vector field self-consistently, we utilize the method of projected characteristics which has its foundation on the wave-decomposition principle of the full MHD system[40]. It has been shown that such method can naturally simulate the transport of magnetic energy and helicity to the corona from below[40,41]. The unit time in the model is set as $\tau=90$ s. By considering that the magnetic evolution at the photosphere is far slower by more than several orders of magnitude than in the corona, we enhance the evolution speed at the bottom boundary of the model by 40 times for the sake of saving the computational time. By this, we assume that 1 h in the HMI data accounts for 1τ in the model. More details of the model can be found in the Method section.

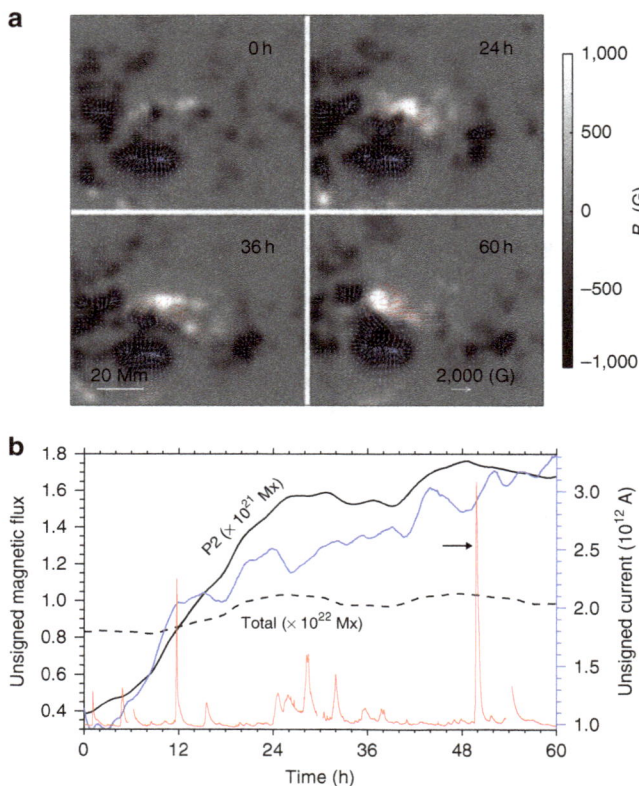

Figure 2 | Magnetic field evolution of the FER at the photosphere.
(**a**) Evolution of magnetic flux distribution B_z and transverse magnetic components (B_x,B_y) shown by the arrows, which are coloured as red (blue) for regions of positive (negative) flux. Transverse field less than 100 G is not shown. The field of view (FoV) for the selected region is displayed in Fig. 1b (dashed box). Time starts from 00:00 UT on 4 September 2011.
(**b**) Evolution of unsigned magnetic flux for the emerging positive polarity P2 (black solid line) and the whole region shown in **a** (black dashed line). The blue line shows unsigned electric current crossing the photospheric surface of P2, which indicates an increase of the non-potentiality of the coronal magnetic field. The GOES soft X-ray (SXR) flux is also shown as the red line. The arrow denotes the M5.3 flare produced by the FER, while the preceding flares recorded are not related with this region.

Energies and magnetic helicity evolution. When monitoring the temporal evolution from $t=0$ to $t=60$ (in unit of τ) for different energies of the MHD system (Fig. 4), we find that its dynamics consist of two distinct phases, a quasi-static evolution phase (from $t=0$ to $t=51$) and an eruption phase (after $t=51$). Furthermore, the onset time $t_c=51$ of the modelled eruption matches that of the observed flare eruption with a lag of $<2\tau$. This suggests that the key transition of dynamics from pre-eruption to eruption is correctly captured by the simulation.

In the first phase from $t=0$ to $t_c=51$, the coronal MHD system evolves stably in response to the changing of the photospheric field. The kinetic energy keeps a rather low value (compared with the magnetic energy) without noticeable variation. On the other hand, there is continuous injection of magnetic energy through the bottom boundary derived from the Poynting flux. Most of this added energy goes to the non-potential energy (that is, the free energy), especially on the second day, when the fast shearing motion of the emerging polarity commences. During this phase, the free magnetic energy, which can be used to power eruptions, is accumulated to an amount close to 10^{32} erg.

From the time $t_c=51$, the kinetic energy begins to rapidly rise resembling an exponential growth, and within a short time interval from $t=51$ to 60, it increases by about 1 order of magnitude. This clearly indicates that the system runs into a loss of quasi-equilibrium, that is, a fast eruptive state, which is confirmed by tracking the evolution of magnetic field configuration (Fig. 5, Supplementary Movies 3 and 4). Note that

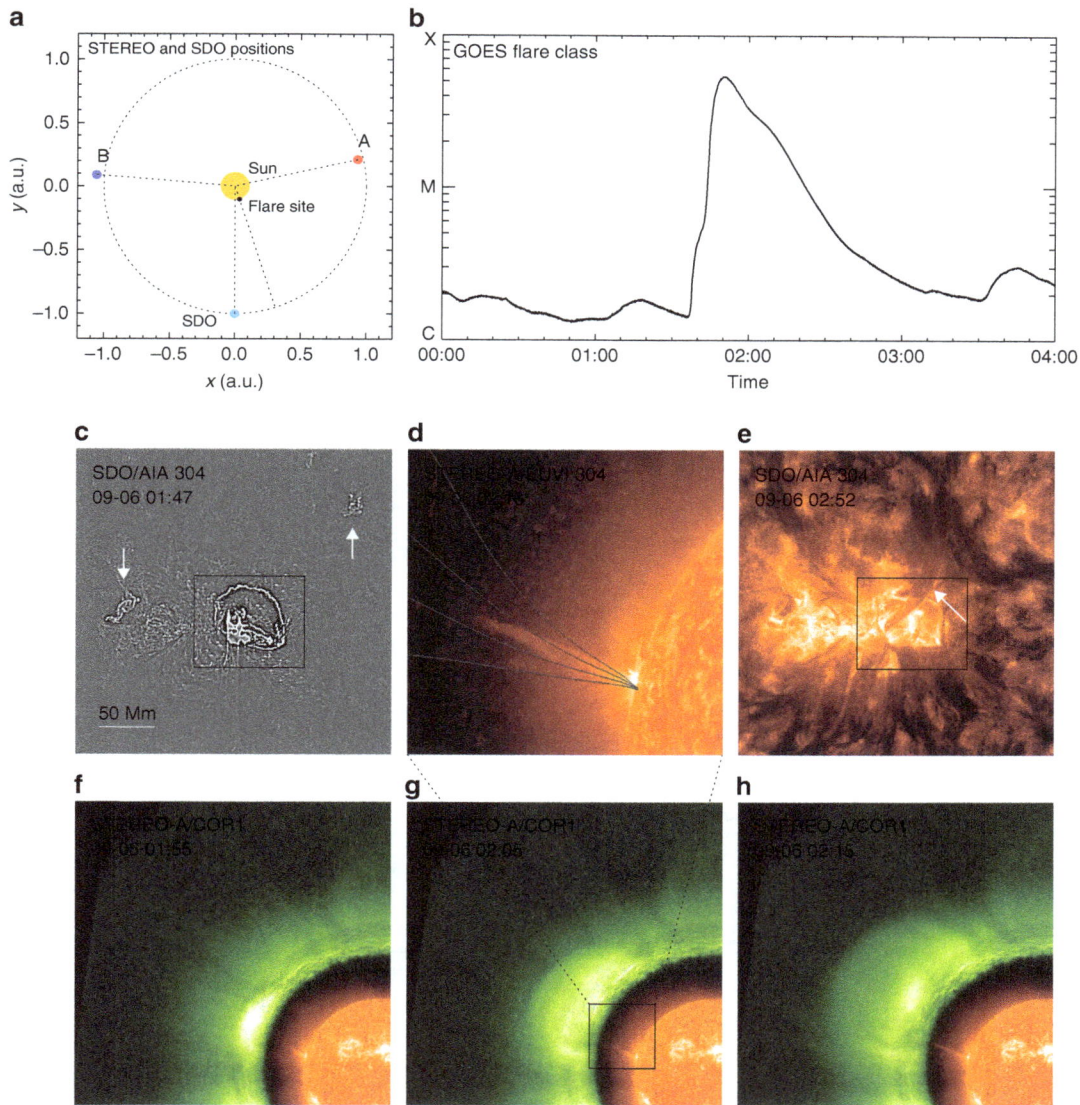

Figure 3 | Observation of the flare and filament ejections leading to CME. (**a**) The positions of the Sun, SDO and STEREO-A/B satellites on 6 September 2011. (**b**) The GOES SXR flux around the flare time with the flare class labelled. (**c**) Enhanced image in SDO/AIA 304 Å channel near the peak time of the M5.3 flare. It shows a central circular flare ribbon and two patches of remote flare brightening (marked by arrows). (**d**) STEREO-A extreme ultraviolet imager (EUVI) 304 Å image of the first filament ejection, which starts at the flare peak time and can also be seen by AIA until 02:20 UT. Overlaid are the open magnetic field lines that are also shown in Fig. 1 but now with the same view angle as STEREO-A. (**e**) Of the same FoV in **c**, AIA observation of the second filament ejection (marked by arrow) from the northwest around the circular ribbon, which can be seen from 2:30 UT to 3:00 UT (Supplementary Movie 2). The boxed regions in **c,e** denote the same FER shown in Fig. 2a. (**f–h**) Combined images of coronagraph (COR1) and EUVI 304 Å observations from STEREO-A showing filament ejection and CME. The boxed region in **g** shows the FoV of **d**.

even through the eruption, the magnetic free energy keeps increasing due to the uninterrupted injection of energy into the volume. In addition, we carried out two experimental runs of the model (Fig. 4b), the first (second) with the photosphere driving ended slightly before (after) t_c. In the first run, the kinetic energy decreases eventually without any sign of eruption, while in the second run it evolves similarly as in the case of full-time driving, indicating that the eruption can only occur with the data driving supplied through the critical time point t_c. In the second run, the magnetic free energy drops as expected during the eruption (Fig. 4d). The released magnetic energy is on the order of 10^{31} erg, which is comparable with the energy budget for M-class flares[42].

Besides the free energy, the relative magnetic helicity is also an important indicator of the non-potentiality of the magnetic field by quantifying the magnetic twist and writhe[43]. Figure 4d shows

that the relative helicity evolves in a similar way as the free energy because of the similar injection of helicity flux through the bottom boundary (Fig. 4e). Some observational and theoretical studies[44,45] indicate that there is a threshold (0.25 ± 0.05) for the ratio of relative helicity to the square of total magnetic flux, and eruption seems to occur only when this threshold is exceeded. The estimated value of this ratio near the eruption is about 0.01, which is far below the aforementioned threshold. Such inconsistence might suggest that here the eruption is not directly driven by magnetic twist (or flux rope), and an analysis of the specific topology is required.

Magnetic topology evolution. In most part of the corona, the plasma is frozen with the magnetic field and so the observed

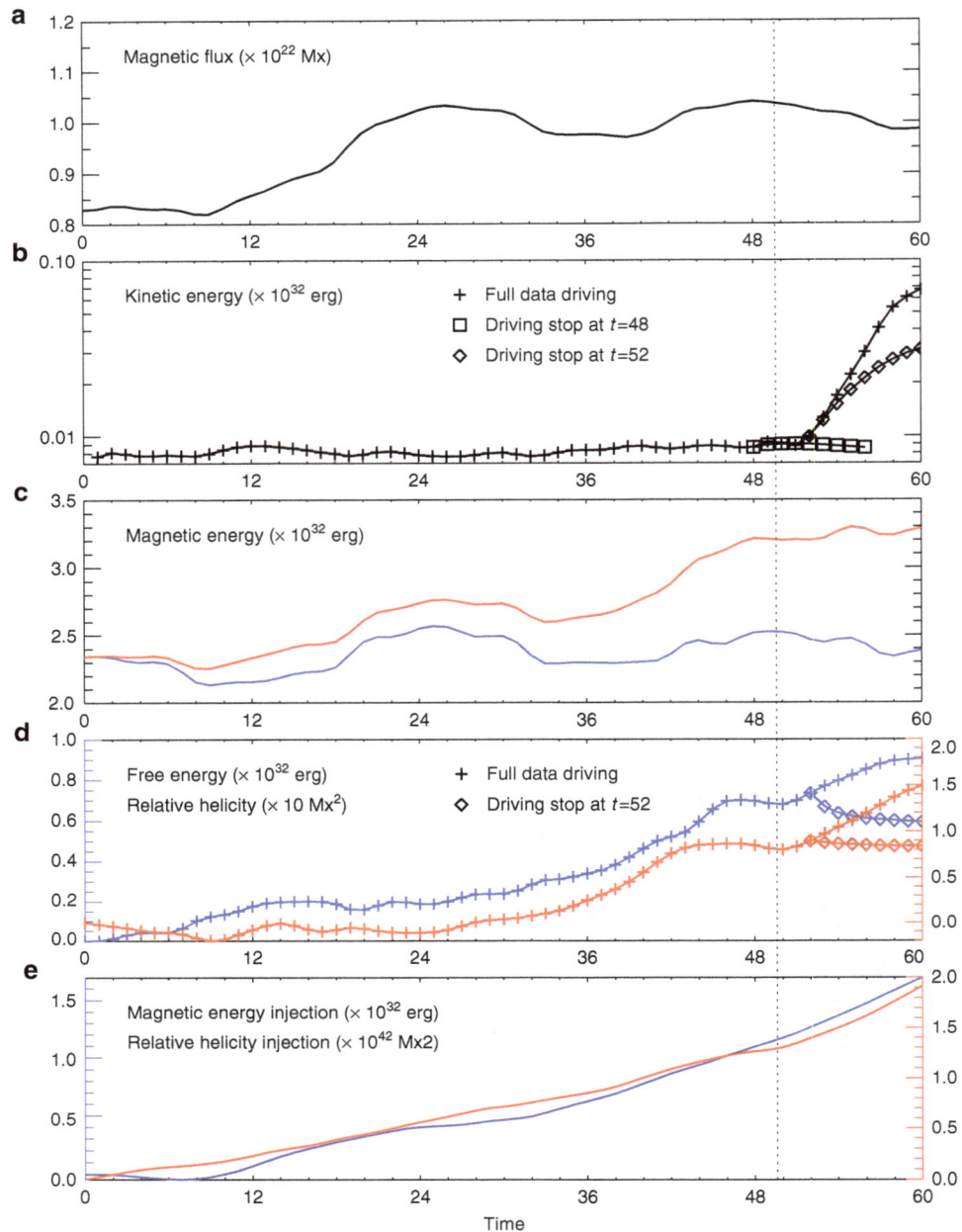

Figure 4 | Temporal evolution of energy and related quantities derived from the data-driven MHD simulation. (**a**) Total unsigned magnetic flux. (**b**) Total kinetic energy. In addition to the run with bottom driving supplied all the time, two experiments are carried out: one (the other) with driving ended at $t = 48$ ($t = 52$), slightly before (after) the eruption onset. (**c**) Magnetic energies derived from the MHD model (red) and potential field model (blue) with the same magnetic flux distribution on the photosphere. (**d**) Free magnetic energy and relative magnetic helicity. (**e**) Estimated amount of injected magnetic energy and helicity through the bottom boundary. All the values are calculated within a sub-volume defined by the FER shown in Fig. 2a with a vertical extent of 100 Mm. The vertical dashed line through the figures denotes the start time of the GOES flare.

filament-like plasma emission outlines well the geometry of the magnetic field lines. Figure 5 and Supplementary Movie 3 show that overall the simulated magnetic configuration and its evolution resemble the AIA images from emergence to eruption. To characterize the magnetic topology, the squashing degree (Q) of the field lines is calculated to locate the important topological structures like separatrices and quasi-separatrix layers (QSLs)[46,47]. By this, we find that the emergence of magnetic polarity P2 results in a topological separatrix like a closed dome separating the new emerging flux from the pre-existing one (Fig. 6). As can be seen, the closed field lines with connection to the newly emerging polarity are encircled within the separatrix,

while outside of it are pre-existing field lines, and with the increasing of the new flux the separatrix expands in both area and height. Quiescent magnetic reconnection should occur at the separatrix for the successive replacement in the corona of the old flux with the new one[48]. Probably as a result of heating by such reconnection, the separatrix location is manifested in the EUV image (AIA 304 Å) as a bright kernel expanding with time (Fig. 5a). Such a distinct evolving feature is usually observed when new flux is emerging into a region of opposite polarity[49–50].

Another important feature of the emerging field is its growing shear. As can be seen along the south part of photospheric polarity inversion line (PIL) separating P2 and N, the

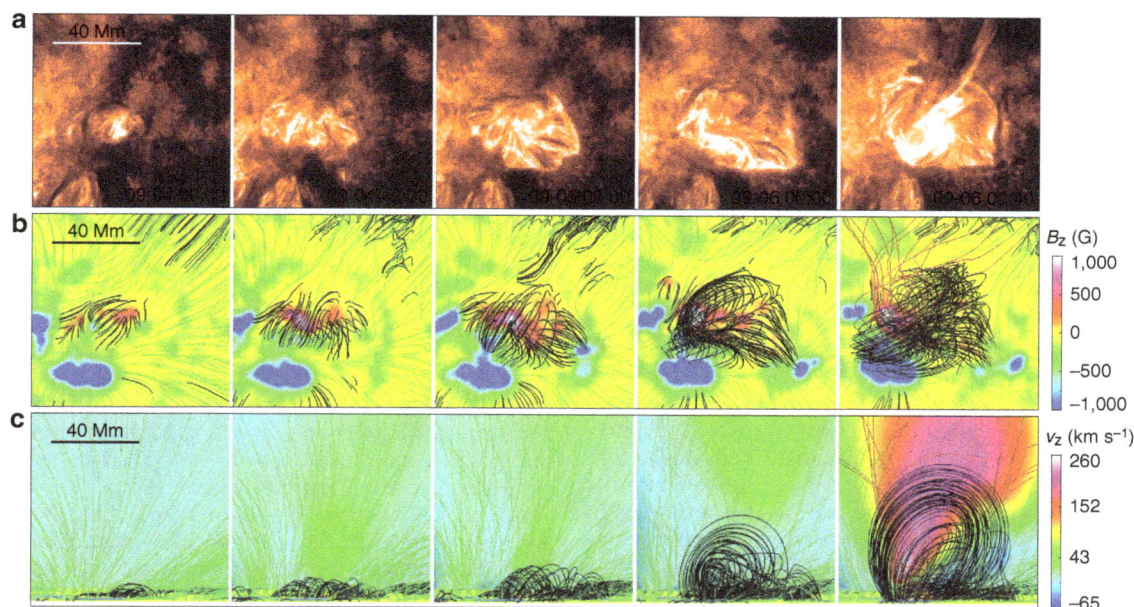

Figure 5 | Comparison of the simulated coronal magnetic field of the FER with the SDO/AIA observations. (**a**) AIA 304 Å images at different times from the initial emergence to the eruption. (**b**) Top view of the corresponding magnetic field evolution at different times ($t = 0, 12, 24, 48$ and 57) from the MHD model. The field lines are traced from footpoints evenly distributed at the bottom surface, which is shown with the photospheric magnetic flux map. Field lines closed (opening) in the box are coloured black (green), while those becoming open from the closed during the eruption are coloured red. (**c**) Side view of the magnetic field lines from south (that is, the horizontal and vertical axes are x and z, respectively). The background shows a 2D central cross-section of the 3D volume and its colour indicates the value of vertical component of velocity.

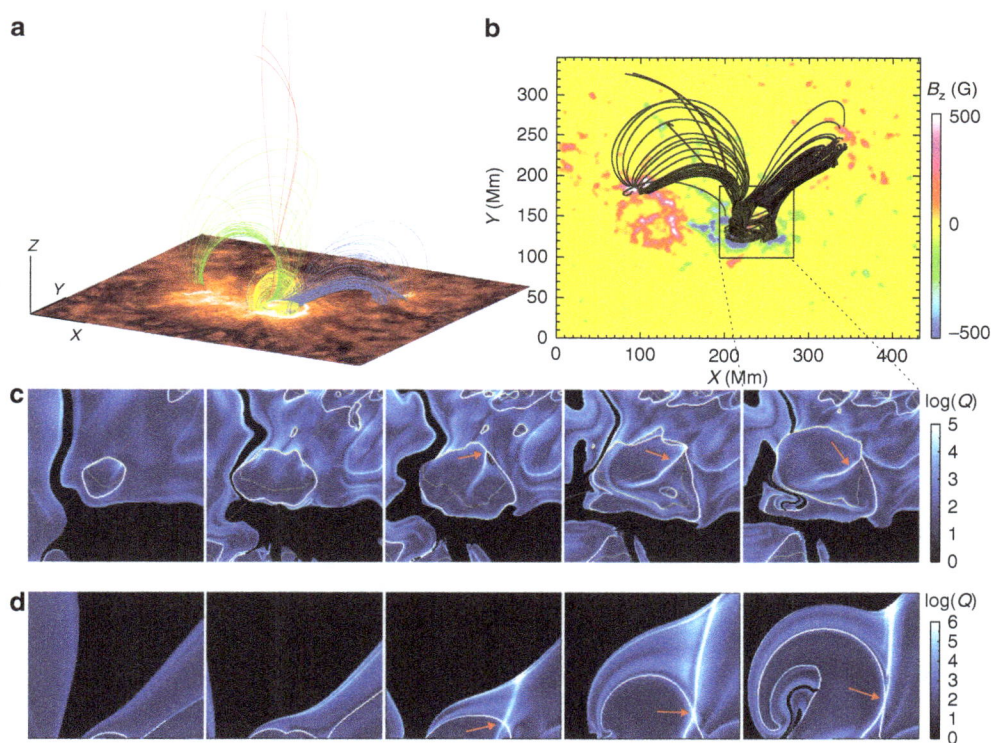

Figure 6 | Magnetic topology evolution. (**a**) Sampled field lines traced from topology separatrix at modelling time $t = 57$ (with different colours denoting different connectivity as shown). They represent the field lines undergoing reconnection during the flare. At the bottom is AIA 304 Å image near the flare peak time to show the flare ribbons. It can be seen that the locations of flare ribbons are matched well by those footpoints of the reconnecting field lines. (**b**) Top view of the field lines with the background image showing the photospheric magnetic flux. Extents shown in **a,b** are identical. (**c**) Magnetic squashing degree $\log(Q)$ at the bottom surface of the FER (FoV is marked in **b** by the box) at different times ($t = 0, 12, 24, 48$ and 57). The separatrix is distinctly revealed by the quasi-circular narrow line with $\log(Q) > 5$. The black regions represent footpoints of open flux. The yellow lines are PILs. The arrows denote the newly-formed QSL. (**d**) Squashing degree maps for vertical cross-sections whose locations are denoted by the oblique lines shown in **c**. The arrows denote the X-like configuration that is formed along with the new QSL.

chromospheric filament threads (and the corresponding magnetic field lines) become more and more co-aligned with the PIL. Such stressing of the magnetic field corresponds to the continuous increasing of its free energy and relative helicity (Fig. 4). Here the shearing process does not produce a fully formed magnetic flux rope in the model. Otherwise there should be a distinct QSL wrapping the rope[51,52], which is not seen in the model (Fig. 7). We further estimate the magnetic twist number of the sheared field, which is found to be lower than a half turn. Thus a rope structure is not yet formed.

With the growing of the newly emerging flux system, part of its edge gets into contact with that of the open flux (Fig. 6c,d, see the changes from $t = 0$ to 12). The Q maps also show that a new QSL is created during the emerging process. Initially the separatrix surface between the emerging flux and pre-existing one is simply a 'bald-patch' type[53], as for the field lines that form the separatrix surface each has one point touching tangentially with the bottom surface. All these points of tangency form a special part of the separatrix at the bottom surface where it coincides with the PIL (see Fig. 8a), and near there its vertical cross-section demonstrates a U shape. After around $t = 24$, the new QSL forms, making the topology surface as a mixed type of a bald-patch and an X-line configuration (see Fig. 8b and the Supplementary Movie 5), of which the vertical cross-section appears as an X-shaped structure similar to the topology at a magnetic null point (Fig. 6d). The emergence of such X-line structure provides a favourable configuration for reconnection. The further development of the shearing of the core field increases magnetic pressure and makes its overlying field expand towards the north in the environment of highly asymmetric magnetic flux distribution. This results in a jet-like configuration (as seen in Fig. 8c), in which reconnection can occur between the newly emerged outer arcade (connecting P2 and N) with the side flux of much longer connection paths to polarities P and P1 and even some open flux. Study of electric current distribution in the model shows that a thin layer of intense current (that is, current sheet) is built up at the X-line slightly before the eruption and grows impulsively, extending to almost the whole separatrix surface during the eruption (Fig. 9a). As a result, the global magnetic topology is fully involved in the reconnection (Fig. 6a,b), which provides a plausible explanation of why there forms the circular flare ribbon and additionally the two remote flaring patches. The linear ejection of filaments around the flare ribbon is most likely a result of the opening of the overlying magnetic field, which is reasonably shown by the

model. As can be seen in Fig. 5 and Supplementary Movies 3 and 4, the field lines whose colour changes from black to red denote the flux becoming open from closed configuration during the eruption. This is also reflected in the map of squashing degree which shows 'holes' corresponding to the open flux cutting into the closed circular separatrix.

Initiation mechanism of eruption. Our modelling results suggest that this eruption is not likely triggered by an unstable flux rope formed prior to the eruption. As mentioned above, the pre-eruptive state is still in sheared-arcade form rather than a well-shaped flux rope. Even if a flux rope exists, it resides far below the critical height for triggering torus instability (Fig. 7), which would occur if the rope axis reaches the height h where the decay index (defined by $n = - d\log(B)/d\log(h)$) of the overlying strapping field B satisfies $n > 1.5$ (ref. 10). The observed features of this eruption are also not consistent with those of flux rope eruptions, for example, the linear shape of the filament ejection does not agree with the eruption of a twisted flux rope, which often demonstrates helical or much more complex structures after being launched[13,54,55]. We also note that it is the filaments around the circular separatrix rather than along the main PIL (that is, the main body of the possible flux rope) that eject. These filaments are activated possibly due to the opening of their overlying flux, and they may further contribute to the eruption.

Based on the analysis of the magnetic topology from the model, the most appropriate mechanism is that the jet-like reconnection triggers the eruption. This is because once the reconnection sets in, naturally a positive feedback is established between the reconnection, which reduces the inward magnetic tension force that confines the flux below, and the consequent outward expansion of the closed arcades, which in turn enhances the reconnection. Such a mechanism is essentially in correspondence with the breakout eruption model[20], and here we demonstrate the magnetic configuration in intrinsic three dimensions (3D)[56]. To characterize how fast this reconnection occurs in our simulation, we locate the current sheet (see Fig. 9a) and estimate its size, as well as the rate of magnetic flux injection into the current sheet (that is, the reconnected magnetic flux). It is found that the rate of reconnection is temporally coupled with the acceleration of the plasma (Fig. 9c), clearly indicating the positive feedback between the reconnection and field expansion.

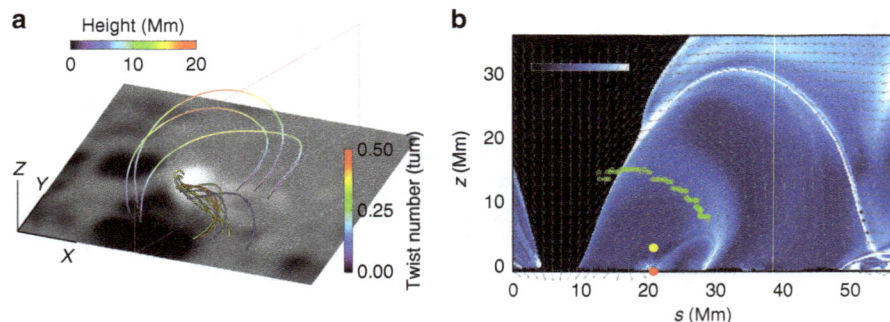

Figure 7 | Magnetic field configuration of the emerging sheared structure at the eruption onset time $t = 52$. (a) Sampled field lines with the low-lying ones as the strongly sheared field, and the overlying ones as its strapping field. Colour of the sheared flux denotes the magnetic twist number of the field lines. Colour of the overlying lines denotes their height. The photospheric magnetogram is shown at the bottom with white dashed line as the PIL. **(b)** A central cross-section of the field whose boundary is denoted by the vertical magenta box in **a**. Its background shows the squashing degree map. The arrows show the direction of magnetic vector components transverse to the cross-section, which form a helical shape centred at the thick yellow dot. Such centre can be regarded as the axis of a magnetic flux rope that may be formed by the twisted field lines. Decay index is computed for a number of paths from the bottom PIL point (the red dot), and a threshold of torus instability is marked by the green diamonds, at which the value of decay index is 1.5.

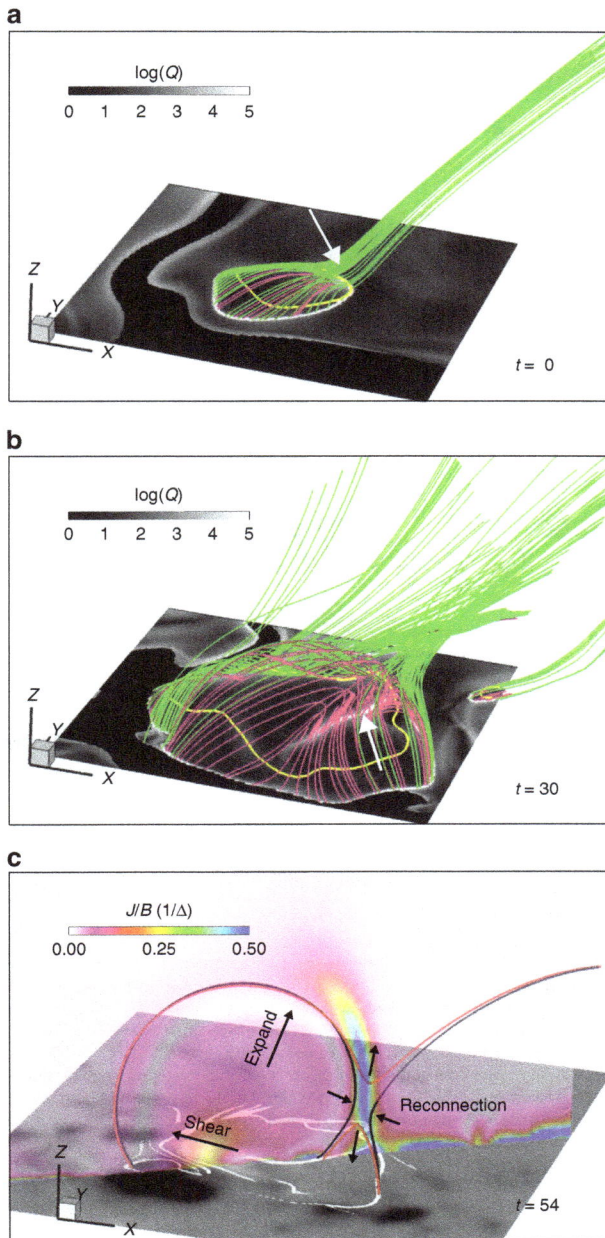

Figure 8 | Formation of a jet-like reconnection structure. (**a**) Local magnetic topology at $t = 0$ for the newly emerging polarity P2. Squashing degree map is shown on the bottom surface, of which the transverse size is 60×60 Mm^2. Magnetic field lines that form the magnetic topology separatrix surface are traced from the circular line with $\log(Q) > 5$. The PIL at the bottom is shown by the yellow line. As indicated by the arrow, these field lines become tangential to the bottom surface at the locations where the PIL coincides with the separatrix. (**b**) Same as **a** but for $t = 30$, when the new QSL (marked by the arrow) has formed. The closed magnetic field within the separatrix is coloured in red. (**c**) Illustration of the jet-like reconnection as eruption trigger mechanism (simulation time at $t = 54$). The transverse size of the bottom surface is 140×120 Mm^2. Field lines in black (red) denote magnetic flux before (after) reconnection. Large arrows denote the bottom shearing and the resulting expansion of the closed arcade. Small arrows indicate the inflow and outflow at the reconnection site. The bottom surface shows the map of B_z overlaid by the white lines showing the trace of separatrix and QSL ($\log(Q) > 5$). The vertical cross-section false-coloured by value of J/B shows distinctly a current sheet at the reconnection site. Note that the reconnecting field lines are not coplanar, thus the configuration is fully 3D.

Questions still arise: what makes such reconnection possible and when is it triggered? First there should be a reconnection-favourable topology and this is fulfilled after the X-line magnetic configuration is formed. A further requirement is the building up of a current sheet so that the resistivity is not negligible there and reconnection might happen. By the stressing of the core field which brings field lines of distinctly different directions close to each other along the X-line, such a thin current layer comes into being there at around $t = 46$ (see Supplementary Movie 6). To finally trigger the reconnection, the profile of magnetic field across the current sheet needs to be steepened sufficiently (that is, the nearly inversely directed magnetic components on both sides of the current sheet are brought to be close enough to each other) for the numerical diffusion to take effect and 'merge' the inverse magnetic field components. By analysing the velocity field near the current layer, we find in the model this reconnection is triggered only after $t_c = 51$, because a clear pattern of reconnection inflow/outflow is not seen before t_c but can be identified shortly afterward. That explains why no eruption occurs when the driving ended at $t = 48$, since the reconnection is not yet triggered. This supports that the eruption can only occur after the reconnection (and feedback) is triggered, and once the feedback is established it can eventually cause the eruption even without further surface driving. Here we note that our interpretation for the triggering of the reconnection is restricted within the context of the present numerical MHD model. The other aspects related to the microscopic processes in space plasmas are beyond the scope of the present work.

Discussion

We have simulated a solar eruption in a realistic and self-consistent way from its origin to onset with a data-driven MHD model. The investigated event consists of a relatively long-duration quasi-equilibrium evolution preceding its eruptive stage of extreme dynamics, and with a single model we are able to calculate the coronal magnetic field evolution for the whole process. The modelled results are supported by the agreement of the magnetic field with EUV images in morphology, the consistency with observation along the timeline from quasi-equilibrium to loss-of-equilibrium, and most importantly, the truly dynamic evolution driven directly by magnetic field data from observation without artificial configuration or constraint.

The modelling offers a reasonable scenario for the eruption. In the background of a multi-polar AR, a small new-flux emergence into the core of the AR leads to the formation of a jet-like configuration that is favourable for reconnection between the newly emerged short arcade and the pre-existing open flux. Meanwhile, the non-potential flux emergence also continuously injects magnetic free energy/helicity into the system due to photospheric shearing motions. Consequently it stresses the field, gradually creating an intense current sheet at the reconnection-favourable site. The system becomes unstable once the reconnection is triggered, since a positive feedback is established between the reconnection and the expansion of the newly emerged arcades. On the other hand, there is no magnetic flux rope fully formed in the modelling, suggesting that a flux rope, although attracting intense interest recently[13,14,31], is not a 'must' for causing a solar eruption. However, 'on-the-fly' flux rope formation might still happen during the eruption, which again, needs reconnection.

In summary, a data-driven MHD modelling like the one shown here, which is able to realistically simulate the whole process from origin to onset of a solar eruption, can be used as a new way for studying the cause of solar eruptions. Furthermore, utilizing the output of such realistic model as the CME initiation input for

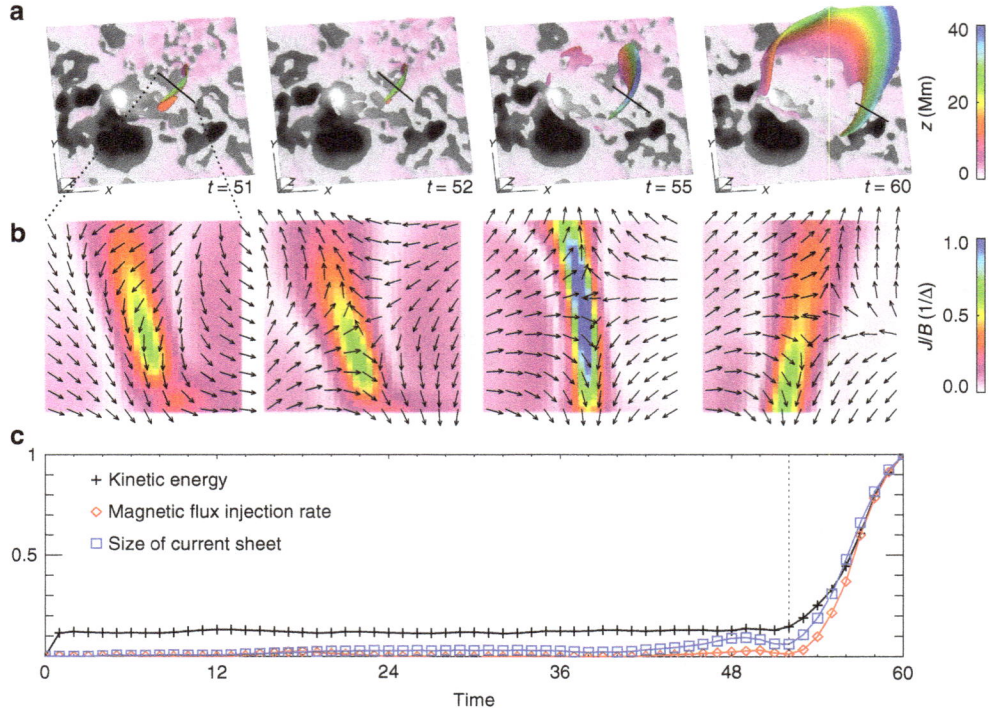

Figure 9 | Current sheet development and magnetic reconnection. (**a**) 3D shape of the current sheet. It is defined as the region with $J/B > 1/(2\Delta)$ (where Δ is the minimal grid size in the model), which consists of intense current layers with width of $\sim \Delta$, thin enough for resistivity to take effect[70]. Its colour denotes the height z from the bottom, and the bottom surface is shown with the photospheric magnetogram of the FER defined in Fig. 2a. (**b**) Flow directions at a vertical cross-section of the current sheet, whose horizontal extent is denoted by the short line in **a**. Reconnection inflow and outflow can be clearly seen after $t = 52$. (**c**) Evolution of the size of the current sheet, magnetic flux injection rate (defined by $\oint_S |\mathbf{B}|\mathbf{v}d\mathbf{S}$ where $\mathbf{v}$ is plasma velocity and $\mathbf{S}$ is the full surface of the current sheet) compared with that of the kinetic energy. All are scaled by their values at $t = 60$.

models of solar storms travelling from the Sun to Earth[57–59] will be, we believe, a step forward in developing sophisticated modelling for space weather.

Methods

MHD model. We numerically solve the full set of time-dependent, 3D MHD equations with the bottom boundary condition driven continuously by the changing photospheric magnetic field from observations. The model does not include the physics of the thin layer (about several Mm) from the photosphere and chromosphere to the transition region. Otherwise it is required to consider the still unknown mechanism of coronal heating to explain how the temperature increases steeply from thousands of degrees to millions. Even more, the ionization degree at the photosphere is extremely low, making the MHD model inappropriate[60]. Instead, we set the bottom boundary of the model at the coronal base (where the temperature is already at a level of 10^6 K) and use the magnetic field measured on the photosphere as a reasonable approximation of the field at the coronal base. The plasma thermodynamics is simplified by an adiabatic energy equation as we focus on the structure and evolution of the coronal magnetic field and its interaction with plasma, which dominates the basic dynamics in the corona. No explicit resistivity is included in the magnetic induction equation, and magnetic reconnection is still allowed due to numerical diffusion if the current sheets are thin enough that their thickness is below the grid resolution[9]. A small kinematic viscosity ν is used with its value corresponding to the viscous diffusion time ($\tau_\nu = L^2/\nu$, where L is the unit length) as $\sim 10^2$ of the Alfvén time ($\tau_A = L/v_A$, where v_A is the Alfvén speed) in the strong-field region. The plasma is initialized as in a hydrostatic, isothermal state $T = 10^6$ K (with sound speed $c_s = 128$ km s^{-1}) with solar gravity. It is configured to make the plasma β as small as 2×10^{-3} (the maximal v_A is 4 Mm s^{-1}) to mimic the coronal low-β (highly tenuous) condition[39]. Here the unit length L is set as 16 arcsec (or 11.5 Mm), double the length of a basic grid block (8 arcsec) used in the model, and the unit time is set as $\tau = L/c_s = 90$ s.

Vector magnetogram data. We use the SDO/HMI observation of the photospheric magnetic field[61]. In particular, the Space-weather HMI Active Region Patches (SHARP) vector magnetogram data product 'hmi.sharp_cea_720 s' (ref. 62) is used to drive the MHD model. With cadence of 12 min and spatial resolution of 1 arcsec, the SHARP data is adequate for simulation of relatively

long-term evolution (hours to days) of eruptive AR magnetic structures from their origin to eruption. Furthermore, the SHARP data includes inverted magnetic field data, projected and re-mapped on the cylindrical equal area (CEA) Cartesian coordinate system centred on the tracked AR, which is well-suited for our simulations performed in the Cartesian coordinate system.

Smoothing is needed when data from observation is involved in a computing scheme based on numerical finite difference. Besides, the lower boundary of the MHD model represents the base of the corona rather than the photosphere and the magnetic structures should be broadened from the photosphere to the coronal base. We simulate such broadening using Gauss smoothing of the data with $\sigma = 2$ arcsec as suggested in ref. 63. We also smooth the data in time with Gaussian window of $\sigma = 4 \times 12$ min to remove short-term temporal oscillations and spikes due to bad pixels (Supplementary Fig. 1 shows comparison of the data before and after being smoothed).

To fully characterize the related magnetic environment for the eruption, we first cut out a large-scale magnetogram (as shown by the full image in Fig. 1b) from a full-disk HMI data observed near the eruption time (at the beginning of day 3) using the same CEA mapping for the SHARP data. This large map is not changed with time as being a fixed background. Then the sub-area of flux emerging (denoted by the dashed box in Fig. 1b) is replaced by the corresponding SHARP data evolving from day 1 to day 3, and finally the combined maps are smoothed. As can be seen in Fig. 2a and the Supplementary Movie 1, we carefully selected this sub-area to avoid significant flux distributions and changes at its borders. The smoothing further mitigates the mismatch of the evolving embedded sub-area and the fixed background.

Numerical scheme and boundary conditions. The model equation is solved using an advanced space-time high-accuracy scheme (AMR–CESE–MHD[64]). The computational volume is sufficiently large to enclose the eruptive region of interest and its surrounding magnetic topology of relevance (see Fig. 1), and at the same time consists of a sufficiently small grid size of $\Delta = 360$ km (equal to 0.5 arcsec on the Sun) matching that of the HMI pixel. This is realized by a non-uniform mesh based on the magnetic flux distribution. The smallest grid is made around the flux-emerging site, where the photospheric field changes most actively. Grid size is increased gradually to 4 arcsec near the side and top boundaries.

When parallelized with a medium number (for example, a hundred) of CPUs (3 GHz), each time-step advancing of the computing code takes about 5 s. We thus face an extremely time-consuming computational task. On the one hand, our model settings require that the time step (that is, the size of iteration step in time) must be smaller than $\Delta/\max(v_A) \approx 0.1$ s due to the Courant–Friedrichs–Lewy stability condition[65]. Accordingly, to update in 1 h of real time needs about 50 h of computing time. On the other hand, the self-consistent modelling of eruption initiation requires us to include the preceding long-term energy buildup process for a time scale of days. Thus, the whole evolution process would require months of computing time. To make the computation manageable, we speed up the cadence of inputting the HMI data by 40 times. This is justified by the fact that the photospheric flow speed in accordance with the photospheric field evolution is about 0.1–1 km s^{-1} (refs 66,67). So in our model settings, the evolution speed of the boundary field, even enhanced by a factor of 40, is still sufficiently small compared with the coronal Alfvén speed ($\sim$ Mm s^{-1}), and the basic reaction of the coronal field to the bottom changes should not be affected in the non-eruptive time duration. As a result, 1 h in the HMI data accounts for 1τ in the simulation. When comparing the simulation with the EUV observations, such scaling also applies to the AIA data in the quasi-static evolution phase, but for the eruptive duration, in principle, time should be scaled according to the ratio of the realistic coronal Alfvén speed to our modelled one. As we have no such data for the real coronal Alfvén speed, we scale the modelling time interval from $t = 49\tau$ to 60τ as being the real 2 h from 01:00 UT to 03:00 UT of day 3, since this gives a reasonable morphological similarity between simulations and observations from AIA for the eruption process.

The continuous input of boundary vector field to the model is implemented by the projected-characteristics method based on the wave-decomposition principle of the full MHD system[40]. The method can naturally simulate the transferring of magnetic energy and helicity to the corona from below[40] by self-consistently calculating the surface flow field[41], which otherwise would have to be derived by local correlation tracking or similar techniques[66,68]. Since the cadence of the input data is 12 min, we linearly interpolate the data in time to produce a data set with cadence matching the time step of the MHD model.

Uncertainty analysis. As being driven directly by data from observations, it is absolutely essential for our modelling that the data are given with good quality and reliability. Here we discuss the possible effects on the modelling results by the known uncertainties and errors of the data.

The SHARP data contains random and systematic errors that may affect our modelling. Estimation of the random errors is included in the data set at each pixel for each magnetic component, that is, s.d. (σ). Conservatively, such uncertainty is as much as 200 G in weak-field regions and as little as 70 G where the field is strong (see Supplementary Fig. 2). Accordingly, we test the performance of our modelling with respect to these uncertainties. Due to the limitation of computational resource, we carried out only two experiments but with the data modified to two extremes (or under two extreme conditions): one (the other) with all the magnetic components plus (minus) their s.d., that is, by σ, then the modified data are smoothed and input into the MHD model as in the original modelling. Undoubtedly, such kind of modifications to the original data can make systematic changes to the modelling, and moreover the effects accumulate during the long-term run. Supplementary Fig. 3 compares the experiment results with the original one for the kinetic and magnetic free energies, which clearly shows quantitative differences between the results. However, the evolution trend from quasi-static to eruptive states is not changed, and in particular, the critical timing of the eruption onset remains accurate with a small uncertainty of $\sim 2\tau$. We also compare the magnetic squashing degree maps derived from the experiment results with the original one in Supplementary Fig. 4. It can be expected that the details of the topology will be changed or its shape will be distorted, since, for example, the PIL is modified in the experiments. In particular, the new-emerging area originally enclosed by the PIL expands if we add σ to the original data, and it shrinks if subtracting σ from the original data. As a result, in the first experiment, it appears that the originally closed separatrix expands and connects to the separatrix in the very weak-field region in the northwest. Nevertheless, the key components constituting the basic topology and their development are still similar to those in the original case. These experiments show that the data uncertainties can quantitatively affect the simulations. However, for the studied event, the main characteristics including the timing of phase transition and the associated dynamic evolution, owing to free energy accumulation and magnetic topology change, remain.

Due to the periodic variation of the SDO orbit, there are daily temporal oscillations of the data that are not removed in the present study. It is estimated[61] that at for the AR strong field (which is of interest in our study), typically such oscillations only cause about $\pm 10 \sim 30$ G change (amounts to ± 1–2%) of the field strength in a period of 24 h. Such systematic error is even smaller than the estimated random error in the strong-field region. Moreover if compared with the significant change of the new-emerging flux (from nearly zero to the order of 10^{21} Mx) in the 2 days for the specific case here, the change by daily oscillations is sufficiently small. However, the impact can still be seen in the results, for example, the line plots in Fig. 4, as manifested by the small-amplitude undulations on top of the overall gradual changes. In future improvement of the model, we will use the data with the daily oscillations removed as reported recently[69].

The HMI data might lose its reliability at the flare time due to anomalous flare-related emissions. To examine the robustness of the model with respect to

such uncertainties, we assume the flare time (from 1:36 to 2:24 UT on day 3) as a data gap and fill the gap by interpolation in time. Supplementary Fig. 5 compares the simulation results driven by this new data set with those by the original data. As can be seen, the change by this data gap is very limited and does not affect our conclusions. This is because such data gap is very close to the simulated eruption onset, and the eruption-favourable magnetic configuration is already formed.

References

1. Wheatland, M. S. The energetics of a flaring solar active region and observed flare statistics. *Astrophys. J.* **679,** 1621–1628 (2008).
2. Savcheva, A. S., McKillop, S. C., McCauley, P. I., Hanson, E. M. & DeLuca, E. E. A new sigmoid catalog from hinode and the solar dynamics observatory: statistical properties and evolutionary histories. *Solar Phys.* **289,** 3297–3311 (2014).
3. McCauley, P. I. et al. Prominence and filament eruptions observed by the solar dynamics observatory: statistical properties, kinematics, and online catalog. *Solar Phys.* **290,** 1703–1740 (2015).
4. Forbes, T. G. et al. CME theory and models. *Space Sci. Rev.* **123,** 251–302 (2006).
5. Shibata, K. & Magara, T. Solar flares: magnetohydrodynamic processes. *Living Rev. Solar Phys.* **8,** 6 (2011).
6. Aulanier, G. in *IAU Symposium,* Vol. 300 (eds Schmieder, B., Malherbe, J.-M. & Wu, S. T.) 184–196 (SAO/NASA Astrophysics Data System, 2014).
7. Janvier, M., Aulanier, G. & Démoulin, P. From coronal observations to MHD simulations, the building blocks for 3D models of solar flares (invited review). *Solar Phys.* **290,** 3425–3456 (2015).
8. Schmieder, B., Aulanier, G. & Vršnak, B. Flare-CME models: an observational perspective (invited review). *Solar Phys.* **290,** 3457–3486 (2015).
9. Török, T. & Kliem, B. Confined and ejective eruptions of kink-unstable flux ropes. *Astrophys. J. Lett.* **630,** L97–L100 (2005).
10. Kliem, B. & Török, T. Torus instability. *Phys. Rev. Lett.* **96,** 255002 (2006).
11. Bateman, G. *MHD Instabilities* 270 (MIT Press, Cambridge, Mass, 1978).
12. Cheng, X., Zhang, J., Ding, M. D., Guo, Y. & Su, J. T. A comparative study of confined and eruptive flares in NOAA AR 10720. *Astrophys. J.* **732,** 87 (2011).
13. Zhang, J., Cheng, X. & Ding, M.-D. Observation of an evolving magnetic flux rope before and during a solar eruption. *Nat. Commun.* **3,** 747 (2012).
14. Wang, H. et al. Witnessing magnetic twist with high-resolution observation from the 1.6-m new solar telescope. *Nat. Commun.* **6,** 7008 (2015).
15. Leka, K. D., Canfield, R. C., McClymont, A. N. & van Driel-Gesztelyi, L. Evidence for current-carrying emerging flux. *Astrophys. J.* **462,** 547 (1996).
16. Okamoto, T. J. et al. Emergence of a helical flux rope under an active region prominence. *Astrophys. J. Lett.* **673,** L215–L218 (2008).
17. Fan, Y. The Emergence of a twisted flux tube into the solar atmosphere, sunspot rotations and the formation of a coronal flux rope. *Astrophys. J.* **697,** 1529–1542 (2009).
18. van Ballegooijen, A. A. & Martens, P. C. H. Formation and eruption of solar prominences. *Astrophys. J.* **343,** 971–984 (1989).
19. Mikic, Z. & Linker, J. A. Disruption of coronal magnetic field arcades. *Astrophys. J.* **430,** 898–912 (1994).
20. Antiochos, S. K., DeVore, C. R. & Klimchuk, J. A. A model for solar coronal mass ejections. *Astrophys. J.* **510,** 485–493 (1999).
21. Moore, R. L., Sterling, A. C., Hudson, H. S. & Lemen, J. R. Onset of the magnetic explosion in solar flares and coronal mass ejections. *Astrophys. J.* **552,** 833–848 (2001).
22. Priest, E. & Forbes, T. *Magnetic Reconnection: MHD Theory and Applications* (Cambridge Univ. Press, 2000).
23. Wu, S. T., Guo, W. P. & Dryer, M. Dynamical evolution of a coronal streamer-flux rope system-II. A self-consistent non-planar magnetohydrodynamic simulation. *Solar Phys.* **170,** 265–282 (1997).
24. Amari, T., Luciani, J. F., Aly, J. J., Mikic, Z. & Linker, J. Coronal mass ejection: initiation, magnetic helicity, and flux ropes. I. boundary motion-driven evolution. *Astrophys. J.* **585,** 1073–1086 (2003).
25. Wu, S. T. et al. Numerical magnetohydrodynamic experiments for testing the physical mechanisms of coronal mass ejections acceleration. *Solar Phys.* **225,** 157–175 (2005).
26. Aulanier, G., Török, T., Démoulin, P. & DeLuca, E. E. Formation of torus-unstable flux ropes and electric currents in erupting sigmoids. *Astrophys. J.* **708,** 314–333 (2010).

27. Roussev, I. I. *et al.* Explaining fast ejections of plasma and exotic x-ray emission from the solar corona. *Nat. Phys.* **8,** 845–849 (2012).

28. Wiegelmann, T. & Sakurai, T. Solar force-free magnetic fields. *Living Rev. Solar Phys.* **9,** 5 (2012).

29. Kliem, B., Su, Y. N., van Ballegooijen, A. A. & DeLuca, E. E. Magnetohydrodynamic modelling of the solar eruption on 2010 April 8. *Astrophys. J.* **779,** 129 (2013).

30. Jiang, C. W., Feng, X. S., Wu, S. T. & Hu, Q. Magnetohydrodynamic simulation of a sigmoid eruption of active region 11283. *Astrophys. J. Lett.* **771,** L30 (2013).

31. Amari, T., Canou, A. & Aly, J. J. Characterizing and predicting the magnetic environment leading to solar eruptions. *Nature* **514,** 465–469 (2014).

32. Inoue, S., Hayashi, K., Magara, T., Choe, G. S. & Park, Y. D. Magnetohydrodynamic simulation of the X2.2 solar flare on 2011 February 15. I. comparison with the observations. *Astrophys. J.* **788,** 182 (2014).

33. Romano, P. *et al.* Recurrent flares in active region NOAA 11283. *Astron. Astrophys.* **582,** A55 (2015).

34. Schou, J. *et al.* Design and ground calibration of the helioseismic and magnetic imager (HMI) instrument on the solar dynamics observatory (SDO). *Solar Phys.* **275,** 229–259 (2012).

35. Schatten, K. H., Wilcox, J. M. & Ness, N. F. A model of interplanetary and coronal magnetic fields. *Solar Phys.* **6,** 442–455 (1969).

36. Jiang, C. W., Wu, S. T., Feng, X. S. & Hu, Q. Formation and eruption of an active region sigmoid. I. a study by nonlinear force-free field modelling. *Astrophys. J.* **780,** 55 (2014).

37. Wang, H. & Liu, C. Circular ribbon flares and homologous jets. *Astrophys. J.* **760,** 101 (2012).

38. Sakurai, T. Computational modelling of magnetic fields in solar active regions. *Space Sci. Rev.* **51,** 11–48 (1989).

39. Gary, G. A. Plasma beta above a solar active region: rethinking the paradigm. *Solar Phys.* **203,** 71–86 (2001).

40. Wu, S. T., Wang, A. H., Liu, Y. & Hoeksema, J. T. Data-driven magnetohydrodynamic model for active region evolution. *Astrophys. J.* **652,** 800–811 (2006).

41. Wang, A. H., Wu, S. T., Liu, Y. & Hathaway, D. Recovering photospheric velocities from vector magnetograms by using a three-dimensional, fully magnetohydrodynamic model. *Astrophys. J. Lett.* **674,** L57–L60 (2008).

42. Saint-Hilaire, P. & Benz, A. O. Thermal and non-thermal energies of solar flares. *Astron. Astrophys.* **435,** 743–752 (2005).

43. Berger, M. A. & Field, G. B. The topological properties of magnetic helicity. *J. Fluid. Mech.* **147,** 133–148 (1984).

44. Démoulin, P. *et al.* What is the source of the magnetic helicity shed by CMEs? the long-term helicity budget of AR 7978. *Astron. Astrophys.* **382,** 650–665 (2002).

45. Jacobs, C., Poedts, S. & van der Holst, B. The effect of the solar wind on CME triggering by magnetic foot point shearing. *Astron. Astrophys.* **450,** 793–803 (2006).

46. Demoulin, P., Henoux, J. C., Priest, E. R. & Mandrini, C. H. Quasi-separatrix layers in solar flares. I. method. *Astron. Astrophys.* **308,** 643–655 (1996).

47. Titov, V. S., Hornig, G. & Démoulin, P. Theory of magnetic connectivity in the solar corona. *J. Geophys. Res.* **107,** 1164 (2002).

48. Tarr, L. A., Longcope, D. W., McKenzie, D. E. & Yoshimura, K. Quiescent reconnection rate between emerging active regions and preexisting field, with associated heating: NOAA AR 11112. *Solar Phys.* **289,** 3331–3349 (2014).

49. Shibata, K. *et al.* Chromospheric anemone jets as evidence of ubiquitous reconnection. *Science* **318,** 1591–1594 (2007).

50. Mandrini, C. H., Schmieder, B., Démoulin, P., Guo, Y. & Cristiani, G. D. Topological analysis of emerging bipole clusters producing violent solar events. *Solar Phys.* **289,** 2041–2071 (2014).

51. Savcheva, A., Pariat, E., van Ballegooijen, A., Aulanier, G. & DeLuca, E. Sigmoidal active region on the sun: comparison of a magnetohydrodynamical simulation and a nonlinear force-free field model. *Astrophys. J.* **750,** 15 (2012).

52. Pariat, E. & Démoulin, P. Estimation of the squashing degree within a three-dimensional domain. *Astron. Astrophys.* **541,** A78 (2012).

53. Titov, V. S., Priest, E. R. & Demoulin, P. Conditions for the appearance of "bald patches" at the solar surface. *Astron. Astrophys.* **276,** 564–570 (1993).

54. Williams, D. R., Török, T., Démoulin, P., van Driel-Gesztelyi, L. & Kliem, B. Eruption of a kink-unstable filament in NOAA active region 10696. *Astrophys. J. Lett.* **628,** L163–L166 (2005).

55. Su, Y. & van Ballegooijen, A. Observations and magnetic field modelling of a solar polar crown prominence. *Astrophys. J.* **757,** 168 (2012).

56. Sun, J. Q. *et al.* Extreme ultraviolet imaging of three-dimensional magnetic reconnection in a solar eruption. *Nat. Commun.* **6,** 7598 (2015).

57. Feng, X. S. *et al.* Three-dimensional solar wind modelling from the Sun to Earth by a SIP-CESE MHD model with a six-componet grid. *Astrophys. J.* **723,** 300–319 (2010).

58. Feng, X. *et al.* Validation of the 3D AMR SIP-CESE solar wind model for four carrington rotations. *Solar Phys.* **279,** 207–229 (2012).

59. Wu, S. T. *et al.* A data-constrained three-dimensional magnetohydrodynamic simulation model for a coronal mass ejection initiation. *J. Geophys. Res.* **121,** 1009–1023 (2016).

60. Vranjes, J., Poedts, S., Pandey, B. P. & de Pontieu, B. Energy flux of Alfvén waves in weakly ionized plasma. *Astron. Astrophys.* **478,** 553–558 (2008).

61. Hoeksema, J. T. *et al.* The helioseismic and magnetic imager (HMI) vector magnetic field pipeline: overview and performance. *Solar Phys.* **289,** 3483–3530 (2014).

62. Bobra, M. G. *et al.* The helioseismic and magnetic imager (HMI) vector magnetic field pipeline, SHARPs—space-weather HMI active region patches. *Solar Phys.* **289,** 3549–3578 (2014).

63. Yamamoto, T. T. & Kusano, K. Preprocessing magnetic fields with chromospheric longitudinal fields. *Astrophys. J.* **752,** 126 (2012).

64. Jiang, C. W., Feng, X. S., Zhang, J. & Zhong, D. K. AMR simulations of magnetohydrodynamic problems by the CESE method in curvilinear coordinates. *Solar Phys.* **267,** 463–491 (2010).

65. Courant, R., Friedrichs, K. & Lewy, H. On the partial difference equations of mathematical physics. *IBM J. Res. Dev.* **11,** 215–234 (1967).

66. Welsch, B. T., Fisher, G. H., Abbett, W. P. & Regnier, S. ILCT: recovering photospheric velocities from magnetograms by combining the induction equation with local correlation tracking. *Astrophys. J.* **610,** 1148–1156 (2004).

67. Liu, Y., Zhao, J. & Schuck, P. W. Horizontal flows in the photosphere and subphotosphere of two active regions. *Solar Phys.* **287,** 279–291 (2013).

68. Schuck, P. W. Tracking vector magnetograms with the magnetic induction equation. *Astrophys. J.* **683,** 1134–1152 (2008).

69. Schuck, P. W., Antiochos, S., Leka, K. D. & Barnes, G. Achieving consistent Doppler measurements from SDO/HMI vector field inversions. Preprint at http://arxiv.org/abs/1511.06500 (2015).

70. Gibson, S. E. & Fan, Y. Coronal prominence structure and dynamics: a magnetic flux rope interpretation. *J. Geophys. Res.* **111,** A12103 (2006).

Acknowledgements

Data from observations are courtesy of NASA SDO/AIA and the HMI science teams. The computation work was carried out on TianHe-1 (A) at the National Supercomputer Center in Tianjin, China. We also thank Dr Murray Dryer and Dr G. Allen Gary for reading the manuscript. This work is supported by 973 program under grant 2012CB825601, the Chinese Academy of Sciences (KZZD-EW-01-4), the National Natural Science Foundation of China (41204126, 41231068, 41274192, 41531073, 41374176, 41574170 and 41574171), the Specialized Research Fund for State Key Laboratories, and Youth Innovation Promotion Association of CAS (2015122). C.W.J., S.T.W. and Q.H. are also supported by NSF-AGS1153323 and AGS1062050. We also acknowledge the support from the International Space Science Institute through an International Team led by Anthony Yeates, Durham University, UK.

Author contributions

C.W.J. developed the model, performed the result analysis and wrote the first draft. S.T.W. contributed to the idea of driving MHD simulation by observed magnetograph data. X.S.F. contributed to the development of the numerical scheme for the model code. All authors participated in discussions and revisions on the manuscript.

Additional information

Competing financial interests: The authors declare no competing financial interests.

22

Multiradionuclide evidence for the solar origin of the cosmic-ray events of AD 774/5 and 993/4

Florian Mekhaldi[1], Raimund Muscheler[1], Florian Adolphi[1], Ala Aldahan[2,3], Jürg Beer[4], Joseph R. McConnell[5], Göran Possnert[6], Michael Sigl[5,7], Anders Svensson[8], Hans-Arno Synal[9], Kees C. Welten[10] & Thomas E. Woodruff[11]

The origin of two large peaks in the atmospheric radiocarbon (^{14}C) concentration at AD 774/5 and 993/4 is still debated. There is consensus, however, that these features can only be explained by an increase in the atmospheric ^{14}C production rate due to an extraterrestrial event. Here we provide evidence that these peaks were most likely produced by extreme solar events, based on several new annually resolved ^{10}Be measurements from both Arctic and Antarctic ice cores. Using ice core ^{36}Cl data in pair with ^{10}Be, we further show that these solar events were characterized by a very hard energy spectrum with high fluxes of solar protons with energy above 100 MeV. These results imply that the larger of the two events (AD 774/5) was at least five times stronger than any instrumentally recorded solar event. Our findings highlight the importance of studying the possibility of severe solar energetic particle events.

[1] Department of Geology—Quaternary Sciences, Lund University, 22362 Lund, Sweden. [2] Department of Geology, United Arab Emirates University, 17551 Al Ain, UAE. [3] Department of Earth Sciences, Uppsala University, 75236 Uppsala, Sweden. [4] Swiss Federal Institute of Aquatic Science and Technology, 8600 Dübendorf, Switzerland. [5] Division of Hydrologic Sciences, Desert Research Institute, Reno, Nevada 89512, USA. [6] Tandem Laboratory, Uppsala University, 75120 Uppsala, Sweden. [7] Laboratory for Radiochemistry and Environmental Chemistry, Paul Scherrer Institut, 5232 Villigen, Switzerland. [8] Center for Ice and Climate, Niels Bohr Institute, University of Copenhagen, 2100 Copenhagen, Denmark. [9] Laboratory of Ion Beam Physics, ETH Zürich, 8093 Zürich, Switzerland. [10] Space Sciences Laboratory, University of California, Berkeley, California 94720, USA. [11] PRIME Laboratory, Purdue University, West Lafayette, Indiana 47907, USA. Correspondence and requests for materials should be addressed to F.M. (email: florian.mekhaldi@geol.lu.se).

The sun irregularly expels large amounts of energetic particles into the interplanetary space and into the vicinity of the Earth which can be observed as so-called solar proton events (SPE). In the context of modern society, this poses a threat to communication, electronic and power systems[1,2]. In addition, SPEs are known to deplete ozone[3,4] and thus possibly affect weather and atmospheric circulation[5]. In consequence, better assessing the relationship between magnitude and occurrence frequency of such events is of substantial importance for solar physics, space technologies, technological infrastructures and climate sciences. The largest known solar flare is considered to be the Carrington event of AD 1859 which is reported to have caused disturbances on telegraph systems and widespread auroral sightings[6]. Typically, SPEs are quantitatively described by their fluence (that is, number of incident particles per cm[2]) of protons with kinetic energies above 30 MeV. It is estimated that the Carrington event was characterized by a fluence ≥ 30 MeV of 1.9×10^{10} protons per cm[2] (ref. 7). However, this estimate of the Carrington solar flare is debated[8].

Even though observational advances in the past decades helped to constrain the Sun's eruptive limits, the record is not long enough to assess the frequency of rare extreme flares as have been observed in solar-type stars[9]. A long-term perspective on frequency, fluence and energy distribution of SPEs can be provided by cosmogenic radionuclides[1] such as beryllium-10, carbon-14 and chlorine-36 (respectively ^{10}Be, ^{14}C and ^{36}Cl) which all arise from the nuclear cascade triggered when cosmic rays reach the atmosphere. Their main production component comes from the incoming galactic cosmic rays as they have, on average, much higher energies than solar particles. The Earth is partially shielded from galactic cosmic rays by the heliomagnetic and the geomagnetic fields, the strengths of which vary from decadal to millennial timescales[10,11]. Nevertheless, large outbursts of solar protons can lead to a rapid increase in the atmospheric production of radionuclides which are subsequently stored in environmental archives such as tree rings and ice cores.

Miyake et al.[12,13] discovered two large natural rapid increases in atmospheric Δ^{14}C (^{14}C/^{12}C ratio corrected for fractionation and decay, relative to a standard) measured in Japanese cedar trees and dated to AD 774/5 and 993/4. The larger of the two increases (AD 775) was characterized by a sharp enhancement in the atmospheric Δ^{14}C of 12‰ over 1 year which corresponds to six times the measurement error and which was estimated to be about 20 times larger than changes attributed to 'ordinary' solar modulation[12]. As a result, the AD 775 and 994 rapid increases in radiocarbon were linked to exceptional cosmic-ray events which have no counterpart in the instrumental records. A number of potential causes for the AD 775 event have been invoked including a gamma-ray burst[14,15], a cometary event[16] or a solar proton event[13,17,18]. Considering that no SPEs including the Carrington solar flare yielded a notable increase in atmospheric radiocarbon concentrations[19], the magnitude of both the hypothesized AD 775 and 994 solar proton events would have been exceptional. Previous studies[18-20] have aimed at estimating the possible fluence of the AD 775 event with results varying by as much as two orders of magnitude. This is partly due to different assumptions concerning the energy spectrum of the incident particles.

Here we include new and annually resolved measurements of ^{10}Be from three ice cores—the North Greenland Ice core Project (NGRIP), the North Greenland Eemian Ice Drilling (NEEM-2011-S1; henceforth NEEM) as well as the Western Antarctic Ice Sheet Divide Core (WDC) in addition to lower-resolved ^{36}Cl data from the Greenland Ice core Project (GRIP)[21]. These radionuclides are produced through different reaction pathways which have different energy dependencies (Fig. 1). This distinguishing feature allows us to better constrain the cause of

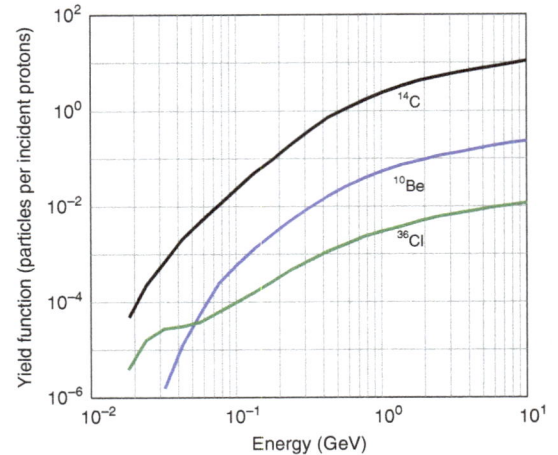

Figure 1 | Yield functions of ^{10}Be,^{14}C, and ^{36}Cl. Globally averaged atmospheric production of each radionuclide per unit flux of incident proton, that is, incoming solar cosmic-ray, as a function of kinetic energy. The yield functions are from refs. 35 and 36.

the events, down to a solar proton event. We further compare the different peaks in production of ^{10}Be and ^{36}Cl to show that the solar proton events in question were both characterized by a very hard fluence spectrum, with high fluxes of protons with kinetic energies ≥ 100 MeV. From this, we can deduce that these events were significantly larger than any solar proton event recorded during the satellite era.

Results
The events shown in ^{10}Be and ^{36}Cl records. The time series of ^{10}Be, ^{14}C and ^{36}Cl which are studied herein are displayed in Fig. 2a–c and Fig. 3a–c where the NGRIP, NEEM and WDC chronologies were adjusted to fit the ^{14}C peaks found by Miyake et al.[12,13] (Methods). The natural background level, that is, the contribution of galactic cosmic rays on the production of radionuclides, was established for each record as the average ^{10}Be and ^{36}Cl flux values as well as the average ^{14}C production rate values prior to and following the peaks. The peaks result from a combination of production and deposition effects leading to an apparent temporal broadening of the measured events. We thus assume that the values exceeding 3σ of the natural background level around the AD 775 (filled areas in Fig. 2a–c) and AD 994 peaks (filled areas in Fig. 3a–c) are related to the two events. The amplitudes which we inferred from the peak areas above these natural background levels are displayed in Fig. 2d–f and Fig. 3d–f where error bars were calculated in order to take in consideration measurement uncertainties as well as a background variability of 1σ (due to noise in the data and the 11-year solar modulation variability).

Our ^{10}Be measurements (Figs 2 and 3, and Table 1) show the existence of the AD 774/5 cosmic-ray event in the Arctic NGRIP and NEEM ice cores as well as in the Antarctic WDC ice core with an average flux enhancement of a factor of 3.4 ± 0.3 (total excess flux related to the average annual background flux). We also report the smaller AD 993/4 event in the ^{10}Be records from NGRIP and NEEM with an average flux enhancement of 1.2 ± 0.2 times the natural background level. The agreement between our stacked ^{10}Be fluxes and modelled ^{14}C production rates (Methods) is comparatively good, especially for the AD 774/5 event which exhibits similar peak amplitudes for both radionuclides. We note, however, that the peak amplitudes in ^{10}Be and ^{14}C differ

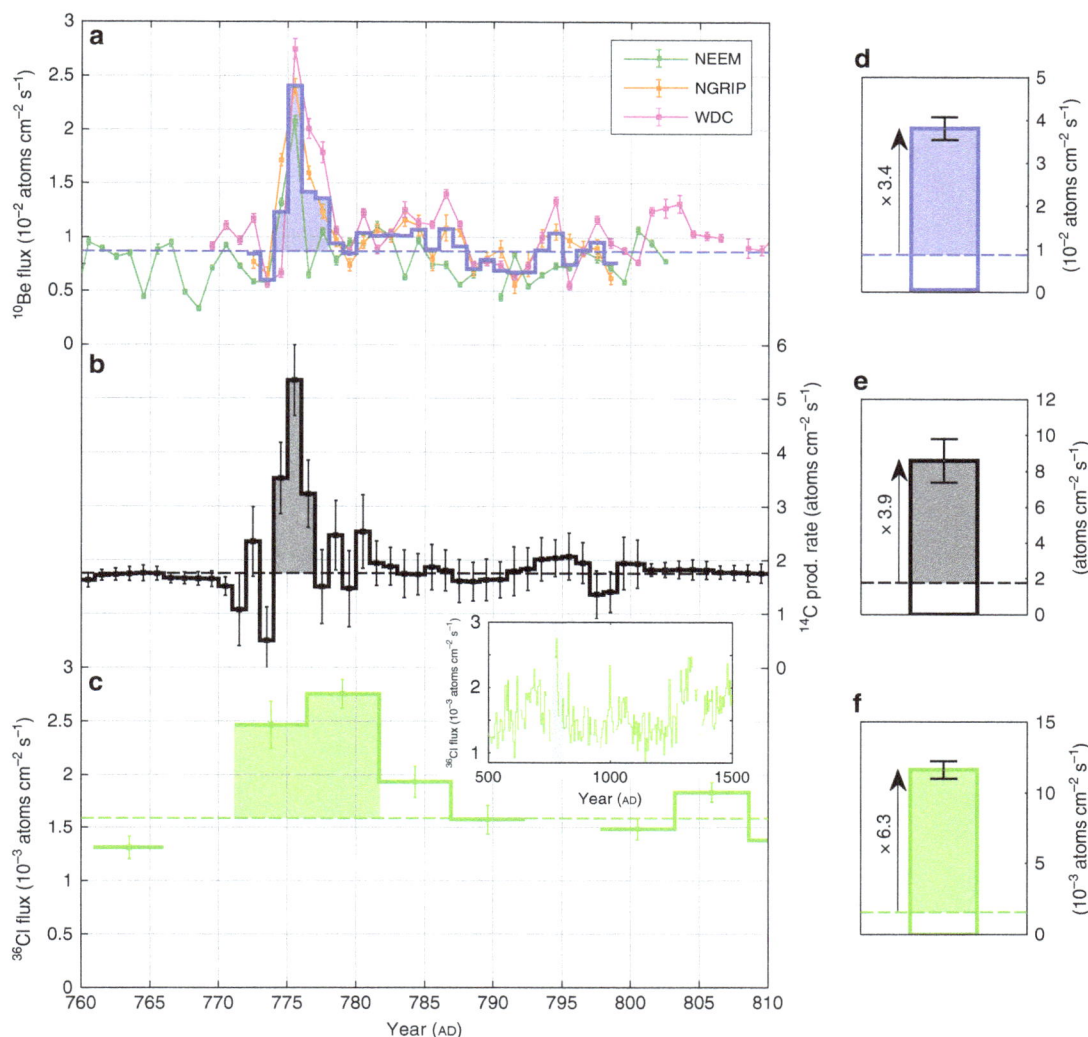

Figure 2 | The AD 774/5 event in view of [10]Be, ^{14}C and [36]Cl. Time series for AD 760–810 (**a**) of [10]Be flux from the NEEM-2011-S1, NGRIP and WDC ice cores in addition to the inferred average [10]Be flux (thick blue curve), (**b**) of modelled ^{14}C production rate based on previously published measurements[12] and (**c**) of [36]Cl flux[21] in addition to an inset with a longer series spanning AD 500–1500 for [36]Cl where the grey rectangle represents the time slice investigated. The dashed lines represent the natural background levels which are set as the average values prior to and following the filled areas. The filled areas represent the estimated production enhancements caused by the cosmic-ray event of AD 774/5. The [10]Be and [36]Cl series have been corrected for a temporal offset between ice-core and tree-ring chronologies (Methods). The right panel shows radionuclide production enhancements caused by the AD 774/5 event in atoms cm^{-2} s^{-1} for 1 year for (**d**) [10]Be, (**e**) ^{14}C and (**f**) [36]Cl. The radionuclide increases are illustrated with arrows corresponding to the ratio between the inferred flux/production enhancements stacked over 1 year (coloured rectangles) and the estimated background levels (white rectangles). Uncertainties are based on error propagation including measurement errors and a background variability of 1σ.

somewhat at AD 993/4 but agree within the margin of errors (Fig. 3d-e, Table 1). The difference likely is due to small uncertainties in the [10]Be measurements (as seen in Fig. 3a with a poorer agreement between the NEEM and NGRIP ice-core data) and/or in the modelled ^{14}C production rates (Methods). Our [10]Be records also indicate that the cosmic-ray event around AD 775 was considerably stronger amounting to a threefold multiple of the AD 994 [10]Be peak, which is consistent with the Δ^{14}C measurements from Miyake et al.[12,13]. The fact that the AD 994 event was weaker leading to a poorer signal-to-noise ratio also could explain the differences between the NGRIP and NEEM [10]Be and the tree ring ^{14}C records for that time period. In general, the agreement with the three [10]Be series around AD 774/5 is very good albeit displaying a slightly different structure of the peak itself. This is likely to be related to the fact that a constant sampling resolution per depth (as was used for NGRIP) translates into a somewhat variable temporal sampling resolution due to fluctuations in the

annual snow accumulation rates and to different sampling in time for the different sites. Our measurements thus provide first unequivocal evidence of a symmetrical production and deposition of [10]Be at both poles (bipolar symmetry) during the AD 774/5 event.

The fact that the [10]Be and ^{14}C increases are imprinted over a time span of 2–3 years despite the probable ephemeral aspect of the cosmic-ray events can be explained by the duration of the transport of the radionuclides from the stratosphere, where they are mostly produced[22,23], to the ground. A smaller fraction of radionuclides also can be produced in the troposphere, which would deposit more rapidly than stratospheric counterparts, given cosmic rays that are sufficiently energetic. The induced complex transport[24,25] of [10]Be, ^{14}C and [36]Cl could thus lead to a temporal broadening of the deposition peaks. Also, deposition fluxes at a specific ice-core site can be differently impacted by the involved scavenging processes (for example, proportion of wet

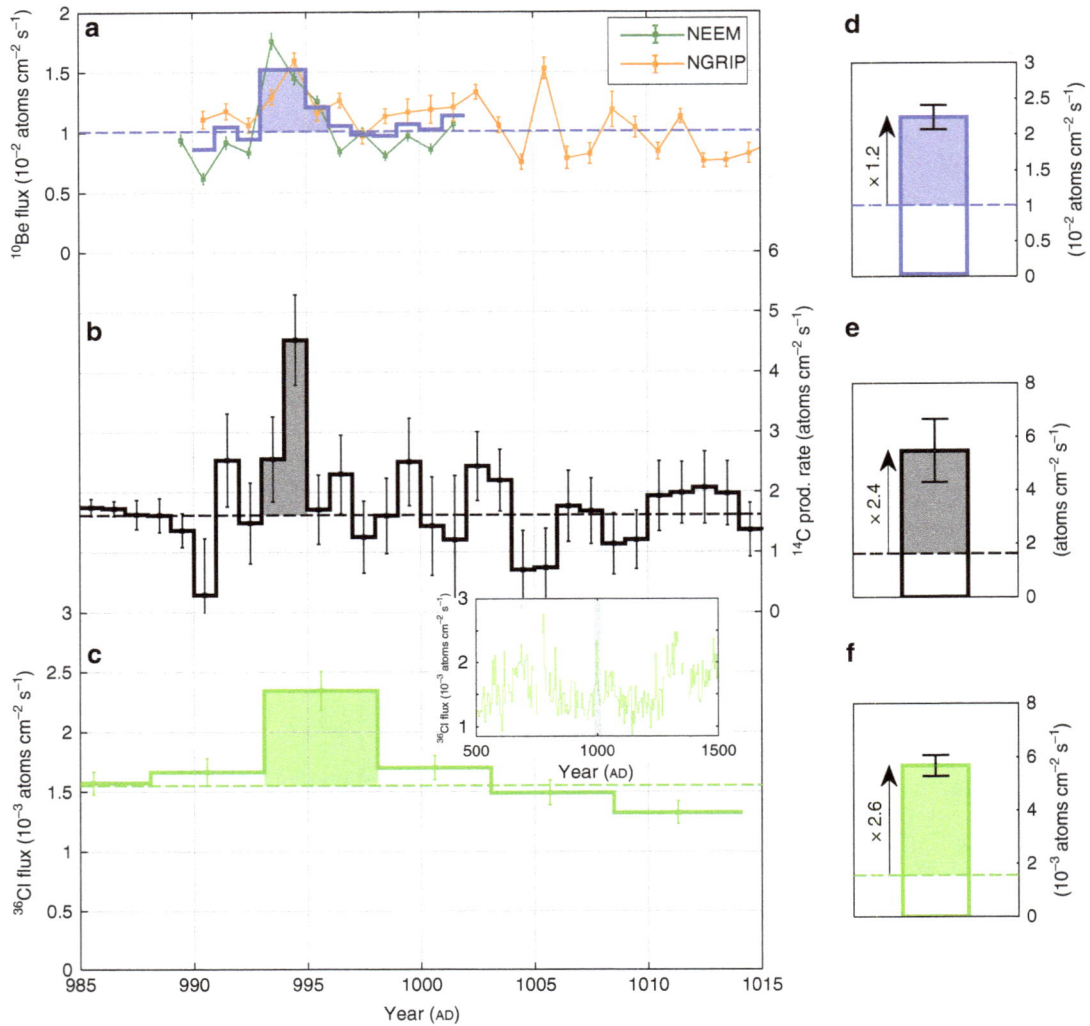

Figure 3 | The AD 993/4 event in view of ^{10}Be, ^{14}C and ^{36}Cl. Time series for AD 985–1015 (**a**) of ^{10}Be flux from the NEEM-2011-S1 and NGRIP ice cores in addition to the inferred average ^{10}Be flux (thick blue curve), (**b**) of modelled ^{14}C production rate based on previously published measurements[13] and (**c**) of ^{36}Cl flux[21] in addition to an inset with a longer series spanning AD 500–1500 for ^{36}Cl where the grey rectangle represents the time slice investigated. The dashed lines represent the natural background level which is set as the average value prior to and following the filled areas. The filled areas represent the estimated production enhancement caused by the cosmic-ray event of AD 993/4. The ^{10}Be and ^{36}Cl series have been corrected for a temporal offset between ice-core and tree-ring chronologies (Methods). The right panel shows radionuclide production enhancements caused by the AD 993/4 event in atoms cm^{-2} s^{-1} for 1 year for (**d**) ^{10}Be, (**e**) ^{14}C and (**f**) ^{36}Cl. The radionuclide increases are illustrated with arrows corresponding to the ratio between the inferred flux/production enhancements stacked over 1 year (coloured rectangles) and the estimated background levels (white rectangles). Uncertainties are based on error propagation including measurement errors and background variability of 1σ.

Table 1 | Summary of results.

	^{10}Be	^{14}C	^{36}Cl
AD 774/5 event			
Peak factor	3.4 ± 0.3	3.9 ± 0.7	6.3 ± 0.4
Flux enhancement (atoms cm^{-2} s^{-1})	$2.94 \pm 0.27 \times 10^{-2}$	6.84 ± 1.16	$10 \pm 0.6 \times 10^{-3}$
AD 993/4 event			
Peak factor	1.2 ± 0.2	2.4 ± 0.7	2.6 ± 0.3
Flux enhancement (atoms cm^{-2} s^{-1})	$1.23 \pm 0.17 \times 10^{-2}$	3.86 ± 1.18	$4.1 \pm 0.4 \times 10^{-3}$

Estimates of the annual production/deposition enhancements of ^{10}Be, ^{14}C and ^{36}Cl caused by the cosmic-ray events of AD 774/5 and 993/4.

and dry deposition) and by atmospheric circulation[26]. We here assume that the relative peak amplitudes at the different sites investigated are not affected by such system effects as supported by the comparable average ^{10}Be flux and similar relative increases at all three locations (Fig. 2a). In addition, ^{36}Cl can be characterized by some mobility in the snowpack[27] as it is deposited to ice caps in the form of gaseous HCl. Although this is not expected to be a major source of uncertainty at high accumulation sites[27], this in turn implies that outgassing and migration of ^{36}Cl to upper snow layers can occur, further broadening the signal. The ^{36}Cl data from the GRIP ice core should consequently be regarded as more uncertain. Nevertheless, there are two large peaks present around AD 775 and 994 which are likely to be attributable to the two cosmic-ray events. The time profile insets in Fig. 2c and Fig. 3c, which span from AD 500 to 1500, emphasize that the two peaks represent the two most conspicuous features of the record during this time slice with flux enhancement factors of 6.3 ± 0.4 and 2.7 ± 0.3 for the AD 774/5 and 993/4 events, respectively. The estimates of the production increases in ^{10}Be, ^{14}C and ^{36}Cl are listed in Table 1. Based on our measured ^{10}Be data, modelled ^{14}C production rates and to the lower-resolved ^{36}Cl data, the production of ^{36}Cl was the most enhanced during the two cosmic-ray events in accordance with the expectations for lower energy particles relative to galactic cosmic rays (Fig. 1).

Supporting a solar origin. It was recently suggested that the radiocarbon peak measured at AD 774/5 was caused by cometary dust from the collision of a bolide into the atmosphere[16]. The authors report a large increase in ^{14}C content in corals from the South China Sea around AD 773. They also note that their measurements are coeval with sightings of a comet and dust event documented in ancient Chinese chronicles. However, it was concluded in other studies that the dimensions needed for a comet to account for this additional injection of radiocarbon would need to be significantly more massive[28,29] than the previous estimates[16]. In consequence, the comet would inevitably have had a considerable and observable impact on the geobiosphere. More problematic for the comet hypothesis is that ^{10}Be and ^{14}C fallouts released from a comet disintegrating in the atmosphere would be, at most, hemispherically redistributed so that the event would only be recorded in either one of the hemispheres[30]. The ^{36}Cl peaks arising from the French nuclear bomb tests which mainly occurred around the 1960s represent a good analogy to this. The related ^{36}Cl fluxes are significantly larger in the southern hemisphere, where the bomb tests had been undertaken[31]. Moreover, the fact that the peaks around AD 774/5 and 993/4 are reported around the globe and in a multitude of radionuclide records[12,13,16,18,30,32,33] in addition to their large amplitude is indicative of an enhanced atmospheric production triggered by an extraordinary influx of cosmic rays in both hemispheres.

It was also suggested that a typical signature of a gamma-ray burst (GRB) on the production of different radionuclides, its 'isotopic footprint', would be a distinct increase in ^{14}C and ^{36}Cl but not in ^{10}Be content[15]. As stated by the authors, the induced secondary neutrons would be at an insufficient energy to initiate spallation reactions on oxygen nuclei and produce detectable amounts of ^{10}Be. Thus, a GRB is inconsistent as a possible astrophysical source for the two events in perspective of our newly obtained ^{10}Be records (Figs 2 and 3). In addition, the two events are rather similar in that they produced abnormal quantities of ^{10}Be, ^{14}C and ^{36}Cl. This leads us to believe that they share the same cause. In purely probabilistic terms, two GRBs striking the Earth within 200 years is unlikely considering the suggested rate of 1 GRB directed at Earth from our galaxy every 125,000 years[34]. Another diagnostic feature is the above-mentioned bipolar symmetry in the production of ^{10}Be, but also in the production of ^{14}C (ref. 30). This implies that the incoming particles must have been affected and redirected by the geomagnetic field and, thus, that they were charged. This rules

out gamma rays (photons) as triggers of the ^{10}Be, ^{14}C and ^{36}Cl peaks at AD 774/5 and 993/4.

Our data, therefore, support the hypothesis that one or several extreme solar proton events are responsible for the radionuclide production peaks measured at AD 774/5 and 993/4 as it is the only option which is in agreement with all available data. Furthermore, the fact that the ^{36}Cl peaks exhibit the largest amplitude mirroring the resonance effect[35] shown in Fig. 1 constitutes further evidence for a solar origin, that is, being caused by solar cosmic rays which generally have lower energies than galactic cosmic rays.

Spectral hardness of the SPEs. The conclusion that one SPE (or a series of SPEs) is responsible for the production increase of ^{10}Be, ^{14}C and ^{36}Cl at AD 774/5 (Fig. 2) is of particular significance because it implies that it must have reached an exceptional magnitude. In fact, no solar phenomena, including the Carrington event, have ever been unequivocally associated with a distinct increase in ^{10}Be concentrations in ice cores. Knowledge of the characteristics of this major solar event, such as its spectral hardness and its fluence ≥ 30 MeV, could consequently help to better estimate the upper limit of the magnitudes of SPEs. The proton fluences of energy ≥ 30 MeV, or F_{30} of an SPE required to yield a given increase in the production rate of a given radio-nuclide is directly bound to the spectral hardness of the SPE (that is, the proportion of high energy protons compared to low energy protons). For instance, Webber et al.[35] have listed the F_{30} of observed SPEs and computed estimates of their impact on the atmospheric production of ^{10}Be and ^{36}Cl. They show that the very hard SPE of February 1956 (SPE56), with a F_{30} of about 1.8×10^9 protons per cm^2 yielded five times more ^{10}Be than the very soft SPE of August 1972 (SPE72) which yet had a F_{30} twice as large. Hence it is crucial to ascertain the spectral hardness of the SPEs around AD 775 and 994 in order to reliably evaluate their F_{30}. To achieve this, one can exploit the different energy sensitivities of the production rates of cosmogenic radionuclides. For instance, the yield functions of ^{10}Be and ^{36}Cl have very different shapes at low energies (Fig. 1). As such, the production of solar-induced ^{36}Cl nuclides is relatively more sensitive to incident protons at about 30 MeV while the production of ^{10}Be nuclides is, compared to ^{36}Cl, more sensitive to solar protons at about 100 MeV (ref. 35). A small ratio of the relative production enhancement of ^{36}Cl relative to ^{10}Be ($^{36}Cl/^{10}Be$) would therefore be expected to be indicative of hard SPEs which are characterized by larger amounts of protons ≥ 100 MeV resulting in a flatter spectrum and vice versa for soft SPEs. As a test, we investigated the relative $^{36}Cl/^{10}Be$ ratios of notable SPEs which occurred between 1956 and 2005, for which the spectral characteristics are known and for which ^{10}Be and ^{36}Cl production yields have been computed[35]. The results are listed in Table 2 while the integral fluence spectra of the related solar proton events are plotted in Fig. 4.

A clear pattern arises from Table 2 with harder SPEs (for example, 1956 and 2005) resulting in lower $^{36}Cl/^{10}Be$ production rate ratios. Inversely, softer SPEs (for example, 1959, 1960, 1972 and 2001) show ratios consistently above 3. By using the different energy dependencies of ^{10}Be and ^{36}Cl and comparing their individual production enhancements, it is thus possible to estimate the spectral shape of paleo-SPEs. From our results, we find the $^{36}Cl/^{10}Be$ ratio to be of about 1.8 ± 0.2 and 2.1 ± 0.4 for the SPEs of AD 774/5 and 993/4, respectively. Despite the uncertainties engendered by the ^{36}Cl record as described earlier, the ratios put the two solar events in the hard spectrum category. By comparing these ratios to Table 2, we find that the best modern analogue would be a very hard spectrum similar to that

Table 2 | Relative $^{36}Cl/^{10}Be$ ratios.

Solar proton event		Relative $^{36}Cl/^{10}Be$ ratio
1	23 February 1956	1.2
2	20 January 2005	1.5
	AD 774/5	1.8 ± 0.2
	AD 993/4	2.1 ± 0.4
3	29 September 1989	2.5
4	29 October 2003	3
5	14 July 2000	3.5
6	19 October 1989	3.6
7	10 July 1959	4
8	12 November 1960	4
9	04 August 1972	6
10	04 November 2001	6

The ratios are based on computations of the annual mean production of ^{10}Be and ^{36}Cl by 10 large solar proton events between 1956 and 2005 (ref. 35). The ratios calculated in this study for the AD 774/5 and 993/4 events are also included. The table is sorted by ascending ratios.

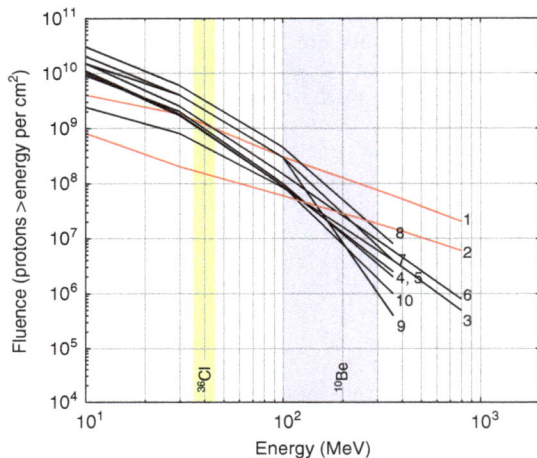

Figure 4 | Event-integrated fluence spectra of recorded SPEs. Integral fluence spectra for 10 notable solar proton events which occurred between 1956 and 2005 (ref. 35). The numbers of the spectra relate to Table 2. The green and blue bands represent the approximate specific peak response energies of ^{36}Cl and ^{10}Be, that is, the incident proton energies at which each radionuclide is mainly produced. The red curves emphasize very hard spectra here defined as leading to a Ground Level Enhancement peak intensity above 1,000% of neutron monitor at sea level on the polar plateau. Modified from ref. 35.

of the SPE of January 2005 (SPE05). This result strengthens SPEs as the cause for the events because it rules out the extremely high F_{30} estimates, based on softer spectra (for example, SPEs of August 1972 and October 1989), required to yield the ^{14}C peak of AD 774/5 and which were associated with moderate to high risks for erythema induced by increased ultraviolet radiation[20].

Discussion

By knowing the approximate spectral hardness of the solar proton events, the specific yield function of ^{10}Be as well as the averaged ^{10}Be flux from the NGRIP, NEEM and WDC ice cores, we can estimate the fluence. More precisely, we applied the multiple of the ^{10}Be increase factor attributable to the SPEs of AD 774/5 and 993/4 (Figs 2d–f and 3d–f and Table 1) relative to that of SPE05 (X_{05}), to the fluence spectrum of the latter (spectrum 2 in Fig. 4).

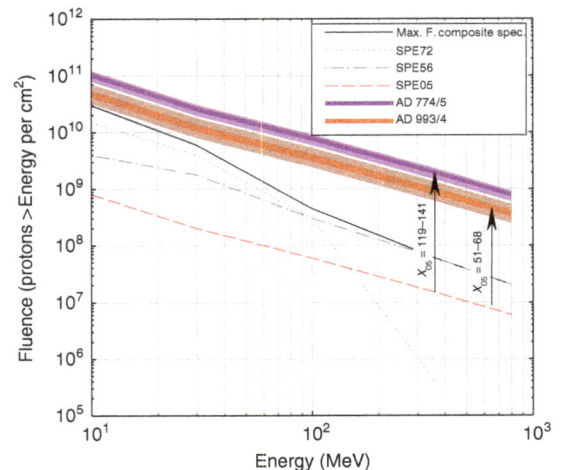

Figure 5 | Event-integrated fluence spectra of the AD 774/5 and 993/4 events. Estimated fluence spectra of the extreme SPEs associated with the AD 774/5 and 993/4 events based on ^{10}Be and ^{14}C and based on the fluence spectrum of SPE05 (red dashed curve) as per ref. 35. The arrows show the scaling factors that were needed to produce the measured ^{10}Be and ^{14}C when assuming a spectral hardness as per SPE05. For perspective, fluence spectra of a very hard (SPE56) and a soft (SPE72) SPE that occurred during the instrumental era are plotted. The black curve represents a composite series of the highest fluences recorded between 1956 and 2005 for protons at $E > 10$, 30, 100, 360 and 800 MeV based on previously published data[35].

Previous computations[35] infer that SPE05 caused an annual ^{10}Be production increase by a factor of about 0.024 under the assumption that complete atmospheric mixing would take place before deposition. This scenario is more realistic than no mixing at all because solar protons mostly would produce ^{10}Be and other radionuclides in the stratosphere which ensures a homogeneous distribution of the cosmogenic radionuclide signature due to a relatively longer mean residence time as opposed to the troposphere[25]. In comparison, our ^{10}Be measurements indicate increase factors of 3.4 and 1.2 (Figs 2d–f and 3d–f) implying a multiple X_{05} of 141 and 51 for the SPEs of AD 774/5 and 993/4, respectively. With a spectral hardness similar to that of SPE05, we therefore find a F_{30} of $2.82 \pm 0.25 \times 10^{10}$ protons per cm^2 for the SPE(s) of AD 774/5 and of $1.02 \pm 0.21 \times 10^{10}$ protons per cm^2 for the SPE(s) of AD 993/4 (Fig. 5). In addition, we performed an alternative estimation by multiplying the specific yield function of ^{14}C (Fig. 1)[36] by a scaled up differential energy spectrum of SPE05 (ref. 37) which could account for the amount of ^{14}C produced during the two events. A multiple X_{05} of 119 and 68 was then needed to reproduce the same amount of radiocarbon than modelled which indicates a F_{30} of $2.38 \pm 0.40 \times 10^{10}$ and $1.36 \pm 0.42 \times 10^{10}$ protons per cm^2 for the SPE(s) of AD 774/5 and 993/4, respectively (Fig. 5). We also performed the same calculation but using a different ^{14}C yield function[38] in order to assess the dependence of our results on the production rate models considered. We found very similar F_{30} estimates of $2.76 \pm 0.46 \times 10^{10}$ protons per cm^2 (AD 774/5) and of $1.56 \pm 0.48 \times 10^{10}$ protons per cm^2 (AD 993/4).

Of course, estimating the fluence ≥ 30 MeV is linked to several assumptions and uncertainties which are difficult to quantify such as the noise inherent to radionuclide data, the true annually resolved ^{36}Cl peaks, the carbon cycle modelling (Methods) as well as the yield functions used. However, the very good agreement between our alternative estimates of the F_{30} for both events based on different and independent methods supports that the assumptions and uncertainties regarding our ^{10}Be fluxes and ^{14}C production rates do not affect our findings significantly. As such, the large amplitudes of the peaks measured in a variety of independent records all are indicative of extreme solar events with fluences ≥ 30 MeV in the order of 10^{10} protons per cm^2 (Fig. 5). Furthermore, the use of two different ^{14}C yield functions[36,38] returned very similar results. In Fig. 5, possible fluence spectra of both events based on the spectrum of SPE05 and on our different estimates of the F_{30} are shown with an envelope representing the uncertainties described above. It emphasizes that the F_{30} of the SPE(s) at AD 774/5 most likely was well above 10^{10} protons per cm^2, even considering the lowest bound of our error margin.

Although our F_{30} for the larger of the two events (AD 774/5) is slightly lower than previously suggested, based on a spectrum assumed to be similar to that of SPE56 (refs 18,20), it remains extremely high and unprecedented. Figure 5 also shows that, for both events, the fluence above any given kinetic energy is much larger than for the soft SPE of August 1972 (dotted curve) and for the hard SPE of February 1956 (dashed dotted curve). The same holds true for a composite spectrum encompassing the highest fluence from the SPEs listed in ref. 35 at kinetic energies above 10, 30, 100, 360 and 800 MeV. In a modern context, such magnitudes would most likely lead to important disruptions of satellite-based technologies and means of communication. In addition, the brief but extensive ozone depletion which would follow[20] could potentially have effects on atmospheric temperatures and circulation.

In conclusion, we have shown that our annually resolved ^{10}Be measurements rule out all suggested sources but, and thereby confirm, a solar cause for the AD 774/5 and 993/4 events. We have also shown that the associated solar proton events were most likely characterized by a very hard spectrum. We conjecture that the strongest event (AD 775) had a fluence above 30 MeV at least five times larger than any observed SPE during the instrumental period between 1956 and 2005. This and previous studies should thus motivate the investigation of similar events in order to better ascertain the occurrence rate of severe solar events.

Methods

^{10}Be data. Beryllium-10 is available at quasi-biennial resolution from the GRIP although the record is presently not continuous[39,40]. Examining the data around AD 775 and 994, it is likely that the events are not yet measured in the record. In addition, ^{10}Be data exist from the Dome Fuji ice core in Antarctica[41] but at a low resolution of 6–15 years. Recently, Miyake et al.[33] performed quasi-annual ^{10}Be measurements on the Dome Fuji ice core though the authors report that the record is affected by multiannual climatic variability. To increase the amount of highly resolved data and decrease the impact of noise on the ^{10}Be data, we conducted new ^{10}Be measurements with annual resolution at depths comprising the two peaks on 3 different ice cores—the NGRIP, the NEEM-2011-S1 as well as the WDC. The NGRIP ice was sampled at a constant spatial resolution of 18.3 cm, resulting in an average temporal resolution of about 1 year. The ^{10}Be nuclides were extracted from the ice samples following the chemical procedure described in Adolphi et al.[42], and measured at the Tandem Laboratory in Uppsala, Sweden. Annual NEEM and WDC ^{10}Be concentrations were determined according to Woodruff et al.[43] and Sigl et al.[44] and measured at the PRIME laboratory in Purdue, USA. Since radionuclides in ice cores are measured as concentrations, deposition fluxes of ^{10}Be were inferred as they might more accurately reflect the production signal by correcting for effects of varying ice accumulation rate[45]. Fluxes also provide a better quantitative estimate of how many atoms were produced and deposited by the AD 774/5 and 993/4 events which can then be compared with computations of the production of ^{10}Be nuclides induced by observed solar proton events.

Timescale adjustments. Age models of ice cores are subject to accumulating uncertainty from the interpretation of ambiguous signatures used for annual layer dating[46], whereas tree ring chronologies have proven calendar-age accuracy[47] (at these timescales). The measurement of the ^{10}Be peaks unravelled a mismatch between ice-core chronologies and tree-ring timescales of 7 years around AD 775 and 994 (ref. 44), which was attributed to a dating bias in previous ice-core chronologies. In consequence, we adjusted the ^{10}Be and ^{36}Cl ice-core records to fit

the tree rings ^{14}C peaks, in agreement with the revised ice-core chronologies proposed for WDC, NEEM and NGRIP[44].

^{14}C production rates. In addition to ^{10}Be, we utilized the tree rings Δ^{14}C data from Miyake et al.[12,13] from which the production rates were derived in order to correct for post-production effects linked to the carbon cycle[48]. Radiocarbon content as retrieved from natural archives does not mirror the atmospheric production rate but rather a damped[49] and time-shifted signal of it due the large active ^{14}C reservoirs in the atmosphere, biosphere and ocean and their interactions. We therefore used a carbon cycle box-diffusion model[50] to rectify this system bias and infer the ^{14}C production signal which can explain the atmospheric ^{14}C concentration, Δ^{14}C (ref. 11). As the model returns normalized production rates, we multiplied the output results by 1.8 atoms cm^{-2} s^{-1}. This corresponds to an average value of different suggestions for the mean global pre-industrial production rate of ^{14}C ranging from 1.6 (ref. 51) to 2.02 atoms cm^{-2} s^{-1} (ref. 23). The applied carbon cycle model is not optimized for short timescales (that is, years). In consequence, it might not accurately reproduce the shape of the ^{14}C increase. Nevertheless, the estimated peak amplitudes should be robust as the model reproduces the integrated longer-term atmospheric ^{14}C signal. In comparison, our estimate for the ^{14}C production rate induced by the AD 774/5 event (6.84 atoms cm^{-2} s^{-1}) is very similar to the findings of Güttler et al.[32] (6.9 atoms cm^{-2} s^{-1}).

^{36}Cl data. Finally, we assessed the existence and amplitude of the peaks in the ^{36}Cl data from the GRIP ice core[21], which has an average uncertainty of 7%, despite its resolution of circa 5 years only. This radionuclide is of particular interest because its production rate is relatively more sensitive to protons with energies below 100 MeV (Fig. 1) compared with other radionuclides. More specifically, its yield function shows an excess for protons with energies around 30 MeV attributed to resonances for ^{36}Cl production arising from the interaction with atmospheric Ar (ref. 35). Similarly to ^{10}Be, deposition fluxes of ^{36}Cl were calculated.

References

1. Schrijver, C. et al. Estimating the frequency of extremely energetic solar events, based on solar, stellar, lunar, and terrestrial records. J. Geophys. Res. 117, A08103 (2012).
2. Shea, M. & Smart, D. Space weather and the ground-level solar proton events of the 23rd solar cycle. Space Sci. Rev. 171, 161–188 (2012).
3. Seppälä, A. et al. Solar proton events of October–November 2003: ozone depletion in the Northern Hemisphere polar winter as seen by GOMOS/Envisat. Geophys. Res. Lett. 31, L19107 (2004).
4. Lopez-Puertas, M. et al. Observation of NOx enhancement and ozone depletion in the Northern and Southern Hemispheres after the October–November 2003 solar proton events. J. Geophys. Res. 110, A09S43 (2005).
5. Calisto, M., Usoskin, I. & Rozanov, E. Influence of a Carrington-like event on the atmospheric chemistry, temperature and dynamics: revised. Environ. Res. Lett. 8, 045010 (2013).
6. Boteler, D. The super storms of August/September 1859 and their effects on the telegraph system. Adv. Space Res. 38, 159–172 (2006).
7. Smart, D., Shea, M. & McCracken, K. The Carrington event: Possible solar proton intensity–time profile. Adv. Space Res. 38, 215–225 (2006).
8. Wolff, E. et al. The Carrington event not observed in most ice core nitrate records. Geophys. Res. Lett. 39, L08503 (2012).
9. Maehara, H. et al. Superflares on solar-type stars. Nature 485, 478–481 (2012).
10. Beer, J. et al. Information on past solar activity and geomagnetism from 10Be in the Camp Century ice core. Nature 331, 675–679 (1988).
11. Muscheler, R. et al. Solar activity during the last 1000yr inferred from radionuclide records. Quat. Sci. Rev. 26, 82–97 (2007).
12. Miyake, F., Nagaya, K., Masuda, K. & Nakamura, T. A signature of cosmic-ray increase in AD 774–775 from tree rings in Japan. Nature 486, 240–242 (2012).
13. Miyake, F., Masuda, K. & Nakamura, T. Another rapid event in the carbon-14 content of tree rings. Nat. Commun. 4, 1748 (2013).
14. Hambaryan, V. & Neuhäuser, R. A Galactic short gamma-ray burst as cause for the ^{14}C peak in AD 774/5. Mon. Not. R. Astron. Soc 430, 32–36 (2013).
15. Pavlov, A. et al. AD 775 pulse of cosmogenic radionuclides production as imprint of a Galactic gamma-ray burst. Mon. Not. R. Astron. Soc 435, 2878–2884 (2013).
16. Liu, Y. et al. Mysterious abrupt carbon-14 increase in coral contributed by a comet. Sci. Rep. 4, 3728 (2014).
17. Melott, A. L. & Thomas, B. C. Causes of an AD 774-775 ^{14}C increase. Nature 491, E1–E2 (2012).
18. Usoskin, I. et al. The AD775 cosmic event revisited: the Sun is to blame. Astron. Astrophys. 552, L3 (2013).
19. Usoskin, I. G. & Kovaltsov, G. A. Occurrence of extreme solar particle events: assessment from historical proxy data. Astron. J. 757, 92 (2012).
20. Thomas, B. C., Melott, A. L., Arkenberg, K. R. & Snyder, B. R. Terrestrial effects of possible astrophysical sources of an AD 774-775 increase in ^{14}C production. Geophys. Res. Lett. 40, 1237–1240 (2013).

21. Wagner, G. *et al.* Reconstruction of the geomagnetic field between 20 and 60 kyr BP from cosmogenic radionuclides in the GRIP ice core. *Nucl. Inst. Meth. Phys. Res.* **172**, 597–604 (2000).

22. Lal, D. & Peters, B. Cosmic ray produced radioactivity on the Earth. *Handb. Phys* **46**, 551–612 (1967).

23. Masarik, J. & Beer, J. Simulation of particle fluxes and cosmogenic nuclide production in the Earth's atmosphere. *J. Geophys. Res.* **104**, 12099–12111 (1999).

24. Stohl, A. *et al.* Stratosphere-troposphere exchange: A review, and what we have learned from STACCATO. *J. Geophys. Res.-Atmos.* **108**, 8516 (2003).

25. Heikkilä, U., Beer, J. & Feichter, J. Meridional transport and deposition of atmospheric [10]Be. *Atmos. Chem. Phys.* **9**, 515–527 (2009).

26. Pedro, J. *et al.* Evidence for climate modulation of the 10Be solar activity proxy. *J. Geophys. Res.-Atmos.* **111**, D21105 (2006).

27. Delmas, R. *et al.* Bomb-test [36]Cl measurements in Vostok snow (Antarctica) and the use of [36]Cl as a dating tool for deep ice cores. *Tellus B* **56**, 492–498 (2004).

28. Overholt, A. C. & Melott, A. L. Cosmogenic nuclide enhancement via deposition from long-period comets as a test of the Younger Dryas impact hypothesis. *Earth Planet. Sci. Lett.* **377**, 55–61 (2013).

29. Usoskin, I. & Kovaltsov, G. A comet could not produce the carbon-14 spike in the 8th century. *Icarus* (2015; in press).

30. Jull, A. *et al.* Excursions in the ^{14}C record at AD 774–775 in tree rings from Russia and America. *Geophys. Res. Lett.* **41**, 3004–3010 (2014).

31. Heikkilä, U. *et al.* [36]Cl bomb peak: comparison of modeled and measured data. *Atmos. Chem. Phys.* **9**, 4145–4156 (2009).

32. Güttler, D. *et al.* Rapid increase in cosmogenic ^{14}C in AD 775 measured in New Zealand kauri trees indicates short-lived increase in ^{14}C production spanning both hemispheres. *Earth Planet. Sci. Lett.* **411**, 290–297 (2015).

33. Miyake, F. *et al.* Cosmic ray event of AD 774-775 shown in quasi-annual [10]Be data from the Antarctic Dome Fuji ice core. *Geophys. Res. Lett.* **42**, 84–89 (2015).

34. Melott, A. L. & Thomas, B. C. Astrophysical ionizing radiation and Earth: a brief review and census of intermittent intense sources. *Astrobiology* **11**, 343–361 (2011).

35. Webber, W., Higbie, P. & McCracken, K. Production of the cosmogenic isotopes ^{3}H, [7]Be, [10]Be, and [36]Cl in the Earth's atmosphere by solar and galactic cosmic rays. *J. Geophys. Res.* **112**, A10106 (2007).

36. Castagnoli, G. & Lal, D. Solar modulation effects in terrestrial production of carbon-14. *Radiocarbon* **22**, 133–158 (1980).

37. Mewaldt, R. *et al.* in Proceedings of the *29th International Cosmic Ray Conference*, 111–114 (Pune, India, 2005).

38. Kovaltsov, G. A., Mishev, A. & Usoskin, I. G. A new model of cosmogenic production of radiocarbon 14 C in the atmosphere. *Earth Planet. Sci. Lett.* **337**, 114–120 (2012).

39. Muscheler, R., Beer, J. & Vonmoos, M. Causes and timing of the 8200yr BP event inferred from the comparison of the GRIP [10]Be and the tree ring Δ^{14}C record. *Quat. Sci. Rev.* **23**, 2101–2111 (2004).

40. Vonmoos, M., Beer, J. & Muscheler, R. Large variations in Holocene solar activity: Constraints from [10]Be in the Greenland Ice Core Project ice core. *J. Geophys. Res.- Space* **111**, A10105 (2006).

41. Horiuchi, K. *et al.* Ice core record of [10]Be over the past millennium from Dome Fuji, Antarctica: a new proxy record of past solar activity and a powerful tool for stratigraphic dating. *Quat. Geochronol.* **3**, 253–261 (2008).

42. Adolphi, F. *et al.* Persistent link between solar activity and Greenland climate during the Last Glacial Maximum. *Nat. Geosci.* **7**, 662–666 (2014).

43. Woodruff, T. E., Welten, K. C., Caffee, M. W. & Nishiizumi, K. Interlaboratory comparison of [10]Be concentrations in two ice cores from Central West Antarctica. *Nucl. Instrum. Methods B* **294**, 77–80 (2013).

44. Sigl, M. *et al.* Timing and climate forcing of volcanic eruptions for the past 2,500 years. *Nature* **523**, 543–549 (2015).

45. Muscheler, R., Beer, J., Wagner, G. & Finkel, R. C. Changes in deep-water formation during the Younger Dryas event inferred from [10]Be and ^{14}C records. *Nature* **408**, 567–570 (2000).

46. Rasmussen, S. O. *et al.* A new Greenland ice core chronology for the last glacial termination. *J. Geophys. Res.* **111**, D06102 (2006).

47. Büntgen, U. *et al.* Extraterrestrial confirmation of tree-ring dating. *Nat. Clim. Change* **4**, 404–405 (2014).

48. Muscheler, R., Beer, J., Kubik, P. W. & Synal, H.-A. Geomagnetic field intensity during the last 60,000 years based on [10]Be and [36]Cl from the Summit ice cores and [14]C. *Quat. Sci. Rev.* **24**, 1849–1860 (2005).

49. Siegenthaler, U., Heimann, M. & Oeschger, H. [14]C variations caused by changes in the global carbon cycle. *Radiocarbon* **22**, 177–191 (1980).

50. Siegenthaler, U. Uptake of excess CO_2 by an outcrop-diffusion model of the ocean. *J. Geophys. Res.* **88**, 3599–3608 (1983).

51. Goslar, T. Absolute production of radiocarbon and the long-term trend of atmospheric radiocarbon. *Radiocarbon* **43**, 743–749 (2001).

Acknowledgements

We thank Anna Sturevik Storm for her help with the chemical preparation of the NGRIP samples for [10]Be measurements. This work was supported by the Swedish Research Council (grant DNR2013-8421 to R.M.). This work was also supported by the U.S. National Science Foundation (NSF) Polar Programs grants including 0839093 and 1142166 to J.R.M. for development of the Antarctic ice-core records and 0909541 and 1204176 to J.R.M. for development of the Arctic ice core records; K.C.W. was funded by NSF grants 0636964 and 0839137; T.W. was funded by NSF grants 0839042 and 0636815. The authors appreciate support of the WAIS Divide Science Coordination Office for collection and distribution of the WAIS Divide ice core; Ice Drilling and Design and Operations for drilling; the National Ice Core Laboratory for curating the core; Raytheon Polar Services for logistics support in Antarctica; and the 109th New York Air National Guard for airlift in Antarctica. NorthGRIP and NEEM are directed and organized by the Center of Ice and Climate at the Niels Bohr Institute and U.S. NSF. It is supported by funding agencies and institutions in Belgium (FNRS-CFB and FWO), Canada (NRCan/GSC), China (CAS), Denmark (SNF, FIST), France (IFRTP, IPEV, CNRS/INSU, CEA and ANR), Germany (AWI), Iceland (RannIs), Japan (MEXT, NIPR), Korea (KOPRI), The Netherlands (NWO/ALW), Sweden (VR, SPRS), Switzerland (SNF), United Kingdom (NERC), and the USA (U.S. NSF Polar Programs).

Author contributions

F.M. performed the analysis in correspondence with R.M., carried out the chemical preparation on the NGRIP samples and wrote the manuscript. R.M. initiated the project and F.A. contributed to the analysis. J.R.M., M.S., K.C.W. and T.E.W. contributed with the [10]Be measurements for NEEM and WDC. A.S. provided with NGRIP material and helped with its sampling. J.B. and H.-A.S provided with [36]Cl data and its analysis. A.A. contributed to the preparation of the NGRIP [10]Be samples while G.P. performed the measurements. All authors were involved in editing the manuscript.

Additional information

23

Laboratory analogue of a supersonic accretion column in a binary star system

J.E. Cross[1], G. Gregori[1], J.M. Foster[1,2], P. Graham[2], J.-M. Bonnet-Bidaud[3], C. Busschaert[4], N. Charpentier[4], C.N. Danson[1,2], H.W. Doyle[1,5], R.P. Drake[6], J. Fyrth[2], E.T. Gumbrell[2], M. Koenig[7,8], C. Krauland[6], C.C. Kuranz[6], B. Loupias[4], C. Michaut[9], M. Mouchet[9], S. Patankar[2], J. Skidmore[2], C. Spindloe[10], E.R. Tubman[11], N. Woolsey[11], R. Yurchak[7] & É. Falize[3,4]

Astrophysical flows exhibit rich behaviour resulting from the interplay of different forms of energy—gravitational, thermal, magnetic and radiative. For magnetic cataclysmic variable stars, material from a late, main sequence star is pulled onto a highly magnetized ($B > 10$ MG) white dwarf. The magnetic field is sufficiently large to direct the flow as an accretion column onto the poles of the white dwarf, a star subclass known as AM Herculis. A stationary radiative shock is expected to form 100–1,000 km above the surface of the white dwarf, far too small to be resolved with current telescopes. Here we report the results of a laboratory experiment showing the evolution of a reverse shock when both ionization and radiative losses are important. We find that the stand-off position of the shock agrees with radiation hydrodynamic simulations and is consistent, when scaled to AM Herculis star systems, with theoretical predictions.

[1] Clarendon Laboratory, University of Oxford, Parks Road, Oxford OX1 3PU, UK. [2] AWE, Aldermaston, Reading, West Berkshire RG7 4PR, UK. [3] Service d'Astrophysique-Laboratoire AIM, CEA/DSM/Irfu, 91191 Gif-sur-Yvette, France. [4] CEA-DAM-DIF, F-91297 Arpajon, France. [5] First Light Fusion Ltd, Unit 10 Oxford Industrial Park, Mead Road, Yarnton Oxfordshire, OX5 1QU, UK. [6] Department of Atmospheric, Oceanic and Space Sciences, University of Michigan, Ann Arbor, Michigan 48109, USA. [7] LULI-CNRS, Ecole Polytechnique, CEA: Université Paris-Saclay; UPMC Univ Paris 06: Sorbonne Universités-F-91128, Palaiseau Cedex, France. [8] Institute for Academic Initiatives, Osaka University, Suita, Osaka 565-0871, Japan. [9] LUTH, Observatoire de Paris, PSL Research University, CNRS, Université Paris Diderot, Sorbonne Paris Cité, 92190 Meudon, France. [10] Target Fabrication Group, Central Laser Facility, Rutherford Appleton Laboratory, Harwell Science and Innovation Campus, Didcot OX11 0QX, UK. [11] York Plasma Institute, Department of Physics, University of York, Heslington, York YO10 5DQ, UK. Correspondence and requests for materials should be addressed to J.E.C. (email: joseph.cross@physics.ox.ac.uk) or to G.G. (email: g.gregori1@physics.ox.ac.uk).

Shocks, waves and jets are important in interstellar and circumstellar regions, and their dynamics can be significantly altered in the presence of radiative losses and magnetic fields[1,2]. Radiative shocks and ionization fronts are examples of such phenomena. By acting as an energy sink, radiative cooling, trapping and ionization modify the shock structure and properties away from that expected in an ideal gas. This is important in star formation[3,4] and in supernovae explosions[5] where a dense shell of material can be formed, with compression much above what would normally be expected from an ideal strong shock.

Laboratory experiments offer an alternative way to study ionizing and radiative shocks, and to probe them in a detailed way that would not be possible in space. High power lasers, such as the Orion Laser Facility, Aldermaston (UK)[6], can produce plasmas of sufficient density, velocity and temperature that are astrophysically relevant[7,8]. This is possible because of the hydrodynamic similarity that can be established between the laboratory and the astrophysical systems, which has been investigated in depth[9-15], and, recently, also include the full combination of magnetohydrodynamics, radiation and quantum effects[16].

One such astrophysical event that shows ionizing and radiative shocks is the white dwarf accretion column in a magnetic cataclysmic variable (MCV) star system—see ref. 17 for a review—where material from a late, main sequence star is pulled off by a highly magnetized white dwarf, a star subclass known as AM Herculis. Instead of the typical accretion disc, the strong magnetic fields ($B > 10$ MG) cause the plasma fluid from the secondary star to follow the magnetic field lines of the white dwarf. When the magnetic field strength is ~ 10–30 MG, bremsstrahlung emission dominates the cooling process[18,19], and, as the flow travels perpendicular to the field, the magnetic field only acts to contain the plasma. The field directs the flow onto the poles of the white dwarf, where it has an impact and a radiative reverse shock is formed, which travels counter to the incoming flow. Theories relating the properties of the star to the shock height[20], and thus the accretion mode[21], have implications for interpreting observational data, such as the ratio of hard to soft X-ray emission[22]. However, the distance of the shock above the white dwarf photosphere is too small for telescopes to resolve.

Here we show a scaled laboratory experiment, building on previously developed platforms[20,23-26], to investigate such astrophysically relevant radiative shocks. An analogue of the astrophysical system is produced and an experimental regime reached such that radiation is beginning to have important effects, (id est, material and radiative energy fluxes are of similar orders of magnitude), which has an impact on the compression across the shock and the shock stand-off distance. Indeed, our measured shock compression ratio is higher than for classical hydrodynamic shocks. We also observe the shock stand-off distance, when scaled to MCVs, is in very good agreement with the expected position of the reverse shock from the white dwarf surface. In future, this platform could be extended to larger laser systems to study hydrodynamics in radiation-dominated environments.

Results

X-ray radiographs. The experimental setup is illustrated in Fig. 1 (see also Methods). Figure 2a shows a comparison of simulated and experimental X-ray radiographs of the plasma having an impact on the obstacle. A radiation-hydrodynamic numerical simulation was performed using a suite of different codes (see Methods) and postprocessed to generate two different synthetic X-ray radiographs, one including the opacity of all the target components, id est, the standard case, and another with

the opacity of the plastic tube artificially set to zero to mimic signal seen through the viewing apertures, cut in the experimental target tube. As shown in Fig. 1, the plastic tube in the experiment had slits cut through it, near the obstacle, to improve contrast by allowing X-rays to go through only the plasma and not to be absorbed by the tube wall. The experimental and simulated X-ray radiographs shown in Fig. 2 are quantitatively similar not only in the morphology of the flow structure—where the position of the reverse shock front is nearly identical—but also in the X-ray transmission values, which are well matched. We can see in the experimental image that the slits were not exactly aligned along the X-ray diagnostic direction by the step down in transmission $\lesssim 2{,}250 \, \mu m$ and again at $\lesssim 2{,}200 \, \mu m$. The transmission values (Fig. 2b) from the synthetic radiograph without the tube wall opacity are consistent with the experimental data near the reverse shock (that is, between 2,300 and 3,000 μm), whereas the simulated transmission with the tube wall included is closer to the data for distances $\lesssim 2{,}300 \, \mu m$. The jump in the simulated transmission around 2,100 μm, which is not seen in the experiment, results from the ablator plasma (which is more transparent to X-rays) immediately following the pusher plasma. It is also worth noting that the experimental results are reproducible: a later shot at the same time as Fig. 2 also shows the same features and shock stand-off distance (see Supplementary Fig. 1 and Supplementary Discussion: Radiograph comparison).

Optical self-emission. Figure 3 shows an image of the optical self-emission streaked in time. Time $t = 0$ ns starts when the drive lasers illuminate the target foils and increases moving up the image. The drive lasers are incident on the foil at 0 μm and distance increases from right to left. The experimental image indicates that the velocity of plasma flowing down the tube is $70 \pm 10 \, km \, s^{-1}$ in the laboratory frame, which agrees with the predicted velocity from numerical simulations. As the flow reaches the obstacle, there is an initial build-up of material, which eventually steepens into a shock at $t \sim 40$ ns. The shock position in the laboratory frame changes very slowly for $t \lesssim 55$ ns and

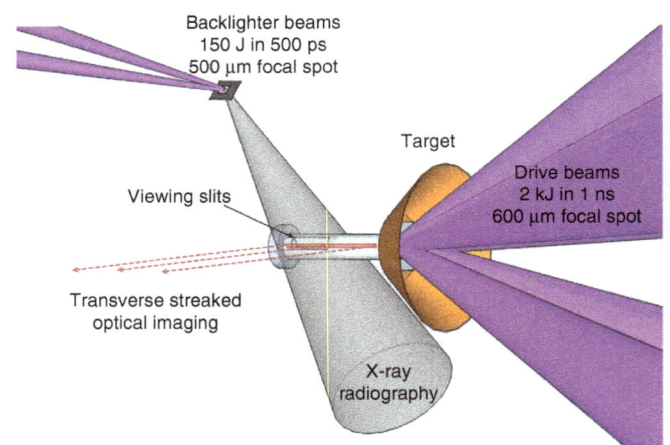

Figure 1 | Schematic of the experimental set-up. The laser beams from the Orion facility are arranged in two cones, which we refer to as 'drive' and 'backlighter'. The drive cone consisted of five beams, which illuminated the target in a 100° cone angle, with each beam supplying ~ 400 J (2 kJ in total) in 1ns onto a 600-μm spot, which gave an average intensity of $7.0 \times 10^{14} \, W \, cm^{-2}$. For the backlighter cone, we used two laser beams of total energy 150 J, in 500 ps onto a 500-μm spot, which gave an average intensity of $1.5 \times 10^{14} \, W \, cm^{-2}$. Both sets of beams were delivered at the third harmonic with a wavelength of 351 nm. Viewing slits, along the X-ray diagnostic line of sight, were cut on either side of the tube at the obstacle end.

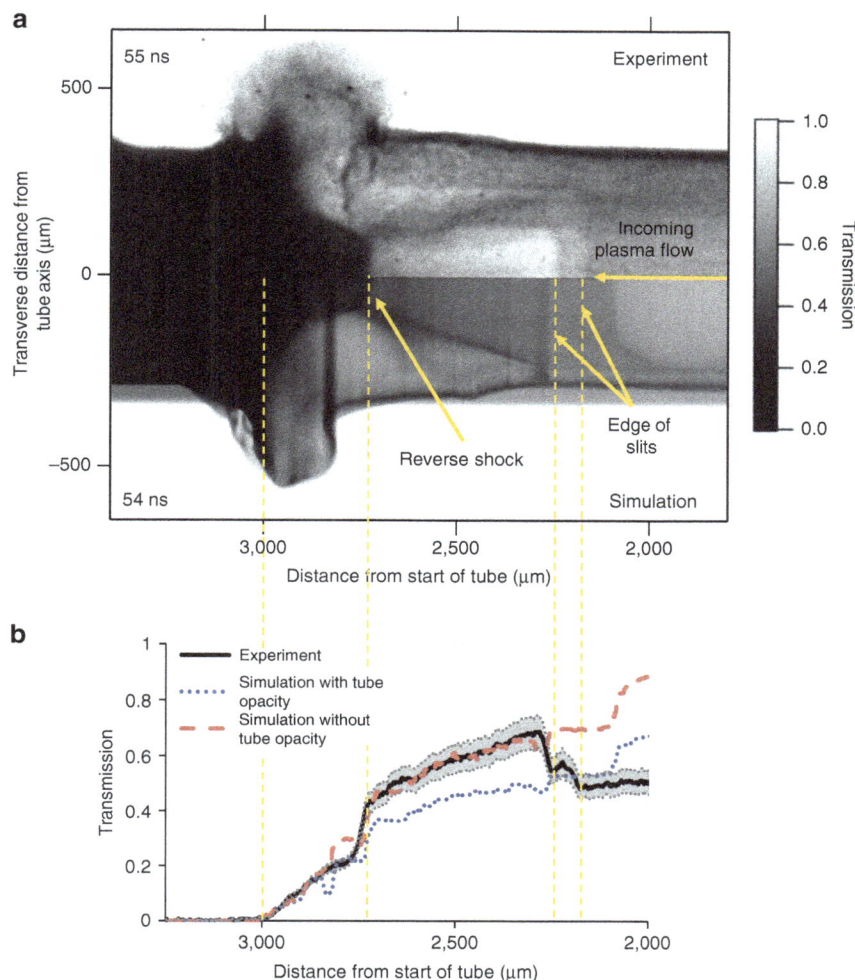

Figure 2 | Comparison of experimental and simulated results. A secondary target was used as an X-ray source for backlit pinhole radiography of the reverse shock region (see Fig. 1 and Methods). The plasma flow enters from the right of the figure and impacts onto the obstacle, originally at 3,000 μm, forming a reverse shock that can be seen around 2,750 μm. The top half of (**a**) shows the experimental radiograph, with the bottom half of (**a**) showing a synthetic radiograph from postprocessing a 2D simulation. The target tube had a slit cut into it from ~2,200 to ~3,100 μm. The same radiative hydrodynamic simulation was postprocessed to give simulated radiographs with and without the effect of the tube opacity, to account for the effect of the tube being absent at the slit position. A line-out of X-ray transmission along the central axis, for the experimental case and postprocessing of the simulated case with and without the opacity of the tube wall, is plotted in (**b**). The transmission is well matched for the experimental case and simulated radiograph without tube opacity from the position of the reverse shock, ~2,750 μm, to the edge of the slit, ~2,200 μm. After this point, the experiment transmission agrees better with the simulated radiograph including the tube opacity.

subsequently begins to move back up the tube. The brightening in the image for distances $\gtrsim 2,200$ μm is due to increased emission where the tube wall is absent (here the slit in the tube faces the optical diagnostics, unlike Fig. 2 where the windows face the X-ray diagnostics). The position of the shock is marked by a cross and the overlaid X-ray radiograph at $t = 55$ ns shows good agreement between the diagnostics. The numerical simulation also predicts a slowly varying shock front stand-off position, albeit at slightly earlier times, between 38 and 50 ns. This is probably due to limitations of a two-dimensional (2D) simulation capturing the full dynamics of the flow.

Numerical simulations. The good agreement between the experimental X-ray images and the synthetic ones obtained from postprocessing the 2D simulations, as well as their prediction of the overall structure of flow gives us confidence in the estimates of the microscopic properties of the plasma. The results for the simulations are shown in Fig. 4. In particular, we notice that

the predicted reverse shock exhibits a sudden jump in both density and ionization fraction.

Assuming that the gold and plastic materials are uniformly mixed, and that the total opacity is simply the weighted sum of the opacity of each component (see Supplementary Fig. 2 and Supplementary Discussion: Compression calculation and Mixing of gold-plastic layers in pusher foil), the density jump in the reverse shock between the upstream and downstream flows is estimated to be $\sim 5.6 \pm 0.6$ at $t = 55$ ns (where the error comes from taking backlighter energies in a range from 3 to 4 keV; the energy of the backlighter was inferred to be ~ 3.75 keV, using transmission through a step wedge. See Supplementary Fig. 3 and Supplementary Methods: X-ray energy). This agrees well with the density jump calculated in the simulations (~ 5.5, as shown in Fig. 4 at 50 ns).

Discussion

Using standard equations for the conservation of mass and momentum across the shock transition region, ignoring the

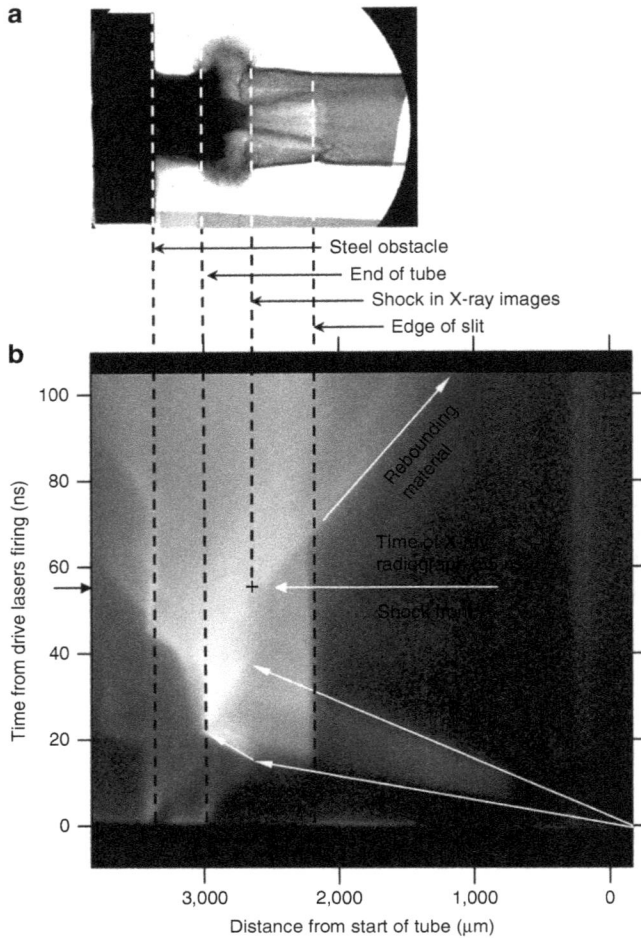

Figure 3 | Streaked optical image of the flow and shock evolution. (a) An X-ray image taken from Fig. 2 aligned with the streaked optical image in **(b)**. Optical diagnostics were fielded in the experiment to record the self-emission from the hot plasma along the entire length of the tube. An imaging system directed the light emitted by the interacting plasma to the streak camera. The streak camera had a spatial field of view of ~ 4 mm and thus the entire tube and obstacle were visible, except the region shielded by the copper cone. An imaging line aligned with the central axis of the tube (along the slit cut-out in the tube), in the horizontal plane, was then streaked in time over ~ 60, 100 or 250 ns as required. This provided a temporal dependence of flow velocities in plasma. Time runs from bottom to top, distance right to left. The time evolution of the plasma can clearly be seen inside the tube. The dotted lines and overlaid radiograph show positions of parts of the target on the streaked image. There is good agreement in the position of features between the X-ray **(a)** and optical **(b)** diagnostics.

radiation pressure term (which is negligible under these conditions), and using the thermal pressure p_i ($i = 1$ and 2 for the upstream flow and downstream plasma, respectively) we obtain[27]:

$$\frac{\rho_2}{\rho_1} = \frac{1}{2a_2^2}\left\{ a_1^2 + u_1^2 \pm \left[\left(a_1^2 + u_1^2\right)^2 - 4a_2^2 u_1^2\right]^{1/2}\right\}, \quad (1)$$

with $a_i = \sqrt{p_i/\rho_i}$ the isothermal speed of sound, ρ_i is the mass density and u_1 the upstream flow velocity.

Equation (1) admits a solution only when $u_1 \geq u_R$ or $u_1 \leq u_D$, where the rarified and dense velocities, respectively, are defined as $u_R = a_2 + \left(a_2^2 - a_1^2\right)^{1/2}$ and $u_D = a_2 - \left(a_2^2 - a_1^2\right)^{1/2}$. If $u_1 > u_R$,

then the flow is supersonic and it results in a compression of the downstream plasma. This requires us to take the positive root in equation (1) and is expected in the reverse shocks occurring in both the MCV star system and in our experiment. The scaling between the two systems is given in Table 1. Values for the incoming flow velocity are taken from our streaked optical data (Fig. 3) and sound speeds are calculated using simulated values for the temperature. The Reynolds number is large in the astrophysical system and the laboratory, indicating that viscous dissipation can be neglected. The thin radiation number—a measure of the incoming material energy flux compared with the radiation flux[16]—is very small in the astrophysical case and by no means large in the laboratory. This implies that radiation losses cannot be ignored and are important in determining the overall evolution of the flow. This is also reflected by the radiation cooling time (τ_c) being of the order of the characteristic dynamical time.

Using the simulated flow velocity ($u_1 \sim 83$ km s^{-1} in the frame of the shock) temperature and ionization fraction (Fig. 4), we estimate, using equation (1), $\rho_2/\rho_1 \approx 5.7 \pm 0.6$, whose value agrees with that inferred from the X-ray radiographs (where the error is calculated by allowing a 10% variation in thermal pressure when calculating the sound speeds). Equation (1) follows from the mass and momentum equations, and is not affected by details relating to ionization and radiation.

The compression ratio, in the strong shock limit (which applies in our experiment), is given by $\rho_2/\rho_1 = (\gamma + 1)/(\gamma - 1)$, where γ is the ratio of specific heats. Taking the compression value as 5.6, from the experiment, we estimate, from the ideal jump condition, an effective $\gamma \approx 1.43^{+0.07}_{-0.05}$.

The adiabatic exponent, taken from the relation of the energy, density and pressure in the ideal case[28], is given by $\gamma = 1 + p/\rho\epsilon$, where ϵ is the internal energy. Using values taken from the simulation gives $\gamma = 1.4$ which agrees with the value of γ taken from the experiment. Ionization of the material at the shock front thus causes a greater densification of the material. The presence of a spike in the simulated electron temperature at the shock front ($\sim 2{,}590\,\mu$m in Fig. 4 at 50 ns) suggests the presence of radiative effects[28]. Following the analysis by ref. 24 in determining whether radiative effects are important, the flow minus the shock velocity must equal or exceed 83 km s^{-1}. From the experimental and simulated numbers, we see that we are just in this regime.

In the astrophysical case, the stand-off position of the shock front with respect to the surface of the white dwarf is estimated by[29,30]

$$h_s = \Delta u_1 \tau_c, \quad (2)$$

where Δ is a value depending on the dimensionless parameters of the system, u_1 is the incoming flow velocity to the shock front and τ_c is the cooling time[11]. The form of this equation is valid in a general sense for both optically thin and thick plasmas. The value of delta has little sensitivity to the dynamics, varying between 0.25 and 0.1 according to the dominant radiative process. In the optical streak image in Fig. 3, the reverse shock appears to maintain a steady distance $h_s \approx 200\,\mu$m from the obstacle between $40\,\text{ns} \lesssim t \lesssim 55\,\text{ns}$. This value is consistent with the prediction of equation (2) and it scales to a stand-off distance of 1,300 km for the MCV system, which is the typical spatial extension as predicted by theory. In the astrophysical case this is mediated by the radiative losses against the incoming mass flux and, hence, radiative effects are very important. However, in the laboratory the radiative and material energy fluxes are of similar magnitude. Owing to the low temperature of the post-shock region in laboratory plasma, a microscopic, scaled model between the laboratory and astrophysics cannot be produced. Indeed, the laboratory plasma does not radiate primarily by bremsstrahlung cooling, but instead by line emission. However,

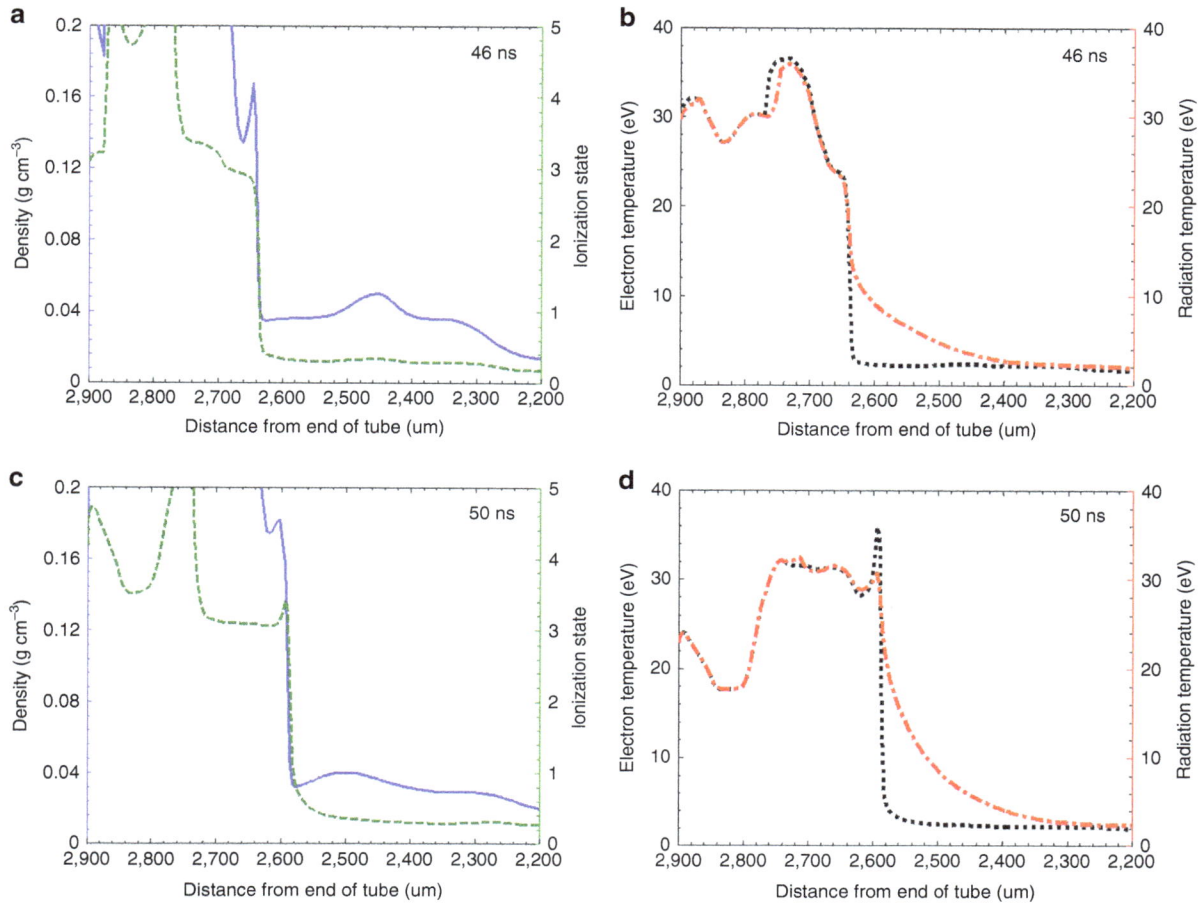

Figure 4 | Prediction from numerical simulations. Lineouts, from the central axis of a 2D simulation, of various plasma properties are shown at two different times. Density and ionization state (**a,c**) are shown as solid blue and dashed green lines, respectively; electron temperature and radiation temperature (**b,d**) are shown as dotted black and dot-dashed red lines, respectively. At the shock position we see an abrupt change in all four of these properties.

Table 1 | Scaling between the laboratory and the astrophysical system.

Characteristic quantity	Astro	Lab
Length	10^6 m	1.5×10^{-4} m
Incoming flow density	10^{-8} g cm^{-3}	0.03 g cm^{-3}
Initial flow velocity	1,000 km s^{-1}	200 km s^{-1}
Incoming flow temperature	0.3 eV	2.2 eV
Post shock temperature	10,000 eV	28 eV
Cooling time (τ_c)	1 s	3.8×10^{-9} s
Reynolds number	10^6	2.2×10^5
Radiation number (Thin)	10^{-16}	2.3
Mach number	>10	8
u_R	1,000 km s^{-1}	57 km s^{-1}
u_D	0.5 km s^{-1}	0.55 km s^{-1}
Shock height (h_s)	~1,300 km	200 μm

The scaling relations and the characteristic dimensionless constants are defined as in ref. 16. Values for the magnetic cataclysmic variable star system are taken from ref. 20.

as the radiative processes modify the macroscopic structure of the flow, as also occurs at astrophysical scales, we can use the parametric similarity introduced by Falize et al.[10,11] to relate the laboratory plasma to the astrophysical flows. Thus, the relation in equation (2) is applicable in the laboratory and astrophysics, and the results can be related to one another.

Our work provides a laboratory platform to investigate the physics of the accretion column near the white dwarf photosphere, which is not currently accessible by observational techniques. Future experiment at larger lasers such as the National Ignition Facility or Laser MegaJoule could increase the radiative nature of the reverse shock and comprehensively test how radiative losses affect the stand-off distance. This can have important consequences for the explanation of the luminosity curve in variable star systems, as the position of the shock above the white dwarf photosphere determines the ratio of hard and soft X-ray emission[22], which itself is dependent on the accretion model[21].

Methods

Laser and target. The drive beams were used to produce an expanding plasma jet, whereas the backlighter beams generated an X-ray source that imaged the plasma flow and the reverse shock formation. The main target consisted of a 550-μm inner diameter, 650-μm outer diameter and a 3.5-mm-long polyimide tube, with foils affixed at one end and an obstacle at the other. The laser spot size was deliberately chosen to be larger than the inner diameter of the tube so as to be less susceptible to misalignment and thus preserve a relatively uniform transverse profile of the plasma flowing down the tube. The drive beams illuminated the end of the target capped with two foils: a 25-μm-thick plastic ablator and a 25-μm-thick gold–plastic layered, pusher foil. The pusher was manufactured by depositing alternating layers of CH and Au, with 6-μm-thick CH and 300-nm-thick Au, giving a total Au concentration in the pusher of 39% by mass. A copper cone was placed just behind the ablator and pusher foils to protect the diagnostics from the direct view of the laser spot and from emission from the blow-off plasma. The other end of the tube was sealed with a steel obstacle, which was inserted ~150 μm into the tube. In some targets, ~1-mm-long slits were cut into the tube near the steel obstacle, at 45° to the horizontal plane. This allowed for clearer imaging of the optical and X-ray emission from the interacting plasma. See also ref. 31 for additional details on the target.

X-ray target. This target consisted of a 400-μm diameter chlorinated plastic (parylene-D) foil, which was then affixed onto a 50-μm CH foil, and finally placed onto a large tantalum foil with a 15-μm pinhole. This target was placed 14.1 mm from the steel obstacle and illuminated by the backlighter beams (see Fig. 1). The backlighter beams overfilled the target, with a focal spot of 500 μm, which caused an expansion of the surrounding plastic and the target material as well. This acted to contain the blow-off and the tantalum pinhole ensured a point-like source of X-rays, which backlit the main target and were recorded on an image plate detector placed 229 mm from the main target[32]. This gave a field of view of ~2 mm and a magnification of ~15. The energy of the X-rays are expected to be mainly from Cl line emission: id est, He-α at 2.78 keV, He-β at 3.27 keV, Ly-α at 2.96 keV and Ly-β at 3.50 keV (see also Supplementary Fig. 4). Appropriate filters were added in front of the image plate: a 12.5-μm polyvinylidene chloride (PVDC) to cut off lower energies and 5 μm Ti to cut off higher energies, giving an effective window of transmission between 2.78 and 5 keV. A step-wedge filter of 50-,100-and 150-μm-thick plastic was also placed over the image plate detector to calibrate the X-ray emission energy on each shot. See Supplementary Fig. 3 and Supplementary Methods: X-ray energy, for more detail.

Simulations. In the calculations, the NYM Lagrangian code[33] was used for the laser-interaction phase, as this allowed fine zoning of the ablator foil, which is essential to resolve light absorption from inverse bremsstrahlung. X-ray emission and absorption were simulated with full multi-group Monte Carlo photonics[34] using CASSANDRA opacities[35] and SESAME equations-of-state[36]. After the laser pulse had ended and the coronal plasma had cooled substantially, the simulation was linked to the PETRA code[37] to continue to late time. This used Eulerian hydrodynamics (run on an orthogonal mesh), which is essential to permit the large shear flows of material along the tube walls. This calculation was restricted to multi-group diffusion for the X-ray exchange but the opacities and equations-of-state were unchanged from NYM. The laser was modelled as a 1-kJ beam normal to the surface and a super-Gaussian spatial intensity profile with 1/e intensity at 300 μm radius. The laser energy was adjusted in order for the predicted flow velocity and the overall flow behaviour to match the experimental case. This gave an effective energy coupling of ~50%, which is reasonable as the experimental beam angle was not modelled. An electron-conduction flux limiter of 0.05 was used, typical of laser-plasma modelling at this irradiance level. See Supplementary Figs 5 and 6, Supplementary Table 1, Supplementary Discussion: Laser intensity in simulation and Supplementary Methods: Numerical simulations, for more detail on the sensitivity of the simulations.

References

1. Hartigan, P. et al. Dynamics of stellar jets in real time: third epoch hubble space telescope images of HH 1, HH 34, and HH 47. Astrophys. J. 736, 29 (2011).
2. Wu, K. Accretion onto magnetic white dwarfs. Space Sci. Rev. 93, 611–649 (2000).
3. Dale, J. E. & Bonnell, I. A. Ionization-induced star formation—III. Effects of external triggering on the initial mass function in clusters. Mon. Not. R. Astron. Soc. 422, 1352–1362 (2012).
4. Tremblin, P. et al. Ionization compression impact on dense gas distribution and star formation probability density functions around H II regions as seen by Herschel. Astron. Astrophys. 564, A106 (2014).
5. Mackey, J. et al. Interacting supernovae from photoionization-confined shells around red supergiant stars. Nature 512, 282–285 (2014).
6. Hopps, N. et al. Overview of laser systems for the Orion facility at the AWE. Appl. Opt. 52, 3597–3607 (2013).
7. Remington, B. A., Arnett, D., Drake, R. P. & Takabe, H. Modeling astrophysical phenomena in the laboratory with intense lasers. Science 284, 1488–1493 (1999).
8. Remington, B. A., Drake, R. P. & Ryutov, D. D. Experimental astrophysics with high power lasers and Z pinches. Rev. Mod. Phys. 78, 755–807 (2006).
9. Falize, É., Bouquet, S. & Michaut, C. Scaling laws for radiating fluids: the pillar of laboratory astrophysics. Astrophys. Space Sci. 322, 107–111 (2009).
10. Falize, É., Michaut, C. & Bouquet, S. Similarity properties and scaling laws of radiation hydrodynamic flows in laboratory astrophysics. Astrophys. J. 730, 96–104 (2011).
11. Falize, É., Dizière, A. & Loupias, B. Invariance concepts and scalability of two-temperature astrophysical radiating fluids. Astrophys. Space Sci. 336, 201–205 (2011).
12. Ryutov, D. D., Drake, R. P. & Kane, J. Similarity criteria for the laboratory simulation of supernova hydrodynamics. Astrophys. J. 518, 821–832 (1999).
13. Ryutov, D. D., Drake, R. P. & Remington, B. A. Criteria for scaled laboratory simulations of astrophysical MHD phenomena. Astrophys. J. 465, 465–468 (2000).
14. Ryutov, D. D., Remington, B. A., Robey, H. F. & Drake, R. P. Magnetohydrodynamic scaling: from astrophysics to the laboratory. Phys. Plasmas 8, 1804–1816 (2001).
15. Ryutov, D. D., Kugland, N. L., Park, H. S., Plechaty, C., Remington, B. A. & Ross, J. S. Basic scalings for collisionless-shock experiments in a plasma without pre-imposed magnetic field. Plasma Phys. Contr. Fusion 54, 105021–105030 (2012).
16. Cross, J. E., Reville, B. & Gregori, G. Scaling of magneto-quantum-radiative hydrodynamic equations: from laser-produced plasmas to astrophysics. Astrophys. J. 795, 59 (2014).
17. Warner, B. Cataclysmic Variables (Cambridge Univ. Press, 1995).
18. Busschaert, C. et al. POLAR project: a numerical study to optimize the target design. New J. Phys. 15, 035020 (2013).
19. Busschaert, C., Falize, É., Michaut, C., Bonnet-Bidaud, J.-M. & Mouchet, M. Quasi-periodic oscillations in accreting magnetic white dwarfs II. The asset of numerical modelling for interpreting observations. Astron. Astrophys. 579, A25 (2015).
20. Falize, É. et al. High-energy density laboratory astrophysics studies of accretion shocks in magnetic cataclysmic variables. High Energ. Dens. Phys. 8, 1–4 (2012).
21. Frank, J., King, A. R. & Lasota, J. -P. The soft X-ray excess in accreting magnetic white dwarfs. Astron. Astrophys. 193, 113–118 (1988).
22. Traulsen, I. et al. X-ray spectroscopy and photometry of the long-period polar AI Tri with XMM-Newton. Astron. Astrophys. 76, 516–527 (2010).
23. Falize, É. et al. The scalability of the accretion column in magnetic cataclysmic variables: the POLAR project. Astrophys. Space Sci. 336, 81–85 (2011).
24. Keiter, P. A. et al. Observation of a hydrodynamically driven, radiative-precursor shock. Phys. Rev. Lett. 89, 165003–165004 (2002).
25. Krauland, C. M. et al. Reverse radiative shock laser experiments relevant to accreting stream-disk impact in interacting binaries. Astrophys. J. Lett. 762, L2 (2013a).
26. Krauland, C. M. et al. Radiative reverse shock laser experiments relevant to accretion processes in cataclysmic variables. Phys. Plasmas 20, 056502–056506 (2013b).
27. Mihalas, D. & Weibel-Mihalas, B. Foundations of Radiation Hydrodynamics (Dover Publications, Inc., 1999).
28. Zel'dovich, Y. B. & Raizer, Y. P. Physics of Shock Waves and High-temperature Hydrodynamic Phenomena (Academic Press, 1966).
29. Lamb, D. Q. & Masters, A. X and UV radiation from accreting magnetic degenerate dwarfs. Astrophys. J. 234, L117–L122 (1979).
30. Wu, K., Chanmugam, G. & Shaviv, G. Structure of steady state accretion shocks with several cooling functions: Closed integral-form solution. Astrophys. J. 426, 664–668 (1994).
31. Spindloe, C. et al. Target fabrication for the POLAR experiment on the Orion laser facility. HPL Sci. Eng. 3, e8 (2015).
32. Kuranz, C. C. et al. Dual, orthogonal, backlit pinhole radiography in OMEGA experiments. Rev. Sci. Instrum. 77, 10E327-1-4 (2006).
33. Roberts, P. D., Rose, S. J., Thompson, P. C. & Wright, R. J. The stability of multiple-shell ICF targets. J. Phys. D 13, 1957 (1980).
34. Fleck, J. A. & Cummings, J. ,D. An implicit Monte Carlo scheme for calculating time and frequency dependent nonlinear radiation transport Jour. Comput. Phys. 8, 313–342 (1971).
35. Crowley, B. J. B. & Harris, J. W. O. Modelling of plasmas in an average-atom local density approximation: the CASSANDRA code. J. Quant. Spec. Rad. Trans. 71, 257–272 (2001).
36. Holian, K. S. T-4 Handbook of Material Properties Data Bases. Vol Ic: Equations of State. Report No. LA-10160-MSLos Alamos National Laboratory, 1984).
37. Youngs, D. L. Numerical simulation of turbulent mixing by Rayleigh-Taylor instability. Phys. D 12, 32–44 (1984).

Acknowledgements
We thank all the Orion technical team at AWE for their support during the experiments. The research leading to these results has received funding from the European Research Council under the European Community's Seventh Framework Programme (FP7/2007–2013)/ERC grant agreement number 256973. The work of JEC was funded by AWE under the Oxford Centre for High Energy Density Science. Partial support from the Science and Technology Facilities Council, and the Engineering and Physical Sciences Research Council of the United Kingdom, and COST action MP1208 is also acknowledged.

Author contributions
The experiment was conceived and designed by C.B., M.K., B.L., C.M., G.G., R.P.D., E.F., J.E.C., J.M.F., P.G., H.W.D., E.R.T., N.W. and R.Y. The work at the Orion laser was carried out by J.E.C., J.M.F., C.B., C.C.K., E.R.T., H.W.D., J.F., E.T.G., S.P., J.S. and R.Y.. The paper was written by J.E.C., G.G., P.G. and R.P.D. The experimental data were analysed by J.E.C. and J.M.F., and the numerical simulations were performed by P.G. Targets were provided by C.S. Additional theoretical support was given by E.F., M.K., C.C.K., R.P.D., C.B., C.N.D., R.Y., J.-M.B.B., N.C., C.K. and M.M.

Additional information

Competing financial interests: The authors declare no competing financial interests.

24

Observational evidence for enhanced magnetic activity of superflare stars

Christoffer Karoff[1,2], Mads Faurschou Knudsen[1], Peter De Cat[3], Alfio Bonanno[4], Alexandra Fogtmann-Schulz[1], Jianning Fu[5], Antonio Frasca[4], Fadil Inceoglu[1], Jesper Olsen[6], Yong Zhang[7], Yonghui Hou[7], Yuefei Wang[7], Jianrong Shi[8] & Wei Zhang[8]

Superflares are large explosive events on stellar surfaces one to six orders-of-magnitude larger than the largest flares observed on the Sun throughout the space age. Due to the huge amount of energy released in these superflares, it has been speculated if the underlying mechanism is the same as for solar flares, which are caused by magnetic reconnection in the solar corona. Here, we analyse observations made with the LAMOST telescope of 5,648 solar-like stars, including 48 superflare stars. These observations show that superflare stars are generally characterized by larger chromospheric emissions than other stars, including the Sun. However, superflare stars with activity levels lower than, or comparable to, the Sun do exist, suggesting that solar flares and superflares most likely share the same origin. The very large ensemble of solar-like stars included in this study enables detailed and robust estimates of the relation between chromospheric activity and the occurrence of superflares.

[1] Department of Geoscience, Aarhus University, Høegh-Guldbergs Gade 2, 8000 Aarhus C, Denmark. [2] Stellar Astrophysics Centre, Department of Physics and Astronomy, Aarhus University, Ny Munkegade 120, 8000 Aarhus C, Denmark. [3] Royal Observatory of Belgium, Ringlaan 3, B-1180 Brussel, Belgium. [4] INAF-Osservatorio Astrofisico di Catania, via S.Sofia 78, 95123 Catania, Italy. [5] Department of Astronomy, Beijing Normal University, 19 Avenue Xinjiekouwai, Beijing 100875, China. [6] AMS, 14C Dating Centre, Department of Physics, Aarhus University, Ny Munkegade 120, 8000 Aarhus C, Denmark. [7] Nanjing Institute of Astronomical Optics and Technology, National Astronomical Observatories, Chinese Academy of Sciences, Nanjing 210042, China. [8] Key Laboratory of Optical Astronomy, National Astronomical Observatories, Chinese Academy of Sciences, Beijing 100012, China. Correspondence and requests for materials should be addressed to C.K. (email: karoff@phys.au.dk).

The largest known solar flare was the Carrington event in AD 1859 (refs 1,2). This flare and the associated coronal mass ejection were so large that they caused world-wide auroras and allowed telegraphs to operate on the currents induced by the accompanying geomagnetic storm[3]. As we have no X-ray measurements of the Carrington event, it is not clear how large it was compared with the largest flares observed during the space age, which are classified according to their peak X-ray flux. Estimates based on magnetometer traces predict that the Carrington event, with a total energy of up to $5 \cdot 10^{32}$ erg, was likely larger than any solar flare observed in the space age[4]. On the other hand, it is clear that the hazard caused by the Carrington event was minimal compared with that potentially posed by, for example, a $2 \cdot 10^{34}$ erg superflare. Fifteen years ago, Schaefer et al.[5] identified nine so-called superflares, defined as flares with energies ranging from 10^{33} to 10^{38} erg, on nine ordinary solar-type stars. We call these stars superflare stars.

Superflares on solar-like stars may arise from at least three different mechanisms, apart from coronal magnetic reconnection: (i) star–star interactions, as is the case for RS CVn systems, where a close binary companion tidally spins up an F or G main-sequence star. It has been suggested that this tidal interaction can be accompanied by large co-rotating flux tubes that temporarily may connect the two stars and thus cause superflares when the interconnecting field lines subsequently disrupt[6]; (ii) star–disk interactions, where the dipole magnetic field of the central star is connected to the co-rotating disk. The twist imposed on the dipole magnetic field can thus lead to disruption and superflares[7]; (iii) star–planet interactions, which can take place in two different ways, either similar to the star–star mechanism through disruption of interconnecting field lines[8,9] or through tidal interaction between the star and the planet, which can lead to enhanced dynamo action[10] and thus greater magnetic reconnection events. One of the main differences between the star–star and the star–planet mechanisms is that where RS CVn-type stars generally have activity levels many times higher than those found in even the most active single solar-like stars[11], stars hosting Jupiter-like planets have activity levels comparable to average solar-like stars[12]. All three mechanisms were analysed theoretically by Shibata et al.[13] along with the possibility for a Sun-like star to generate a superflare through coronal magnetic reconnection. It was found that it would take the Sun 40 years to generate a sunspot large enough and with a magnetic field strong enough to produce a 10^{35} erg superflare. Although Shibata et al.[13] conclude that the coronal magnetic reconnection hypothesis is the most like scenario, they were not able to rule out other scenarios, such as enhanced dynamo action through tidal interactions.

Using 120 days of high-precision, high-cadence photometric observations from NASA's Kepler mission[14], Maehara et al.[15] identified 365 superflares on 148 solar-type stars. This study was updated by Shibayama et al.[16] using 550 days of observations to include 1,547 superflares on 279 G-type stars. The nature of these has been investigated by a number of authors who analysed periodic brightness modulations in their Kepler light curves[15–19]. All these studies show that fast rotating stars are more likely to host superflares, but in general, the measured rotation periods have to be handled with great caution: it has been shown using simulations that Kepler observations cannot be used to extract rotation periods for stars where the spot life-time is comparable to or shorter than the rotation period, as is the case for the Sun[20]; the Kepler light curves are heavily affected by artefacts on time scales longer than 20 days[21]; a large fraction of the apparently rapid rotating (a few days) stars are likely ellipsoidal binaries[22,23].

Instead of looking at the rotation rates, the activity levels can be measured as emission in the Ca II H and K spectral lines at 396.8 and 393.4 nm, respectively. The intensity of the emission scales with the amount of non-thermal heating in the chromosphere, making these lines a useful proxy for the strength of, and fractional area covered by, magnetic fields. This was first suggested by Eberhard and Schwarzschild[24] and has subsequently been used extensively to measure stellar activity since Olin Wilson started regular observations at the Mount Wilson Observatory[25]. The emission in the Ca II H and K lines is measured by the canonical S index, which intensive research over the last 4 decades has shown is related to magnetic activity[26,27].

Stellar activity can also be measured using, for example, the Ca II infrared triplet or the Hα line[28–31]. Notsu et al.[31] used the Subaru telescope to measure activity levels, using the Ca II line at 8,542 Å, of 27 superflare stars, which all show activity levels larger than the Sun. A strong correlation was observed between the photometric modulation of the Kepler light curves over the rotation period and the measured activity levels. It is, however, not clear from these observations if all superflare stars have activity levels higher than the Sun, as the Ca II line at 8,542 Å is insensitive to activity at activity levels lower than or compared with the activity level of the Sun[31,32].

Here, we use observations from the Large Sky Area Multi-Object Fibre Spectroscopic Telescope (LAMOST, also called the Guoshoujing Telescope at Xinglong observatory, China)[33] to show that superflare stars are generally characterized by larger chromospheric emissions than other stars, including the Sun. However, superflare stars with activity levels lower than, or comparable to, the Sun do exist, suggesting that solar flares and superflares most likely share the same origin. The very large ensemble of solar-like stars included in this study enables detailed and robust estimates of the relation between chromospheric activity and the occurrence of superflares.

Results

Superflare stars are characterized by enhanced activity. Using observations from LAMOST, we have measured the S index of 5,648 main-sequence, solar-like stars with effective surface temperatures between 5,100 and 6,000 K in the field of view of the Kepler mission (Methods). Figure 1 shows four examples of spectra of the Ca II H and K lines for solar-like stars with different activity levels. In all, 48 of these 5,648 stars have been categorized as superflare stars by Shibayama et al.[16]. We compare the distribution of the measured activity levels of all the 5,648 solar-like stars with the distribution of the 48 stars that show superflares (Fig. 2). The two distributions are clearly different, with the activity levels of the superflare stars shifted to higher values. A Kolmogorov–Smirnov test shows that the two distributions are different at the 99.999999983% significance level, comparable to 6σ. This result shows that superflare stars generally exhibit stronger activity levels than other solar-like stars, indicating that the superflare stars have larger dynamo actions and thus that the superflares may be caused by a mechanism similar to the one generating solar flares. This is also supported by a bootstrap test, which shows that the mean S index of the superflare stars falls within the top 0.002% of the distribution of mean S indices when 48 stars are repeatedly selected at random from the 5,648 main-sequence stars (Methods).

If the activity distribution of the superflare stars is compared with the activity level of the Sun, which has an S index that varies between 0.169 and 0.205 from minimum to maximum of the 11-year solar cycle[34], we observe that superflare stars generally have activity levels larger than what is observed on the Sun. In all, 42 out of 48 superflare stars have activity levels

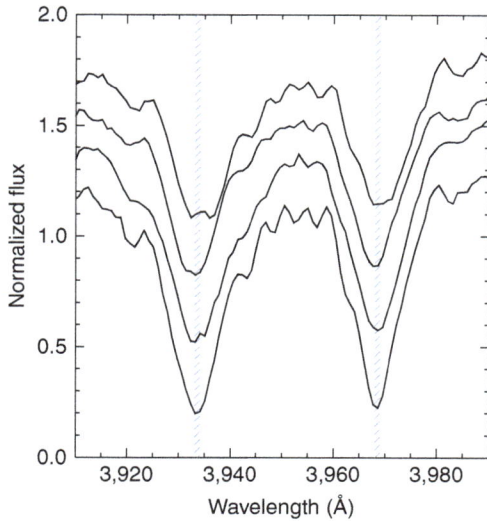

Figure 1 | **Examples of spectra showing different chromospheric emission levels in four solar-like stars.** The spectra show the Ca II H and K absorption lines at 396.8 and 393.4 nm, respectively. The four stars shown here are, bottom to top, KIC 8493735, KIC 9025370, KIC 8552540 and KIC 8396230. The spectra for these four stars have been normalized to one in the spectral range shown, but for clarity we have applied an offset of 0.2 between each spectrum. The measured S index for these four stars are: 0.15, 0.23, 0.30 and 0.34, respectively. The difference in S index can be seen as the increase in emission in the core of the absorption lines. The blue shaded regions show the wavelength bands used to measure the S index.

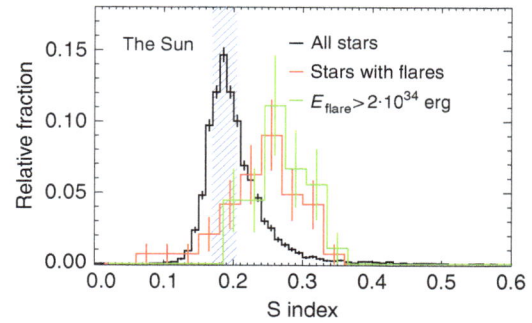

Figure 2 | **Histograms of the activity distribution of solar-like stars.** (main-sequence stars with effective temperatures between 5,100 and 6,000 K) in black compared with those of these stars that show superflares in their Kepler light-curve in the analysis of Shibayama et al.[16] in red. The blue shaded region marks the range of the S index of the Sun between solar cycle minima and maxima[34]. The two distributions are different at a 6σ level, clearly showing that the superflare stars generally have higher activity levels than average solar-like stars. The green curve shows the distribution of superflare stars with total energies larger than $2 \cdot 10^{34}$ erg in the analysis of Shibayama et al.[16] It is seen that this distribution is shifted to even higher activity levels. The error bars are based on the number of stars in each bin, assuming Poisson statistics. The uncertainties on the measured S indicies are smaller than 0.03 (see Supplementary Fig. 2 for details). The error bars in the figure represent a single standard deviation.

larger than 0.169, whereas 36 out of 48 superflare stars have activity levels larger than 0.205. If we only consider superflares with total energies larger than $2 \cdot 10^{34}$ erg, we observe that 30 out of 30 stars have activity levels larger than 0.169, whereas 26 out of 30 have activity levels larger than 0.205 (Methods). The distribution of stars hosting superflares with total energies larger than $2 \cdot 10^{34}$ erg is shifted to higher activity levels, and none of these superflare stars display activity levels significantly lower than the Sun (Fig. 2). However, it is noteworthy that 12 of 48 of superflare stars and 4 of 30 superflare stars with superflares above $2 \cdot 10^{34}$ erg have chromospheric emission levels comparable to the Sun.

Comparison to the Subaru observations. In this study, we only analyse stars with a signal-to-noise ratio higher than 10 in the blue part of the spectrum, as spectra with lower signal-to-noise ratio do not allow the S index to be measured with the required accuracy. The signal-to-noise ratio is measured as described in Luo et al.[35] and shown in Fig. 3. The risk associated with using signal-to-noise ratios down to 10 is that the activity measurements may become unreliable. To test this, we have compared our activity measurements (S index) with activity measurements based on the residual flux in the core of the infrared Ca II 8,542 line (r_0) by Notsu et al.[31] (Fig. 4). The clear correlation between the measured activities indicates that the S index we measure with the LAMOST observations is a reliable measure of chromospheric emission. We also tested the reliability of the measured activity levels by only analysing spectra with a signal-to-noise ratio higher than 30. Although this lowered the number of analysed stars dramatically, it did not change any of the conclusions.

Notsu et al.[31] also calculated the magnetic flux $<fB>$ using the residual flux in the core of the infrared Ca II line. Using the formulation of Schrijver et al.[27], we also calculate the magnetic

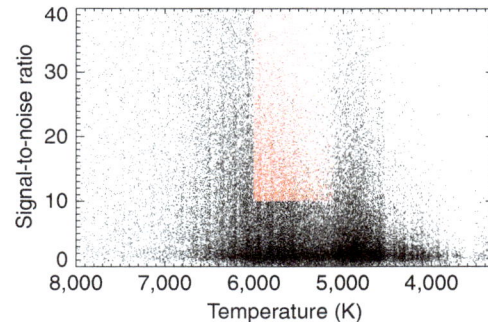

Figure 3 | **Relationship between the signal-to-noise ratio in the blue part of the spectra and effective temperature.** The effective temperatures are from Brown et al.[58] and the signal-to-noise ratio calculated as described by Luo et al.[35] The 5,648 main-sequence stars with signal-to-noise ratio higher than 10, which we analyse in this study, are marked with red. A clear separation between main-sequence (around 6,000 K) and evolved stars (around 5,000 K) can be seen as a function of effective temperature.

flux based on our measured S index (Methods). These fluxes agree with those of Notsu et al.[31] within the uncertainties, although the uncertainties associated with the estimated magnetic fluxes are relatively large in both cases (Methods). The uncertainties are in both cases calculated based on the transformation to magnetic flux $<fB>$ using solar data. It is therefore not clear how the accuracy of the two sets of activity measurements compare with one another. On the one hand, the Subaru spectra have higher resolution and higher signal-to-noise ratio than the LAMOST spectra, but, on the other hand, the S index is a better activity indicator than the residual flux in the core of the infrared Ca II line. The main difference between the Subaru and the LAMOST observations is the much larger (5,648) sample size of the LAMOST observations, which allow us to conclude that superflare stars are generally characterized by higher activity levels than other solar-like stars.

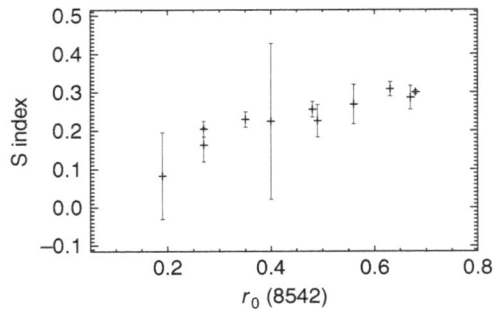

Figure 4 | Relation between residual flux in the core of the infrared Ca II 8542 line and the S index for 11 stars that have been observed with both the Subaru and LAMOST telescope. A clear correlation is seen between the two activity indicators. The clear correlation between the measured activities indicates that the S index we obtain with LAMOST observations is a reliable measure of chromospheric emission. The error bars are calculated using the relation in Supplementary Fig. 2. The error bars in the figure represent a single standard deviation.

Comparison to star spot coverage. Chromospheric emission, as measured with the S index, is one manifestation of stellar magnetic fields. Another manifestation is star spots. Notsu et al.[31] compared the so-called mean stellar brightness variation with the residual flux in the core of the infrared Ca II line. The stellar brightness variation is calculated as the difference between the maximum and minimum of the stellar flux measured with Kepler within a given 3-months period (after outlier removal)[31]. The comparison shows a strong correlation between these two quantities. In Fig. 5, we thus compare our measured chromospheric emission with the stellar brightness variation measured by Notsu et al.[31] Again, we see a strong correlation for the 11 stars that have been observed with both the Subaru and LAMOST telescopes. Unfortunately, 9 out of these 11 stars show both larger chromospheric emission, residual flux in the core of the infrared Ca II line and stellar brightness variation than the Sun. In order to test the relation between chromospheric emission and star spot coverage over a larger parameter range, we compare the measured S indices with the periodic photometric variability amplitude (R_{per}) measured by McQuillan et al.[36] (Fig. 6). The periodic photometric variability amplitude is calculated as the range between the 5th and 95th percentile of the normalized flux[36]. Indications of a correlation between the S index and the periodic photometric variability amplitude are observed down to periodic photometric variability amplitudes around 1,000 p.p.m. (Fig. 6), which is the level of the amplitudes observed for the Sun. Below 1,000 p.p.m., no clear indications of a correlation is seen, indicating that the relation between spot coverage and chromospheric emission breaks down when the magnetic activity is low. The comparison between the measured chromospheric emission and the photometric variability in both the studies by Notsu et al.[31] and McQuillan et al.[36] also indicate that the S index we measure with LAMOST is a reliable measure of chromospheric emission.

Discussion

Based on activity measurements of 5,648 solar-like stars, including 48 superflare stars, we show that superflare stars are generally characterized by higher activity levels than other stars, including the Sun. However, superflare stars with activity levels lower than, or comparable to, the Sun do exist, but none of the stars hosting the largest superflares ($>2\cdot10^{34}$ erg) show activity levels lower than the Sun. As discussed above, superflares on

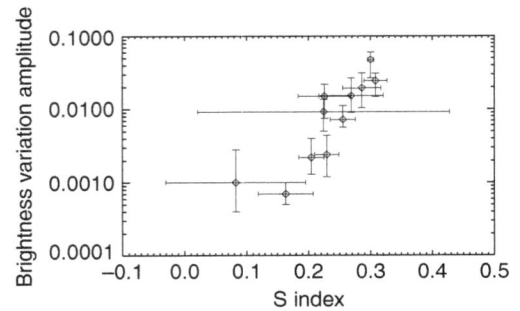

Figure 5 | Relation between the S index and the stellar brightness variation. The brightness variation amplitude measurements are from Notsu et al.[31] for the same 11 stars as in Fig. 4. A strong correlation is seen, which agrees well with the strong correlation seen between the residual flux in the core of the infrared Ca II 8542 line and the brightness variation amplitude by Notsu et al.[31] The error bars are calculated using the relation in Supplementary Fig. 2. The error bars in the figure represent a single standard deviation.

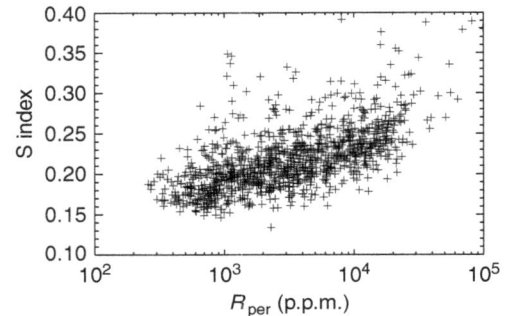

Figure 6 | Relation between the periodic photometric variability amplitude and the S index. The periodic photometric variability amplitudes are from McQuillan et al.[36] for 1,400 stars that also have LAMOST spectra that allow reliable measurements of the chromospheric emission. Indication of a correlation is seen down to periodic photometric variability amplitudes around 1,000 p.p.m., which is the level of the amplitudes seen in the Sun. Below 1,000 p.p.m. no clear indication of a correlation is seen, indicating that the relation between spot coverage and chromospheric emission breaks down when the magnetic activity is low.

solar-like stars may result from at least three different mechanisms apart from coronal magnetic reconnection: (i) star–star interactions, (ii) star–disk interactions and (iii) star–planet interactions. Our activity measurements show that superflare stars generally have S indices between 0.15 and 0.35. Although this is higher than solar-like stars in general and higher than the Sun, it is lower than typical RS CVn-type stars, which usually have S indices higher than 1.00 (ref. 11). It is therefore unlikely that a large fraction of the superflares can be explained by RS CVn-type star–star interaction.

Although we cannot evaluate if the star–disk and the star–planet hypotheses may be correct, it is likely that the magnetic field of the superflare stars should be significantly larger than what we observe on the Sun for these mechanisms to operate[8]. The fact that the lower part of the distribution of the measured activity of the superflare stars overlap with the range observed for the Sun indicates that these mechanisms most likely are not the main cause of superflares. Also, given the transit probability for a hot Jupiter in a 4-day circular orbit[37] and because none of the superflare stars are known to host hot

Jupiters, the star–planet mechanism is unlikely to be the main mechanism responsible for superflares[15].

Our study provides observational (although non-exclusive) support for the coronal magnetic reconnection hypothesis, as we show that superflare stars generally exhibit higher chromospheric emissions than the Sun and other solar-like stars, although there is an overlap between the two distributions. The coronal magnetic reconnection hypothesis can explain the observations via the notion that superflares and solar flares share the same origin and that the two activity distributions therefore are within similar range, but that superflares mainly take place on stars with activity levels larger than the Sun. If the observations had shown two different and non-overlapping ranges of activity levels in superflare (red curve) and non-superflare stars (black curve), for example, centred on S indices around 1.00 and 0.18, respectively, it would have favoured the star–star, star–disk or star–planet mechanisms, but this is not the case.

The observations also make it possible to renew the evaluation of the frequency of solar superflares (Fig. 7). Here, we have compared the frequency of solar flares with the frequency of superflares on solar-like stars as a function of the flare energy (Methods). If only the superflare stars with activity levels comparable to or smaller than the Sun are considered, then the frequency of $2 \cdot 10^{34}$ erg superflares is reduced with almost an order-of-magnitude compared with solar-like stars in general.

This result confirms the result by Shibayama et al.[16], who found that 1,150 out of 1,547 superflares occurred on fast rotating stars. Combined, the two results indicate that although superflares on solar-like stars with Sun-like rotation and chromospheric emission are an order-of-magnitude less likely than superflares on solar-like stars in general, they still occur. We do, however, have to be careful with the definition of Sun-like stars here. Shibata et al.[13] suggested that a sunspot with a radius around 30% of the solar radius would be needed in order for the Sun to produce a superflare with an energy above 10^{35} erg. Although such a large spot has never be observed on the Sun, it is clear that the Sun would likely fall outside our definition of Sun-like, if such a large

spot was observed, as such a large spot would likely result in large chromospheric emission.

The downward adjustment of the flare frequency based on the activity measurements leads to a better agreement between astrophysical activity observations of superflare stars and the frequency of solar flares recorded in geological archives (see Fig. 7 for details). Geological archives, in particular cosmogenic nuclides (^{10}Be and ^{14}C) in ice cores and tree rings, can be used to evaluate the flare frequency through so-called solar particle events, where protons are accelerated in connection with large solar flares to energies sufficiently high to produce cosmogenic nuclides when they reach the Earth's atmosphere (see Schrijver et al.[38] for a recent review).

The raw flare frequency estimated from the Kepler data, calculated without weighting for activity, seems to roughly follow the power-law distribution of solar flares[16,17,39]. To some extent, this contradicts results from a number of geological archives, which indicate a break, or roll-over, in the distribution of solar particle events at energies around 10^{33} erg (refs 39–42). The discussion of this roll-over effect has received renewed attention with recent discoveries based on studies of ^{14}C in Japanese tree rings, indicating that the Sun hosted a superflare with an energy larger than 10^{33} erg in AD 775 (refs 43–48). The solar origin of the AD 775 event was recently questioned due to the lack of simultaneous aurora observations[49], but evidence from a new study based on multi-radionuclides suggest a solar origin[50]. A similar event might have taken place in AD 993 (ref. 51). In Fig. 7, the AD 775 and the AD 993 events are shown with a diamond, assuming that such events take place every 620 years. The upper-limit flare frequency weighted for activity lies over the AD 775 and AD 993 events. This does not contradict the roll-over scenario, but it does not place tight constraints on the occurrence of such events either. The indications of a roll-over effect in the flare frequency weighted for magnetic activity is also in agreement with recent theoretical estimates, which predict that sunspot groups larger than historically reported would be needed in order for the Sun to produce a superflare with an energy larger than $\sim 6 \cdot 10^{33}$ erg (ref. 52).

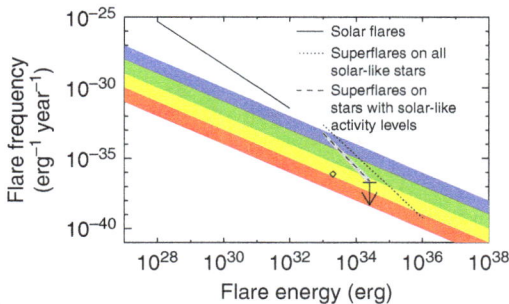

Figure 7 | Comparison of occurrence rates of flares on the Sun and of superflares on solar-like stars. The solid line shows the power-law distribution of solar flares[64], which is compared with the power-law distribution of superflares on the 279 G-type main-sequence stars found by Shibayama et al.[16] (dotted line) and the power-law distribution of those of these stars that have activity levels lower than the Sun at solar cycle maximum (dashed line, Methods). The last power-law assumes that the Sun's chromospheric emission does not change beyond what have been observed so far. The grey region marks the 1σ error range. The diamond marks the probability of an AD 775 or AD 993 event, assuming that such an event takes place every 620 y. We have extended the power-law down to energies of 10^{35} erg, although the observations suffer from detection limit effects below energies of $5 \cdot 10^{34}$ erg. The extension of the power-law to energies below 5×10^{34} erg is, however, supported by Maehara et al.[68] The coloured regions mark where one flare per year (blue), decade (green), century (yellow) or millennium (red) is expected.

Methods

The observations. The magnetic activity measurements from LAMOST enable the first direct comparisons of magnetic activity levels in superflare stars to the activity levels of solar-like stars in general. This is only possible because of the unique combination of a large aperture, large field-of-view and multi-object fibre spectroscopy.

Calculation of the S index. The LAMOST-Kepler project is described by De Cat et al.[53] For the present study, we analysed 71,733 low-resolution spectra ($R \sim 1,800$). These spectra were reduced as described in Luo et al.[35] This reduction included standard bias and dark-frame subtraction, flat-field correction and extraction of a one-dimensional spectrum. All spectra were cross-correlated with a solar spectrum to place the observed spectra on a reference wavelength grid with velocities zeroed.

The emission in the Ca II H and K lines is measured by the canonical S index defined as[54]:

$$S = \alpha \cdot \frac{H+K}{R+V} \tag{1}$$

where H and K are the recorded counts in a 1.09 Å full-width at half-maximum triangular bandpasses centred on the H and K lines at 396.8 and 393.4 nm, respectively. V and R are two 20 Å wide reference bandpasses centred on 390.1 and 400.1 nm. α is a normalization constant.

As our observations are done with a spectrograph and not with a spectrophotometer, it is convenient to rewrite S as a function of mean flux per wavelength interval ($\tilde{X} = X/\Delta\lambda_X$):

$$S = \alpha \cdot 8 \cdot \frac{\Delta\lambda_{HK}}{\Delta\lambda_{RV}} \cdot \frac{\tilde{H}+\tilde{K}}{\tilde{R}+\tilde{V}} = \alpha \cdot 8 \cdot \frac{1.09\text{Å}}{20\text{Å}} \cdot \frac{\tilde{H}+\tilde{K}}{\tilde{R}+\tilde{V}} \tag{2}$$

The factor of 8 comes from the fact that the Mt Wilson spectrophotometer used a rapidly rotating slit mask that exposed the H and K channels eight times longer than the reference channels. The value of $\alpha = 1.8$ is obtained from Hall, Lockwood and Skiff[55]. Other values of α can be found in the literature and these values are sometimes obtained by comparing S indices for the same stars measured with different instruments. Unfortunately, we have not been able to find other measurements of S indices for any of the Kepler stars observed by LAMOST. The main reason for this is likely that most surveys, especially on the northern hemisphere, focus on relatively bright stars, which saturate the LAMOST charge-coupled devices (CCDs)[53]. Instead, we have compared the distributions of the S indices from the LAMOST observations with the distribution of the S indices from Isaacson and Fischer[56] (Supplementary Fig. 1). Here, it is seen that although the LAMOST observations do not see as many high-active stars with an S index above 0.5, the two distributions are identical around 0.20. If we only consider main-sequence stars with effective temperatures between 5,100 and 6,000 K and S indices between 0.1 and 0.3, the mean value of the LAMOST observations is 0.192 ± 0.03 compared with 0.192 ± 0.04 for the Isaacson and Fischer sample. This is confirmed by a Kolmogorov–Smirnov test, which shows that the two distributions are identical between 0.1 and 0.3 at the 93% significance level. For the stars in the Isaacson and Fischer[56] sample, we define main-sequence stars as stars with absolute visual magnitude within 1 magnitude of the main sequence. The effective temperatures are calculated from the B-V colour index using the formulation by Alonso et al.[57] The ensemble thus consists of 60 main-sequence stars with effective temperatures between 5,100 and 6,000 K and with S indices between 0.1 an 0.3. Based on this analysis, we conclude that $\alpha = 1.8$ is an appropriated choice.

The uncertainties associated with the S indices is found by comparing the standard deviation of different measurements of the S index for 16,900 main-sequence stars with effective temperatures below 6,200 K that are located in more than one LAMOST field[53]. In Supplementary Fig. 2, we show the standard deviation as a function of the mean signal-to-noise ratio in the blue part of the spectrum of these stars. We can use this relation to calculate the uncertainties of the measurements of the S indices as a function of the signal-to-noise ratio in the blue part of the spectrum as: $\log \sigma(S) = -\log S/N - 0.5$. The calculation was carried out using χ^2 minimization, yelling a reduced χ^2 of 0.45. In the calculation, we assume that the measurements are not independent measurements of the same quantity, in which case we should have used the uncertainty of the mean value of the S indices as the uncertainty instead of the standard deviation of the different measurements of the S indices. This is done to achieve conservative estimates of the uncertainty and because we expect that the intrinsic chromospheric emission of the stars change between observations. This implies that the uncertainty we report also includes a stellar variability component. We do, however, note that the scatter is large in Supplementary Fig. 2 and that the uncertainties therefore should be handled with caution.

For each star, we made a cross-identification to the Kepler Input Catalog[58] (KIC) and restricted the analysis to stars on the main sequence with effective temperatures between 5,100 and 6,000 K (the same range as used by Shibayama et al.[16]). The main sequence was defined by the green line in Supplementary Fig. 3. The measured S indices for the superflare stars are given in Supplementary Table 1 together with the effective temperatures and surface gravity.

Huber et al.[59] collected fundamental stellar parameters for stars in the KIC from, for example, spectroscopic studies, binaries and planet studies and asteroseismology. Unfortunately, none of the stars analysed in our study were identified in any of these studies. Huber et al.[59] also updated the fundamental parameters for 196,468 stars in the Kepler FOV using isochrones fitted to the collected fundamental stellar parameters. All the stars in our study thus have updated parameters by Huber et al.[59] We show the updated values of the effective temperatures and surface gravity in Supplementary Fig. 4. Although both the effective temperature and surface gravity do change for some stars, it does not change any of the conclusions in our study and we have thus chosen to use the KIC values, as Huber et al.[59] explicitly stress that their updated values are only accurate for large ensembles and not suitable for scientific analysis on a star-by-star basis. Nevertheless, we want to note that KIC 11197517 and KIC 11241343 could have evolved off the main-sequence, based on the updated surface gravities from Huber et al.[59] Asteroseismic studies have previously shown that superflares are not uncommon on stars that have evolved off the main-sequence[60]. The updated effective temperatures imply that 13 out of the 48 superflare stars fall outside the 5,100–6,000 K effective temperature range when using the effective temperatures from Huber et al.[59] In Supplementary Fig. 5, we thus show how Fig. 2, showing the histogram of the activity distributions, would look if effective temperature and surface gravity values were taken from Huber et al.[59] It is clear from this figure that this would not change the conclusions.

Chromospheric flux. The S index is known to be colour dependent[61]. Also, the S index is a purely empirical quantity and it may be advantageous to calculate the chromospheric flux instead. The chromospheric flux can be calculated from the S index and some colour information like B-V and or effective temperature. The precision on the calculated chromospheric flux therefore depends on the precision of the colour information. We use the effective temperatures from KIC to calculate the B-V colour index using the formulation by Alonso et al.[57]

The chromospheric flux is calculated as R_{HK}^+ and F_{HK}^+, using the formulation by Mittag et al.[62] Here, the chromospheric emission, the S index, is converted into physical units and corrected for both photospheric flux and a so-called basal flux, in order to calculate a pure activity-related quantity. The uncertainties are found by combining the uncertainty of the S index with 100 K uncertainty on the effective temperature from KIC. We also show the same histogram of the distribution of

R_{HK}^+ and F_{HK}^+ in solar-like stars in general, in superflare stars and in stars with superflares with total energies larger than $2 \cdot 10^{34}$ erg in Supplementary Figs 6 and 7, as we showed for the S index in Fig. 2. These two figures clearly confirm the results based on the S index, that is, superflare stars are generally characterized by higher activity levels than other stars, but that superflare stars with activity levels lower than, or comparable to, other solar-like stars do exist. Supplementary Figs 6 and 7 do, however, indicate that the noise in the chromospheric flux measurements is larger than in the S index. This is likely due to both noise in the effective temperature from KIC and in the conversion from effective temperature to the B-V colour index. A test showed that using the effective temperatures from Huber et al.[59] did not change the results.

Supplementary Data 1 contains S, R_{HK}^+ and F_{HK}^+, as well as uncertainties, for all the 5,648 main-sequence solar-like stars.

Correlation between stellar activity and flare energy. In Supplementary Fig. 8, we show the relation between the total bolometric flare energy and stellar activity. Except for the absence of any high-energy flares with weak stellar activity, no clear correlation is seen. This is not unexpected, as stars with strong stellar activity are expected to host medium and larger superflares.

The Kepler photometry of the three stars with S index below 0.13 (KIC 2850378, KIC 11197517 and KIC 11241343) could be contaminated by background stars. If this is the case, it would explain both the low value of the total bolometric flare energy (as the absolute luminosity of the superflare stars and thus the total bolometric flare energy would be larger) and the S index (as the light entering the fibre would likely be contaminated and the observed emission in the H and K lines would therefore be relatively small). On the other hand, we do not have any prime indications that these stars are contaminated by background stars. Also, the spectra did not show any signs of binarity or artefacts, so although these three superflare stars with low S index are hard to explain with any of the proposed scenarios, we do not find any observational bias that can remove them from the sample population. The low activity of these three stars could indicate that these stars are in a grand activity minimum, similar to what the Sun went through during the Maunder Minimum[63], but the observations at hand do not allow us to confirm or reject this hypothesis. However, as discussed above, the explanation could also be that at least KIC 11197517 and KIC 11241343 have evolved off the main-sequence, based on the updated surface gravities from Huber et al.[59]

The magnetic flux density. Notsu et al.[31] calculated the magnetic flux $<fB>$ using the residual flux in the core of the infrared Ca II line. Using the formulation in Schrijver et al.[27], we have also calculated the magnetic flux using our measured S indices. Here, the Ca II H and K excess flux density (ΔF_{CaII}^*), calculated as F_{HK}^+ in the formulation by Mittag et al.[62], is given as:

$$\Delta F_{CaII}^* = 0.055 <fB>^{0.62 \pm 0.14} \qquad (3)$$

11 stars have been observed with both the Subaru and the LAMOST telescopes. Of these, one (KIC 11197517) has an S index so low that it results in a Ca II H and K flux density smaller than the basal flux and is therefore not used in the comparison shown in Supplementary Fig. 9. This comparison shows that, although the uncertainties are large on both types of estimates, the magnetic fluxes calculated from the Subaru observations generally seem to be larger than the magnetic fluxes calculated from LAMOST observations. The reason for this could be that the magnetic fluxes calculated based on the Subaru observations are not corrected for basal flux.

Calculation of the flare rates. After we have calculated the S indices for the 5,648 main-sequence stars based on the LAMOST spectra with signal-to-noise ratios higher than 10 in the blue part of the spectrum, including the subset of 48 superflare stars, it is possible to calculate the flare rates. We calculate these flare rates for S indices of 0.169, 0.179 and 0.205 corresponding to the mean value of the Sun during solar cycle minimum, the mean value of the Sun during the whole solar cycle, and the mean value during solar cycle maximum[34]. The flare rates are given below, together with the associated uncertainties, assuming that the flare rates follow a Poisson distribution (Table 1).

We also performed the same calculation using only the stars with superflares having a total bolometric flare energy larger than $2 \cdot 10^{34}$ erg (Table 2).

Similar flare rates were calculated by Maehara et al.[15] and Shibayama et al.[16] using rotation instead of chromospheric emission as the criterion for selecting

Table 1 | Fraction of superflare stars as a function of chromospheric emission.

S index smaller than	Fraction
0.169	13 ± 5%
0.179	13 ± 5%
0.205	25 ± 7%

Table 2 | The same as Table 1, but only bolometric flare energies larger than $2 \cdot 10^{34}$ erg.

S index smaller than	Fraction
0.169	0%
0.179	0%
0.205	8 ± 4%

Table 3 | Mean fraction of random stars as a function of chromospheric emission.

S index smaller than	Fraction
0.169	17.5134% ± 0.0005%
0.179	29.6214% ± 0.0007%
0.205	63.1812% ± 0.0007%

Sun-like stars with magnetic activity levels similar to the Sun. This only works for active stars with large spots characterized by long lifetimes. For Sun-like stars with magnetic activity levels similar to the Sun, one should be very careful with such analysis. The reason is that Kepler would not be able to measure the rotation period of the Sun, as sunspots have lifetimes much shorter than the rotation period. We can therefore not assume that the stars that do not show rotational modulation in the Kepler observations rotate as slow or slower than the Sun. It could also be that the reason why we cannot see any rotational modulation in the Kepler observations is because the starspots of these stars have lifetime much shorter than their rotation periods.

The significance of the flare fractions. To test the significance of the occurrence of flares, we performed a Monte Carlo simulation, where we randomly removed 43 stars from the 5,648 main-sequence stars and calculated the distribution of the S indices of these stars. This process was repeated 10,000,000 times and the resulting fractions were calculated as the mean values, and the uncertainty as the standard deviations (Table 3).

When these fractions are compared with the flare fractions, it is clear that the flare fractions are significantly different from the fractions obtained from randomly selecting 48 stars from the 5,648 main-sequence stars.

We also performed a bootstrapping test, where we compared the mean S index of the superflare stars with the mean S index of 48 stars randomly selected from the 5,648 stars. This process was also repeated 10,000,000 times. The mean S index of the superflare stars (0.2399) falls within the top 0.002% of the distribution of mean S indices based on 48 stars randomly selected from the 5,648 main-sequence stars (Supplementary Fig. 10).

Calculation of the flare frequency. The frequency of solar flares in Fig. 7 (that is, the power-law distribution) is obtained from Fig. 6 in Crosby, et al.[64] By noting that the power-law distribution in that figure passes through the point (10^{31} erg; 10^{-32} erg^{-1} day^{-1}), we obtain the following equation for the power-law:

$$\log\frac{dN}{dE} = 17.7 - 1.52 \pm 0.02 \log E \qquad (4)$$

where E is measured in erg and dN/dE is measured in erg^{-1} year^{-1}. The frequency of superflares is taken from Fig. 5b in Shibayama et al.[16] Here, we have the exact values of the fit (Shibayama, T., personnel communication):

$$\log\frac{dN}{dE} = 40 - 2.2 \log E \qquad (5)$$

In order to include the contributions from stellar activity in this equation, we construct a new power-law relationship that should be a straight line in the log–log plot, going through a point at 10^{33} and one at $2 \cdot 10^{34}$ erg. Using the values from Shibayama et al.[16], these two points should be: (10^{33} erg; $10^{-32.6}$ erg^{-1} year^{-1}) and ($10^{34.30}$ erg; $10^{-35.5}$ erg^{-1} year^{-1}). We then calculate the new points by taking 25% of the value at 10^{33} erg and 8% of the value at $2 \cdot 10^{34}$ erg (these values are found above). In this way, we get (10^{33} erg; $10^{-33.20}$ ergs^{-1} year^{-1}) and ($10^{34.30}$ erg; $10^{-36.6}$ erg^{-1} year^{-1}). The straight line that passes through these two points in a log–log plot, can be described by the following equation:

$$\log\frac{dN}{dE} = 121 - 2.62 \log E \qquad (6)$$

This relation is an upper limit (therefore, the arrow in Fig. 7), as we only calculate it for stars less active than the Sun at solar cycle maximum (0.205). It would arguably have been better to calculate the upper limit for stars less active than the mean activity level of the Sun over a solar cycle (0.179). However, the problem is that we do not have any stars hosting large superflares with total bolometric energies larger than $2 \cdot 10^{34}$ erg, which have magnetic activity levels less than 0.179.

The slope of -2.62 can be compared with the slope of -2.0 found for slowly rotating stars by Shibayama et al.[16] It is clear that the smaller slope obtained from the chromospheric emission measurements agrees better with the idea from the geological archives about a break, or roll-over, in the power-law distribution. The discrepancy between the two slopes suggests that slowly rotating stars can have large chromospheric emissions and likely large spots too. In other words, the chromospheric emission measurements suggest that it is less likely for the Sun to have a superflare compared with what is estimated based on the rotation measurements. When we use the slope of -2.62 to estimate the likelihood of the

Sun hosting a superflare, we therefore implicitly assume that the Sun's chromospheric emission does not change dramatically. If the Sun, on the other hand, is capable of producing the large ($0.3R_\odot$) spot suggested by Shibata et al.[13], then this assumption is likely violated, as the chromospheric emission of the Sun, in connection with such a large spot, would be dramatically larger than what we have observed so far.

Contamination of the Kepler photometry. Each pixel on the Kepler photometer spans four times 4 arcs on the sky and each star is observed with an optimal aperture that consists of several pixels[65]. There is, therefore, a significant risk that multiple stars will be located in a given optimal aperture and thus that a given star is contaminated by a number of background stars. This again leaves the possibility that the observed superflares do not originate from the stars we assigned them to, but from background stars. This problem was discussed by Balona[66], who used an analysis of the total bolometric flare energy to conclude that it is unlikely that the observation of superflares on A-type stars can be attributed to contamination. Notsu et al.[31], on the other hand, obtained high-resolution spectra of 50 superflare stars with the Subaru telescope. These observations revealed that 16 out of these 50 superflare stars show signs of binarity. These signs were either radial velocity variations (1 star), Hα line profile variability (2 stars), double-lined profiles (9 stars) or visual binarity seen in the slit viewer images (4 stars).

Of the 48 superflare stars analysed in this study, 6 are identified as binaries by Notsu et al.[31] These are: KIC 4045215, KIC 5445334, KIC 8226464, KIC 9653110, KIC 9764192 and KIC 11073910. In general, these stars all have S indices higher than the average for superflare stars. This is in agreement with the idea that binarity, through tidal coupling, can lead to increased magnetic activity[67].

We have also examined the Hα line of the 48 superflare stars in our study to search for any signs of binarity. All 48 superflare stars show nice narrow Hα absorption lines, with no indication of binarity. Owing to the low resolution of the LAMOST spectra, this does not rule out the possibility that more than six of these stars are binaries. It does, however, show that none of the 48 superflare stars are T Tauri stars. T Tauri stars are pre-main-sequence stars that show strong activity, especially in the Hα line and are known to host large flares[69].

In order to test if our conclusions are affected by contamination of the Kepler photometry, we have repeated the whole analysis excluding the six binary stars. This did not change any of the conclusions presented in this study. The distribution of the measured activity in the now 42 superflare star ensemble is still significantly different from the distribution of the measured activity levels of all the 5,648 solar-like stars at a 6σ significance level. 36 out of 42 superflare stars have activity levels larger than 0.169 and 30 out of 42 have activity levels larger than 0.205. If we only look at superflares with total energies larger than $2 \cdot 10^{34}$ erg, we observe that 26 out of 26 stars have activity levels larger than 0.169, whereas 22 out of 26 have activity levels larger than 0.205. We thus conclude that the conclusions presented here are not affected by contamination of the Kepler photometry.

References

1. Carrington, R. C. Description of a singular appearance seen in the Sun on September 1, 1859. *Mon. Not. R. Astron. Soc.* **20**, 13–15 (1859).
2. Hodgson, R. On a curious appearance seen in the Sun. *Mon. Not. R. Astron. Soc.* **20**, 15–16 (1859).
3. Muller, C. The carrington solar flares of 1859: consequences on life. *Orig. Life Evol. Biosph.* **44**, 185–195 (2014).
4. Cliver, E. W. & Dietrich, W. F. The 1859 space weather event revisited: limits of extreme activity. *J. Space Weather Space Clim* **3**, AA31 (2013).
5. Schaefer, B. E., King, J. R. & Deliyannis, C. P. Superflares on ordinary solar-type stars. *Astrophys. J.* **529**, 1026–1030 (2000).
6. Simon, T., Linsky, J. L. & Schiffer, F. H. IUE spectra of a flare in the RS Canum Venaticorum-type system UX ARIETIS. *Astrophys. J.* **239**, 911–918 (1980).
7. Hayashi, M. R., Shibata, K. & Matsumoto, R. X-Ray flares and mass outflows driven by magnetic interaction between a protostar and its surrounding disk. *Astrophys. J.* **468**, L37–L40 (1996).
8. Rubenstein, E. P. & Schaefer, B. E. Are superflares on solar analogues caused by extrasolar planets? *Astrophys. J.* **529**, 1031–1033 (2000).
9. Ip, W.-H., Kopp, A. & Hu, J.-H. 2004, On the star-magnetosphere interaction of close-in exoplanets. *Astrophys. J.* **602**, L53–L56 (2004).

10. Cuntz, M., Saar, S. ~ H. & Musielak, Z. ~ E. On stellar activity enhancement due to interactions with extrasolar giant planets. *Astrophys. J.* **533**, L151–L154 (2000).

11. Buccino, A. P. & Mauas, P. J. D. Long-term chromospheric activity of non-eclipsing RS CVn-type stars. *A&A* **495**, 287–295 (2009).

12. Hartman, J. ~ D. A correlation between stellar activity and the surface gravity of hot Jupiters. *Astrophys. J* **717**, L138–L142 (2010).

13. Shibata, K., Isobe, H. & Hillier, A. Can superflares occur on our Sun? *Publ. Astron. Soc. Jpn.* **65**, 49 (2013).

14. Koch, D. G. *et al.* Kepler mission design, realized photometric performance, and early science. *Astrophys. J.* **713**, L79–L86 (2010).

15. Maehara, H. *et al.* Superflares on solar-type stars. *Nature* **485**, 478–481 (2012).

16. Shibayama, T. *et al.* Superflares on solar-type stars observed with Kepler. I. Statistical properties of superflares. *Astrophys. J. Ser.* **209**, 5 (2013).

17. Notsu, Y. *et al.* Superflares on solar-type stars observed with Kepler II. Photometric variability of superflare-generating stars: a signature of stellar rotation and starspots. *Astrophys. J.* **771**, 127 (2013).

18. Candelaresi, S., Hillier, A., Maehara, H., Brandenburg, A. & Shibata, K. Superflare occurrence and energies on G-, K-, and M-type dwarfs. *Astrophys. J.* **792**, 67 (2014).

19. Nogami, D. *et al.* Two sun-like superflare stars rotating as slow as the Sun. *Publ. Astron. Soc. Jpn* **66**, L4 (2014).

20. Nielsen, M. B. & Karoff, C. Starspot modeling, Kepler, surface rotation. *AN* **333**, 1036–1039 (2012).

21. Christiansen, J. L. *et al.* Kepler Data Characteristics Handbook. KSCI-19040-004. Available online: https://archive.stsci.edu/kepler/documents.html (2013).

22. Basri, G. *et al.* Photometric Variability in Kepler Target Stars. II. An Overview of Amplitude, Periodicity, and Rotation in First Quarter Data. *AJ* **141**, 20 (2011).

23. Meibom, S. *et al.* A spin-down clock for cool stars from observations of a 2.5-billion-year-old cluster. *Nature* **517**, 589–591 (2015).

24. Eberhard, G. & Schwarzschild, K. On the reversal of the calcium lines H and K in stellar spectra. *ApJ* **38**, 292–295 (1913).

25. Wilson, O. C. Flux measurements at the centers of stellar H- and K-lines. *Astrophys. J.* **153**, 221–234 (1968).

26. Linsky, J. L., McClintock, W., Robertson, R. M. & Worden, S. P. Stellar model chromospheres. X - High-resolution, absolute flux profiles of the CA II H and K lines in stars of spectral types F0-M2. *Astrophys. J.* **41**, 481–500 (1979).

27. Schrijver, C. J., Cote, J., Zwaan, C. & Saar, S. H. Relations between the photospheric magnetic field and the emission from the outer atmospheres of cool stars. *I - The solar CA II K Line Core Emission* **337**, 964–976 (1989).

28. Notsu, S. *et al.* High-Dispersion Spectroscopy of the Superflare Star KIC 6934317. *Publ. Astron. Soc. Jpn* **65**, 112 (2013).

29. Wichmann, R., Fuhrmeister, B., Wolter, U. & Nagel, E. Kepler super-flare stars: what are they? *A&A* **567**, A36 (2014).

30. Notsu, Y. *et al.* High-dispersion spectroscopy of solar-type superflare stars. I. Temperature, surface gravity, metallicity, and vsin i. *Publ. Astron. Soc. Jpn.* **67**, 3224 (2015).

31. Notsu, Y. *et al.* High dispersion spectroscopy of solar-type superflare stars. II. Stellar rotation, starspots, and chromospheric activities. *Publ. Astron. Soc. Jpn.* **67**, 3314 (2015).

32. Takeda, Y., Honda, S., Kawanomoto, S., Ando & Sakurai, T. Behavior of Li abundances in solar-analog stars. II. Evidence of the connection with rotation and stellar activity. *A&A* **515**, A93 (2010).

33. Cui, X. Q. *et al.* The Large Sky Area Multi-Object Fiber Spectroscopic Telescope (LAMOST). *Res. Astron. Astrophys.* **12**, 1197–1242 (2012).

34. Henry, T. J., Soderblom, D. R., Donahue, R. A. & Baliunas, S. L. A survey of Ca II H and K chromospheric emission in southern solar-type stars. *AJ* **111**, 439–465 (1996).

35. Luo, A.-L. *et al.* Data release of the LAMOST pilot survey. *Res. Astron. Astrophys.* **12**, 1243–1246 (2012).

36. McQuillan, A., Mazeh, T. & Aigrain, S. Rotation Periods of 34,030 Kepler main-sequence stars: the full autocorrelation sample. *Astrophys. J. Suppl. Ser.* **211**, 14 (2014).

37. Kane, S. R. & von Braun, K. Constraining orbital parameters through planetary transit monitoring. *Astrophys. J.* **689**, 492–498 (2008).

38. Schrijver, C. J., Beer, J. & Baltensperger, U. *et al.* Estimating the frequency of extremely energetic solar events, based on solar, stellar, lunar, and terrestrial records. *JGR* **117**, 8103 (2012).

39. Lingenfelter, R. E. & Hudson, H. S. in *The Ancient Sun: Fossil Record in the Earth, Moon and Meteorites* (eds Pepin, R. O., Eddy, J. A. & Merrill, R. B.) 69–79 (Pergamon Press, 1980).

40. Kucera, T. A., Dennis, B. R., Schwartz, R. A. & Shaw, D. Evidence for a cutoff in the frequency distribution of solar flares from small active regions. *Astrophys. J.* **475**, 338–347 (1997).

41. Usoskin, I. G. & Kovaltsov, G. A. Occurrence of extreme solar particle events: assessment from historical proxy data. *Astrophys. J.* **757**, 92 (2012).

42. Kovaltsov, G. A. & Usoskin, I. G. Occurrence probability of large solar energetic particle events: assessment from data on cosmogenic radionuclides in lunar rocks. *Solar Phys.* **289**, 211–220 (2014).

43. Miyake, F., Nagaya, K., Masuda, K. & Nakamura, T. A signature of cosmic-ray increase in AD 774-775 from tree rings in Japan. *Nature* **486**, 240–242 (2012).

44. Melott, A. L. & Thomas, B. C. Causes of an AD 774-775 ^{14}C increase. *Nature* **491**, E1–E2 (2012).

45. Usoskin, I. ~ G. *et al.* The AD775 cosmic event revisited: the Sun is to blame. *A&A* **552**, L3 (2013).

46. Cliver, E. W., Tylka, A. J., Dietrich, W. F. & Ling, A. G. On a solar origin for the cosmogenic nuclide event of 775A.D. *Astrophys. J.* **781**, 32 (2014).

47. Jull, A. J. T. *et al.* Excursions in the ^{14}C record at A.D. 774-775 in tree rings from Russia and America. *GRL* **41**, 3004–3010 (2014).

48. Miyake, F. *et al.* Cosmic ray event of A.D. 774-775 shown in quasi-annual ^{10}Be data from the Antarctic Dome Fuji ice core. *GRL* **42**, 84–89 (2015).

49. Neuhäuser, R. & Neuhäuser, D. L. Solar activity around AD 775 from aurorae and radiocarbon. *AN* **336**, 225–248 (2015).

50. Mekhaldi, F. *et al.* Multiradionuclide evidence for the solar origin of the cosmic-ray events of AD 774/5 and 993/4. *Nat. Commun.* **6**, 8611 (2015).

51. Miyake, F., Masuda, K. & Nakamura, T. Another rapid event in the carbon-14 content of tree rings. *Nat. Commun.* **4**, 1748 (2013).

52. Aulanier, G. *et al.* The standard flare model in three dimensions. II. Upper limit on solar flare energy. *A&A* **549**, A66 (2013).

53. De Cat, P. *et al.* LAMOST observations in the Kepler field. *Astrophys. J. Suppl. Ser.* **220**, 19 (2015).

54. Vaughan, A. H., Preston, G. W. & Wilson, O. C. Flux measurements of Ca II H and K emission. *Astron. Soc. Pac.* **90**, 267–274 (1978).

55. Hall, J. C., Lockwood, G. W. & Skiff, B. A. The activity and variability of the Sun and Sun-like stars. *I. Synoptic Ca II H and K Observations* **133**, 862–881 (2007).

56. Isaacson, H. & Fischer, D. Chromospheric Activity and Jitter Measurements for 2630 Stars on the California Planet Search. *Astrophys. J.* **725**, 875–885 (2010).

57. Alonso, A., Arribas, S. & Martinez-Roger, C. The empirical scale of temperatures of the low main sequence (F0V-K5V). *A&A* **313**, 873–890 (1996).

58. Brown, T. M., Latham, D. W., Everett, M. E. & Esquerdo, G. A. Kepler Inout Catalog: Photometric Calibratic and Stellar Classification. *Astron. J.* **142**, 112 (2011).

59. Huber, D. *et al.* Revised stellar properties of Kepler targets for the quarter 1-16 transit detection run. *Astrophys. J. Suppl. Ser.* **211**, 2 (2015).

60. Karoff, C. Comparison of photometric variability before and after stellar flares. *Astrophys. J.* **781**, L22 (2014).

61. Noyes, R. W. *et al.* Rotation, convection, and magnetic activity in lower main-sequence stars. *Astrophys. J.* **279**, 763–777 (1984).

62. Mittag, M., Schmitt, J. H. M. M. & Schröder, K. P. Ca II H + K fluxes from S-indices of large samples: a reliable and consistent conversion based on PHOENIX model atmospheres. *A&A* **549**, A117 (2013).

63. Eddy, J. A. The maunder minimum. *Science* **192**, 1189–1202 (1976).

64. Crosby, N. B., Aschwanden, M. J. & Dennis, B. R. Frequency distributions and correlations of solar X-ray flare parameters. *Solar Phys.* **143**, 275–299 (1993).

65. Jenkins, J. M. *et al.* Overview of the Kepler science processing pipeline. *Astrophys. J.* **713**, L87 (2010).

66. Balona, L. A. Kepler observations of flaring A-F type stars. *Mon. Not. R. Astron. Soc.* **423**, 3420–3429 (2012).

67. Strassmeier, K. G. *et al.* Chromospheric CA II H and K and H-alpha emission in single and binary stars of spectral types F6-M2. *Astrophys. J. Suppl. Ser.* **72**, 191–230 (1990).

68. Maehara, H. *et al.* Statistical properties of superflares on solar-type stars based on 1-min cadence data. *Earth Planets Space* **67**, 59 (2015).

69. Herbig, G. H. The properties of T Tauri stars and related objects. *Adv. Astron. Astrophys.* **1**, 47–103 (1962).

Acknowledgements

Guoshoujing Telescope (the Large Sky Area Multi-Object Fiber Spectroscopic Telescope, LAMOST) is a National Major Scientific Project built by the Chinese Academy of Sciences. Funding for the project has been provided by the National Development and Reform Commission. LAMOST is operated and managed by the National Astronomical Observatories, Chinese Academy of Sciences. Funding for the Stellar Astrophysics Centre is provided by the Danish National Research Foundation (Grant agreement No.: DNRF106). The project has been supported by the Villum Foundation. JNF acknowledges the support from the Joint Fund of Astronomy of National Natural Science Foundation of China (NSFC) and Chinese Academy of Sciences through the Grant U1231202, and the support from the National Basic Research Program of China (973 Program 2014CB845700 and 2013CB834900).

Author contributions

C.K. was responsible for the planning, coordination and data analysis. C.K. also wrote the majority of the text. M.F.K., A.B. and A.F. helped with data analysis, theoretical interpretations and writing of the paper. P. de C. and J.F. were responsible for the planning, coordination and reduction of the LAMOST observations. A.F.-S., F.I. and J.O. provided theoretical interpretation of the observations and gave advice on the paper's content. Y.Z., Y.H., Y.W., J.S. and W.Z. were responsible for producing the LAMOST data.

Additional information

Competing financial interests: The authors declare no competing financial interest.

25

Aerosol influence on energy balance of the middle atmosphere of Jupiter

Xi Zhang[1], Robert A. West[2], Patrick G.J. Irwin[3], Conor A. Nixon[4] & Yuk L. Yung[5]

Aerosols are ubiquitous in planetary atmospheres in the Solar System. However, radiative forcing on Jupiter has traditionally been attributed to solar heating and infrared cooling of gaseous constituents only, while the significance of aerosol radiative effects has been a long-standing controversy. Here we show, based on observations from the NASA spacecraft Voyager and Cassini, that gases alone cannot maintain the global energy balance in the middle atmosphere of Jupiter. Instead, a thick aerosol layer consisting of fluffy, fractal aggregate particles produced by photochemistry and auroral chemistry dominates the stratospheric radiative heating at middle and high latitudes, exceeding the local gas heating rate by a factor of 5–10. On a global average, aerosol heating is comparable to the gas contribution and aerosol cooling is more important than previously thought. We argue that fractal aggregate particles may also have a significant role in controlling the atmospheric radiative energy balance on other planets, as on Jupiter.

[1] Department of Earth and Planetary Sciences, University of California Santa Cruz, Santa Cruz, California 95064, USA. [2] Jet Propulsion Laboratory, California Institute of Technology, 4800 Oak Grove Drive, Pasadena, California 91109, USA. [3] Atmospheric, Oceanic and Planetary Physics, University of Oxford, Clarendon Laboratory, Parks Road, Oxford OX1 3PU, UK. [4] NASA Goddard Space Flight Center, Greenbelt, Maryland 20771, USA. [5] Division of Geological and Planetary Sciences, California Institute of Technology, Pasadena, California 91125, USA. Correspondence and requests for materials should be addressed to X.Z. (email: xiz@ucsc.edu).

As on Earth, Jupiter's atmospheric temperature profile exhibits a strong inversion above the tropopause[1], implying that its middle atmosphere, or the 'stratosphere', is convectively inhibited. Therefore, the energy budget should be dominated by radiation and the stratified middle atmosphere is in global radiative equilibrium. A first-order question is: which constituents in the atmosphere control this energy balance? About half of the incoming solar radiation on Jupiter penetrates deep into the troposphere and one third is reflected back to space (Fig. 1)[2]. The bulk constituents, hydrogen and helium, are not radiatively active except via H_2–H_2 and H_2–He collisional-induced absorption (CIA) at pressures >10 hPa (refs 3,4). The next most abundant gas, methane (CH_4), diffuses upward from the deep atmosphere and heats the stratosphere by absorbing the near-infrared solar flux[3-8]. The methane photochemical products acetylene (C_2H_2) and ethane (C_2H_6), together with H_2–H_2 and H_2–He CIA, absorb the upward mid-infrared radiation from the troposphere and re-radiate it to space, resulting in an efficient net cooling of the middle atmosphere[3-9] to compensate the solar heating.

The global maps of temperature and C_2 hydrocarbons were recently retrieved from the Jupiter flyby data from Cassini and Voyager-1 spacecraft in 2000 (refs 4,10-12) and 1979 (refs 4,12), respectively. On the basis of a state-of-the-art radiative transfer model (see Methods section), we investigate the global energy balance of Jupiter[4]. Surprisingly, the global average cooling flux by gaseous constituents in the middle atmosphere is estimated to be ~ 1.4 W m^{-2}, about 1.5 times larger than solar flux absorbed by the stratospheric CH_4 (~ 0.9 W m^{-2}; Fig. 1). Vertically, the gas solar heating rate is substantially smaller than the gas thermal cooling at pressures >10 hPa (ref. 4). The energy imbalance consistently revealed by the Voyager and Cassini data is not a seasonal effect because Jupiter has nearly zero obliquity. The Jupiter–Sun distance was different for the two flybys, varying

from northern fall equinox (Voyager) to the northern summer solstice (Cassini), but the global average heating is not altered significantly. Long-term ground-based observations from 1980 to 2000 also show that the global average temperature at 20 hPa does not substantially vary with time[10], and thus neither does the thermal radiative cooling. The violation of the radiative energy equilibrium thereby suggests the presence of an additional strong heat source other than CH_4 in the middle atmosphere of Jupiter, which absorbs the missing ~ 0.5 W m^{-2}.

Here we show that the missing heat source is aerosols, the end product of atmospheric chemistry on Jupiter. As a result of photochemistry, with the help of auroral chemistry, especially at high latitudes (or the 'auroral zone') where high-energy particles penetrate into the atmosphere, complex hydrocarbon compounds can form and eventually coagulate and condense as aerosols, or haze particles[13,14]. On the basis of Cassini imaging science subsystem (ISS) observations[15], here we derive the globally averaged solar flux absorbed by the stratospheric aerosols of ~ 0.5 to 0.7 W m^{-2}, more than half of the amount due to CH_4 (Fig. 1). The aerosol heating is predominant at middle and high latitudes, exceeding the local gas heating rate by a factor of 5–10. For the first time, we estimate the possible aerosol cooling effect, which might be as important as the cooling via hydrocarbons at high latitudes. We conclude that the aerosols maintain the atmospheric energy balance and must be partially responsible for the stratospheric temperature inversion. That the photochemistry and auroral chemistry control the atmospheric energetics via the production of aerosols suggests that Jupiter exhibits a new regime of atmospheric energy balance that is different from that of the Earth.

Results

Aerosol heating effect. The global aerosol map has been revealed from images acquired by the ISS during its Jupiter flyby[15]. At low latitudes (40° S to 25° N), an optically thin layer composed of compacted particles with radii ~ 0.2–0.5 µm is found to be concentrated at ~ 50 hPa. The rest of the atmosphere is covered by an optically thick aerosol layer at 10–20 hPa composed of fluffy, fractal particles aggregated from hundreds to thousands of ten-nanometre size monomers, similar to the haze particles on Titan[16,17]. The fractal dimension of these aggregates is assumed as 2, meaning that their geometric structure of the aggregate particles lies between a long linear chain (fractal dimension of 1) and a fully compacted cluster (fractal dimension of 3). This type of fractal aggregates is consistent with the ISS observations and the polarization observations of Jupiter, whereas spherical particles are not[16,17]. Fractal aggregates are known to be much more absorbing than spherical particles in the ultraviolet and visible wavelengths[18]. For instance, from 0.2 to 1 µm, an aggregate particle composed of a thousand ten-nanometre monomers can absorb twice as much of the solar flux as an ensemble of 0.07 µm individual spherical particles (assumed in ref. 19) with the same extinction optical depth, because the former is less scattering than the latter. Figure 2b shows that, at high latitudes, the opacity of the fractal aggregates on Jupiter is considerably larger than the CH_4 opacity in the ultraviolet and visible ranges, indicating that these particles can absorb a significant fraction of solar energy at wavelengths where the solar blackbody radiation peaks (Fig. 2a). In addition, long-term observations suggest that the seasonal variation of the Jovian north–south polarization asymmetry is only $\sim 0.5\%$ (ref. 20). The lack of strong temporal variation implies that the fractal aggregates constitute a steady heat source.

The radiative heating calculations demonstrate that the middle atmosphere of Jupiter is heated by two components: aerosols in

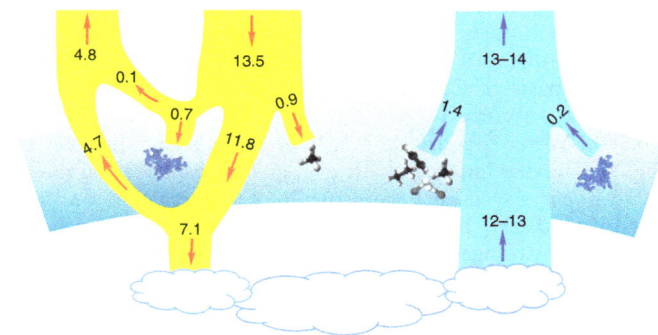

Figure 1 | Globally averaged heating and cooling fluxes on Jupiter. The heating (yellow branch) and cooling (cyan branch) fluxes are in units of W m^{-2}. The stratosphere is shaded. The heating flux is associated with the incoming solar radiation and the cooling flux is related to the outgoing thermal radiation. Of the 13.5 W m^{-2} of solar radiation incident to Jupiter's atmosphere, 0.1 W m^{-2} is reflected back to space and 11.8 W m^{-2} is transmitted to the troposphere. Tropospheric hazes and clouds absorbed 7.1 W m^{-2} and 4.7 W m^{-2} is reflected back to space[2]. The remainder of the solar energy is absorbed in the middle atmosphere by fractal haze particles (0.7 W m^{-2}) and CH_4 gas molecules (0.9 W m^{-2}). The total outgoing thermal radiation from our radiative calculation is ~ 13–14 W m^{-2}, consistent with that from Cassini and Voyager observations[9]. The thermal cooling flux is mainly emitted from the troposphere (12–13 W m^{-2}). In the middle atmosphere, the net cooling flux is 1.4 W m^{-2} emitted by gas molecules H_2, CH_4, C_2H_2, and C_2H_6 (black and white molecule diagrams). The upper limit of the outgoing thermal flux from the fractal aggregates (blue diagrams) is ~ 0.2 W m^{-2} as determined in this study.

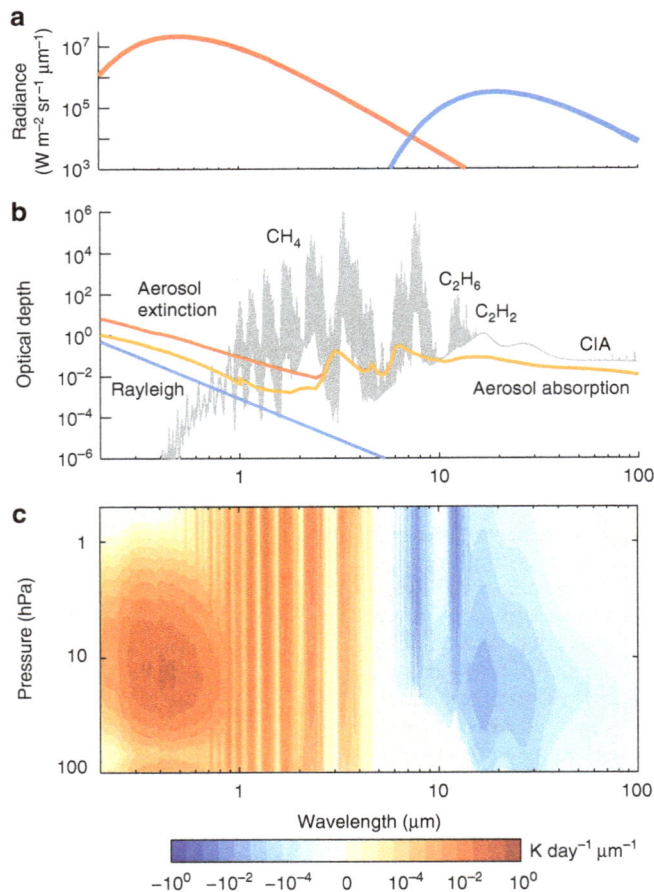

Figure 2 | Spectrally resolved heating and cooling rates and corresponding energy fluxes and opacity. (**a**) Globally averaged solar radiation received by Jupiter, approximated by a blackbody of 5,778 K (red) and Jupiter thermal radiation in the stratosphere approximated by a blackbody of 150 K (blue). (**b**) Total optical depth from the top of the atmosphere to 100 hPa as a function of wavelength at 60° S. The gas optical depth (grey) includes H_2-H_2 and H_2-He CIA and CH_4, C_2H_2 and C_2H_6 absorption. The non-gas components include Rayleigh scattering (blue), fractal aggregate aerosol extinction (red) and the aerosol absorption (orange). (**c**) Spectrally resolved zonally averaged solar heating (0.2-5 μm) and cooling (5-100 μm) map at 60° S. Absolute values of the heating/cooling rates that are $<10^{-6}$ K per day per μm are not shown. Solar heating dominates shortwards of 5 μm while Jupiter thermal cooling (shown in negative values here) dominates longwards of 5 μm. Contributions from the H_2-H_2 and H_2-He CIA and gas vibrational-rotational bands are shown. Aerosol heating is important in the ultraviolet and visible regions and aerosol cooling is important in the mid-infrared region beyond 11 μm.

the ultraviolet and visible wavelengths and CH_4 in the near-infrared. Figure 2c shows the spectrally resolved zonally averaged heating rate as a function of wavelength and pressure at 60° S. At near-infrared wavelengths longer than 0.9 μm, strong CH_4 bands completely dominate the heating with minor contributions from H_2–H_2 and H_2–He CIA (Fig. 2b). At shorter wavelengths, heating by the fractal aggregates is predominant. The maximum aerosol heating occurs near the wavelength of the solar spectrum peak ($\sim$0.5 μm), but slightly shortwards owing to the increasing absorption of aerosols towards the ultraviolet. At 60° S, the integrated aerosol heating rate over all wavelengths can reach $\sim$0.2 K per day at 10 hPa, where it exceeds the CH_4 heating rate by a factor of 10 (Fig. 3a). The aerosol heating rate appears to

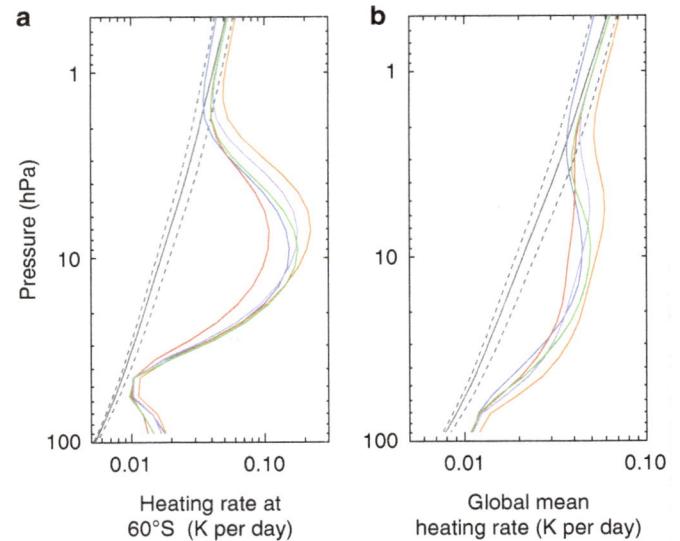

Figure 3 | Vertical heating rate profiles. (**a**) Zonally averaged heating rates at 60° S; (**b**) Globally averaged heating rates. The gas-only calculations are shown in black. The dashed black lines show the possible gas heating rates due to the uncertainty of CH_4 profiles. The coloured lines represent different aerosol retrieval solutions. Cases H1–H5 correspond to the green, red, purple, blue and orange curves, respectively. See Table 1 for detailed input information of the cases.

decrease rapidly above 5 hPa and does not contribute significantly to the total heating rate in the upper stratosphere. However, this conclusion is not certain since the resolution of the near-infrared spectra[21] is not sufficient to fully characterize the upper stratosphere. On the basis of higher-resolution observations[22], another aerosol layer was found above 5 hPa at the poles that might contribute to the local heating rate in the upper stratosphere, though not to the total energy budget of the middle atmosphere due to its lower density at those levels.

On the basis of Cassini observations, the globally averaged heat flux absorbed by the stratospheric aerosols is $\sim$0.5-0.7 W m^{-2}, more than half of the amount due to CH_4 (Fig. 1). The maximum globally averaged aerosol heating rate of $\sim$0.03 K per day occurs at $\sim$10 hPa, comparable to the total near-infrared CH_4 heating rate at the same pressure level. But the globally averaged heating rate including aerosols is about twice that of the gas-only heating rate at pressures $>$20 hPa (Fig. 3b). Spatially, the aerosol heating is predominant at middle and high latitudes, which is attributable to the optically thick fractal aggregates layer (Fig. 4b). Therefore, the aerosol heating naturally, if not coincidentally, compensates for the energy deficit due to gas heating and cooling at the appropriate pressure levels (Fig. 4e).

Owing to the existence of degenerate solutions in the interpretation of ISS observations[15], we performed sensitivity tests to estimate the uncertainty ranges of the aerosol heating rate. The tropospheric haze and cloud are treated as an effective cloud layer in the troposphere[15]. Our tests show that the effective cloud albedo and phase function within the retrieved uncertainties has insignificant effect on the stratospheric heating rate. The heating rate is more sensitive to the total optical depth and single scattering albedo of the stratospheric aerosols. Given the constraints from the Cassini ISS observations, testing the sensitivity of the heating rate to each individual aerosol parameter is inappropriate. However, multiple solutions still exist in the aerosol retrieval[15]. Therefore, we adopted five typical retrieval solutions for the middle and high latitudes, namely cases

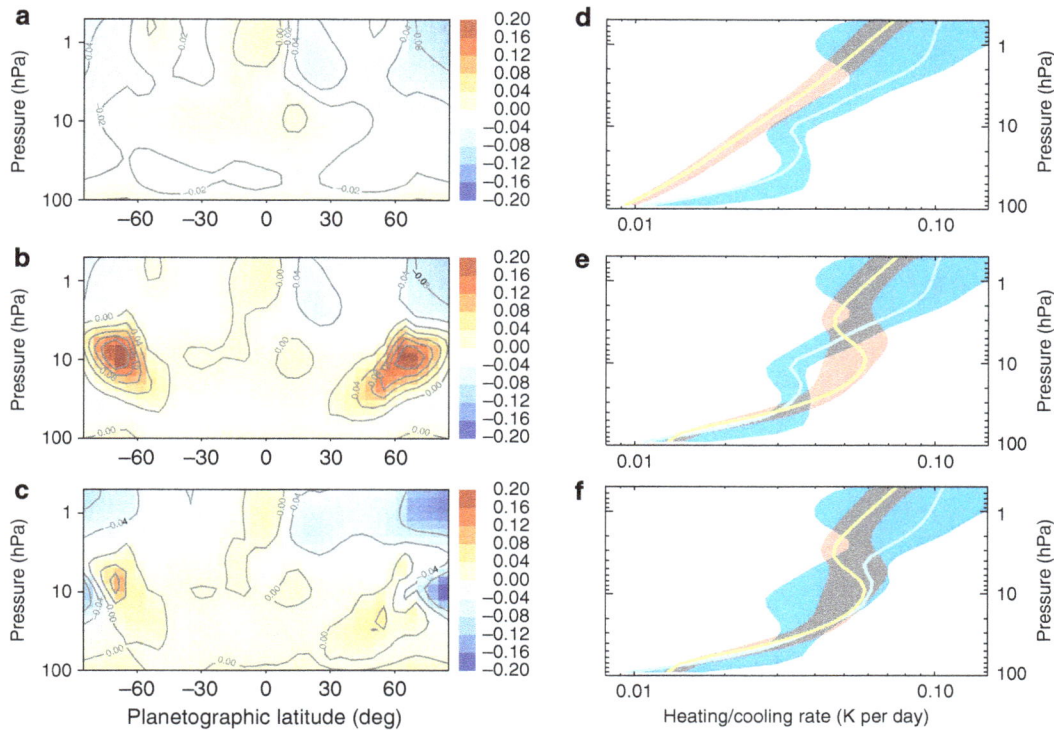

Figure 4 | Radiative balance calculation results in the middle atmosphere of Jupiter based on Cassini flyby observations. (**a**) Net radiative heating rate map (in units of K per Earth day) without aerosols; (**b**) net radiative heating rate map with aerosol heating; (**c**) net radiative heating rate map with aerosol heating and cooling; (**d**) globally averaged heating (yellow with pink shading) and cooling (cyan with blue shading) profiles without aerosols; (**e**) globally averaged heating and cooling profiles with aerosol heating; and (**f**) globally averaged heating and cooling profiles with aerosol heating and cooling. The uncertainty ranges are shaded.

Table 1 | Sensitivity cases for ISS retrieval and heating rate calculation.

Case	CH_4 mixing ratio	k (UV1)	k (CB3/MT3)	Colour
H1 (Nominal)	1.8×10^{-3}	2×10^{-2}	1×10^{-3}	Green
H2	1.8×10^{-3}	6×10^{-3}	1×10^{-4}	Red
H3	1.8×10^{-3}	8×10^{-2}	4×10^{-3}	Purple
H4	1.5×10^{-3}	2×10^{-2}	1×10^{-3}	Blue
H5	2.5×10^{-3}	2×10^{-2}	1×10^{-3}	Orange

H1–H5 in Table 1. Those cases were designed to explore the parameter space within the uncertainties of the CH_4 mixing ratio[4,23] and the imaginary part of the refractive indices (hereafter k values) of the UV1 and CB3/MIT3 channels of Cassini ISS[15]. All cases provide good fits to the limb-darkening observations from Cassini ISS (Table 1).

The heating rate calculation at 60° S (Fig. 3a) shows that the maximum heating rate including aerosols is about a factor of 2 larger than the minimum at ~ 10 hPa. The maximum (H3) and minimum (H2) occurs when the k value of the UV1 channel reaches the upper and lower bound, respectively, outside which the UV1 limb-darkening profiles cannot be explained[15]. On a global average (Fig. 3b), the maximum heating rate including aerosols is about a factor of 1.5 larger than the minimum (shaded in red in Fig. 4e,f).

Aerosol cooling effect. Aerosols could also cool the middle atmosphere of Jupiter but this effect has not been explored in previous studies. The mid-infrared optical properties of aerosols produced in a hydrogen-dominated environment have not been

measured experimentally. But at wavelengths shorter than 2.5 µm, previous laboratory experiments found that the k values of aerosols produced in a CH_4/H_2 gas mixture could be either larger or smaller than their counterpart in the CH_4/N_2 mixture, depending on the chemical composition and environmental pressure[24,25]. Therefore, the fractal aggregates on Jupiter might have non-negligible opacity in the mid-infrared compared with H_2–H_2 and H_2–He CIA if their optical constants behave like the aerosols on Titan, which are strongly absorbing with almost no scattering beyond 5 µm (Fig. 2b)[26,27]. To estimate the possible cooling effect from aerosols, we included aerosol absorption in a non-linear inversion model to fit the spectra from Cassini CIRS[28]. However, owing to insufficient sensitivity of CIRS observations to the Jovian aerosol opacity, a pure-gas (that is, non-detection of aerosols) model is also able to fit the spectra[4,10–12]. Future analysis on the possible C–H bending vibrational features of aerosols at 1,380 and 1,460 cm^{-1} that have been detected on Titan[27] might provide more constraints on the infrared opacity and chemical structure of aerosol particles on Jupiter. Here we aim to derive the upper limit of aerosol opacity from the CIRS spectra and estimate the upper bound of the aerosol thermal infrared cooling.

For each latitude, we included aerosols in the Non-linear optimal Estimator for MultivariatE spectral analySIS (NEMESIS) model[28] and retrieved the temperature profile and the mixing ratios of C_2H_2 and C_2H_6 following the procedure detailed in ref. 4. Owing to the lack of Jupiter-analogue aerosol measurements in mid-infrared, we tested several k values based on the laboratory tholin results[26] and recently derived k values from Cassini observations on Titan[27]. The latter shows cooling 2–3 times smaller than the former and exhibits different wavelength dependence. We gradually increased the k values

Table 2 | Sensitivity cases for CIRS retrieval and cooling rate calculation.

Case	k values in the mid-infrared	χ^2/N (600–850 cm^{-1})	χ^2/N (1,225–1,325 cm^{-1})	Colour
C1	Pure gas, no aerosol	0.544	1.094	Red
C2	Titan tholin experiment[26]	0.543	1.042	Blue
C3	Titan CIRS observations[27]	0.541	1.090	Orange
C4 (Nominal)	C2 values divided by 2	0.543	1.051	Green
C5	Five times of C2 values	0.607	1.964	Brown

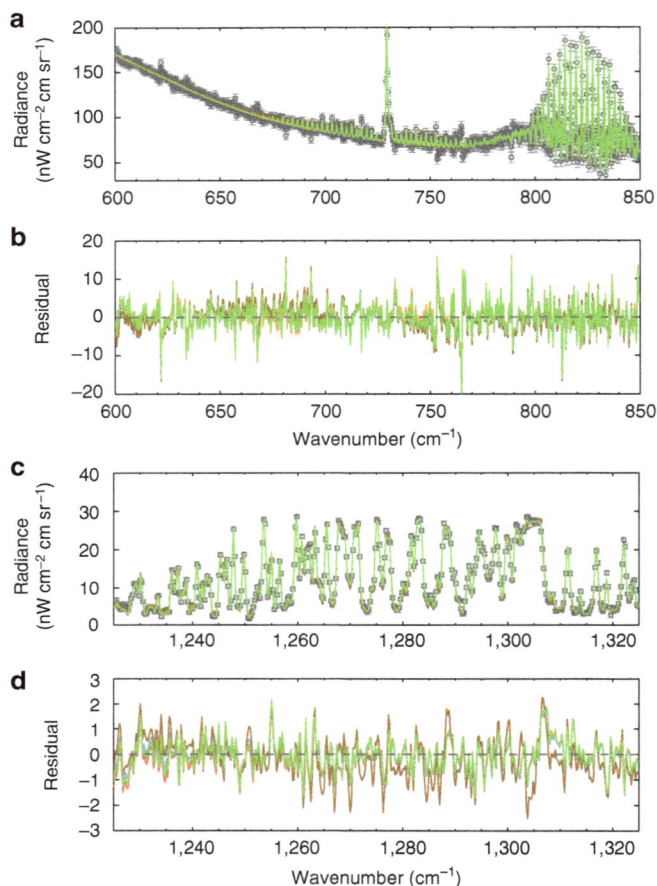

Figure 5 | Spectral inversion results at 57° S. (a) Spectra at 600–850 cm^{-1} region (H$_2$–H$_2$ and H$_2$–He CIA, C$_2$H$_2$ and C$_2$H$_6$ bands); **(b)** fitting residual at 600–850 cm^{-1} region; **(c)** spectra at 1,225–1,325 cm^{-1} region (CH$_4$ bands); **(d)** fitting residual at 1,225–1,325 cm^{-1} region. CIRS observations are shown as black circles. The red, blue, orange, green and brown colours represent NEMESIS retrieval cases C1–C5, respectively. The goodness of fit (χ^2/N where N is the number of measurements) in the 600–850 cm^{-1} region is ~0.5–0.6 for each case. In the CH$_4$ band, the goodness of fit is around unity for each case except for the brown case ($\chi^2/N = 1.96$), which does not fit the CIRS spectra. See Table 2 for detailed information of the cases.

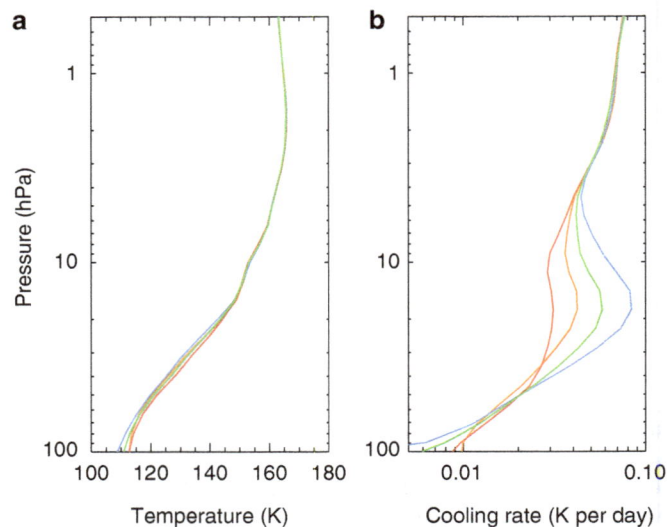

Figure 6 | Vertical temperature and cooling rate profiles at 57° S.
(a) Retrieved temperature profiles; **(b)** corresponding zonally averaged cooling rates. The red, blue, orange and green colours represent cases C1–C4, respectively. The C5 case is not used because it cannot explain the CIRS observations.

until the CIRS spectra cannot be fitted within the measurement uncertainties. Table 2 summarized five typical tests. For each choice of optical constants, we performed an atmospheric retrieval of the Cassini CIRS spectra using NEMESIS model to constrain the aerosol opacity. Figure 5 illustrates the retrieval fitting results and residual values at 57° S and the values of goodness of fit are shown in Table 2. The retrieved temperature profiles are shown in Fig. 6a. If we enhance the optical constants

from ref. 25 by a factor of 5 (case C5), we are not able to fit the CH$_4$ emission spectra (Fig. 5). Overall, we estimated the upper limit of the aerosol optical depth in the mid-infrared wavelengths to be ~0.1 at 100 hPa at high latitudes. This places an upper limit on the aerosol contribution to the globally averaged cooling flux of ~0.2 W m^{-2} in the stratosphere (Fig. 1).

The aerosol cooling effect could be significant at middle and high latitudes where the particles are abundant but negligible at low latitudes. When the aerosol opacity is included, the cooling rate increases in the aerosol layer (Fig. 6b). At 57° S, with a moderate choice of k (case C4), the zonally averaged cooling rate including aerosols is ~0.06 K per day. This is about two times larger than the gas-only cooling rate (case C1) at the pressures where the aerosol mixing ratio peaks (~10–20 hPa, Fig. 6b). The globally averaged aerosol cooling rate could be comparable to the gas cooling rate at 20 hPa and partially compensate for the aerosol heating effect (Fig. 4f). On the other hand, the cooling rates from the aerosol cases are smaller than the pure-gas case at pressure levels below the aerosol layer (Fig. 6b), a result of energy conservation as constrained by the total emission observed by CIRS. Indeed, stronger aerosol absorption leads to a colder retrieved temperature profile (Fig. 6a). For instance, the temperature profile at 57° S from the C2 case is ~3 K colder than the case without aerosols (C1 case) below ~40 hPa. This colder temperature leads to a smaller cooling effect than the pure-gas case (C1) at pressures > ~40 hPa where the cooling is dominated by H$_2$–H$_2$ and H$_2$–He CIA (Fig. 6b).

We estimate the possible aerosol cooling rate for each latitude and their influences on the globally averaged cooling rate by combining the 'ensemble uncertainty'[4] of the temperature and gas abundances and the aerosol cooling rate tests. The uncertainty range including the aerosol contribution (Fig. 4f) is larger than that without aerosols (Fig. 4d) because the aerosols remain elusive from the CIRS spectra. Global radiative equilibrium is achievable when both aerosol heating and cooling are included (Fig. 4f).

Discussion
Owing to the lack of sufficient observational evidence before, the importance of aerosol heating in the middle atmosphere of Jupiter has long been a controversial question. It was suggested since the 1970s that the Jovian aerosols might absorb solar radiation and heat the atmosphere[6–8]. In the 1990s, based on the International Ultraviolet Explorer and Voyager-2 data, aerosol heating was shown to have a large impact on the atmospheric circulation[29]. However, a later analysis using Hubble Space Telescope (HST) images found that the aerosols have relatively insignificant effects[19]. With better constraints on the spatial distribution and optical properties of aerosols, Cassini observations confirm a pronounced aerosol heating at high latitudes on Jupiter. Aside from the aerosol cooling effect, our conclusion is qualitatively consistent with previous estimate from the International Ultraviolet Explorer and Voyager-2 data[29] but with a much better spatial coverage. However, this work disagrees with heating rates derived solely on the basis of HST data[3,19].

The major difference between the current work and previous studies is probably attributable to the fractal nature of the aerosol and aerosol spatial distribution. Through the multi-channel–multi-phase retrieval on Cassini images, we can characterize the fractal aggregates in great detail, including the optical depth, single scattering albedo and phase function of the particles. On the basis of low-phase-angle images alone[19], the polar aerosols were assumed as tiny spherical particles of $\sim 0.07\,\mu m$ in radius instead of the submicron size aggregate particles. The latter have lower single scattering albedo than the spherical particles. Furthermore, with a larger particle size, the fractal aggregates have less backscattering than the spherical particles. Owing to the above two reasons, a larger total optical depth of the fractal aggregates is required to explain the low-phase-angle I/F observations, leading to a larger heating rate in this study, than that in ref. 19, at the south pole. Note that the k values used in ref. 19 are lower than our nominal model but still within the sensitivity test range in our study.

The vertical distributions of aerosols in previous studies[19,29] are based on microphysical simulations that are inconsistent with the near-infrared observations[21]. Previous studies adopted a haze layer located above 10 hPa at polar regions, while the near-infrared spectra reveal a main haze layer at 10–20 hPa (ref. 20). West et al.[29] only sampled two latitudes and estimated the other latitude information by scaling. The HST images[19] have a good latitudinal coverage, but the aerosol heating appears influential only at the south pole, not at middle latitudes or at the north pole. Moreno et al.[19] reported the aerosol optical depth at the north pole one order of magnitude less than that at the south pole. This result is inconsistent with other observations. For example, recent high-resolution ground-based near-infrared spectra[22] concluded that the near-infrared haze optical depth at the northern pole is comparable to that at the southern pole. The Cassini images in low and high phase angles also revealed that the haze ultraviolet optical depth at north high latitudes is not significantly less than its south counterpart[15]. Furthermore, Cassini images show that the haze optical depth in the ultraviolet channel can approach unity down to 100 hPa at middle latitudes[15]. Including the haze

contribution at those latitudes would greatly enhance the aerosol heating. This will not only influence the local heat balance but also on the global energy equilibrium, especially at 10–20 hPa, as shown in our study.

Several other factors might also attribute to the difference between the current work and previous studies. For instance, we have a much better global coverage of the temperature, hydrocarbons and aerosols based on Cassini observations. The cooling rate in ref. 3 was likely to be underestimated because the temperature profile from Galileo entry probe is shown to be colder than the globally averaged temperature profile from Voyager and Cassini observations[4], albeit the cooling rate in ref. 3 is still slightly larger than the gas heating rate at pressures >10 hPa. The spectroscopic data of CH_4, C_2H_2 and C_2H_6 have been greatly improved in the last decade (see Methods section for gas opacity). The state-of-the-art line data allow us to adopt the line-by-line approach to resolve the vibrational–rotational line shape of hydrocarbons[4], the most accurate radiative transfer method to estimate the gas heating and cooling rate.

Another possible heating mechanism in the middle atmosphere is energy dissipation of upward propagating gravity waves from the troposphere. However, as per previous studies[30,31], this hypothesis has several defects. First, there is little evidence of stratospheric gravity waves at middle and high latitudes. Second, there is no direct evidence of wave breaking in the lower stratosphere of Jupiter. Third, gravity wave breaking could either heat or cool the middle atmosphere but the net effect is difficult to quantify[30].

Unlike the Earth, on which the photochemical product (ozone) only dominates the atmospheric radiative heating, Jupiter might exhibit a different regime of atmospheric energy balance where both the heating and cooling are significantly controlled by the photochemistry and auroral chemistry via the production of aerosols and C_2 hydrocarbons. Aside from the first-order global energy balance, aerosol heating and cooling on Jupiter also influence the spatial distributions of radiative forcing, which has a significant impact on the large-scale dynamical circulation in the middle atmosphere[5,19,29]. The NASA JUNO spacecraft, arriving at Jupiter in 2016, will provide more insights on the aurora processes and aerosol formation in the polar region.

Jupiter is the second planetary body and the first hydrogen-dominated planet that shows evidence of hydrocarbon aerosols playing a significant role in regulating the radiation flux and most probably the circulation of its middle atmosphere. The other one is Titan[32]. Although Jupiter's atmosphere is primarily dominated by hydrogen, Titan's is dominated by nitrogen, both of these atmospheres produce fluffy, fractal aggregate particles, suggesting that fractal aggregates might be a ubiquitous result of hydrocarbon chemistry. In view of the existence of hydrocarbon aerosols in many other atmospheres dominated by hydrogen or nitrogen, such as those of Saturn[33], Uranus[34] and Neptune[35], the early Earth[18], and possibly some exoplanets[36–40], we hypothesize that the heating and cooling from fractal aggregates could also be important for determining the radiative energy distribution and climate evolution on these planets. Owing to their strong heating effects, fractal aggregates play a significant role in creating the temperature inversion in the lower stratosphere of Jupiter. They might also be partially responsible for the temperature inversions observed on the other giant planets in the Solar System, but neglected in previous studies[41]. A typical feature of the fluffy aggregate particle is its extremely low density, which might help to explain the existence of very high and thick haze layers at pressures <0.1 hPa, as seen on the Neptune and sub-Neptune size planets GJ436b (ref. 39) and GJ1214b (ref. 40). A thorough study of fractal aggregates will shed light on how to characterize these particles in future photometry and polarization observations.

Methods

Radiative heating and cooling model. For heating rate calculations between 0.20 and 0.94 μm where the aerosol contribution is significant, we use a multiple scattering model based on the C version of the discrete ordinates radiative transfer code (DISORT Program for a multi-layered plane-parallel medium)[42]. The phase function and cross sections of low-latitude particles were calculated based on Mie theory, while that of the fractal aggregates at middle and high latitudes were computed using a parameterization method for the aggregates with a fractal dimension of two[17]. The parameterization is based on electromagnetic scattering computations using the multi-sphere method[43]. In the heating rate calculations, we use 32 streams to characterize the intensity angular distribution, which displays almost no difference from the 64-stream case. A Gaussian quadrature method with 10 zenith angles is used to average the heating rates longitudinally. The spectral resolution is 0.001 μm. The effective cloud albedo in the troposphere is interpolated between 0.20 and 0.94 μm based on the retrieved albedo from the UV1 channel and CB3/MT3 channels of Cassini ISS[15].

At longer wavelengths (0.94–200 μm), our calculations adopt the line-by-line approach, based on a state-of-the-art high-resolution radiative heating and cooling model for the stratosphere of Jupiter, which has been rigorously validated against simple but realistic analytical solutions[4]. The CH_4 heating rate calculation from 0.94 to 10 μm is performed with a spectral resolution of 0.005 cm^{-1} to resolve the CH_4 spectral line shape using the most current CH_4 line lists (see discussion for gas opacity below). The thermal cooling rate from 50 to 2,500 cm^{-1} (4–200 μm) is calculated with a spectral resolution of 0.001 cm^{-1}.

We calculate heating and cooling rates for every latitude to produce the latitude-pressure two-dimensional maps. The specific heat of Jupiter's atmosphere is taken as $1.0998 \times 10^4 \, J \, kg^{-1} \, K^{-1}$ (ref. 44). The globally averaged profiles are obtained via an area-weighted mean from 90° N to 90° S. Since the data quality at latitudes north of 70° N and south of 70° S is not sufficiently good for rigorous atmospheric retrieval, we do not derive the atmospheric characteristics from the ISS images and CIRS spectra. Instead, we assume that the vertical profiles of aerosol, temperature and gases are identical to their values at 70° north and south, respectively.

This assumption might introduce some uncertainty in our estimate of the global energy balance because the heating and cooling in the polar region, especially in the infrared aurora region, are not negligible. The polar aurora is known to be highly variable both temporally and spatially[1]. According to our limited data on the polar regions from Galileo[45] and Cassini[46] spacecraft as well as ground-based observations[47,48], both regions poleward of 70° N and 70° S could be different from that ~65°–70°. For example, the Cassini CIRS instrument[46] detected a variation of thermal emission over the C_2H_2 and C_2H_6 bands at regions poleward of 65°, within about a factor of 2. However, because the surface area poleward of 70° amounts to merely 6% of the total surface area of Jupiter, increasing the polar cooling rate by a factor of 2 will probably introduce an uncertainty of only ~6% of the total cooling rate, which is still well-located within our estimated uncertainty range (Fig. 4f). On the other hand, the pressure level of infrared aurora source has not been precisely determined. If the emission originates from the upper stratosphere (for example, above 0.1 mbar level), the aurora and its variability might have little impact on the thermal cooling rate at pressures we focus here. We should also point out that the aurora might also be associated with some heating mechanisms in the polar region that have not been considered in this study. Future analysis of the Jovian polar region will provide more information on the local energetics.

The aerosol and gas opacities used in the radiative calculations are discussed below.

Aerosol opacity sources. In the visible and near-infrared wavelengths, the early laboratory studies[24–26] measured the optical properties of Titan-analogue aerosols (that is, in the nitrogen environment) and Jupiter-analogue aerosols (that is, in the hydrogen environment), respectively. Using the radio frequency glow discharge in a CH_4/H_2 gas mixture, it was found that the refractive indices of these aerosols are consistent with high-phase-angle photometry data of Uranus by Voyager-2 at 0.55 μm (ref. 24). Compared with aerosols produced in the CH_4/N_2 mixture[26], the imaginary part of the refractive index of the Jupiter-analogue aerosols could be either larger or smaller, depending on the chemical composition. Unfortunately, ref. 24 has been the only laboratory experiment of the CH_4/H_2 gas mixture to date. It should be noted that the subsequent Titan-analogue laboratory measurements show that the aerosol properties are significantly influenced by the gas composition and environmental pressure[25].

Zhang et al.[15] combined the ground-based infrared spectra with the Cassini ISS observations at both low and high phase angles, and determined the k values for the UV1 channel (0.258 μm) and the infrared channels (CB3 at 0.938 μm and MT3 at 0.889 μm). Owing to the existence of degeneracy, the k value at UV1 channel varies from 0.008 to 0.02 and that at 0.9 μm varies from 0.0001 to 0.004. Different choices of k would imply different solutions to fit the ISS data, such as the radius of the monomers from 10 to 40 nm, the number of monomers per aggregate particle from 100 to 1,000, and the abundance of particles. In all possible solutions, the total aerosol opacity only changes by ~30% among all solutions. The corresponding aerosol heating rate does not change significantly.

For the heating rate calculations, we choose the model parameters for the k values from ref. 15 as our nominal case (H1 in Table 1), with a careful sensitivity study in the radiatiave heating calculation. The k value is ~0.02 at 0.258 μm and

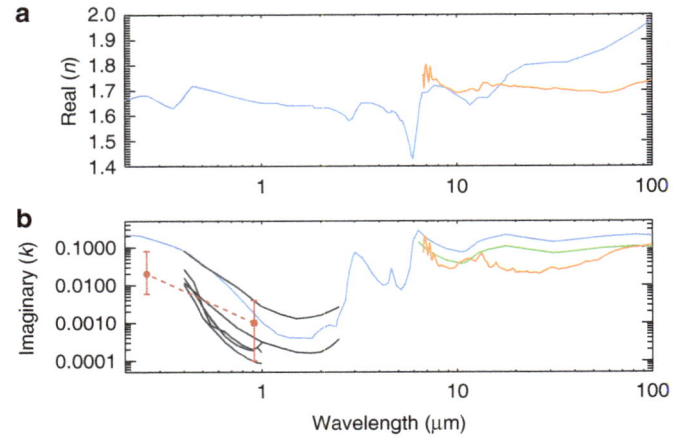

Figure 7 | Refractive indices of aerosols from 0.2 to 100 μm. (**a**) Real part of the refractive index; (**b**) imaginary part. The blue one is from the Titan tholin experiment[26]; orange is from Titan CIRS observations[27]; green is same as the blue one but reduced by a factor of 2 (case C4); black is the Jupiter-analogue aerosol experiment[24]; red is the derived from Cassini ISS observations[15], with an interpolation in the coordinate of linear wavelength and logarithmic k value. The red dashed line is used in the nominal case for the heating rate calculation (case H1) in this study.

0.001 at ~0.9 μm. We performed an interpolation for the wavelengths between 0.20 and 0.94 μm in the coordinate of linear wavelength and logarithmic k value. The interpolation can be justified by the approximate same trend shown in the laboratory measurements (Fig. 7). As in ref. 15, we adopt the real part of refractive index (n) from ref. 26.

Gas opacity sources. From 0.20 to 0.94 μm, the CH_4 opacity is based on ref. 49 and Rayleigh scattering optical depth is taken from ref. 14. From 0.94 to 10 μm, a line-by-line calculation is performed. We compared several CH_4 line databases, including the HITRAN2012 (refs 50,51), the database from ref. 22, and the M5 database[52]. For the CH_4 broadening width, it has been suggested that line widths in the Jovian atmosphere (H_2–He mixture) are similar to those in the Earth's atmosphere (N_2–O_2 mixture)[22]. All the above three CH_4 opacity sources result in a consistent heating rate in our calculation. However, none of the above opacity sources covers the CH_4 band between 0.94 and 1.1 μm. Recently, a '10 to 10' CH_4 line database is computed from first principles[53]. This database is shown to be roughly consistent with the HITRAN2012 CH_4 data in their overlapping near-infrared wavelengths (personal communication with J. Bailey) and therefore is helpful to fill the gap between 0.94 and 1.1 μm in our calculation. For this CH_4 band, due to the lack of laboratory measurements of line shape parameters, we adopted an average pressure-broadened half-width of 0.06 cm^{-1} for foreign-broadening and 0.077 cm^{-1} for self-broadening at 1 bar pressure, and the temperature dependence exponent is ~0.85 (refs 52,54). The contribution from this band to the total heating rate is negligible (Fig. 2c). For the heating rate calculation, we obtain the near-infrared H_2–H_2 CIA absorption from refs 55,56 and H_2–He CIA absorption from refs 57–60.

The thermal cooling rate is calculated from 50 to 2,500 cm^{-1} (4–200 μm). The opacity sources of CH_4, C_2H_2 and C_2H_6 are obtained from HITRAN2012 with hydrogen-broadening widths[4,61]. Fractal aggregates are treated as pure absorbers in the thermal wavelengths due to their negligible single scattering albedo in the mid-infrared. The mid-infrared H_2–H_2 and H_2–He CIAs are obtained from ref. 62. Figure 2c shows the spectrally resolved heating/cooling rate as a function of wavelength and pressure, in which one can see the contributions from the CIAs and different vibrational–rotational bands from gases in Fig. 2b.

Spectral inversion model. The temperature and hydrocarbon distributions are simultaneously retrieved from the Cassini CIRS spectra[4,10–12] using the NEMESIS algorithm[28]. This inversion model has been used in previous studies involving CIRS data retrieval[4,11,12]. In this study we extended the previous retrieval framework to include aerosol absorption in the mid-infrared wavelengths. No scattering calculations were needed as the fractal aggregate particles are significant absorbers and have negligible single scattering albedos at these wavelengths.

References

1. Moses, J. I. et al. in Jupiter-The Planet, Satellites and Magnetosphere (eds Bagenal, F., Dowling, T. & McKinnon, W.) 129–157 (Cambridge University Press, 2004).

2. Hanel, R., Conrath, B., Herath, L., Kunde, V. & Pirraglia, J. Albedo, internal heat, and energy balance of Jupiter: Preliminary results of the Voyager infrared investigation. *J. Geophys. Res.* **86**, 8705–8712 (1981).

3. Yelle, R. V., Griffith, C. A. & Young, L. A. Structure of the Jovian stratosphere at the Galileo probe entry site. *Icarus* **152**, 331–346 (2001).

4. Zhang, X. *et al.* Radiative forcing of the stratosphere of Jupiter, part I: atmospheric cooling rates from Voyager to Cassini. *Planet. Space Sci.* **88**, 3–25 (2013).

5. Conrath, B. J., Gierasch, P. J. & Leroy, S. S. Temperature and circulation in the stratosphere of the outer planets. *Icarus* **83**, 255–281 (1990).

6. Wallace, L., Prather, M. & Belton, M. J. S. The thermal structure of the atmosphere of Jupiter. *Astrophys. J.* **193**, 481–493 (1974).

7. Cess, R. & Chen, S. The influence of ethane and acetylene upon the thermal structure of the Jovian atmosphere. *Icarus* **26**, 444–450 (1975).

8. Appleby, J. F. & Joseph, S. H. Radiative-convective equilibrium models of Jupiter and Saturn. *Icarus* **59**, 336–366 (1984).

9. Li, L. *et al.* Emitted power of Jupiter based on Cassini CIRS and VIMS observations. *J. Geophys. Res. Planets* **117**, E11 (2012).

10. Simon-Miller, A. A. *et al.* Jupiter's atmospheric temperatures: from Voyager IRIS to Cassini CIRS. *Icarus* **180**, 98–112 (2006).

11. Nixon, C. A. *et al.* Meridional variations of C_2H_2 and C_2H_6 in Jupiter's atmosphere from Cassini CIRS infrared spectra. *Icarus* **188**, 47–71 (2007).

12. Nixon, C. A. *et al.* Abundances of Jupiter's trace hydrocarbons from Voyager and Cassini. *Planet. Space Sci.* **58**, 1667–1680 (2010).

13. Wong, A. S., Yung, Y. L. & Friedson, A. J. Benzene and haze formation in the polar atmosphere of Jupiter. *Geophys. Res. Lett.* **30**, 1–4 (2003).

14. West, R. A. *et al.* in *Jupiter-The Planet, Satellites and Magnetosphere* (eds Bagenal, F., Dowling, T. & McKinnon, W.) 79–104 (Cambridge University Press, 2004).

15. Zhang, X., West, R. A., Banfield, D. & Yung, Y. L. Stratospheric aerosols on Jupiter from Cassini observations. *Icarus* **226**, 159–171 (2013).

16. West, R. A. & Smith, P. H. Evidence for aggregate particles in the atmospheres of Titan and Jupiter. *Icarus* **90**, 330–333 (1991).

17. Tomasko, M. G. *et al.* A model of Titan's aerosols based on measurements made inside the atmosphere. *Planet. Space Sci.* **56**, 669–707 (2008).

18. Wolf, E. T. & Toon, O. B. Fractal organic Hazes provided an ultraviolet shield for early Earth. *Science* **328**, 1266–1268 (2010).

19. Moreno, F. & Sedano, J. Radiative balance and dynamics in the stratosphere of Jupiter: results from a latitude-dependent aerosol heating model. *Icarus* **130**, 36–48 (1997).

20. Starodubtseva, O. M., Akimov, L. A. & Korokhin, V. V. Seasonal variations in the north–south asymmetry of polarized light of Jupiter. *Icarus* **157**, 419–425 (2002).

21. Banfield, D., Conrath, B. J., Gierasch, P. J. & Nicholson, P. D. Near-IR spectrophotometry of Jovian aerosols—meridional and vertical distributions. *Icarus* **134**, 11–23 (1998).

22. Kedziora-Chudczer, L. & Bailey, J. Modelling the near-IR spectra of Jupiter using line-by-line methods. *Mon. N. Astron. Soc.* **414**, 1483–1492 (2011).

23. Wong, M. H., Mahaffy, P. R., Atreya, S. K., Niemann, H. B. & Owen, T. C. Updated Galileo probe mass spectrometer measurements of carbon, oxygen, nitrogen, and sulfur on Jupiter. *Icarus* **171**, 153–170 (2004).

24. Khare, B. N., Sagan, C., Thompson, W., Arakawa, E. & Votaw, P. Solid hydrocarbon aerosols produced in simulated Uranian and Neptunian stratospheres. *J. Geophys. Res.* **92**, 15067–15082 (1987).

25. Imanaka, H. *et al.* Laboratory experiments of Titan tholin formed in cold plasma at various pressures: implications for nitrogen-containing polycyclic aromatic compounds in Titan haze. *Icarus* **168**, 344–366 (2004).

26. Khare, B. N. *et al.* Optical constants of organic tholins produced in a simulated Titanian atmosphere: from soft X-ray to microwave frequencies. *Icarus* **60**, 127–137 (1984).

27. Vinatier, S. *et al.* Optical constants of Titan's stratospheric aerosols in the 70–1,500 cm^{-1} range constrained by Cassini/CIRS observations. *Icarus* **219**, 5–12 (2012).

28. Irwin, P. G. J. *et al.* The NEMESIS planetary atmosphere radiative transfer and retrieval tool. *J. Quant. Spectrosc. Radiat. Transfer* **109**, 1136–1150 (2008).

29. West, R. A., Friedson, A. J. & Appleby, J. Jovian large-scale stratospheric circulation. *Icarus* **100**, 245–259 (1992).

30. Young, L. A., Yelle, R. V., Young, R., Seiff, A. & Kirk, D. B. Gravity waves in Jupiter's stratosphere, as measured by the Galileo ASI experiment. *Icarus* **173**, 185–199 (2005).

31. Watkins, C. & Cho, J. Y. K. The vertical structure of Jupiter's equatorial zonal wind above the cloud deck, derived using mesoscale gravity waves. *Geophys. Res. Lett.* **40**, 472–476 (2013).

32. Tomasko, M. G. *et al.* A model of Titan's aerosols based on measurements made inside the atmosphere. *Planet. Space Sci.* **56**, 669–707 (2008).

33. West, R. A., Baines, K. H., Karkoschka, E. & Sánchez-Lavega, A. in *Saturn from Cassini-Huygens* (eds Dougherty, M. K., Esposito, L. W. & Krimigis, S. S. M.) 161–179 (Springer, 2009).

34. Pollack, J. B. *et al.* Nature of stratospheric haze on Uranus: evidence for condensed hydrocarbons. *J. Geophys. Res.* **92**, 15037–15065 (1987).

35. Baines, K. H. & Hammel, H. B. Clouds, hazes and stratospheric methane abundance ratio in Neptune. *Icarus* **109**, 20–39 (1994).

36. Sing, D. K. *et al.* Hubble Space Telescope transmission spectroscopy of the exoplanet HD 189733b: high-altitude atmospheric haze in the optical and near-ultraviolet with STIS. *Mon. Not. R. Astron. Soc.* **416**, 1443–1455 (2011).

37. Sing, D. K. *et al.* HST hot-Jupiter transmission spectral survey: evidence for aerosols and lack of TiO in the atmosphere of WASP-12b. *Mon. Not. R. Astron. Soc* **436**, 2956–2973 (2013).

38. Deming, D. *et al.* Infrared transmission spectroscopy of the exoplanets HD 209458b and XO-1b using the wide field camera-3 on the Hubble Space Telescope. *Astrophys. J.* **774**, 95 (2013).

39. Knutson, H. A., Benneke, B., Deming, D. & Homeier, D. A featureless transmission spectrum for the Neptune-mass exoplanet GJ436b. *Nature* **505**, 66–68 (2014).

40. Kreidberg, L. *et al.* Clouds in the atmosphere of the super-Earth exoplanet GJ1214b. *Nature* **505**, 69–72 (2014).

41. Robinson, T. D. & Catling, D. C. Common 0.1-bar tropopause in thick atmospheres set by pressure-dependent infrared transparency. *Nat. Geosci.* **7**, 12–15 (2014).

42. Buras, R., Dowling, T. & Emde, C. New secondary-scattering correction in DISORT with increased efficiency for forward scattering. *J. Quant. Spectrosc. Radiat. Transfer* **112**, 2028–2034 (2011).

43. Mishchenko, M. I., Travis, L. D. & Mackowski, D. W. T-matrix computations of light scattering by nonspherical particles: A review. *J. Quant. Spectrosc. Radiat. Transfer* **55**, 535–575 (1996).

44. Irwin, P. G. J. *Giant Planets of Our Solar System: Atmospheres, Composition, and Structure* (Springer, 2009).

45. Rages, K., Beebe, R. & Senske, D. Jovian stratospheric hazes: the high phase angle view from Galileo. *Icarus* **139**, 211–226 (1999).

46. Kunde, V. G. *et al.* Jupiter's atmospheric composition from the Cassini thermal infrared spectroscopy experiment. *Science* **305**, 1582–1586 (2004).

47. Kostiuk, T., Romani, P. N., Espenak, F., Livengood, T. A. & Goldstein, J. J. Temperature and abundances in the Jovian auroral stratosphere, 2. Ethylene as a probe of the microbar region. *J. Geophys. Res. Planets* **98**, 18823–18830 (1993).

48. Livengood, T. A., Kostiuk, T., Espenak, F. & Goldstein, J. J. Temperature and abundances in the Jovian auroral stratosphere, 1. Ethane as a probe of the millibar region. *J. Geophys. Res. Planets* **98**, 18813–18822 (1993).

49. Karkoschka, E. & Tomasko, M. G. Methane absorption coefficients for the Jovian planets from laboratory, Huygens, and HST data. *Icarus* **205**, 674–694 (2010).

50. Brown, L. R. *et al.* Methane line parameters in the HITRAN2012 database. *J. Quant. Spectrosc. Radiat. Transfer* **130**, 201–219 (2013).

51. Rothman, L. S. *et al.* The HITRAN2012 molecular spectroscopic database. *J. Quant. Spectrosc. Radiat. Transfer* **130**, 4–50 (2013).

52. Sromovsky, L. A., Fry, P. M., Boudon, V., Campargue, A. & Nikitin, A. Comparison of line-by-line and band models of near-IR methane absorption applied to outer planet atmospheres. *Icarus* **218**, 1–23 (2012).

53. Yurchenko, S. N., Tennyson, J., Bailey, J., Hollis, M. D. & Tinetti, G. Spectrum of hot methane in astronomical objects using a comprehensive computed line list. *Proc. Natl Acad. Sci. USA* **111**, 9379–9383 (2014).

54. Nikitin, A. V. *et al.* GOSAT-2009 methane spectral line list in the 5,550–6,236 cm^{-1} range. *J. Quant. Spectrosc. Radiat. Transfer* **111**, 2211–2224 (2010).

55. Borysow, A. Collision-induced absorption coefficients of H_2 pairs at temperatures from 60 K to 1,000 K. *Astron. Astrophys.* **390**, 779–782 (2002).

56. Borysow, A., Trafton, L., Frommhold, L. & Birnbaum, G. Modeling of pressure-induced far-infrared absorption spectra: Molecular hydrogen pairs. *Astrophys. J.* **296**, 644–654 (1985).

57. Borysow, J., Frommhold, L. & Birnbaum, G. Collison-induced rototranslational absorption spectra of H_2-He pairs at temperatures from 40 to 3000 K. *Astrophys. J.* **326**, 509–515 (1988).

58. Borysow, A. & Frommhold, L. Collision-induced infrared spectra of H_2-He pairs at temperatures from 18 to 7,000 K. II-Overtone and hot bands. *Astrophys. J.* **341**, 549–555 (1989).

59. Borysow, A., Frommhold, L. & Moraldi, M. Collision-induced infrared spectra of H_2-He pairs involving 0-1 vibrational transitions and temperatures from 18 to 7,000 K. *Astrophys. J.* **336**, 495–503 (1989).

60. Borysow, A. New model of collision-induced infrared absorption spectra of H_2-He pairs in the 2-2.5 μm range at temperatures from 20 to 300 K: An update. *Icarus* **96**, 169–175 (1992).

61. Orton, G. S. *et al.* Semi-annual oscillations in Saturn's low-latitude stratospheric temperatures. *Nature* **453**, 196–199 (2008).

62. Orton, G. S., Gustafsson, M., Burgdorf, M. & Meadows, V. Revised ab initio models for H_2-H_2 collision-induced absorption at low temperatures. *Icarus* **189**, 544–549 (2007).

Acknowledgements

We thank E. Karkoschka, L. Brown, G. Orton, J. Bailey, T. Kostiuk, A. Showman and L. Li for useful discussions and comments. Special thanks to M. Gerstell, P. Gao, R. Hu, P. Kopparla, C. Li, M.C. Liang, S. Newman, R.L. Shia, M. Wong, X. Xi and Q. Zhang for proofreading the manuscript. The early phase of this research was supported by the Outer Planets Research program via NASA Grant JPL 1452240 to the California Institute of Technology. R.A.W. and C.A.N. are supported by the NASA Cassini project. P.G.J.I. acknowledges the support of the UK Science and Technology Facilities Council.

Author contributions

X.Z. carried out the radiative modelling and CIRS spectral retrieval; R.A.W. provided the ISS data; C.A.N. provided the CIRS data; R.A.W. and Y.L.Y. helped with radiative modelling; P.G.J.I. and C.A.N. helped with spectral inversion modelling; all authors contributed to the paper writing.

Additional information

Competing financial interests: The authors declare no competing financial interests.

26

Ultra-low-frequency wave-driven diffusion of radiation belt relativistic electrons

Zhenpeng Su[1,2], Hui Zhu[1,3], Fuliang Xiao[4], Q.-G. Zong[5], X.-Z. Zhou[5], Huinan Zheng[1,2], Yuming Wang[1,2,6], Shui Wang[1,2], Y.-X. Hao[5], Zhonglei Gao[1,3], Zhaoguo He[7], D.N. Baker[8], H.E. Spence[9], G.D. Reeves[10], J.B. Blake[11] & J.R. Wygant[12]

Van Allen radiation belts are typically two zones of energetic particles encircling the Earth separated by the slot region. How the outer radiation belt electrons are accelerated to relativistic energies remains an unanswered question. Recent studies have presented compelling evidence for the local acceleration by very-low-frequency (VLF) chorus waves. However, there has been a competing theory to the local acceleration, radial diffusion by ultra-low-frequency (ULF) waves, whose importance has not yet been determined definitively. Here we report a unique radiation belt event with intense ULF waves but no detectable VLF chorus waves. Our results demonstrate that the ULF waves moved the inner edge of the outer radiation belt earthward 0.3 Earth radii and enhanced the relativistic electron fluxes by up to one order of magnitude near the slot region within about 10 h, providing strong evidence for the radial diffusion of radiation belt relativistic electrons.

[1] CAS Key Laboratory of Geospace Environment, Department of Geophysics and Planetary Sciences, University of Science and Technology of China, Hefei, Anhui 230026, China. [2] Collaborative Innovation Center of Astronautical Science and Technology, University of Science and Technology of China, Hefei, Anhui 230026, China. [3] Mengcheng National Geophysical Observatory, School of Earth and Space Sciences, University of Science and Technology of China, Hefei, Anhui 230026, China. [4] School of Physics and Electronic Sciences, Changsha University of Science and Technology, Changsha Hunan 410004, China. [5] Institute of Space Physics and Applied Technology, Peking University, Beijing 100871, China. [6] Synergetic Innovation Center of Quantum Information and Quantum Physics, University of Science and Technology of China, Hefei, Anhui 230026, China. [7] Harbin Institute of Technology Shenzhen Graduate School, Shenzhen, Guangdong 518055, China. [8] Laboratory for Atmospheric and Space Physics, University of Colorado Boulder, Boulder, Colorado 80303-7814, USA. [9] Institute for the Study of Earth, Oceans, and Space, University of New Hampshire, Durham, New Hampshire 03824-3525, USA. [10] Space Science and Applications Group, Los Alamos National Laboratory, Los Alamos, New Mexico 87544, USA. [11] The Aerospace Corporation, Los Angeles, California 90245-4609, USA. [12] School of Physics and Astronomy, University of Minnesota, Minneapolis, Minnesota 55455, USA. Correspondence and requests for materials should be addressed to Z.S. (email: szpe@mail.ustc.edu.cn).

The geomagnetic field geometry allows three quasi-periodic motions of outer Van Allen radiation belt[1] relativistic electrons over distinct timescales: gyration about the magnetic field line on a timescale of milliseconds; bounce along the magnetic field line between two magnetic mirror points on a timescale of seconds; drift circling the Earth on a timescale of kiloseconds. Each periodicity gives rise to an approximate constant of motion defined as the adiabatic invariant. Surrounding the highly dynamic outer radiation belt[2], one outstanding question has been how electrons are accelerated to relativistic energies of several million electron volts. Two invariant-violating processes have been proposed: local acceleration by VLF (~kHz) chorus waves[3,4] and radial diffusion by ULF (~mHz) waves[5–9]. On 30 August 2012, National Aeronautics and Space Administration launched a twin-spacecraft mission, Van Allen Probes (formerly known as the Radiation Belt Storm Probes (RBSP))[10], to make the measurements for the identification of acceleration mechanisms. Recent analyses[11,12] of the Van Allen Probes data have presented compelling evidence for the local acceleration in the heart of the

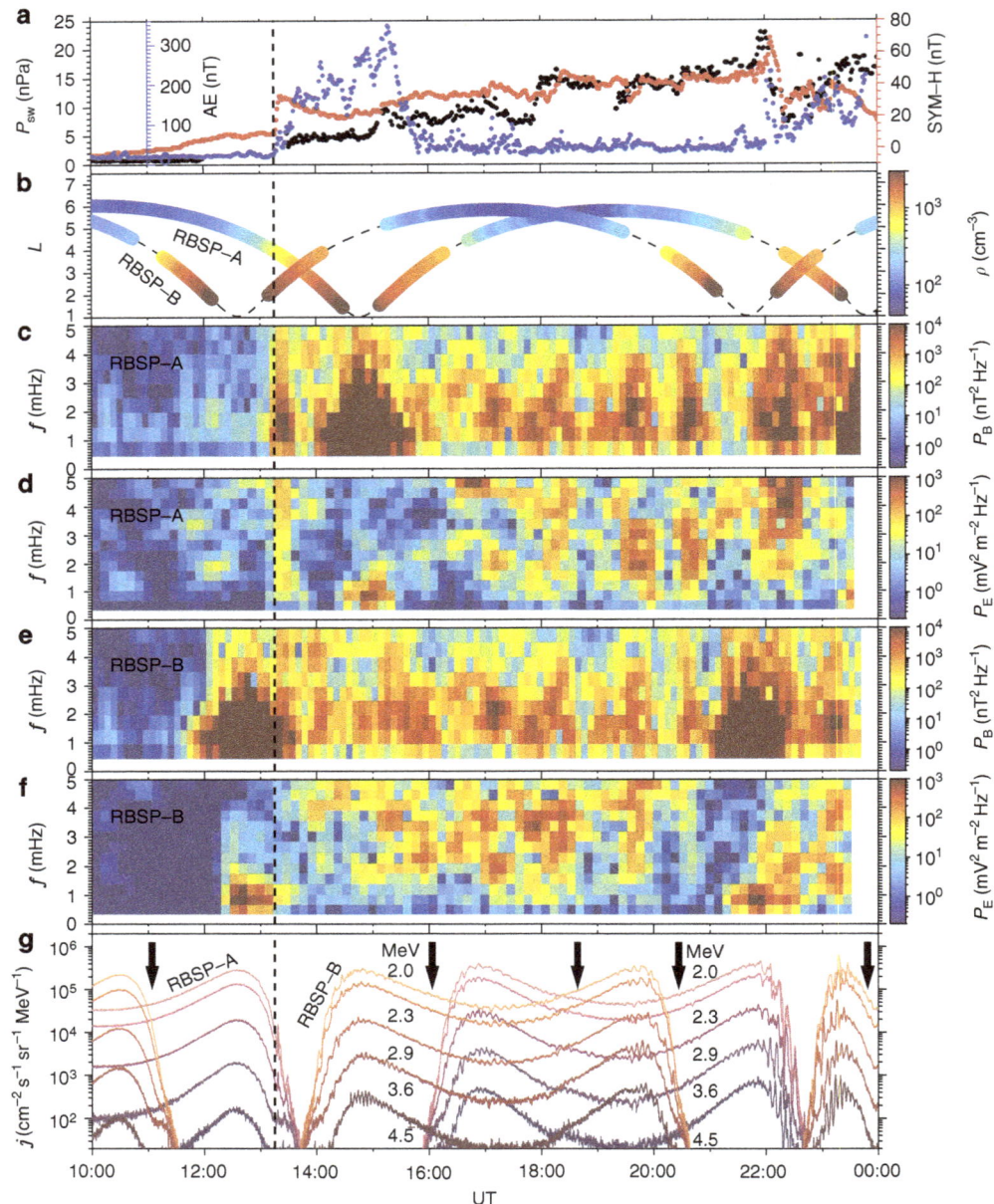

Figure 1 | An overview of the 15 February 2014 radiation belt event. (**a**) Solar wind dynamic pressure P_{sw}, geomagnetic activity indices AE (measuring the substorm intensity) and SYM-H (measuring the storm intensity and partially reflecting the variation of solar wind dynamic pressure). (**b**) Cold electron number density ρ from the EFW instrument. The density were always beyond $40\,cm^{-3}$, indicating the locations of Van Allen Probes in the plasmasphere. (**c,e**) Power spectral density P_B of the compressional ULF wave magnetic field in the MFA coordinate system from the EMFISIS magnetometer. (**d,f**) Power spectral density P_E of the y component ULF wave electric field in the mGSE coordinate system from the EFW instrument. In the outer belt, the ULF wave power were intensified obviously by the solar wind variation after 13:15 UT (vertical-dashed line). (**g**) Spin-averaged differential electron fluxes j (colour-coded according to energy) in the outer radiation belt from the REPT instrument. Black arrows mark the times around which the regular oscillations of fluxes were relatively weak.

outer radiation belt. However, the importance of radial diffusion for the radiation belt evolution has not yet been determined definitively.

ULF waves have been thought to effectively violate the third adiabatic invariant L^* under the drift-resonance condition[5] $\omega = m\omega_d$ (with the wave frequency ω, the azimuthal wave mode number m and the electron drift frequency ω_d). The quantity L^* is the Roederer's drift-shell parameter[13], equal to the equatorial radial distance (made dimensionless by Earth radii R_E) of the adiabatically equivalent electron drift orbit in the dipole field. Violation of the third invariant (but conservation of the first two invariants) causes the radial migration of resonant electrons and consequently the variation of their energies and pitch angles. The electrons in drift resonance with broadband ULF waves move stochastically along the radial direction, which is described by the radial diffusion theory[5,7,8]. This theory was indirectly supported by the strong correlation between ULF wave power and radiation belt electron flux enhancement[14-18], but due to the limitations in previous particle/field observations, the *in situ* wave–particle interaction characteristics were lacking to clarify the associated physical process. Particularly, previous works[7,14-16,19,20] often concentrated on the radiation belt reformation events during the geomagnetic storms. These geomagnetic storms did cause marked dynamics of energetic electrons, but the strongly disturbed magnetosphere very likely allowed the concurrence[20-22] of ULF wave-driven radial diffusion and VLF chorus wave-driven local acceleration. The superposition of two processes is not conducive to isolating the contribution of radial diffusion.

Here we report a unique radiation belt event serendipitously observed by the Van Allen Probes in the plasmasphere during non-storm times. The plasmasphere is a torus-shaped region of cold and dense plasma surrounding the Earth, where the VLF chorus wave-driven local acceleration seldom occurs. By analysing the high-resolution data and performing the detailed simulation, we demonstrate that the radial diffusion by intense ULF waves was responsible for the radiation belt relativistic electron evolution in this event.

Results

Wave modulation of relativistic electrons. The radiation belt event of interest occurred on 15 February 2014, with the required particle/field data collected by the Relativistic Electron–Proton Telescope (REPT)[23] and the Magnetic Electron Ion Spectrometer (MagEIS)[24] of the Energetic particle, Composition and Thermal plasma (ECT) instrument suite[25], the Electric Fields and Waves (EFW) instruments[26] and the Electric and Magnetic Field Instrument Suite and Integrated Science (EMFISIS) instrument suite[27] on board the Van Allen Probes. During the event, the magnetosphere was free from magnetic storms but experienced some weak and short-duration substorms (Fig. 1a). Such magnetospheric conditions favoured the expansion of plasmasphere to at least $L = 6.5$ (Figs 1b and 2a,c; Supplementary Table 1). The quantity L is the McIlwain's drift-shell parameter[28], equal to the equatorial radial distance (made dimensionless by Earth radii R_E) of the drift orbit of the electron having the same mirror

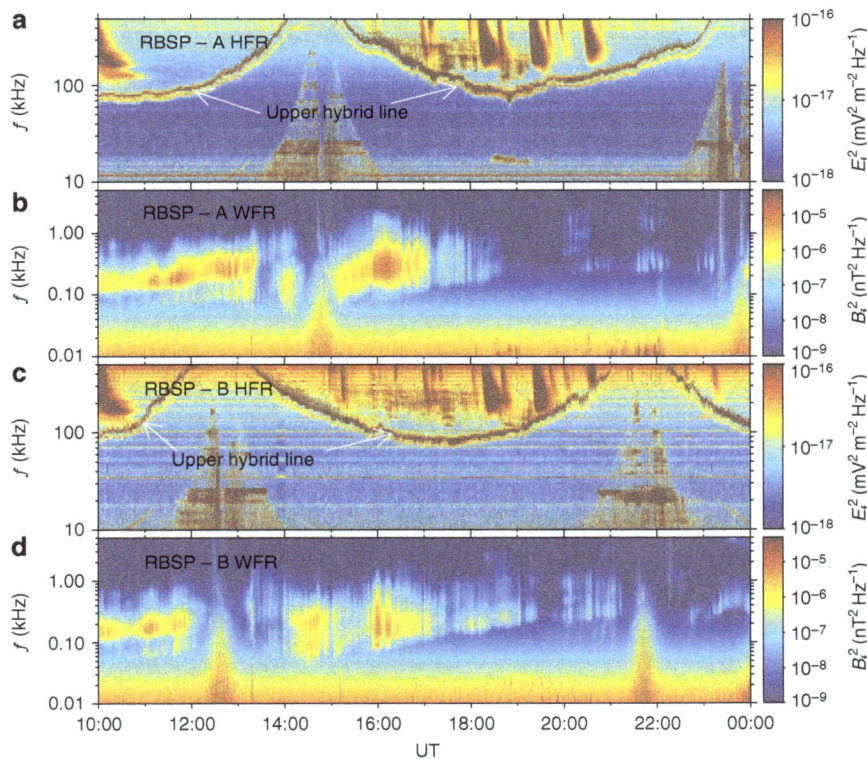

Figure 2 | Electromagnetic power spectral densities of high-frequency and very-low-frequency waves. (**a,c**) Wave electric power spectral density from the high-frequency receiver (HFR) of the EMFISIS Suite and Integrated Science suite. The upper hybrid resonance bands (bright lines) had frequencies (positively correlated with the background electron density[59]) beyond 70 kHz, indicating that the twin spacecrafts stayed in the high-density plasmasphere for the entire orbits. (**b,d**) Wave magnetic power spectral density from the Waveform Receiver (WFR) of the EMFISIS suite. There were VLF hiss waves in the frequency range 0.1–1.0 kHz contributing to the slow loss of relativistic electrons, but no VLF chorus waves responsible for the local acceleration of relativistic electrons.

field, second invariant and energy in the dipole field. In the plasmasphere, there were no VLF chorus waves contributing to the local acceleration (Fig. 2b,d). In the time range from 13:15 to 24:00 UT, the solar wind dynamic pressure (Fig. 1a) exhibited continuous fluctuations especially around 13:15 and 22:00 UT. The solar wind variation drove the generation of ULF waves over a wide frequency range in the outer radiation belt (Fig. 1c–f). Correspondingly, the relativistic (2.0–4.5 MeV) electron fluxes oscillated periodically (Fig. 1g), which is one of the expected drift-resonance characteristics[29]. In contrast to the previously reported ULF modulation[30,31] of electron fluxes at the relatively low energies and/or in a localized spatiotemporal region, this event presented the ULF modulation of highly relativistic electron fluxes for about 10 h throughout the outer radiation belt, serving as an observational evidence for the global and long-lasting drift resonance between ULF waves and relativistic electrons.

The azimuthal mode number m of ULF waves is required to quantify the drift-resonance process. Wavelet transforms[32] have been performed on the electron fluxes at different energy channels to identify their dominant oscillation periods (Fig. 3). During some time intervals (10:00–13:15, 15:30–16:00, 17:30–19:30, 20:20–21:00 and 23:30–24:00 UT), the regular oscillations of fluxes were so weak (as marked by the black arrows in Fig. 1g) that the wavelet transforms mainly captured the power associated with the irregular noise and/or the radial variation of fluxes. Most of the time the dominant oscillation periods of electron fluxes are found to change with energy and location but

remain close to the corresponding electron drift periods, which can be considered as the result of the $m=1$ mode drift-resonance between relativistic electrons and broadband ULF waves[29,33]. It should be mentioned that recent magnetohydrodynamic simulation[34] for a magnetic storm had suggested the dominance of $m=1$ mode ULF waves in the radiation belt region.

Global evolution of relativistic electrons. The Van Allen Probes passed through the radiation belt six times in the time range of interest. The comparison of the L-dependent electron fluxes observed during each passage is made to identify the global evolution of radiation belt electrons (Fig. 4). Within about 10 h, the inner edge of the outer belt moved inward about 0.3 R_E at 2.0 MeV and 0.8 R_E at 4.5 MeV. Around the inner edge, the electron fluxes increased by an average factor of up to 10. To exclude the possible adiabatic effect associated with the compression of magnetosphere, these electron fluxes are transformed into phase space density (PSD) in the adiabatic invariant coordinate system (Fig. 5a; Supplementary Figs 1–3; Supplementary Note 1). For the fixed L^* near the slot region, the variation of mapped L with time was below 0.05 throughout the event (Fig. 6). The initial PSD peaked at the large L^* (that is, the external source region) and monotonically decreased with the decreasing L^*. As time went on, the external source showed some fluctuations. At the centre of the outer belt ($L^* \approx 4.3$), the PSD remained unchanged, once again demonstrating the absence of local

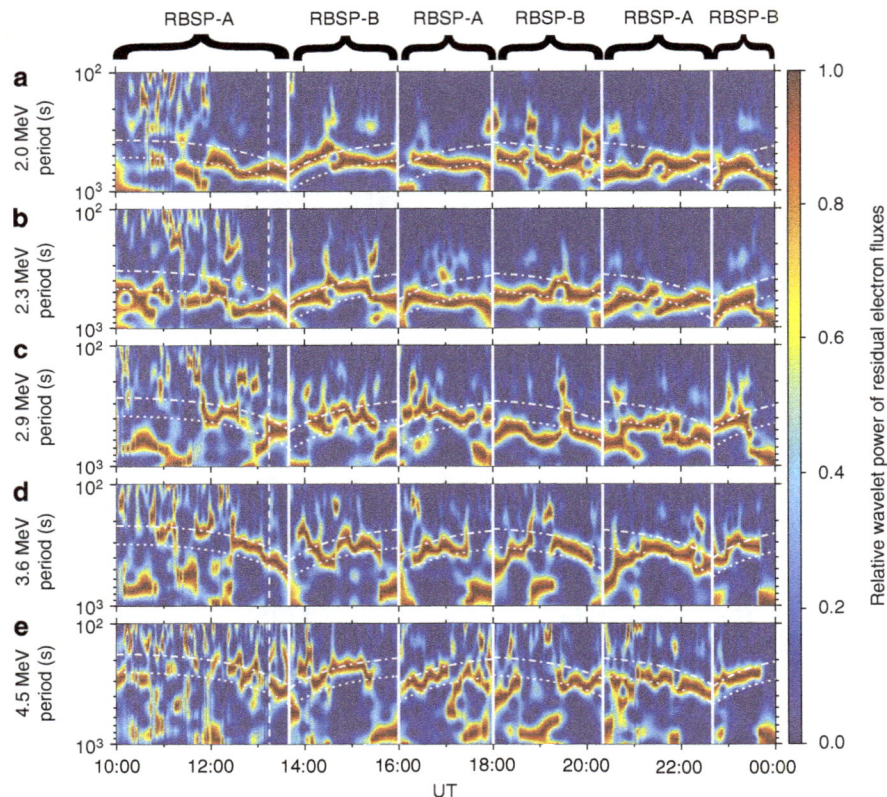

Figure 3 | Relative wavelet power of residual electron fluxes. (a-e) Relative wavelet power at the energy channels 2.0, 2.3, 2.9, 3.6 and 4.5 MeV, respectively. Residual flux is defined as $(j - j_0)/j_0$ with j the spin-averaged differential flux from the REPT instrument and j_0 the 1,000 s running averaged j, which reflects the oscillation of electron flux at each energy channel. Relative wavelet power is defined as the wavelet power of residual flux normalized by the maximum power with oscillation period <1,000 s. The obtained relative wavelet power (colour-coded scale) of the twin spacecrafts is plotted alternatively to characterize the oscillation of the outer radiation belt electron fluxes. The superposed dotted and dot-dashed lines represent the drift periods of electrons with the equatorial pitch angles 0° and 90° at the corresponding energy channels in the dipole field. The vertical-dashed line denotes the beginning of ULF wave enhancement induced by the solar wind dynamic pressure variation.

12:24–13:34   19:13–20:30
13:50–15:05   21:15–22:35
16:01–17:17   22:46–23:50

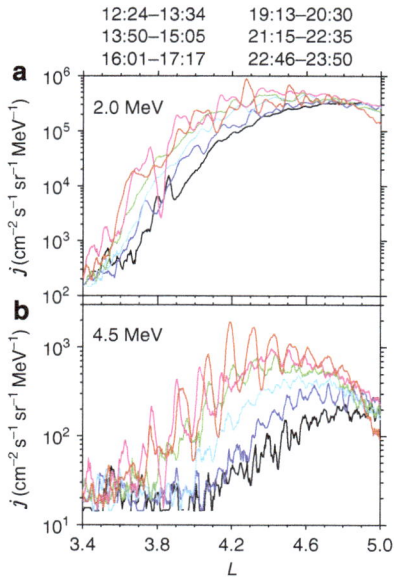

Figure 4 | Radial profiles of relativistic electron fluxes for six outer radiation belt passages. Each colour-coded profile shows the L-dependent differential flux j from REPT instrument during a passage. (**a**,**b**) Differential fluxes of electrons with the equatorial pitch angles 55° at the energy channels 2.0 and 4.5 MeV, respectively. Because of the latitudinal variation of twin spacecrafts, the corresponding local pitch angles varied between 55° and 90° in the TS04D geomagnetic field model[60]. Electron fluxes at the other energy channels 2.3, 2.9 and 3.6 MeV exhibited the similar characteristics.

μ=700 MeV G^{-1} K=0.15 R_EG$^{1/2}$

12:06–13:32   18:52–20:28
13:52–15:26   20:54–22:32
16:03–17:39   22:48–23:44

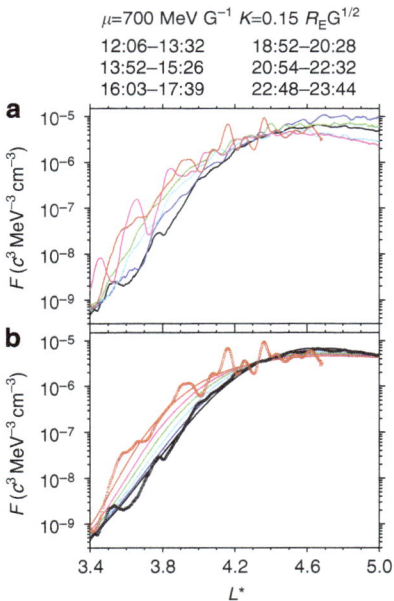

Figure 5 | Radial profiles of relativistic electron phase space densities for six outer radiation belt passages. Each colour-coded profile shows the L^*-dependent phase space density F at the fixed first adiabatic invariant $\mu = 700$ MeV G^{-1} and second adiabatic invariant $K = 0.15$ R_EG$^{1/2}$ during a passage. The corresponding energies were about 1.0 MeV at $L^* = 5.5$ and 2.5 MeV at $L^* = 3.4$, and the corresponding local pitch angles were 40°–70°. (**a**) Observations (lines) from MagEIS and REPT instruments in the TS04D geomagnetic field model. (**b**) Simulations (lines) from the radial diffusion equation, overplotted with the observations (circles) for ease of comparison.

11:05–13:32   16:03–19:29   19:30–22:32
13:52–17:30   17:30–20:28   22:48–23:44

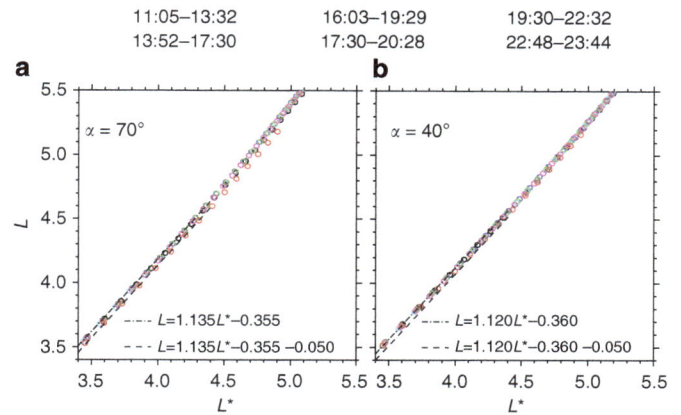

Figure 6 | Mapping relations between L and L^* for six outer radiation belt passages. Mapping relations (circles colour-coded according to the time) are calculated every 300 s in the TS04D geomagnetic field model. (**a**,**b**) Mapping relations at two different local pitch angles 70° and 40°, respectively. The dot-dashed and dashed lines are overplotted to identify the variation of the mapped L with time for the fixed L^* near the slot region. Similar characteristics can be found in the mapping relations at the other local pitch angles.

acceleration. In the slot region, the inner edge of the outer belt was transported inward about 0.3 R_E. Specifically, at $L^* \approx 3.6$, the PSD gradually increased by one order of magnitude within 10 h. These global evolution characteristics were qualitatively consistent with the prediction of radial diffusion theory[35,36]. It should be mentioned that the very strong fluctuations of electron PSDs after 22:00 UT were perhaps associated with the coherent transport process[33].

The radial diffusion of radiation belt electrons is simulated by solving the equation[37]

$$\frac{\partial F}{\partial t} = L^{*2} \frac{\partial}{\partial L^*} \left(\frac{D_{L^*L^*}}{L^{*2}} \frac{\partial F}{\partial L^*} \right) \qquad (1)$$

with the electron PSD F and the radial diffusion rate $D_{L^*L^*}$ calculated from the observed ULF waves[9] (Fig. 7). The simulations (Fig. 5b; Supplementary Fig. 4; Supplementary Note 2) well reproduce the average rate and extent of the observed PSD variations. The radial diffusion process can effectively reduce the electron PSD gradients[35]. In the region $L^* > 4.3$, the electron PSD behaved smoothly and consequently presented minor changes. In contrast, the electron PSD possessed a very steep gradient in the region $L^* < 4.3$ and then allowed the significant enhancement.

Discussion

There was an average enhancement of relativistic electron fluxes by up to one order of magnitude within about 10 h on 15 February 2014, which could be easily misinterpreted in the framework of the VLF chorus wave-driven local acceleration. The high-resolution data and the detailed simulation clearly show that, in the absence of VLF chorus waves, the ULF waves could radially diffuse and effectively accelerate the radiation belt electrons in this event. The ULF waves[38–44], as well as the VLF chorus waves[45–54], are commonly observed in the magnetosphere. To provide a more comprehensive picture, we additionally analyse two more radiation belt events with the concurrence of ULF and VLF chorus waves (Supplementary Figs 5–14; Supplementary Tables 2 and 3; Supplementary Note 3). During the 18 January 2013 event, there were moderate ULF waves but quite weak VLF chorus waves. The radial diffusion, dominating over the local acceleration, caused the earthward

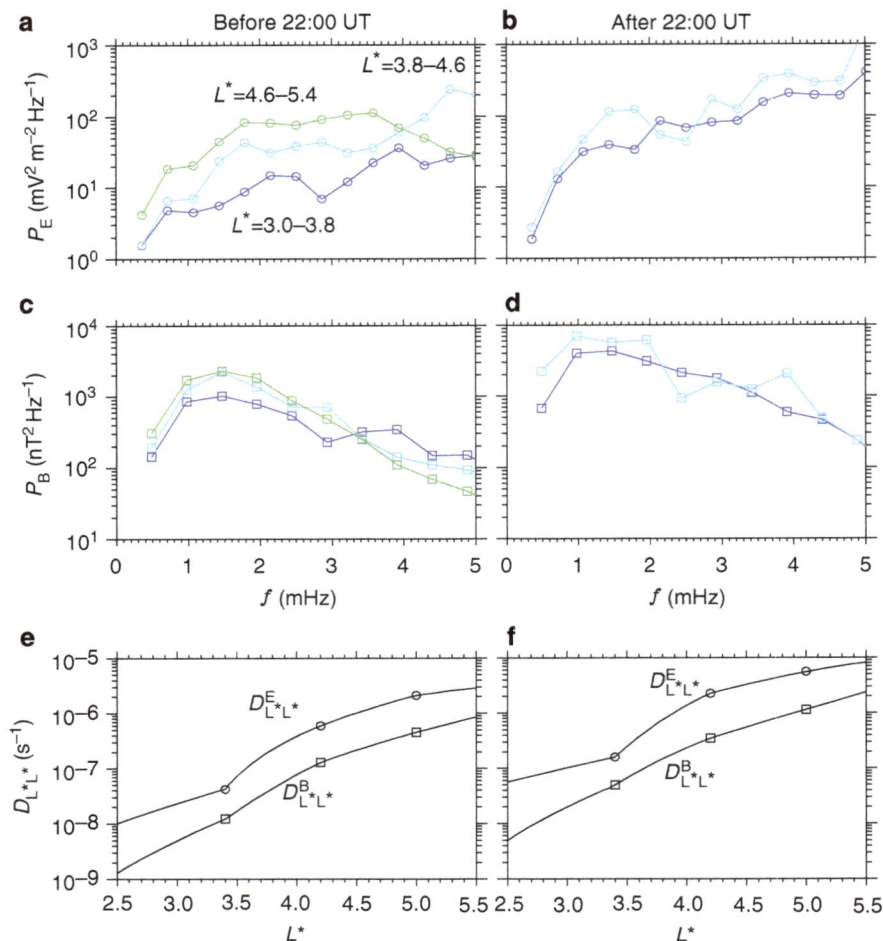

Figure 7 | Electromagnetic power spectral densities of ultra-low-frequency waves and radial diffusion rates. (a–d) Averaged electric P_E (circles) and magnetic P_B (squares) power spectral densities in three different regions before and after 22:00 UT. Solid lines are introduced to guide the eye. (e,f) Radial diffusion rates (lines) from the electric $D^E_{L^*L^*}$ and magnetic $D^B_{L^*L^*}$ perturbations. Symbols are drawn to mark the diffusion rates at the centres of the three spatial regions. The total radial diffusion rates $D_{L^*L^*} = D^E_{L^*L^*} + D^B_{L^*L^*}$ were the sum of the electric and magnetic radial diffusion rates.

movement of the inner edge of the outer radiation belt. In contrast, both moderate ULF waves and strong VLF chorus waves occurred during the 22 September 2014 event. The local acceleration produced the relativistic electron PSD peaks, and the radial diffusion redistributed the electrons along the radial direction to reduce the PSD gradients.

Our results present evidence for the importance of ULF wave-driven radial diffusion in the outer radiation belt dynamics. Under differing magnetospheric conditions, the accurate contributions of local acceleration and radial diffusion to the radiation belt evolution should be carefully examined. The radial diffusion can play a dominant role in some radiation belt electron acceleration events (such as the one discussed here). Compared with the local acceleration in the previously reported marked events[11,12,21], the radial diffusion in the present event yielded a smaller acceleration rate and a smaller relativistic electron flux enhancement. However, no matter how strong the action of local acceleration is, the additional redistribution of relativistic electrons by the radial diffusion would constitute an important part of the outer radiation belt dynamics[11,33,35,55–57].

Methods

Calculation of wave power spectral density. We use the level-3 magnetic field data with the time resolution of 1 s in the Geocentric Solar Magnetospheric coordinate system to obtain the magnetic power spectral density of ULF waves.

A three-step procedure is performed as follows. First, the magnetic field data are running averaged over 11 s to reduce the spin modulation of Van Allen Probes[44]. Then, these magnetic field data are projected on the so-called mean field aligned (MFA) coordinates[58]. In the MFA coordinate system, the parallel direction is determined by the 2,048-s running average of the instantaneous magnetic field, the azimuthal direction is obtained by the cross product of the parallel vector and the satellite position vector, and the radial direction completes the triad. This system is used to separate magnetic field perturbations into the toroidal (azimuthal), poloidal (radial) and compressional (parallel) components[26]. Finally, an overlapped (80%) 2,048-point fast Fourier transform is adopted to calculate the power spectral densities of each magnetic field components. The obtained power spectra densities have the time resolution of 6.83 min and the frequency resolution of 0.49 mHz. We use the level-3 electric field data with the time resolution of 10.9 s in the modified geocentric solar ecliptic (mGSE) coordinate system to obtain the electric power spectral density of ULF waves. The mGSE unit vectors ($\mathbf{X}_{mGSE}$, $\mathbf{Y}_{mGSE}$ and $\mathbf{Z}_{mGSE}$) can be generally expressed in terms of the GSE unit vectors ($\mathbf{X}_{GSE}$, $\mathbf{Y}_{GSE}$ and $\mathbf{Z}_{GSE}$) and the spin axis unit vector $\mathbf{S}_{GSE}$: $\mathbf{X}_{mGSE} = \mathbf{S}_{GSE}$, $\mathbf{Y}_{mGSE} = -(\mathbf{S}_{GSE} \times \mathbf{Z}_{GSE})/\|\mathbf{S}_{GSE} \times \mathbf{Z}_{GSE}\|$, and $\mathbf{Z}_{mGSE} = \mathbf{X}_{mGSE} \times \mathbf{Y}_{mGSE}$. When the spin axis points towards the sun ($\mathbf{S}_{GSE} = \mathbf{X}_{GSE}$), the mGSE system is exactly the same as the GSE system. The EFW instrument only measured the two electric field components in the spin plane, and the third component has to be derived from the assumption $\mathbf{E} \cdot \mathbf{B} = 0$. For this event after 22:00 UT, the magnetic field component B_x of RBSP-B in the mGSE system was nearly zero and the electric field component E_x was unconstrained by the $\mathbf{E} \cdot \mathbf{B} = 0$ condition, not allowing the calculation of electric field vectors in the MFA system. Note that the most significant ULF waves happened to be generated after 22:00 UT. Hence, we calculate the power spectral densities of the electric field components in the mGSE system (rather than in the MFA system) to ensure the continuous monitoring of ULF waves throughout this event. Specifically, an overlapped (80%) 256-point fast Fourier transform is performed on the electric field data (with the preliminary substraction of the co-rotation electric field and the $V_{sc} \times B$ electric field induced by the satellite motion through the Earth's magnetic field), and the obtained electric field power spectral densities possess the time resolution of 9.3 min and the frequency resolution of 0.36 mHz. It should be mentioned the power spectral densities exhibited unphysical intensification[44] associated with the steep variation of the background magnetic field during the averaging time of 2,048 s in the region $L < 3$ or the contamination by the large $V_{sc} \times B$ electric field in the region $L < 2.5$.

Calculation of radial diffusion rates. The radial diffusion is generally produced by both the ULF electric and magnetic perturbations. In this event, the $m = 1$ mode symmetric drift resonance is suggested to be dominant. The corresponding radial diffusion rates are written as[9]

$$D_{L^*L^*}^{E} = \frac{1}{8 B_0^2 R_E^2} L^{*6} P_E(\omega_d) \qquad (2)$$

$$D_{L^*L^*}^{B} = \frac{\mu^2}{8 e^2 \gamma^2 B_0^2 R_E^4} L^{*4} P_B(\omega_d) \qquad (3)$$

with the equatorial magnetic field magnitude at the Earth's surface $B_0 = 0.312$ G, the Lorentz factor of electrons γ, the Earth's radii R_E, and the power spectral densities of the azimuthal wave electric field P_E and the compressional wave magnetic field P_B at the resonant frequency $\omega = \omega_d$ in the MFA system. As discussed in the previous section, the azimuthal wave electric field cannot be obtained continuously. In this study, P_E in the MFA system is approximated as the power spectral density of the y component electric field in the mGSE system. We calculate the radial diffusion rates in two time ranges (14:00–22:00 and 22:00–24:00 UT). In each time range, we average the wave power in three spatial regions ($L^* = 3.4 \pm 0.4$, 4.2 ± 0.4 and 5.0 ± 0.4) and then obtain the L^*-dependent ULF wave power distribution through the linear interpolation. In the inner region $L^* < 3.4$, the wave power is assumed to equal that at $L^* = 3.4$; in the outer region $L^* > 5.0$ before 22:00 UT or $L^* > 4.2$ after 22:00 UT, the wave power is assumed to equal that at $L^* = 5.0$ or 4.2. Note that there were no ULF wave data in the region $L^* = 5.0 \pm 0.4$ after 22:00 UT. These assumptions would not significantly affect our conclusions, since the ULF wave-driven variation in electron PSDs mainly occurred in the region $L^* = 3.4$–4.2. In the frequency range 1–5 mHz, the obtained electromagnetic power spectral densities (Fig. 7a–d) were comparable to the previous statistical results[38–41,43]. The obtained electric diffusion rates dominated over the magnetic diffusion rates (Fig. 7e,f), consistent with the previous calculations[38,40,43].

Simulation of radiation belt evolution. The simulation code is extracted from our previously developed STEERB model[37]. The fully implicit finite difference method is adopted to solve the radial diffusion equation with the computational domain $2.5 \leq L^* \leq 5.5$. The initial condition is given by the observed electron PSD during the first passage (12:06–13:32 UT) of the outer radiation belt. The electron PSD is assumed to be fixed at the outer boundary $L^* = 5.5$ and to be zero at the inner boundary $L^* = 2.5$.

Software availability. The software for wavelet analysis is available at http://paos.colorado.edu/research/wavelets/.

References

1. Van Allen, J. A. & Frank, L. A. Radiation around the Earth to a radial distance of 107,400 km. *Nature* **183**, 430–434 (1959).
2. Friedel, R. H. W., Reeves, G. D. & Obara, T. Relativistic electron dynamics in the inner magnetosphere - a review. *J. Atmos. Sol. Terr. Phys* **64**, 265–282 (2002).
3. Horne, R. B. & Thorne, R. M. Potential waves for relativistic electron scattering and stochastic acceleration during magnetic storms. *Geophys. Res. Lett.* **25**, 3011–3014 (1998).
4. Summers, D., Thorne, R. M. & Xiao, F. Relativistic theory of wave-particle resonant diffusion with application to electron acceleration in the magnetosphere. *J. Geophys. Res.* **103**, 20487 (1998).
5. Fälthammar, C.-G. Effects of time-dependent electric fields on geomagnetically trapped radiation. *J. Geophys. Res.* **70**, 2503–2516 (1965).
6. Elkington, S. R., Hudson, M. K. & Chan, A. A. Acceleration of relativistic electrons via drift-resonant interaction with toroidal-mode Pc-5 ULF oscillations. *Geophys. Res. Lett.* **26**, 3273–3276 (1999).
7 Hudson, M. K., Elkington, S. R., Lyon, J. G. & Goodrich, C. C. Increase in relativistic electron flux in the inner magnetosphere: ULF wave mode structure. *Adv. Space Res.* **25**, 2327–2337 (2000).
8 Elkington, S. R., Hudson, M. K. & Chan, A. A. Resonant acceleration and diffusion of outer zone electrons in an asymmetric geomagnetic field. *J. Geophys. Res.* **108**, 1116 (2003).
9. Fei, Y., Chan, A. A., Elkington, S. R. & Wiltberger, M. J. Radial diffusion and MHD particle simulations of relativistic electron transport by ULF waves in the september 1998 storm. *J. Geophys. Res.* **111**, A12209 (2006).
10. Mauk, B. H. et al. Science objectives and rationale for the radiation belt storm probes mission. *Space Sci. Rev.* **179**, 3–27 (2013).
11. Reeves, G. D. et al. Electron acceleration in the heart of the Van Allen radiation belts. *Science* **341**, 991–994 (2013).
12. Thorne, R. M. et al. Rapid local acceleration of relativistic radiation-belt electrons by magnetospheric chorus. *Nature* **504**, 411–414 (2013).
13. Roederer, J. G. *Dynamics of Geomagnetically Trapped Radiation* (Springer-Verlag, 1970).
14. Baker, D. N. et al. Coronal mass ejections, magnetic clouds, and relativistic magnetospheric electron events: ISTP. *J. Geophys. Res.* **103**, 17279–17292 (1998).
15. Rostoker, G., Skone, S. & Baker, D. N. On the origin of relativistic electrons in the magnetosphere associated with some geomagnetic storms. *Geophys. Res. Lett.* **25**, 3701–3704 (1998).
16. Mathie, R. A. & Mann, I. R. A correlation between extended intervals of ULF wave power and storm-time geosynchronous relativistic electron flux enhancements. *Geophys. Res. Lett.* **27**, 3261–3264 (2000).
17. O'Brien, T. P. et al. Energization of relativistic electrons in the presence of ULF power and MeV microbursts: Evidence for dual ULF and VLF acceleration. *J. Geophys. Res.* **108**, 1329 (2003).
18. Ukhorskiy, A. Y., Anderson, B. J., Takahashi, K. & Tsyganenko, N. A. Impact of ULF oscillations in solar wind dynamic pressure on the outer radiation belt electrons. *Geophys. Res. Lett.* **33**, L06111 (2006).
19. Reeves, G. D. et al. The global response of relativistic radiation belt electrons to the January 1997 magnetic cloud. *Geophys. Res. Lett.* **25**, 3265–3268 (1998).
20. Loto'Aniu, T. M. et al. Radial diffusion of relativistic electrons into the radiation belt slot region during the 2003 Halloween geomagnetic storms. *J. Geophys. Res.* **111**, 4218 (2006).
21. Horne, R. B. et al. Wave acceleration of electrons in the Van Allen radiation belts. *Nature* **437**, 227–230 (2005).
22. Shprits, Y. Y. et al. Acceleration mechanism responsible for the formation of the new radiation belt during the 2003 halloween solar storm. *Geophys. Res. Lett.* **33**, L05104 (2006).
23. Baker, D. N. et al. The relativistic electron-proton telescope (REPT) instrument on board the radiation belt storm probes (RBSP) spacecraft: Characterization of Earth's radiation belt high-energy particle populations. *Space Sci. Rev.* **179**, 337–381 (2013).
24. Blake, J. B. et al. The magnetic electron ion spectrometer (MagEIS) instruments aboard the radiation belt storm probes (RBSP) spacecraft. *Space Sci. Rev.* **179**, 383–421 (2013).
25. Spence, H. E. et al. Science goals and overview of the energetic particle, composition, and thermal plasma (ECT) suite on NASAs Radiation Belt Storm Probes (RBSP) mission. *Space Sci. Rev.* **179**, 311–336 (2013).
26. Wygant, J. et al. The electric field and waves instruments on the radiation belt storm probes mission. *Space Sci. Rev.* **179**, 183–220 (2013).
27. Kletzing, C. A. et al. The electric and magnetic field instrument suite and integrated Science (EMFISIS) on RBSP. *Space Sci. Rev.* **179**, 127–181 (2013).
28. McIlwain, C. E. Coordinates for mapping the distribution of magnetically trapped particles. *J. Geophys. Res.* **66**, 3681–3691 (1961).
29. Southwood, D. J. & Kivelson, M. G. Charged particle behavior in low-frequency geomagnetic pulsations. II - Graphical approach. *J. Geophys. Res.* **87**, 1707–1710 (1982).
30. Zong, Q. et al. Ultralow frequency modulation of energetic particles in the dayside magnetosphere. *Geophys. Res. Lett.* **34**, L12105 (2007).
31. Claudepierre, S. G. et al. Van Allen Probes observation of localized drift resonance between poloidal mode ultra-low frequency waves and 60 keV electrons. *Geophys. Res. Lett.* **40**, 4491–4497 (2013).
32. Torrence, C. & Compo, G. P. A practical guide to wavelet analysis. *Bull. Amer. Meteor. Soc.* **79**, 61–78 (1998).
33. Mann, I. R. et al. Discovery of the action of a geophysical synchrotron in the Earth's Van Allen radiation belts. *Nat. Commun.* **4**, 2795 (2013).
34. Tu, W., Elkington, S. R., Li, X., Liu, W. & Bonnell, J. Quantifying radial diffusion coefficients of radiation belt electrons based on global MHD simulation and spacecraft measurements. *J. Geophys. Res.* **117**, 10210 (2012).
35. Green, J. C. & Kivelson, M. G. Relativistic electrons in the outer radiation belt: differentiating between acceleration mechanisms. *J. Geophys. Res.* **109**, A03213 (2004).
36. Chen, Y., Reeves, G. D. & Friedel, R. H. W. The energization of relativistic electrons in the outer Van Allen radiation belt. *Nat. Phys.* **3**, 614–617 (2007).
37. Su, Z., Xiao, F., Zheng, H. & Wang, S. STEERB: a three-dimensional code for storm-time evolution of electron radiation belt. *J. Geophys. Res.* **115**, A09208 (2010).
38. Brautigam, D. H. et al. CRRES electric field power spectra and radial diffusion coefficients. *J. Geophys. Res.* **110**, A02214 (2005).

39. Huang, C.-L., Spence, H. E., Singer, H. J. & Hughes, W. J. Modeling radiation belt radial diffusion in ULF wave fields: 1. Quantifying ULF wave power at geosynchronous orbit in observations and in global MHD model. *J. Geophys. Res.* **115**, A06215 (2010).

40. Ozeke, L. G. *et al.* ULF wave derived radiation belt radial diffusion coefficients. *J. Geophys. Res.* **117**, A04222 (2012).

41. Rae, I. J. *et al.* Ground-based magnetometer determination of in situ Pc4-5 ULF electric field wave spectra as a function of solar wind speed. *J. Geophys. Res.* **117**, A04221 (2012).

42. Mann, I. R. *et al. Dynamics of the Earth's Radiation Belts and Inner Magnetosphere* **199**, 69–92 (2012).

43. Ali, A. F. *et al.* Magnetic field power spectra and magnetic radial diffusion coefficients using CRRES magnetometer data. *J. Geophys. Res.* **120**, 973–995 (2015).

44. Takahashi, K. *et al.* Externally driven plasmaspheric ULF waves observed by the Van Allen Probes. *J. Geophys. Res.* **120**, 526–552 (2015).

45. Meredith, N. P., Horne, R. B. & Anderson, R. R. Substorm dependence of chorus amplitudes: Implications for the acceleration of electrons to relativistic energies. *J. Geophys. Res.* **106**, 13165–13178 (2001).

46. Summers, D. *et al.* Model of the energization of outer-zone electrons by whistler-mode chorus during the October 9, 1990 geomagnetic storm. *Geophys. Res. Lett.* **29**, 2174 (2002).

47. Horne, R. B. *et al.* Timescale for radiation belt electron acceleration by whistler mode chorus waves. *J. Geophys. Res.* **110**, A03225 (2005).

48. Thorne, R. M. *et al.* Refilling of the slot region between the inner and outer electron radiation belts during geomagnetic storms. *J. Geophys. Res.* **112**, A06203 (2007).

49. Li, W. *et al.* Global distribution of whistler-mode chorus waves observed on the THEMIS spacecraft. *Geophys. Res. Lett.* **36**, L09104 (2009).

50. Meredith, N. P. *et al.* Global model of lower band and upper band chorus from multiple satellite observations. *J. Geophys. Res.* **117**, 10225 (2012).

51. Horne, R. B. *et al.* A new diffusion matrix for whistler mode chorus waves. *J. Geophys. Res.* **118**, 6302–6318 (2013).

52. Su, Z. *et al.* Nonstorm time dynamics of electron radiation belts observed by the Van Allen Probes. *Geophys. Res. Lett.* **41**, 229–235 (2014).

53. Su, Z. *et al.* Intense duskside lower band chorus waves observed by Van Allen Probes: Generation and potential acceleration effect on radiation belt electrons. *J. Geophys. Res.* **119**, 4266–4273 (2014).

54. Xiao, F. *et al.* Wave-driven butterfly distribution of Van Allen belt relativistic electrons. *Nat. Commun.* **6**, 8590 (2015).

55. Albert, J. M., Meredith, N. P. & Horne, R. B. Three-dimensional diffusion simulation of outer radiation belt electrons during the october 9, 1990, magnetic storm. *J. Geophys. Res.* **114**, A09214 (2009).

56. Shprits, Y. Y., Subbotin, D. & Ni, B. Evolution of electron fluxes in the outer radiation belt computed with the VERB code. *J. Geophys. Res.* **114**, A11209 (2009).

57. Su, Z., Xiao, F., Zheng, H. & Wang, S. Radiation belt electron dynamics driven by adiabatic transport, radial diffusion, and wave-particle interactions. *J. Geophys. Res.* **116**, A04205 (2011).

58. Takahashi, K., McEntire, R. W., Lui, A. T. Y. & Potemra, T. A. Ion flux oscillations associated with a radially polarized transverse Pc 5 magnetic pulsation. *J. Geophys. Res.* **95**, 3717–3731 (1990).

59. Kurth, W. S. *et al.* Electron densities inferred from plasma wave spectra obtained by the Waves instrument on Van Allen Probes. *J. Geophys. Res.* **120**, 904–914 (2015).

60. Tsyganenko, N. A. & Sitnov, M. I. Modeling the dynamics of the inner magnetosphere during strong geomagnetic storms. *J. Geophys. Res.* **110**, A03208 (2005).

Acknowledgements

This work was supported by the National Natural Science Foundation of China grants 41422405, 41274169 and 41274174, the Chinese Academy of Sciences grants KZCX2-EW-QN510 and KZZD-EW-01-4, the National Key Basic Research Special Foundation of China grant 2011CB811403, the Fundamental Research Funds for the Central Universities WK2080000077. We acknowledge the University of Iowa as the source for the EMFISIS data in this study (this acknowledgement does not imply endorsement of the publication by the University of Iowa or its researchers), acknowledge J.H. King, N. Papatashvili and CDAWeb for providing interplanetary parameters and magnetospheric indices, and acknowledge C. Torrence and G. Compo for providing the wavelet analysis software.

Author contributions

Z.S. designed this study, performed most of the data analysis and simulation and wrote the paper. H.Z., Q.Z., X.Z. and Y.H. contributed to the ULF wave analysis. F.X. assisted with the simulations and the English improving. H.Z., Y.W. and S.W. contributed to the discussions. Z.G. and Z.H. assisted with the electron flux analysis. D.N.B., H.E.S., G.D.R. and J.B.B. provided the ECT-REPT and ECT-MagEIS data. J.R.W. provided the EFW data.

Additional information

27

Wave-driven butterfly distribution of Van Allen belt relativistic electrons

Fuliang Xiao[1], Chang Yang[1], Zhenpeng Su[2], Qinghua Zhou[1], Zhaoguo He[3], Yihua He[1], D.N. Baker[4], H.E. Spence[5], H.O. Funsten[6] & J.B. Blake[7]

Van Allen radiation belts consist of relativistic electrons trapped by Earth's magnetic field. Trapped electrons often drift azimuthally around Earth and display a butterfly pitch angle distribution of a minimum at 90° further out than geostationary orbit. This is usually attributed to drift shell splitting resulting from day–night asymmetry in Earth's magnetic field. However, direct observation of a butterfly distribution well inside of geostationary orbit and the origin of this phenomenon have not been provided so far. Here we report high-resolution observation that a unusual butterfly pitch angle distribution of relativistic electrons occurred within 5 Earth radii during the 28 June 2013 geomagnetic storm. Simulation results show that combined acceleration by chorus and magnetosonic waves can successfully explain the electron flux evolution both in the energy and butterfly pitch angle distribution. The current provides a great support for the mechanism of wave-driven butterfly distribution of relativistic electrons.

[1] School of Physics and Electronic Sciences, Changsha University of Science and Technology, 2nd Section, South Wanjiali Road #960, Yuhua District, Changsha, Hunan 410004, China. [2] Chinese Academy of Sciences Key Laboratory for Basic Plasma Physics, University of Science and Technology of China, Hefei, Anhui 230026, China. [3] Center for Space Science and Applied Research, Chinese Academy of Sciences, Beijing 100190, China. [4] Laboratory for Atmospheric and Space Physics, University of Colorado, Boulder, Colorado 80303, USA. [5] Institute for the Study of Earth, Oceans, and Space, University of New Hampshire, Durham, New Hampshire 03824-3525, USA. [6] ISR Division, Los Alamos National Laboratory, Los Alamos, New Mexico 87545, USA. [7] The Aerospace Corporation, Los Angeles, California 90245-4609, USA. Correspondence and requests for materials should be addressed to F.X. (email: flxiao@126.com).

The Earth's outer Van Allen radiation belt is composed of trapped electrons with relativistic energy ($E_k \geq 1$ MeV). Fluxes and pitch angle distributions of relativistic electrons often exhibit dramatic and highly dynamic changes during geomagnetic storms or substorms, which are associated with different physical mechanisms. Those mechanisms include transport, energization and loss processes. Wave–particle interaction plays an important role in energy exchange between various modes of plasma waves and Van Allen radiation belt relativistic electrons. Two types of plasma waves, chorus and magnetosonic (MS) wave (also known as 'equatorial noise'), can efficiently accelerate electrons up to relativistic energies. Relativistic electrons can pose serious damage to satellites and astronauts in space, it is therefore important to understand and ultimately predict Van Allen radiation belt electron dynamic variations.

Radiation belt trapped electrons experience azimuthal drift-motion around Earth when they bounce between northern and southern magnetic mirror points. Due to the day–night azimuthal asymmetry in Earth's magnetic field, their drift shells which are traced out by their guiding centres can separate radially for different pitch angles, which is known as drift shell splitting[1]. Significant drift shell splitting occurs only in the outer radiation belt $L > 6$ (with L being the radial distance in Earth radii R_E), where the asymmetry becomes substantial. Previous theoretical[1,2] and observational[3-5] works have long confirmed drift shell splitting.

Drift shell splitting separates the high and low pitch angle particles in nightside injections as they move to the dayside magnetosphere. The higher pitch angle particles drift to larger radial distance beyond the magnetopause on the dayside and may consequently be lost from the distribution[6]. Therefore, drift shell splitting can be interpreted in terms of butterfly pitch angle distributions, namely, a sharp dropout in the flux of 90° electrons. Another drift shell splitting is in combination with a negative radial flux gradient as equatorially mirroring particles drift around Earth at locations $L = 6$-12 (refs 7,8). Drift shell splitting will most easily affect trapped particle populations further out than geosynchronous orbit. Occurrence of drift shell splitting inside of geosynchronous orbit would need a very large magnetopause compression during very large geomagnetic storms.

A fundamental problem, both theoretically and observationally, is that there is any butterfly pitch angle distribution of relativistic electrons below $L = 5$ associated with a new mechanism instead of drift shell splitting. Previous studies have analysed the characteristics and evolution of pitch angle distributions of the outer radiation belt electrons[9,10]. Chorus and MS waves were proposed to produce butterfly distributions by preferentially accelerating off-equator electrons[11,12]. Adiabatic processes could potentially yield the formation of butterfly distributions at locations $L \geq 6$ (refs 13,14). Direct confirmation of wave-driven butterfly distributions below $L = 5$ requires simultaneous high-resolution data but this is generally unavailable before the launch of NASA's Van Allen Radiation Belt Storm Probes in 2012 (ref. 15). Fortunately, the unique events on the 28 June 2013 geomagnetic storm observed from Van Allen Probes provide an excellent opportunity to identify such mechanisms.

Here we report the formation of a unusual butterfly pitch angle distribution of relativistic electrons around $L = 4.8$, corresponding to the occurrence of strong chorus and MS waves at the same time. Using two-dimensional simulation, we show that flux enhancements of relativistic electrons are most pronounced between the medium pitch angles 30° and 60° due to the dominant acceleration process for combined chorus and MS waves. Meanwhile, the pitch angle distributions close to 90° are increased due to relatively larger acceleration process for chorus

alone. The combined acceleration by chorus and MS waves substantially modifies the whole population of relativistic electrons, producing the butterfly distribution. Our detailed modelling, together with the correlated Van Allen Probes data, presents a further understanding on how chorus and MS waves play different roles in Earth's outer radiation belts.

Results

Correlated Van Allen Probe data. On 28 June 2013, a moderate storm with Dst ≈ -100 nT (Fig. 1a) was triggered by an interplanetary coronal mass ejection[16]. This is a relatively minor magnetopause compression, thus drift shell splitting is not expected to occur inside of geosynchronous orbit. A large negative B_z (the z component of the interplanetary magnetic field) occurs during the period 1100 hours on June 28 to 1200 UT on June 29 (Fig. 1b), leading to efficient coupling with Earth's magnetosphere and prolonged geomagnetic activity[17]. Distinct whistler-mode chorus and MS emissions (Fig. 1d) were observed by the Electric and Magnetic Field Instrument Suite and Integrated Science (EMFISIS) Waves instrument[18,19] onboard Van Allen Probe A. Chorus and MS waves are right hand polarized electromagnetic waves. Chorus waves are excited by the injection of energetic (tens of keV) plasma sheet electrons into the inner magnetosphere during the period of enhanced plasma convection associated with the intervals of negative B_z (refs 20–22). MS waves are generated by a ring distribution of low-energy (~ 10 keV) protons at frequencies close to the harmonics of the proton gyrofrequency[23].

The fluxes of relativistic (2–3.6 MeV) electrons (Fig. 1e–h) from the Relativistic Electron-Proton Telescope (REPT) instrument[24] onboard Van Allen Probes decreased during the main phase of the storm when the magnetopause was compressed down to $L = 7.6$, and dramatically increased during the recovery phase after 29 June when the solar wind dynamic pressure decreased and the magnetopause moved beyond $L = 10$ (Fig. 1c).

In Fig. 2, we show data during the period 000 to 1600 UT on 29 June 2013 (corresponding to the shaded area in Fig. 1) from the EMFISIS instrument for magnetic and electric spectral intensity of chorus and MS waves (Fig. 2a,b), wave normal angle θ (Fig. 2c) and wave ellipticity (Fig. 2d). The observed chorus wave has a low normal angle ($\theta \approx 0°$) and circular polarization (ellipticity ≈ 0)[25]. Meanwhile, the observed MS wave is highly oblique ($\theta \approx 90°$) and linearly polarized (ellipticity ≈ 0)[26]. This indicates that the MS or chorus wave **k** vector is approximately perpendicular or parallel to the ambient magnetic field direction[12].

A very interesting feature here is that distinct butterfly distributions of fluxes of relativistic (2–3.6 MeV) electrons from the REPT instrument occurred around $L = 4.8$ over the interval 1228-1312 UT (Fig. 2e–h), corresponding to the occurrence of enhanced chorus and MS waves. Electron fluxes have minima around pitch angle 90°, and peaks around the pitch angle range of 30°–60° or 120°–150°. Observations of butterfly distributions at locations $L < 5$ are unlikely due to drift shell splitting as Earth's magnetic field is more dipolar and only distorted/stretched under extreme geomagnetic conditions. As this was a relatively minor compression and only a moderate geomagnetic storm, it is unlikely that drift shell splitting is the cause of these observations. Here we perform a two-dimensional simulation, together with the high-resolution observations from the Van Allen Probes, to examine whether chorus and MS waves can be responsible for evolution of both the energy and butterfly pitch angle distribution of the observed relativistic electron flux increase.

Numerical modelling. To model the temporal evolution of the electron distribution function, we need to solve the

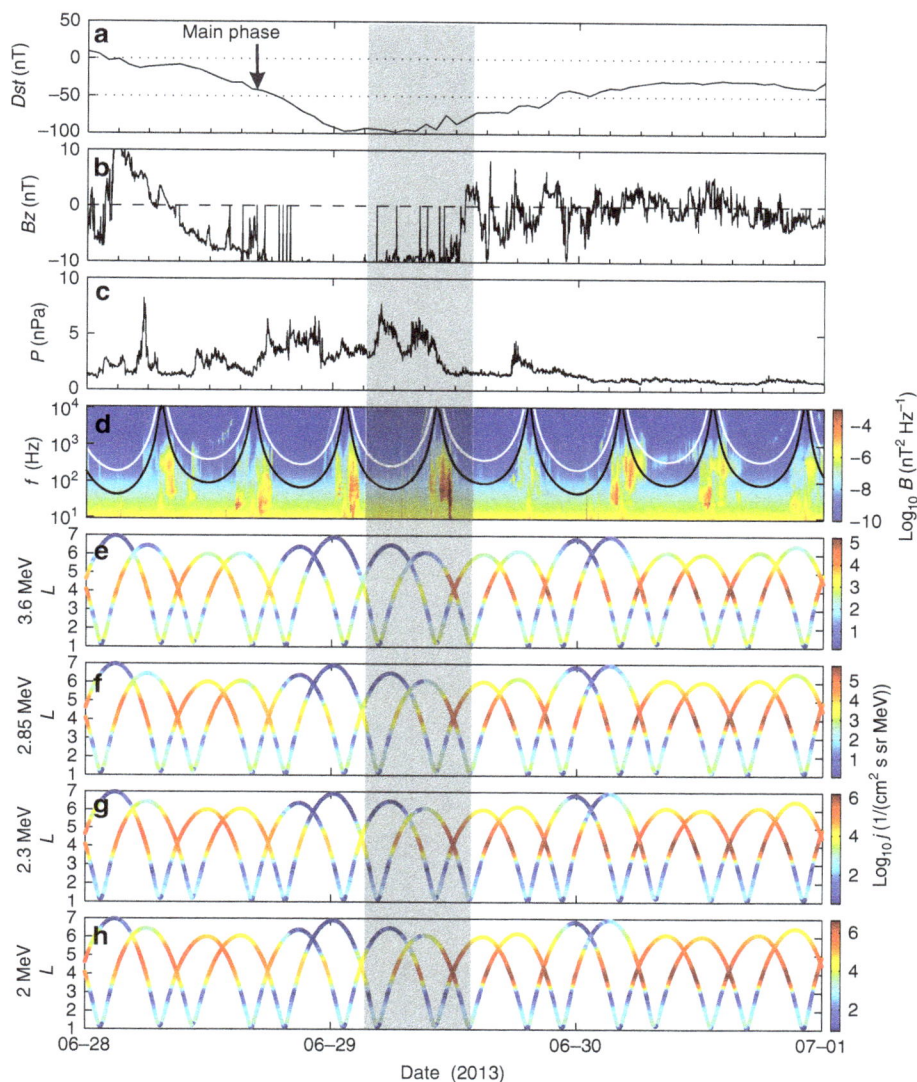

Figure 1 | Van Allen Probe data during the 28–30 June 2013 storm. (a) The *Dst* index. **(b)** The interplanetary magnetic field B_z. **(c)** Solar wind dynamic pressure. **(d)** Magnetic spectral intensity of chorus and MS waves from the EMFISIS instrument on Van Allen Probe A. The black and white lines denote the lower hybrid resonance frequency f_{lhr} and $0.1f_{ce}$ (f_{ce} being the electron gyrofrequency). **(e–h)** Relativistic electron differential fluxes from the REPT instrument onboard the Van Allen Probe A as a function of *L* showing a rapid increase in the radial range $3.5 < L < 5.5$ where strong chorus and MS waves occur. The light grey shading area indicates the simulation period.

Fokker–Planck diffusion equation, which is associated with pitch-angle and momentum diffusion coefficients driven by wave–particle interaction. Then evaluation of diffusion coefficients needs specific information of wave amplitudes and spectral properties. In this event, due to the brief UT intervals of Van Allen Probes, the *in situ* measurements of chorus or MS waves were confined to a limited range of magnetic local time (MLT) and *L*-shells: $L = 4$–6 and MLT $= 22$–24 for chorus; $L = 2$–3.5 and MLT $= 10$–18 for MS.

Previous works have demonstrated that MS waves can propagate over a broad region of MLT and *L*-shells[23,27,28]. Here using the previously developed programmes[29,30], we present ray-tracing results of MS waves based on the wave data. MS waves are launched at $L = 5.6$ for different parameters and MLT regions at the geomagnetic equator (Supplementary Table 1). The modelled results (Supplementary Fig. 1) confirm that MS waves can propagate either into or out of the plasmasphere through the plasmapause, covering a broad region of $L = 2$–5.6, particularly the observed butterfly distribution location $L = 4.8$ and MLT $= 19$.

In this event, Van Allen Probes stayed relatively deep inside the plasmasphere in the day–evening sector. Consequently, those MS waves which potentially occurred in the day–evening sector can not be directly observed by Van Allen Probes in their orbits. Since relativistic electrons drift eastward around Earth approximately in a circular orbit, the observed butterfly distribution should come from the whole contribution of resonances with those MS waves in different MLT regions.

We have also checked the Van Allen Probes data on 28 July 2013 and found that MS waves were indeed observed at $L = 4.8$ and MLT $= 19$. However, Van Allen Probes only passed the observed location in a brief UT interval every day, they may not observe the MS waves each time. In addition, using the Gaussian fitting method (not shown for brevity), we find that those MS waves display a similar Gaussian distribution to those at $L = 3$ and MLT $= 17.3$ on 29 July. It is therefore possibly reasonable to expect that the MS waves potentially existing at $L = 4.8$ and MLT $= 19$ on 29 July to follow the similar Gaussian distribution to that at $L = 3$ and MLT $= 17.3$.

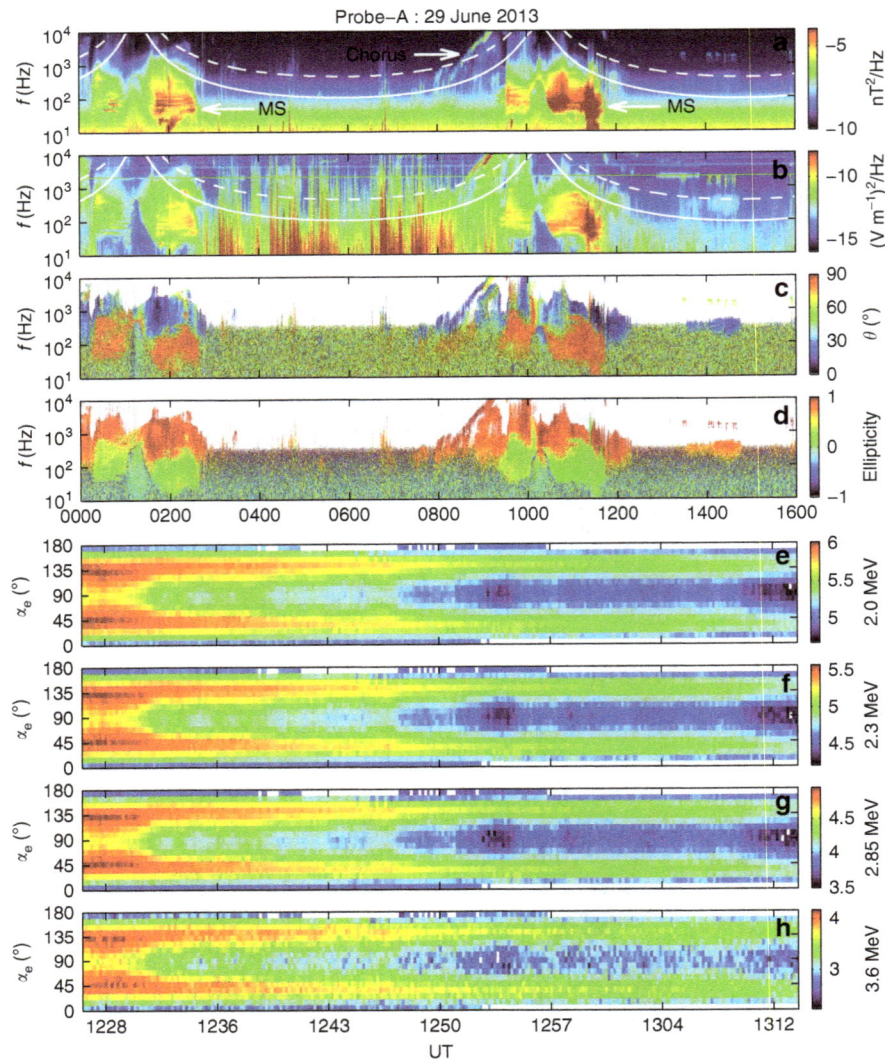

Figure 2 | Formation of butterfly distribution of relativistic electrons. Data from the EMFISIS instrument for magnetic and electric spectral intensity (in unit of $\log_{10}$) of chorus and MS waves (**a,b**), wave normal angle (**c**), the angle between Earth's magnetic field and the normal to the plane of the wave, and wave ellipticity (**d**), the degree of elliptical polarization. Chorus wave occurs above $0.1f_{ce}$ (the white dashed line). MS wave occur as a series of narrow tones, spaced at multiples of the proton gyrofrequency f_{cp} up to f_{lhr} (the white solid line). (**c,d**) The observed MS or chorus wave has a high ($\theta \approx 90°$) or low ($\theta \approx 0°$) normal angle and a high (ellipticity ≈ 0) or low (ellipticity ≈ 1) degree of elliptical polarization. (**e–h**) Pitch angle distribution for different indicated energies (2–3.6 MeV) over ~40 min duration from the REPT instrument. Fluxes (in the same unit as that in Fig. 1) of relativistic electrons have minima around pitch angle 90°, and peaks around the pitch angle range of 30°–60° or 120°–150°.

Meanwhile, for calculating the diffusion coefficients of MS waves, we assume that MS waves propagate very highly oblique ($\theta = 86°$–$89°$), distributed in a standard Gaussian form ($X = \tan\theta$) with a peak value $X_m = \tan 89°$. We noted that a concise modelling method by using maxima wave intensities was adopted in the previous work[26].

Moreover, chorus occurs over a broad range of MLT from the nightside through dawn to the dayside[31,32], potentially producing efficient scattering precipitation of energetic (tens of keV) electrons into the atmosphere. Considering that the ratio between the precipitated and trapped electron fluxes measured by Polar Orbiting Environmental Satellites (POES) is approximately proportional to the chorus power spectral intensity[33], previous studies have developed a novel technique to obtain a dynamic global model of chorus wave amplitudes as a function of L, MLT and time by converting the measured POES flux ratios at different MLT[34,35]. Here we use the same approach

(described in Methods), together with the previous parametric study[36], to obtain the global distribution of chorus wave amplitude as a function of L, MLT in this event. Specifically, we remove the proton contamination by the same method in the previous work[37]. The model parameters listed in Table 1 are adopted to calculate diffusion coefficients for chorus waves. However, the current POES model does not incorporate other loss mechanisms and further improvements are needed in the future. Considering that observations of a global distribution of chorus waves are still very limited, this model should move a step forward in constructing an event-dependent global dynamic model of chorus waves.

In general, chorus waves which consist of substructures are coherent at the equator but become less coherent off the equator, resonating with relativistic electrons[38]. The consequences of coherent wave–particle interactions involving relativistic electrons and chorus waves off the equator have been presented

Table 1 | Adopted model parameters for chorus[a].

MLT	f_m/f_{ce}	$\delta f/f_{ce}$	λ_m	N_e (cm^{-3})	f_{pe}/f_{ce}	B_t (pT)
00–04	0.25	0.05	15°	12.2	3.9	63
04–08	0.23	0.05	25°	14.0	4.3	73
08–12	0.21	0.04	45°	20.7	5.2	$10^{(0.75 + 0.04\lambda)}$
12–16	0.20	0.04	40°	25.6	5.7	36
20–24	0.35	0.05	40°	17.1	4.7	81

MLT, magnetic local time.
No realistic data available either in 08-12 or in 16-20 MLT.
[a]Chorus wave parameters either obtained from direct observations by the Van Allen Probes or inferred from POES data or from the previous parametric result[36].

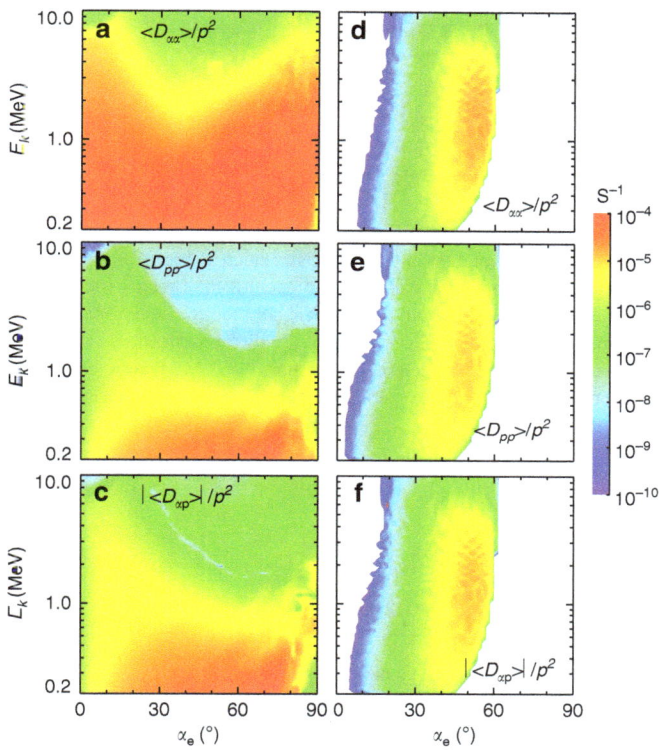

Figure 3 | Diffusion rates. Bounce-averaged pitch angle ($\langle D_{\alpha\alpha} \rangle$, (a,d)), momentum ($\langle D_{pp} \rangle$, (b,e)) and cross ($\langle D_{\alpha p} \rangle$, (c,f)) diffusion rates (in unit of s^{-1}) for resonant interactions between chorus (left) or MS (right) wave with electrons. Diffusion rates cover an entire region of pitch angles 0°–90° for chorus but a limited region 30°–60° for MS waves. Combination of chorus and MS waves leads to efficient acceleration of electrons between 2 and 3.6 MeV at medium pitch angles 30°–60° and high pitch angles up to 90°, yielding the resultant butterfly distribution on a timescale comparable to 10 h.

in the previous work[39]. Here the quasi-linear theory to treat wave–particle interaction is adopted and the corresponding conditions will be discussed below. We assume chorus waves to obey the same Gaussian distributions in wave frequency and wave normal angle at and off the equator. Considering that relativistic electrons move along the geomagnetic field line and bounce back between mirror points, we consider wave–particle interaction at each location and calculate bounce-averaged diffusion coefficients.

The obtained bounce-averaged diffusion coefficients (Fig. 3) cover an entire region of pitch angles 0°–90° for chorus but a limited region 30°–60° for MS waves. This can allow for significant increases in fluxes of relativistic (2–3.6 MeV) electrons

at medium pitch angles 30°–60° by MS waves[12] and at high pitch angles up to 90° by chorus waves[40], leading to the resultant butterfly distribution on a timescale comparable to 10 h.

Using aforementioned bounce-averaged diffusion coefficients, we calculate phase space density (PSD) f_t evolution of electrons in solving a two-dimensional Fokker–Planck diffusion equation[41]. The evolution of differential flux j is then simulated in the pitch angle region 0°–90° by the subsequent conversion $j = p^2 f_t$ and the results are extended to the range 90°–180° due to the mirror symmetry. We show in Fig. 4a remarkable agreement between the simulated pitch angle distribution (Fig. 4c,d) and the REPT observation (Fig. 4a,b) during the acceleration interval. The most pronounced flux enhancement occurs over the medium pitch angles 30°–60°, where momentum diffusion rates for combined chorus and MS waves dominate (Fig. 3). Furthermore, the pitch angle distributions at higher pitch angles close to 90° are enhanced due to larger momentum diffusion rates of chorus waves. Finally, the combined acceleration by chorus and MS waves significantly alters the whole population of relativistic electrons, yielding the butterfly distribution in about 9 h.

Discussion

The present modelling provides a further confirmation of wave-driven butterfly pitch angle distribution of relativistic electrons observed by the REPT instrument at lower locations ($L < 5$) of the outer radiation belt. This is in contrast to the formation of butterfly distributions at higher locations ($L > 6$) induced by drift shell splitting due to local magnetic field asymmetry. The associated physical processes are schematically presented in Fig. 5. Relativistic electrons at $L < 5$ drift azimuthally around Earth inside the magnetosphere without loss to the magnetopause, continuously resonating with chorus and MS waves. The combined acceleration by chorus and MS waves occurs preferably between the medium pitch angles 30° and 60°, leading to the unusual butterfly distribution. The combined acceleration process described here is a universal physical process, which should also be effective in the magnetospheres of Jupiter, Saturn and other magnetized plasma environments in the cosmos.

It should be mentioned that, in a departure from the coherent chorus–electron interaction approach[38], we use the quasi-linear theory of wave–particle interaction, which has been frequently adopted by radiation belt research community to treat wave–particle interaction[12,40,42]. Previous work[43] has shown that energetic electron pitch angle scattering by coherent chorus waves is 3 orders more rapid than by incoherent chorus waves. Coherent chorus waves can produce particle diffusion, phase trapping and/or phase bunching, which is determined primarily by the competing effects of wave amplitude and ambient magnetic field inhomogeneity at resonance. However, application of coherent wave–particle interaction to the storm-time global evolution of energetic particle distributions has not

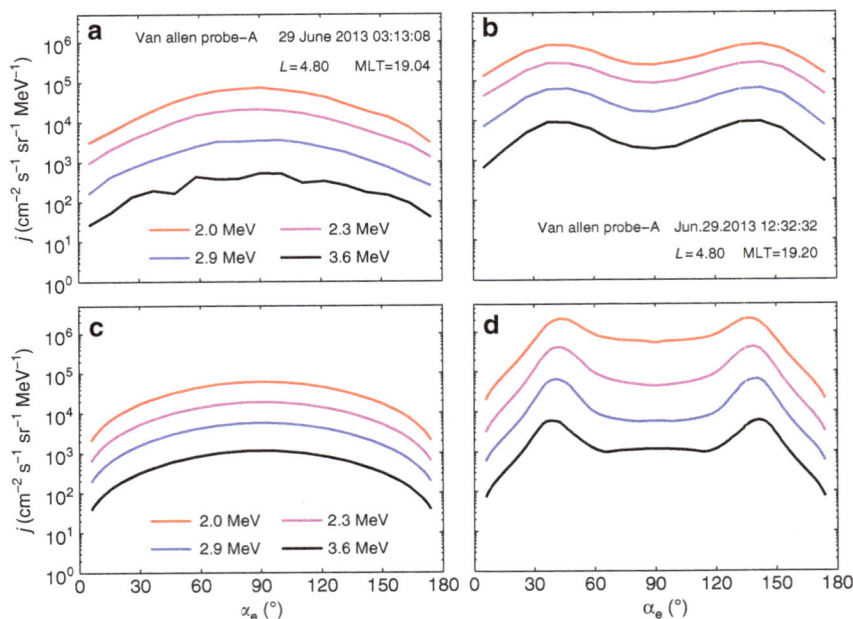

Figure 4 | Comparison of Fokker–Planck simulation results with observations. (a,b) Observed evolution of the relativistic electron fluxes as a function of pitch angle at the same location (shown inside panels) during the simulation. **(c,d)** Starting with an initial condition representative of relativistic electrons **(a)**, we show the relativistic (2–3.6 MeV) electron fluxes due to acceleration by chorus and MS waves from a numerical solution to the two-dimensional Fokker–Planck diffusion equation. Acceleration is most pronounced within the medium pitch angle region 30°–60° or 120°–150° produced by both chorus and MS waves, and noticeable around the high pitch angles induced by chorus waves. The combined acceleration by chorus and MS waves leads to the butterfly distribution both in the magnitude and the time scale **(d)** comparable to the observation **(b)**. The current simulation provides a further support for wave-driven butterfly distribution of relativistic electrons during this storm.

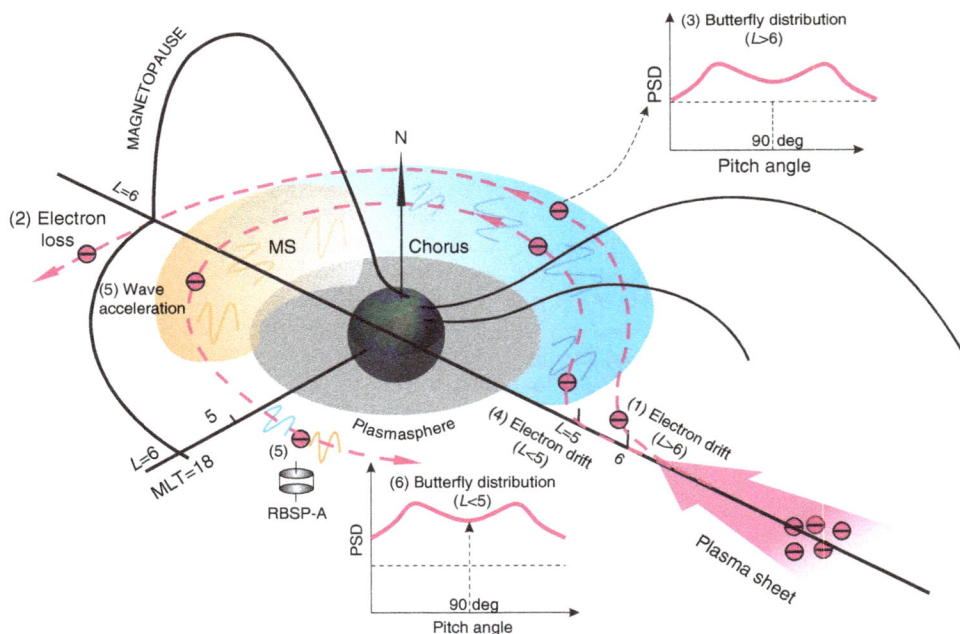

Figure 5 | Schematic diagram of formation of butterfly distributions. The long black curves indicate Earth's magnetic field lines. The grey area denote the region of dense cold plasma known as plasmasphere. The blue and yellow areas stand for primary occurrences of chorus and MS waves. The wavy line represents chorus (blue) and MS (yellow) waves. Relativistic electrons originating from the plasma sheet azimuthally drift eastward around Earth. (1) Higher pitch angle electrons at $L>6$ drift to larger radial distance beyond the dayside magnetopause and are consequently lost due to the day–night magnetic field asymmetry (2), yielding the regular butterfly distribution (3). (4) Electrons at $L<5$ drift inside the magnetosphere without loss to the dayside magnetopause because of a very small day–night magnetic field asymmetry, continuously resonating with chorus and MS waves. (5) Wave acceleration can enhance electron PSD primarily within the medium pitch angles, leading to the formation of the unusual butterfly distribution (6).

yet been established so far. We therefore expect the adopted quasi-linear theory here to be a basis for comparison with future developments in nonlinear modelling.

The basic conditions for the quasi-linear theory are that each individual particle randomly moves in velocity space, resonates with a succession of uncorrelated and small amplitude waves, and is scattered in a small amount in pitch angle and energy each time. Those conditions are basically satisfied in Earth's radiation belts for naturally generated MS and chorus waves, where the MS wave bandwidth is generally above the proton gyrofrequency up to the lower hybrid frequency and the chorus wave roughly lies in the frequency range 0.1–0.8 f_{ce} (f_{ce} being the electron gyrofrequency).

In general, there could be a distorted/stretched geomagnetic field on the nightside during a geomagnetic storm. We have examined the geomagnetic field data from Van Allen Probes and found that the observed data are close to the dipole field model during this storm period. The largest distortion of the dipole field is about 20% in the simulation period. Moreover, we perform a test-particle simulation of trajectories of trapped relativistic (2 MeV) electrons by the TS04 magnetic model[44] for different high pitch angles. Simulation results (Fig. 6) show that, starting at the location $L = 4.8$ and MLT = 24, relativistic electrons drift eastward around Earth and approach a farthest location $L \approx 5.1$ on the dayside. Then they pass the observed location without loss to the magnetopause because the magnetopause locates above $L = 7.6$ in the simulation period. Simulations for different energies or pitch angles (not shown for brevity) indeed show similar results. This indicates that drift shell splitting is unlikely to play a role in the observed butterfly electron distribution in this event.

A basic assumption is adopted here that the cold plasma density remains constant and unchanging during the 9-h simulation period. Using the upper hybrid resonance frequency observation from Van Allen Probes, we infer the ambient electron density at locations along Van Allen Probes' orbits and find that the electron density is comparable to the adopted plasma density and does not change much in 9 h. Analogous to the previous work[40], this assumption is probably reasonable in the absence of

simultaneous and continuous local time observations of cold plasma density in the 9-h period.

To check whether the plasmaspheric plume exists, we analyse the potential data from the Time History of Events and Macroscale Interactions During Substorms (THEMIS) spacecraft at $L = 4.8$, and find no distinct presence of plasmaspheric plume except a slightly higher electron density in the region around/after MLT = 16 in the simulation period. The plasmaspheric plume resulting from time-varying convection generally forms with a more structured distribution of plasma in the dusk region in the storm main phase. Meanwhile, the plume moves eastward (towards later MLTs) in Earth's rotation eastward, and gradually fades/erodes as the strength of convection decreases. As shown in Fig. 1, the simulation period is in the storm recovery phase on 29 June, > 24 h after the onset of the storm on 28 June. Hence, the plume probably already faded and moved to the later MLT region. However, further research is still required in the future because there are numerous uncertainties in accurate determination of either the upper hybrid resonance frequency from Van Allen Probes or the THEMIS spacecraft potential.

Methods

Ray tracing of MS waves. We perform the ray tracing of MS waves with wave vector **k** and frequency ω by the following standard Hamiltonian equations[45]:

$$\frac{d\mathbf{R}}{dt} = -\frac{\partial D}{\partial \mathbf{k}} \bigg/ \frac{\partial D}{\partial \omega} \qquad (1)$$

$$\frac{d\mathbf{k}}{dt} = \frac{\partial D}{\partial \mathbf{R}} \bigg/ \frac{\partial D}{\partial \omega}, \qquad (2)$$

where **R** is the position vector of a point on the ray path, t is the group time, D represents the standard wave dispersion relation obeying $D(\mathbf{R}, \omega, \mathbf{k}) = 0$ at every point along the ray path. The spatial variation in D can be written:

$$\frac{\partial D}{\partial \mathbf{R}} = \frac{\partial D}{\partial \mathbf{B}_0}\frac{\partial \mathbf{B}_0}{\partial \mathbf{R}} + \frac{\partial D}{\partial N_c}\frac{\partial N_c}{\partial \mathbf{R}} + \frac{\partial D}{\partial k}\frac{\partial \mathbf{k}}{\partial \mathbf{R}} \qquad (3)$$

where $\mathbf{B}_0$ is the ambient magnetic field and N_c is the background plasma density. We adopt a dipole magnetic field model and the MLT-dependent plasmatrough density model[46]. The Earth-centred Cartesian and local Cartesian coordinate systems for the ray-tracing calculation are described in Supplementary Note 1.

Calculation of the global chorus wave amplitude. Recent works[34,35] have constructed a global chorus wave model based on the data of low-altitude electron population collected by multiple POES satellites. The Medium Energy Proton and Electron Detector (MEPED) onboard POES has two electron solid-state detector (0° and 90°) telescopes to measure electron fluxes in three energy channels (> 30 keV, > 100 keV and > 300 keV)[47]. The 90° telescope largely measures the trapped flux over the invariant latitude range between 55° and 70°, and the 0° telescope measures precipitating flux inside the bounce loss cone at $L > 1.4$ (ref. 48). Using the electron distribution function near the loss cone[33], the chorus wave amplitude can be calculated from the ratio between the measured precipitated and trapped electron fluxes (30–100 keV and 100–300 keV). The electron energy spectrum is assumed to follow a kappa-type function[49,50]. The basic equation for linking the ratio of electron count rates measured by the 0° and 90° telescopes to chorus wave amplitude is shown in Supplementary Note 2. The ratios of precipitated and trapped electron fluxes measured by multiple POES satellites are used to construct the global chorus wave model over a broad range in L and MLT, and the ratios obtained in four distinct MLT sectors are shown in Supplementary Fig. 2.

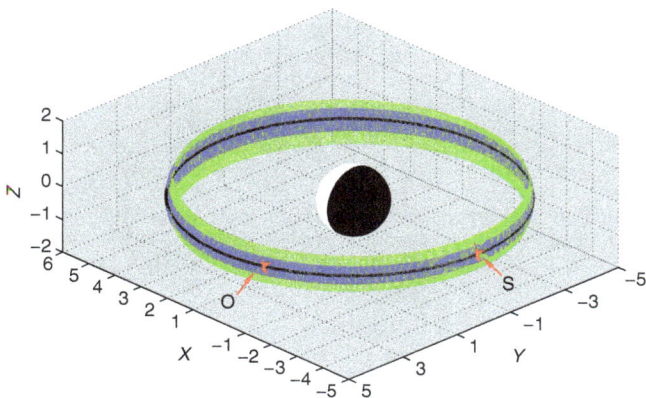

Figure 6 | Trajectories of trapped electrons. Test-particle simulations of relativistic (2 MeV) electron trajectories by the TS04 magnetic model for different pitch angles 70° (green), 80° (blue) and 90° (black). The input parameters for TS04 magnetic model are based on the observations. Starting at the location (the symbols S): $L = 4.8$ and MLT = 24, relativistic electrons drift eastward around Earth and approach a farthest location $L \approx 5.1$ on the dayside. Then they pass the observed location (the symbol O) without loss to the magnetopause because the magnetopause locates above $L = 7.6$ in the simulation period.

Calculation of diffusion coefficients due to chorus and MS waves. We assume that the wave spectral density B_f^2 follows a typical Gaussian frequency distribution with a center f_m, a half-width δf, a band between f_1 and f_2 (ref. 51).

$$B_f^2 = \frac{2B_t^2}{\sqrt{\pi}\delta f}\left[\mathrm{erf}\left(\frac{f_2 - f_m}{\delta f}\right) + \mathrm{erf}\left(\frac{f_m - f_1}{\delta f}\right)\right]^{-1}\exp\left[-\frac{(f - f_m)^2}{(\delta f)^2}\right] \qquad (4)$$

here B_t^2 is the wave amplitude in units of Tesla and erf is the error function. To allow the data modelling to be more reliable, we average the observed wave magnetic field intensity over the indicated time period in this event and then apply the corresponding Gaussian fit for MS waves as shown in Supplementary Fig. 3.

We choose the wave normal angle distribution to also satisfy a standard Gaussian form:

$$g(X) \propto \begin{cases} \exp[-(X - X_m)/X_\omega]^2 & \text{for } X_1 \leq X \leq X_2, \\ 0 & \text{otherwise,} \end{cases} \quad (5)$$

where $X = \tan\theta$ ($\theta_1 \leq \theta \leq \theta_2$, $X_{1,2} = \tan\theta_{1,2}$), with a half-width X_ω and a peak X_m. Based on the observation, we choose $X_m = \tan 89°$, $X_\omega = \tan 86°$, $X_1 = X_m - X_\omega$, $X_2 = X_m + X_\omega$; and the maximum latitude for the presence of MS wave $\lambda_m = 10°$. Based on the ray-tracing results (Supplementary Fig. 1), we assume the MS wave spectral intensity at $L = 4.8$ and MLT $= 19$ hours to follow the similar Gaussian distribution to that as shown in Supplementary Fig. 3, which are then used to calculate the MS-driven bounce-averaged diffusion coefficients. For the chorus waves, we adopt the wave parameters as shown in Table 1 to calculate the chorus-driven bounce-averaged diffusion coefficients at the location $L = 4.8$.

We consider contribution from harmonic resonances up to $n = \pm 5$ for both chorus and MS waves. The ambient plasma density is obtained by the MLT-dependent plasmatrough density model[46] and further assumed to remain latitudinally constant and unchanged during the 9-h simulation period. Note that there is no realistic data available either in 08–12 or 16–20 MLT. We assume the chorus wave power in 08–12 MLT to follow the latitude-dependent model[36]. The corresponding adopted wave amplitude, wave spectrum and latitudinal occurrence of chorus waves, the ambient electron density and the equatorial ratio of plasma frequency to gyrofrequency (f_{pe}/f_{ce}) are shown in Table 1.

Evaluation of the relativistic electron PSD evolution. The evolution of the electron PSD f_t is calculated by solving the bounce-averaged pitch angle and momentum diffusion equation

$$\frac{\partial f_t}{\partial t} = \frac{1}{Gp} \frac{\partial}{\partial \alpha_e} \left[G \left(\langle D_{\alpha\alpha} \rangle \frac{1}{p} \frac{\partial f_t}{\partial \alpha_e} + \langle D_{\alpha p} \rangle \frac{\partial f_t}{\partial p} \right) \right]$$
$$+ \frac{1}{G} \frac{\partial}{\partial p} \left[G \left(\langle D_{p\alpha} \rangle \frac{1}{p} \frac{\partial f_t}{\partial \alpha_e} + \langle D_{pp} \rangle \frac{\partial f_t}{\partial p} \right) \right] \quad (6)$$

here p is the electron momentum, $G = p^2 T(\alpha_e) \sin\alpha_e \cos\alpha_e$ with α_e being the equatorial pitch angle, the normalized bounce time $T(\alpha_e) \approx 1.30 - 0.56\sin\alpha_e$; $\langle D_{\alpha\alpha} \rangle$, $\langle D_{pp} \rangle$, and $\langle D_{\alpha p} \rangle = \langle D_{p\alpha} \rangle$ denote bounce-averaged diffusion coefficients in pitch angle, momentum and cross pitch angle–momentum. The explicit expressions of those bounce-averaged diffusion coefficients can be found in the previous work.[41]

The initial condition is modelled by a bi-modal kappa-type distribution function of energetic electrons[49,50]:

$$f_0^\kappa(p, \sin\alpha_e) = a_1 f^{l_1}(p, \sin\alpha_e) + a_2 f^{l_2}(p, \sin\alpha_e) \quad (7)$$

Each component, with variable weighting parameters a_1 and a_2, is expressed as:

$$f^l(p, \sin\alpha_e) = \frac{n_h \Gamma(\kappa + l + 1)}{\pi^{3/2} \theta_\kappa^3 \kappa^{(l+3/2)} \Gamma(l+1) \Gamma(\kappa - 1/2)} \left(\frac{p \sin\alpha_e}{\theta_\kappa} \right)^{2l} \left[1 + \frac{p^2}{\kappa \theta_\kappa^2} \right]^{-(\kappa + l + 1)} \quad (8)$$

here n_h is the number density of energetic electrons, l indicates the loss-cone index, Γ is the gamma function, κ and θ_κ^2 are the spectral index and effective thermal energy scaled by the electron rest mass energy $m_e c^2$ (~ 0.5 MeV).

For the pitch angle boundary condition, $f_t = 0$ at the loss-cone $\alpha_e = \alpha_L (\sin\alpha_L = L^{-3/2}(4 - 3/L)^{-1/4})$ to simulate a rapid precipitation of electrons inside the loss cone (Fig. 2e–h), and $\partial f_t / \partial \alpha_e = 0$ at $\alpha_e = 90°$. For the energy diffusion boundary conditions, $f_t = f_0^\kappa$ (0.2 MeV) = const at the lower boundary 0.2 MeV, and $f_t = f_0^\kappa$ (10 MeV) = const at the upper boundary 10 MeV.

Based on the observation, we choose the following values of parameters: $\theta_\kappa = 0.15$ (~ 75 keV), $\kappa = 4$, $n_h = 0.16$ cm^{-3}; $a_1 = 0.6$, $l_1 = 0.6$; $a_2 = 0.4$, $l_2 = 1.5$. We solve the diffusion equation(6) using the recently developed hybrid difference method[41], which is efficient, stable and easily parallel programmed. The numerical grid is set to be 91×101 and uniform in pitch angle and natural logarithm of momentum. The drifting average is taken 25% for MS wave and 1/6 for chorus wave in each aforementioned MLT sector (Table 1).

References

1. Roederer, J. G. On the adiabatic motion of energetic particles in a model magnetosphere. *J. Geophys. Res.* **72**, 981–992 (1967).
2. Stone, E. C. The physical significance of and application of L, B_0 and R_0 to geomagnetically trapped particles. *J. Geophys. Res.* **68**, 4157–4166 (1963).
3. West, H. I., Buck, Jr R. M. & Walton, J. R. Electron pitch angle distributions throughout the magnetosphere as observed on Ogo 5. *J. Geophys. Res.* **78**, 1064–1081 (1973).
4. Baker, D. N., Higbie, P. R., Hones, Jr E. W. & Belian, R. D. High-resolution energetic particle measurements at 6.6 R_E. Low-energy electron anisotropies and short-term substorm predictions. *J. Geophys. Res.* **83**, 4863–4868 (1978).
5. Takahashi, K. *et al.* Drift-shell splitting of energetic ions injected at pseudo-substorm onsets. *J. Geophys. Res.* **102**, 22117–22130 (1997).
6. Selesnick, R. S. & Blake, J. B. Relativistic electron drift shell splitting. *J. Geophys. Res.* **1079**, 1265 (2002).

7. Garcia, H. A. & Spjeldvik, W. N. Anisotropy characteristics of geomagnetically trapped ions. *J. Geophys. Res.* **90**, 347–358 (1985).
8. Sibeck, D. G., McEntire, R. W., Lui, A. T. Y., Lopez, R. E. & Krimigis, S. M. Magnetic field drift shell splitting: Cause of unusual dayside particle pitch angle distributions during storms and substorms. *J. Geophys. Res.* **92**, 13345–13497 (1987).
9. Horne, R. B. *et al.* Evolution of energetic electron pitch angle distributions during storm time electron acceleration to megaelectronvolt energies. *J. Geophys. Res.* **108**, SMP 11-1-SMP 11-13 (2003).
10. Gannon, J. L., Li, X. & Heynderickx, D. Pitch angle distribution analysis of radiation belt electrons based on combined release and radiation effects satellite medium electrons A data. *J. Geophys. Res.* **112**, A05212 (2007).
11. Horne, R. B. *et al.* Timescale for radiation belt electron acceleration by whistler mode chorus waves. *J. Geophys. Res.* **110**, A03225 (2005).
12. Horne, R. B. *et al.* Electron acceleration in the Van Allen radiation belts by fast magnetosonic waves. *Geophys. Res. Lett.* **34**, L17107 (2007).
13. Su, Z., Xiao, F., Zheng, H. & Wang, S. Combined radial diffusion and adiabatic transport of radiation belt electrons with arbitrary pitch angles. *J. Geophys. Res.* **115**, A10249 (2010).
14. Su, Z., Xiao, F., Zheng, H. & Wang, S. Radiation belt electron dynamics driven by adiabatic transport, radial diffusion, and wave–particle interactions. *J. Geophys. Res.* **116**, A04205 (2011).
15. Mauk, B. H. *et al.* Science objectives and rationale for the radiation belt storm probes mission. *Space Sci. Rev.* **179**, 3–27 (2012).
16. Gonzalez, W. D., Tsurutani, B. T. & Gonzalez, A. L. C. Interplanetary origin of geomagnetic storms. *Space Sci. Rev.* **88**, 529–562 (1999).
17. Gonzalez, W. D. *et al.* What is a geomagnetic storm. *J. Geophys. Res.* **99**, 5771–5792 (1994).
18. Kletzing, C. A. *et al.* The electric and magnetic field instrument suite and integrated science (EMFISIS) on RBSP. *Space Sci. Rev.* **179**, 127–181 (2013).
19. Wygant, J. R. *et al.* The electric field and waves (EFW) instruments on the radiation belt storm probes mission. *Space Sci. Rev.* **179**, 183–220 (2013).
20. Papadopoulos, K. *Waves and Instabilities in Space Plasmas.* (eds Palmadesso, P. & Papadopoulos, K.) (D. Reidel Co., 1979).
21. Xiao, F., Zhou, Q., Zheng, H. & Wang, S. Whistler instability threshold condition of energetic electrons by kappa distribution in space plasmas. *J. Geophys. Res.* **111**, A08208 (2006).
22. Summers, D., Tong, R. & Thorne, R. M. Limit on stably trapped particle fluxes in planetary magnetospheres. *J. Geophys. Res.* **114**, A10210 (2009).
23. Chen, L., Thorne, R. M., Jordanova, V. K., Thomsen, M. F. & Horne, R. B. Magnetosonic wave instability analysis for proton ring distributions observed by the LANL magnetospheric plasma analyser. *J. Geophys. Res* **116**, A03223 (2011).
24. Baker, D. N. *et al.* The relativistic electron-proton telescope (REPT) instrument on board the Radiation Belt Storm Probes (RBSP) spacecraft: Characterization of Earth's radiation belt high-energy particle populations. *Space Sci. Rev.* **179**, 337–381 (2012).
25. Tsurutani, B. T., Verkhoglyadova, O. P., Lakhina, G. S. & Yagitani, S. Properties of dayside outer zone chorus during HILDCAA events: loss of energetic electrons. *J. Geophys. Res.* **114**, A03207 (2009).
26. Tsurutani, B. T. *et al.* Extremely intense ELF magnetosonicwaves: A survey of polar observations. *J. Geophys. Res. Space Phys.* **119**, 964–977 (2014).
27. Xiao, F., Zhou, Q., He, Z. & Tang, L. Three-dimensional ray tracing of fast magnetosonic waves. *J. Geophys. Res.* **117**, A06208 (2012).
28. Ma, Q., Li, W., Thorne, R. M. & Angelopoulos, V. Global distribution of equatorial magnetosonic waves observed by THEMIS. *Geophys. Res. Lett.* **40**, 1895–1901 (2013).
29. Horne, R. B. Path-integrated growth of electrostatic waves: The generation of terrestrial myriametric radiation. *J. Geophys. Res.* **94**, 8895–8909 (1989).
30. Xiao, F., Chen, L., Zheng, H. & Wang, S. A parametric ray tracing study of superluminous auroral kilometric radiation wave modes. *J. Geophys. Res.* **112**, A10214 (2007).
31. Meredith, N. P. *et al.* Global models of lower band and upper band chorus from multiple satellite observations. *J. Geophys. Res.* **117**, A10225 (2012).
32. Li, W. *et al.* Global distribution of whistler-mode chorus waves observed on the THEMIS spacecraft. *Geophys. Res. Lett.* **36**, L09104 (2009).
33. Kennel, C. F. & Petschek, H. E. Limit on stably trapped particle fluxes. *J. Geophys. Res.* **71**, 1–28 (1966).
34. Li, W. *et al.* Constructing the global distribution of chorus wave intensity using measurements of electrons by the POES satellites and waves by the Van Allen Probes. *Geophys. Res. Lett.* **40**, 4526–4532 (2013).
35. Ni, B. *et al.* A novel technique to construct the global distribution of whistler mode chorus wave intensity using low-altitude POES electron data. *J. Geophys. Res. Space Phys.* **119**, 5685–5699 (2014).
36. Shprits, Y. Y., Meredith, N. P. & Thorne, R. M. Parameterization of radiation belt electron loss timescales due to interactions with chorus waves. *Geophys. Res. Lett.* **34**, L11110 (2007).

37. Lam, M. M. *et al.* Origin of energetic electron precipitation >30 keV into the atmosphere. *J. Geophys. Res.* **115**, A00F08 (2010).

38. Tsurutani, B. T. *et al.* Quasi-coherent chorus properties: 1. Implications for wave–particle interactions. *J. Geophys. Res.* **116**, A09210 (2011).

39. Tsurutani, B. T., Lakhina, G. S. & Verkhoglyadova, O. P. Energetic electron (>10 keV) microburst precipitation, ~5-15 s X-ray pulsations, chorus, and wave–particle interactions: A review. *J. Geophys. Res. Space Phys.* **118**, 2296–2312 (2013).

40. Thorne, R. M. *et al.* Rapid local acceleration of relativistic radiation-belt electrons by magnetospheric chorus. *Nature* **504**, 411–414 (2013).

41. Xiao, F., Su, Z., Zheng, H. & Wang, S. Modelling of outer radiation belt electrons by multidimensional diffusion process. *J. Geophys. Res.* **114**, A03201 (2009).

42. Horne, R. B. *et al.* Wave acceleration of electrons in the Van Allen radiation belts. *Nature* **437**, 227–230 (2005).

43. Lakhina, G. S., Tsurutani, B. T., Verkhoglyadova, O. P. & Pickett, J. S. Pitch angle transport of electrons due to cyclotron interactions with the coherent chorus subelements. *J. Geophys. Res.* **115**, A00F15 (2010).

44. Tsyganenko, N. A. & Sitnov, M. I. Modelling the dynamics of the inner magnetosphere during strong geomagnetic storms. *J. Geophys. Res.* **110**, A03208 (2005).

45. Suchy, K. Real Hamilton equations of geomagnetic optics for media with moderate absorption. *Radio Sci.* **16**, 1179–1182 (1981).

46. Sheeley, B. W., Moldwin, M. B., Rassoul, H. K. & Anderson, R. R. An empirical plasmasphere and trough density model: CRRES observations. *J. Geophys. Res.* **106**, 25631–25641 (2001).

47. Evans, D. S. & Greer, M. S. *NOAA Technical Memorandum.* 93 (Space Weather Prediction Center, Boulder, 2004).

48. Rodger, C. J., Clilverd, M. A., Green, J. C. & Lam, M. M. Use of POES SEM-2 observations to examine radiation belt dynamics and energetic electron precipitation into the atmosphere. *J. Geophys. Res.* **115**, A04202 (2010).

49. Xiao, F., Shen, C., Wang, Y., Zheng, H. & Wang, S. Energetic electron distributions fitted with a relativistic kappa-type function at geosynchronous orbit. *J. Geophys. Res.* **113**, A05203 (2008).

50. Vasyliunas, V. M. A survey of low-energy electrons in the evening sector of the magnetosphere with ogo 1 and ogo 3. *J. Geophys. Res.* **73**, 2839–2884 (1968).

51. Lyons, L. R., Thorne, R. M. & Kennel, C. F. Pitch angle diffusion of radiation belt electrons within the plasmasphere. *J. Geophys. Res.* **77**, 3455–3474 (1972).

Acknowledgements

This work was supported by 973 Program 2012CB825603, the National Natural Science Foundation of China grants 41531072, 41274165, 41204114, the Aid Program for Science and Technology Innovative Research Team in Higher Educational Institutions of Hunan Province and the Construct Program of the Key Discipline in Hunan Province. All the Van Allen Probes data are publicly available at https://emfisis.physics.uiowa.edu/data/index by the EMFISIS suite and at http://www.rbsp-ect.lanl.gov/data_pub/ by the REPT and MagEIS instrument. The OMNI data are obtained from http://omni-web.gsfc.nasa.gov/form/dx1.html. Work at Los Alamos was performed under the auspices of the US Department of Energy and supported by the Los Alamos LDRD program. We would like to thank G.D. Reeves for making available the data used in this work.

Author contributions

F.X. led the idea, modelling and was responsible for writing the paper. C.Y. and Z.S. contributed to data analysis, interpretation and the Fokker–Planck simulations. Q.Z. contributed to ray-tracing modelling. Z.H. and Y.H. contributed to data analysis. D.N.B., H.E.S., H.O.F. and J.B.B. provided the RBSP-ECT data.

Additional information

PERMISSIONS

LIST OF CONTRIBUTORS

P. Merino and J. Cernicharo
Centro de Astrobiología INTA-CSIC, Carretera de Ajalvir, km.4, ES-28850 Madrid, Spain

& J.A. Martin-Gago
Centro de Astrobiología INTA-CSIC, Carretera de Ajalvir, km.4, ES-28850 Madrid, Spain
Instituto Ciencia de Materiales de Madrid-CSIC, c/. Sor Juana Inés de la Cruz, 3, ES-28049 Madrid, Spain

M. Švec and P. Jelinek
Institute of Physics, Academy of Sciences of the Czech Republic, Cukrovarnicka 10, CZ-16200 Prague, Czech Republic

J.I. Martinez
Instituto Ciencia de Materiales de Madrid-CSIC, c/. Sor Juana Inés de la Cruz, 3, ES-28049 Madrid, Spain

P. Lacovig, M. Dalmiglio and S. Lizzit
Elettra-Sincrotrone Trieste S.C.p.A., Area Science Park, S.S. 14, Km 163.5, I-34149 Trieste, Italy

P. Soukiassian
Commissariat à l'Energie Atomique et aux Energies Alternatives, SIMA, DSM-IRAMIS-SPEC, Bât. 462, 91191 Gif sur Yvette, France
Synchrotron SOLEIL, L'Orme des Merisiers, Saint-Aubin, 91192 Gif sur Yvette, France

J.Q. Sun, X. Cheng, M.D. Ding, Y. Guo, P.F. Chen and C. Fang
School of Astronomy and Space Science, Nanjing University, Nanjing 210093, China

E.R. Priest and C.E. Parnell
School of Mathematics and Statistics, University of St Andrews, Fife, KY16 9SS Scotland, UK

S.J. Edwards
Department of Mathematical Sciences, Durham University, Durham DH1 3LE, UK

J. Zhang
School of Physics, Astronomy and Computational Sciences, George Mason University, Fairfax, Virginia 22030, USA

Benhui Yang and P.C. Stancil
Department of Physics and Astronomy and the Center for Simulational Physics, The University of Georgia, Athens, Georgia 30602, USA

P. Zhang
Department of Chemistry, Duke University, Durham, North Carolina 27708, USA

X. Wang and J.M. Bowman
Department of Chemistry, Emory University, Atlanta, Georgia 30322, USA

N. Balakrishnan
Department of Chemistry, University of Nevada Las Vegas, Las Vegas, Nevada 89154, USA

R.C. Forrey
Department of Physics, Penn State University, Berks Campus, Reading, Pennsylvania 19610, USA

Haimin Wang, Wenda Cao, Chang Liu and Yan Xu
Space Weather Research Laboratory, New Jersey Institute of Technology, University Heights, Newark, New Jersey 07102-1982, USA
Big Bear Solar Observatory, New Jersey Institute of Technology, 40386 North Shore Lane, Big Bear City, California 92314-9672, USA

Zhicheng Zeng
Big Bear Solar Observatory, New Jersey Institute of Technology, 40386 North Shore Lane, Big Bear City, California 92314-9672, USA

Rui Liu
Department of Geophysics and Planetary Sciences, CAS Key Laboratory of Geospace Environment, University of Science and Technology of China, Hefei 230026, China
Collaborative Innovation Center of Astronautical Science and Technology, China

Jongchul Chae
Department of Physics and Astronomy, Astronomy Program, Seoul National University, Seoul 151-747, Korea

Haisheng Ji
Purple Mountain Observatory, Chinese Academy of Sciences, Nanjing 210008, China

A.V. Artemyev
LPC2E/CNRS, 3A, Avenue de la Recherche Scientifique, 45071 Orleans Cedex 2, France.
Space Research Institute (IKI) 117997, 84/32 Profsoyuznaya Street, Moscow, Russia. (A.V.A.)
Astronomy and Space Physics Department, National Taras Shevchenko University of Kiev, 2 Glushkova Street, 03222 Kiev, Ukraine (O.V.A.)

V.V. Krasnoselskikh
LPC2E/CNRS, 3A, Avenue de la Recherche Scientifique, 45071 Orleans Cedex 2, France

F.S. Mozer
Space Sciences Laboratory, University of California, 7 GaussWay, Berkeley, California 94720, USA

O.V. Agapitov
Space Sciences Laboratory, University of California, 7 GaussWay, Berkeley, California 94720, USA
Space Research Institute (IKI) 117997, 84/32 Profsoyuznaya Street, Moscow, Russia. (A.V.A.); Astronomy and Space Physics Department, National Taras Shevchenko University of Kiev, 2 Glushkova Street, 03222 Kiev, Ukraine (O.V.A.)

D. Mourenas
CEA, DAM, DIF, F-91297 Arpajon Cedex, France
Space Research Institute (IKI) 117997, 84/32 Profsoyuznaya Street, Moscow, Russia. (A.V.A.)
Astronomy and Space Physics Department, National Taras Shevchenko University of Kiev, 2 Glushkova Street, 03222 Kiev, Ukraine (O.V.A.)

A. Wallner
Department of Nuclear Physics, Australian National University, Canberra, Australian Capital Territory 0200, Australia.
VERA Laboratory, Faculty of Physics, University of Vienna, Währinger Strasse 17, A-1090 Vienna, Austria

J. Feige, W. Kutschera and P. Steier
VERA Laboratory, Faculty of Physics, University of Vienna, Währinger Strasse 17, A-1090 Vienna, Austria

F. Quinto
VERA Laboratory, Faculty of Physics, University of Vienna, Währinger Strasse 17, A-1090 Vienna, Austria
Institute for NuclearWaste Disposal (INE), Hermann-von Helmholtz-Platz 1, D-76344 Eggenstein-Leopoldshafen, Germany (F.Q.)
Helmholtz-Zentrum Dresden-Rossendorf, Helmholtz Institute Freiberg for Resource Technology, Halsbruecker Strasse 34, 09599 Freiberg, Germany (G.R.)

T. Faestermann and G. Korschinek
Physik Department, Technische Universität München, D-85747 Garching, Germany

G. Rugel
Physik Department, Technische Universität München, D-85747 Garching, Germany
Institute for NuclearWaste Disposal (INE), Hermann-von-Helmholtz-Platz 1, D-76344 Eggenstein-Leopoldshafen, Germany (F.Q.)

Helmholtz-Zentrum Dresden-Rossendorf, Helmholtz Institute Freiberg for Resource Technology, Halsbruecker Strasse 34, 09599 Freiberg, Germany (G.R.)

C. Feldstein, A. Ofan and M. Paul
Racah Institute of Physics, Hebrew University, Jerusalem 91904, Israel

K. Knie
Physik Department, Technische Universität München, D-85747 Garching, Germany
GSI Helmholtz-Zentrum für Schwerionenforschung GmbH, Planckstrasse 1, 64291 Darmstadt, Germany

Uğur Ural
Leibniz Institute fu¨r Astrophysik Potsdam, An der Sternwarte 16, Potsdam 14482, Germany

Mark I. Wilkinson
Department of Physics and Astronomy, University of Leicester, University Road, Leicester LE1 7RH, UK

Justin I. Read
Astrophysics Research Group, Faculty of Engineering and Physical Sciences, University of Surrey, Guildford GU2 7XH, UK

Matthew G. Walker
Department of Physics, McWilliams Center for Cosmology, Carnegie Mellon University, 5000 Forbes Avenue, Pittsburgh, Pennyslvania 15213, USA

Christian Möstl
Space Research Institute, Austrian Academy of Sciences, A-8042 Graz, Austria
IGAM-Kanzelhöhe Observatory, Institute of Physics, University of Graz, A-8010 Graz, Austria

Tanja Rollett
Space Research Institute, Austrian Academy of Sciences, A-8042 Graz, Austria

Martin A. Reiss, Manuela Temmer and Peter Boakes
IGAM-Kanzelhöhe Observatory, Institute of Physics, University of Graz, A-8010 Graz, Austria

Rudy A. Frahm
Southwest Research Institute, 6220 Culebra Road, San Antonio, Texas 78238, USA

Ying D. Liu
State Key Laboratory of Space Weather, National Space Science Center, Chinese Academy of Sciences, Beijing 100190, China

David M. Long
Mullard Space Science Laboratory, University College London, Holmbury St Mary, Dorking RH5 6NT, UK

Robin C. Colaninno
Space Science Division, Naval Research Laboratory, Washington, District of Columbia 20375, USA

Charles J. Farrugia
Space Science Center, Department of Physics, University of New Hampshire, Durham, New Hampshire 03824, USA

Arik Posner
NASA Headquarters,Washington, District of Columbia 20546, USA.

Mateja Dumbović and Bojan Vršnak
Hvar Observatory, Faculty of Geodesy, University of Zagreb, 10 000 Zagreb, Croatia

Miho Janvier
Department of Mathematics, University of Dundee, Dundee DD1 4HN, Scotland

Pascal Démoulin
Observatoire de Paris, LESIA, UMR 8109 (CNRS), F-92195 Meudon Principal, France

Andy Devos and Emil Kraaikamp
Solar-Terrestrial Center of Excellence - SIDC, Royal Observatory of Belgium, 1180 Brussels, Belgium

Mona L. Mays
Catholic University of America, Washington, District of Columbia 20064, USA
Heliophysics Science Division, NASA Goddard Space Flight Center, Greenbelt, Maryland 20771, USA

Michael A. Balikhin, Simon N. Walker and Keith H. Yearby
Department of Automatic Control and Systems Engineering, University of Sheffield, Mappin Street, Sheffield S1 3JD, UK

Yuri Y. Shprits
Department of Earth Planetary and Space Sciences, UCLA, 595 Charles Young Drive East, Box 951567, Los Angeles, California 90095-1567, USA
Department of Earth Atmospheric and Planetary Sciences, MIT, 77 Massachusetts Avenue, Cambridge, Massachusetts 02139-4307, USA

Benjamin Weiss
Department of Earth Atmospheric and Planetary Sciences, MIT, 77 Massachusetts Avenue, Cambridge, Massachusetts 02139-4307, USA

Lunjin Chen
W.B. Hanson Center for Space Sciences, Department of Physics, The University of Texas at Dallas, 800 West Campbell Road, Richardson, Texas 75080-3021, USA

Nicole Cornilleau-Wehrlin
LPP, CNRS, École Polytechnique, Palaiseau 1128, France. 6 LESIA, Observatoire de Paris, Section de Meudon, 5, Place Jules Janssen, Meudon 92195, France

Iannis Dandouras
CNRS, IRAP, 9, Avenue du Colonel Roche, Toulouse BP 44346-31028, France. 8 UPS-OMP, IRAP, 14, Avenue Edouard Belin, Toulouse 31400, France

Ondrej Santolik
Department of Space Physics, Institute of Atmospheric Physics ASCR, Bocni II/1401, 14131 Praha 4, Czech Republic. 10 Faculty of Mathematics and Physics, Charles University in Prague, V Holesovickach 2, 18000 Praha 8, Czech Republic

Christopher Carr
Blackett Laboratory, Imperial College London, South Kensington Campus, London SW7 2AZ, UK

Mauricio Bustamante
Center for Cosmology and AstroParticle Physics (CCAPP), The Ohio State University, 191 W. Woodruff Avenue, Columbus, Ohio 43210, USA
DESY, Platanenallee 6, D-15738 Zeuthen, Germany
Institut für Theoretische Physik und Astrophysik, Universität Würzburg, Am Hubland, D-97074 Würzburg, Germany

Walter Winter
DESY, Platanenallee 6, D-15738 Zeuthen, Germany. Princeton, New Jersey 08540, USA

Philipp Baerwald
Department of Astronomy and Astrophysics, Center for Particle and Gravitational Astrophysics, Institute for Gravitation and the Cosmos, Pennsylvania State University, 525 Davey Lab, University Park, Pennsylvania 16802, USA
Department of Physics, Pennsylvania State University, 525 Davey Lab, University Park, Pennsylvania 16802, USA

Kohta Murase
Department of Astronomy and Astrophysics, Center for Particle and Gravitational Astrophysics, Institute for Gravitation and the Cosmos, Pennsylvania State University, 525 Davey Lab, University Park, Pennsylvania 16802, USA

Department of Physics, Pennsylvania State University, 525 Davey Lab, University Park, Pennsylvania 16802, USA
Institute for Advanced Study

David H. Brooks
College of Science, George Mason University, 4400 University Drive, Fairfax, Virginia 22030, USA
Present address: Hinode Team, ISAS/JAXA, 3-1-1 Yoshinodai, Chuo-ku, Sagamihara, Kanagawa 252-5210, Japan

Ignacio Ugarte-Urra
College of Science, George Mason University, 4400 University Drive, Fairfax, Virginia 22030, USA

Harry P. Warren
Space Science Division, Naval Research Laboratory, 4555 Overlook Avenue SW, Washington, District Of Columbia 20375, USA

R.J. Morton
Department of Mathematics and Information Sciences, Northumbria University, Newcastle Upon Tyne NE1 8ST, UK
High Altitude Observatory, National Center for Atmospheric Research, Boulder, Colorado 80307-3000, USA

S. Tomczyk
High Altitude Observatory, National Center for Atmospheric Research, Boulder, Colorado 80307-3000, USA

R. Pinto
UPS-OMP, IRAP, Universitéde Toulouse, 14 Avenue Edouard Belin -314000 Toulouse, France
CNRS, IRAP, 9 Avenue colonel Roche, BP 44346, F-31028 Toulouse, France

Dainis Dravins
Lund Observatory, Lund University, Box 43, Lund SE-22100, Sweden

Tiphaine Lagadec
Lund Observatory, Lund University, Box 43, Lund SE-22100, Sweden
ESTEC, European Space Research and Technology Centre, Keplerlaan 1, NL-2200 AG Noordwijk, The Netherlands (T.L.)
JPL, Jet Propulsion Laboratory, California Institute of Technology, 4800 Oak Grove Drive, Pasadena, California 91109-8099, USA (P.D.N.)

Paul D. Nuñez
Collège de France, 11 Place Marcelin Berthelot, Paris FR-75005, France

Laboratoire Lagrange, Observatoire de la Coˆte d'Azur, BP 4229, Nice FR-06304, France
ESTEC, European Space Research and Technology Centre, Keplerlaan 1, NL-2200 AG Noordwijk, The Netherlands (T.L.)
JPL, Jet Propulsion Laboratory, California Institute of Technology, 4800 Oak Grove Drive, Pasadena, California 91109-8099, USA (P.D.N.)

Timothy D. Glotch and Jessica A. Arnold
Department of Geosciences, Stony Brook University, Stony Brook, New York 11794-2100, USA

Joshua L. Bandfield
Space Science Institute, 4750 Walnut St #205, Boulder, Colorado 80301, USA

Paul G. Lucey
Hawaii Institute of Geophysics and Planetology, University of Hawaii, Honolulu, Hawaii 96822, USA

Paul O. Hayne and Benjamin T. Greenhagen
Jet Propulsion Laboratory, M/S 183-301, 4800 Oak Grove Drive, Pasadena, California 91109, USA

Rebecca R. Ghent
Department of Earth Sciences, University of Toronto, Toronto, Ontario, Canada M5S 3B1

David A. Paige
University of California Los Angeles, Box 951567, Los Angeles, California 90095-1567, USA

Bin Yang and Xiang Li
Key Laboratory of Dark Matter and Space Astronomy, Purple Mountain Observatory, Chinese Academy of Sciences, Nanjing 210008, China.

University of Chinese Academy of Sciences, Yuquan Road 19, Beijing 100049, China

Yi-Zhong Fan
Key Laboratory of Dark Matter and Space Astronomy, Purple Mountain Observatory, Chinese Academy of Sciences, Nanjing 210008, China

Collaborative Innovation Center of Modern Astronomy and Space Exploration, Nanjing University, Nanjing 210046, China

Zhi-Ping Jin, Xian-Zhong Zheng and Da-Ming Wei
Key Laboratory of Dark Matter and Space Astronomy, Purple Mountain Observatory, Chinese Academy of Sciences, Nanjing 210008, China

Stefano Covino
INAF/Brera Astronomical Observatory, via Bianchi 46, I-23807 Merate, Italy

Kenta Hotokezaka and Tsvi Piran
Racah Institute of Physics, The Hebrew University, Jerusalem 91904, Israel

Scott W. McIntosh
High Altitude Observatory, National Center for Atmospheric Research, PO Box 3000, Boulder, Colorado 80307, USA

Robert J. Leamon
Department of Physics, Montana State University, Bozeman, Montana 59717, USA

Larisza D. Krista
Cooperative Institute for Research in Environmental Sciences, University of Colorado, Boulder, Colorado 80205, USA

Alan M. Title
Lockheed Martin Advanced Technology Center, 3251 Hanover Street, Building 252, Palo Alto, Colorado 94304, USA

Hugh S. Hudson
Space Sciences Laboratory, University of California, Berkeley, California 94720, USA

Pete Riley
Predictive Science Inc., 9990 Mesa Rim Road, Suite 170, San Diego, California 92121, USA

Jerald W. Harder, Greg Kopp, Martin Snow and Thomas N. Woods
Laboratory for Atmospheric and Space Physics, University of Colorado, 1234 Innovation Drive, Boulder, Colorado 80303, USA

Michael L. Stevens
Harvard-Smithsonian Center for Astrophysics, 60 Garden Street, Cambridge, Massachusetts 02138, USA

Justin C. Kaspe
Harvard-Smithsonian Center for Astrophysics, 60 Garden Street, Cambridge, Massachusetts 02138, USA
Department of Atmospheric, Oceanic and Space Sciences, University of Michigan, Ann Arbor, Michigan 48109, USA

Roger K. Ulrich
Division of Astronomy and Astrophysics, University of California, Los Angeles, Colorado 90095, USA

Ketron Mitchell-Wynne and Asantha Cooray
Department of Physics & Astronomy, University of California, Irvine, California 92697, USA

Yan Gong
Department of Physics & Astronomy, University of California, Irvine, California 92697, USA
National Astronomical Observatories, Chinese Academy of Sciences, 20A Datun Road, Chaoyang District, Beijing 100012, China

Matthew Ashby
Harvard-Smithsonian Center for Astrophysics, 60 Garden St., Cambridge, Massachusetts 02138, USA

Timothy Dolch
Department of Astronomy, Cornell University, Ithaca, New York 14853, USA

Henry Ferguson, Norman Grogin and Anton Koekemoer
Space Telescope Science Institute, 3700 San Martin Dr., Baltimore, Maryland 21218, USA

Steven Finkelstein
Department of Astronomy, The University of Texas at Austin, Austin, Texas 78712, USA

Dale Kocevski
Department of Physics and Astronomy, University of Kentucky, Lexington, Kentucky 40506, USA

Joel Primack
Physics Department, University of California Santa Cruz, Santa Cruz, California 95064, USA

Joseph Smidt
Theoretical Division, Los Alamos National Laboratory, Los Alamos, New Mexico 87545, USA

Xiaoli Yan, Li Zhao, Zhong Liu, Yi Bi and Yongyuan Xiang
Yunnan Observatories, Chinese Academy of Sciences, Kunming, Yunnan 650216, China

Zhike Xue and Liheng Yang
Yunnan Observatories, Chinese Academy of Sciences, Kunming, Yunnan 650216, China
Key Laboratory of Solar Activity, National Astronomical Observatories, Chinese Academy of Sciences, Beijing 100012, China

Bernhard Kliem
Yunnan Observatories, Chinese Academy of Sciences, Kunming, Yunnan 650216, China
Institute of Physics and Astronomy, University of Potsdam, Potsdam 14476, Germany

Jun Zhang
Key Laboratory of Solar Activity, National Astronomical Observatories, Chinese Academy of Sciences, Beijing 100012, China

Kai Yang and Xin Cheng
School of Astronomy and Space Science, Nanjing University, Nanjing, Jiangsu 210093, China

Yingna Su
Key Laboratory for Dark Matter and Space Science, Purple Mountain Observatory, Chinese Academy of Sciences, Nanjing, Jiangsu 210008, China

N. Booth, A.P.L. Robinson, R.J. Clarke, B. Li and P.P. Rajeev
Central Laser Facility, STFC Rutherford Appleton Laboratory, Didcot OX11 0QX, UK

P. Hakel
Department of Physics, College of Science, University of Nevada, Reno, Nevada 89557-0208, USA
Division of Computational Physics, Los Alamos National Laboratory, Los Alamos, New Mexico 87545, USA (P.H.)
ELI Beamlines, Fyzikalni Ustav AV CR Vvi, 182 21 Prague, Czech Republic (T.L.)

J. Pasley
Central Laser Facility, STFC Rutherford Appleton Laboratory, Didcot OX11 0QX, UK
Department of Physics, York Plasma Institute, University of York, Heslington York YO10 5DD, UK

R.C. Mancini
Department of Physics, College of Science, University of Nevada, Reno, Nevada 89557-0208, USA

R.J. Dance, E. Wagenaars, J.N. Waugh and N.C. Woolsey
Department of Physics, York Plasma Institute, University of York, Heslington York YO10 5DD, UK

D. Doria, M. Makita and D. Riley
School of Mathematics and Physics, Queen's University Belfast, Belfast BT1 4NN, UK

L.A. Gizzi, P. Koester and L. Labate
Intense Laser Irradiation Laboratory, Istituto Nazionale di Ottica, Area della Ricerca del CNR, 56124 Pisa, Italy

T. Levato
Intense Laser Irradiation Laboratory, Istituto Nazionale di Ottica, Area della Ricerca del CNR, 56124 Pisa, Italy
Division of Computational Physics, Los Alamos National Laboratory, Los Alamos, New Mexico 87545, USA (P.H.)
ELI Beamlines, Fyzikalni Ustav AV CR Vvi, 182 21 Prague, Czech Republic (T.L.)

G. Gregori
Department of Physics, University of Oxford, Oxford OX4 3PU, UK

X. Yi, K. Vahala and J. Li
Department of Applied Physics and Materials Science, Pasadena, California 91125, USA

S. Diddams and G. Ycas
National Institute of Standards and Technology, 325 Broadway, Boulder, Colorado 80305, USA
Department of Physics, University of Colorado, 2000 Colorado Avenue, Boulder, Colorado 80309, USA

P. Plavchan
Department of Physics, Missouri State University, 901 S National Avenue, Springfield, Missouri 65897, USA

S. Leifer, J. Sandhu, G. Vasisht and P. Chen
Jet Propulsion Laboratory, California Institute of Technology, 4800 Oak Grove Drive, Pasadena, California 91109, USA

P. Gao
Division of Geological and Planetary Sciences, California Institute of Technology, Pasadena, California 91125, USA

J. Gagne
Department of Terrestrial Magnetism, Carnegie Institution of Washington, 5241 Broad Branch Road, Washington, District of Columbia 20015, USA

E. Furlan and C. Beichman
NASA Exoplanet Science Institute, California Institute of Technology, Pasadena, California 91125, USA

M. Bottom
Department of Astronomy, California Institute of Technology, Pasadena, California 91125, USA

E.C. Martin and M.P. Fitzgerald
Department of Physics and Astronomy, University of California Los Angeles, Los Angeles, California 90095, USA

G. Doppmann
W.M. Keck Observatory, Kamuela, Hawaii 96743, USA

Xuesheng Feng
SIGMA Weather Group, State Key Laboratory for Space Weather, National Space Science Center, Chinese Academy of Sciences, No.1 Nan-Er-Tiao, Zhong-Guan-Cun, Hai-Dian District, Beijing 100190, China

Chaowei Jiang
SIGMA Weather Group, State Key Laboratory for Space Weather, National Space Science Center, Chinese Academy of Sciences, No.1 Nan-Er-Tiao, Zhong-Guan-Cun, Hai-Dian District, Beijing 100190, China
Center for Space Plasma & Aeronomic Research, The University of Alabama in Huntsville, Huntsville, Alabama 35899, USA

S.T. Wu and Qiang Hu
Center for Space Plasma & Aeronomic Research, The University of Alabama in Huntsville, Huntsville, Alabama 35899, USA

Florian Mekhaldi, Raimund Muscheler and Florian Adolphi
Department of Geology—Quaternary Sciences, Lund University, 22362 Lund, Sweden

Ala Aldahan
Department of Geology, United Arab Emirates University, 17551 Al Ain, UAE
Department of Earth Sciences, Uppsala University, 75236 Uppsala, Sweden

Jürg Beer
Swiss Federal Institute of Aquatic Science and Technology, 8600 Dübendorf, Switzerland

Joseph R. McConnell
Division of Hydrologic Sciences, Desert Research Institute, Reno, Nevada 89512, USA

Göran Possnert
Tandem Laboratory, Uppsala University, 75120 Uppsala, Sweden

Michael Sigl
Division of Hydrologic Sciences, Desert Research Institute, Reno, Nevada 89512, USA
Laboratory for Radiochemistry and Environmental Chemistry, Paul Scherrer Institut, 5232 Villigen, Switzerland

Anders Svensson
Center for Ice and Climate, Niels Bohr Institute, University of Copenhagen, 2100 Copenhagen, Denmark

Hans-Arno Synal
Laboratory of Ion Beam Physics, ETH Zu¨rich, 8093 Zürich, Switzerland

Kees C. Welten
Space Sciences Laboratory, University of California, Berkeley, California 94720, USA

Thomas E. Woodruff
PRIME Laboratory, Purdue University, West Lafayette, Indiana 47907, USA

J.E. Cross and G. Gregori
Clarendon Laboratory, University of Oxford, Parks Road, Oxford OX1 3PU, UK.

J.M. Foster and C.N. Danson
Clarendon Laboratory, University of Oxford, Parks Road, Oxford OX1 3PU, UK

AWE, Aldermaston, Reading, West Berkshire RG7 4PR, UK

H.W. Doyle
Clarendon Laboratory, University of Oxford, Parks Road, Oxford OX1 3PU, UK
First Light Fusion Ltd, Unit 10 Oxford Industrial Park, Mead Road, Yarnton Oxfordshire, OX5 1QU, UK

P. Graham, J. Fyrth, E.T. Gumbrell, S. Patankar and J. Skidmore
AWE, Aldermaston, Reading, West Berkshire RG7 4PR, UK

J.-M. Bonnet-Bidaud
Service d'Astrophysique-Laboratoire AIM, CEA/DSM/Irfu, 91191 Gif-sur-Yvette, France

É. Falize
Service d'Astrophysique-Laboratoire AIM, CEA/DSM/Irfu, 91191 Gif-sur-Yvette, France
CEA-DAM-DIF, F-91297 Arpajon, France

C. Busschaert, N. Charpentier and B. Loupias
CEA-DAM-DIF, F-91297 Arpajon, France

R.P. Drake, C. Krauland and C.C. Kuranz
Department of Atmospheric, Oceanic and Space Sciences, University of Michigan, Ann Arbor, Michigan 48109, USA

R. Yurchak
LULI-CNRS, Ecole Polytechnique, CEA: Université Paris-Saclay; UPMC Univ Paris 06: Sorbonne Universités-F-91128, Palaiseau Cedex, France

M. Koenig
LULI-CNRS, Ecole Polytechnique, CEA: Université Paris-Saclay; UPMC Univ Paris 06: Sorbonne Universités-F-91128, Palaiseau Cedex, France
Institute for Academic Initiatives, Osaka University, Suita, Osaka 565-0871, Japan

C. Michaut and M. Mouchet
LUTH, Observatoire de Paris, PSL Research University, CNRS, Université Paris Diderot, Sorbonne Paris Cité, 92190 Meudon, France

C. Spindloe
Target Fabrication Group, Central Laser Facility, Rutherford Appleton Laboratory, Harwell Science and Innovation Campus, Didcot OX11 0QX, UK

E.R. Tubman and N. Woolsey
York Plasma Institute, Department of Physics, University of York, Heslington, York YO10 5DQ, UK

Mads Faurschou Knudsen, Alexandra Fogtmann-Schulz and Fadil Inceoglu
Department of Geoscience, Aarhus University, Høegh-Guldbergs Gade 2, 8000 Aarhus C, Denmark

Christoffer Karoff
Department of Geoscience, Aarhus University, Høegh-Guldbergs Gade 2, 8000 Aarhus C, Denmark
Stellar Astrophysics Centre, Department of Physics and Astronomy, Aarhus University, Ny Munkegade 120, 8000 Aarhus C, Denmark

Peter De Cat
Royal Observatory of Belgium, Ringlaan 3, B-1180 Brussel, Belgium

Alfio Bonanno and Antonio Frasca
INAF-Osservatorio Astrofisico di Catania, via S.Sofia 78, 95123 Catania, Italy

Jianning Fu
Department of Astronomy, Beijing Normal University, 19 Avenue
Xinjiekouwai, Beijing 100875, China

Jesper Olsen
AMS, 14C Dating Centre, Department of Physics, Aarhus University, Ny Munkegade 120, 8000 Aarhus C, Denmark

Yong Zhang, Yonghui Hou and Yuefei Wang
Nanjing Institute of Astronomical Optics and Technology, National Astronomical Observatories, Chinese Academy of Sciences, Nanjing 210042, China

Jianrong Shi and Wei Zhang
Key Laboratory of Optical Astronomy, National Astronomical Observatories, Chinese Academy of Sciences, Beijing 100012, China

Xi Zhang
Department of Earth and Planetary Sciences, University of California Santa Cruz, Santa Cruz, California 95064, USA

Robert A. West
Jet Propulsion Laboratory, California Institute of Technology, 4800 Oak Grove Drive, Pasadena, California 91109, USA

Patrick G.J. Irwin
Atmospheric, Oceanic and Planetary Physics, University of Oxford, Clarendon Laboratory, Parks Road, Oxford OX1 3PU, UK

Conor A. Nixon
NASA Goddard Space Flight Center, Greenbelt, Maryland 20771, USA

Yuk L. Yung
Division of Geological and Planetary Sciences, California Institute of Technology, Pasadena, California 91125, USA

Zhenpeng Su, Huinan Zheng and Shui Wang
CAS Key Laboratory of Geospace Environment, Department of Geophysics and Planetary Sciences, University of Science and Technology of China, Hefei, Anhui 230026, China
Collaborative Innovation Center of Astronautical Science and Technology, University of Science and Technology of China, Hefei, Anhui 230026, China

Hui Zhu and Zhonglei Gao
CAS Key Laboratory of Geospace Environment, Department of Geophysics and Planetary Sciences, University of Science and Technology of China, Hefei, Anhui 230026, China
Mengcheng National Geophysical Observatory, School of Earth and Space Sciences, University of Science and Technology of China, Hefei, Anhui 230026, China

Yuming Wang
CAS Key Laboratory of Geospace Environment, Department of Geophysics and Planetary Sciences, University of Science and Technology of China, Hefei, Anhui 230026, China
Collaborative Innovation Center of Astronautical Science and Technology, University of Science and Technology of China, Hefei, Anhui 230026, China
Synergetic Innovation Center of Quantum Information and Quantum Physics, University of Science and Technology of China, Hefei, Anhui 230026, China

Fuliang Xiao
School of Physics and Electronic Sciences, Changsha University of Science and Technology, Changsha Hunan 410004, China

Q.-G. Zong, X.-Z. Zhou and Y.-X. Hao
Institute of Space Physics and Applied Technology, Peking University, Beijing 100871, China

Zhaoguo He
Harbin Institute of Technology Shenzhen Graduate School, Shenzhen, Guangdong 518055, China

D.N. Baker
Laboratory for Atmospheric and Space Physics, University of Colorado Boulder, Boulder, Colorado 80303-7814, USA

H.E. Spence
Institute for the Study of Earth, Oceans, and Space, University of New Hampshire, Durham, New Hampshire 03824-3525, USA

G.D. Reeves
Space Science and Applications Group, Los Alamos National Laboratory, Los Alamos, New Mexico 87544, USA

J.B. Blake
The Aerospace Corporation, Los Angeles, California 90245- 4609, USA

J.R. Wygant
School of Physics and Astronomy, University of Minnesota, Minneapolis, Minnesota 55455, USA

Fuliang Xiao, Chang Yang, Qinghua Zhou and Yihua He
School of Physics and Electronic Sciences, Changsha University of Science and Technology, 2nd Section, South Wanjiali Road #960, Yuhua District, Changsha, Hunan 410004, China

Zhenpeng Su
Chinese Academy of Sciences Key Laboratory for Basic Plasma Physics, University of Science and Technology of China, Hefei, Anhui 230026, China

Zhaoguo He
Center for Space Science and Applied Research, Chinese Academy of Sciences, Beijing 100190, China

D.N. Baker
Laboratory for Atmospheric and Space Physics, University of Colorado, Boulder, Colorado 80303, USA

H.E. Spence
Institute for the Study of Earth, Oceans, and Space, University of New Hampshire, Durham, New Hampshire 03824-3525, USA

H.O. Funsten
ISR Division, Los Alamos National Laboratory, Los Alamos, New Mexico 87545, USA

J.B. Blake
The Aerospace Corporation, Los Angeles, California 90245-4609, USA

Index

www.ingramcontent.com/pod-product-compliance
Lightning Source LLC
Chambersburg PA
CBHW080518200326
41458CB00012B/4253